WELL LOGGING FOR PHYSICAL PROPERTIES

WELL LOGGING FOR PHYSICAL PROPERTIES

Joseph R Hearst, Ph.D.
University of California
Lawrence Livermore National Laboratory

and

Philip H. Nelson, Ph.D.
Sohio Petroleum Company

McGraw-Hill Book Company

New York St. Louis San Francisco Auckland Bogotá Hamburg Johannesburg London Madrid Mexico Montreal New Delhi Panama Paris São Paulo Singapore Sydney Tokyo Toronto

Library of Congress Cataloging in Publication Data

Hearst, Joseph R
Well logging for physical properties.

Includes bibliographies and index.
1. Geophysical well logging. 2. Petroleum—Geology. 3. Gas, Natural—Geology. 4. Prospecting. I. Nelson, Philip H. II. Title.
TN871.35.H38 1985 662'.1828 85-115
ISBN 0-07-027698-6

1234567890 DOC/DOC 898765

ISBN 0-07-027698-6

The editors for this book were Betty Sun and Galen H. Fleck, the book was designed by M.R.P. Design, and the production supervisor was Teresa F. Leaden. It was set in Caledonia by University Graphics, Inc.

Printed and bound by R. R. Donnelley & Sons Company.

We dedicate this book to the scientists and engineers who, over the years, have made modern logging technology what it is today.

Contents

Preface

Books on formation evaluation tell the reader how to use logging methods to locate oil and gas. We recognize that there are many uses of logs other than that; the aim of this book is to teach basic understanding of the methods themselves. There are large numbers of pitfalls in obtaining data from logs; many will be described. We shall discuss the physical principles behind the logging methods and the assumptions used in the conversion of the signals obtained from the logging equipment into values of the properties of the formation being logged. We are not, however, trying to tell our readers how to design logging equipment.

We have tried to aim at readers with degrees in technical subjects other than geophysics: geology, engineering, or basic science. We have included a fair amount of mathematics. The math is included to justify the verbal statements and convince the reader of their validity; we have made no attempt at rigor. Generally we use the physics of the situation to develop simple versions of the important equations, describe some of the complexities that arise, and then state the result without proof.

We have drawn extensively on the American logging literature and in particular on publications of the Society of Professional Well Log Analysts (SPWLA), the Society of Petroleum Engineers (SPE), and the

Society of Exploration Geophysicists (SEG). We have used the Soviet literature when translations were available.

The book is in three parts. Part 1 is an introduction; it outlines the problems to be solved by logging, describes the drilling of a hole, and explains the logging process. Part 2, the longest, discusses the basic physics behind the common logging methods. It describes the standard logging equipment and explains some of the data reduction techniques. Part 3 uses many of the ordinary, and some of the not so ordinary, applications of logging equipment to illustrate the ways in which the basic equipment is used to obtain properties of the formation beyond the specific quantities that the equipment can measure directly. It discusses the assumptions needed to infer the properties when the methods work—and when they do not. We hope that, with the aid of this book, the reader will be able to use *understanding* of the capabilities of the equipment to critically analyze and evaluate the data from well logs.

Many authors have been kind enough to permit us to reproduce figures from their work. We would particularly like to thank R. A. Broding and T. J. Taylor of Amoco, F. L. Paillet and W. S. Keys of the U.S. Geological Survey, D. K. Smith of Halliburton, W. M. Sturdevant of Birdwell, B. F. Swanson of Shell, and A. Youmans of Dresser Atlas, who sent us original photographs to permit better reproduction.

Several people helped us by reviewing parts of this book. We are especially grateful to Michael King of the University of California and Donald Hill of Chevron Overseas Petroleum, Inc., each of whom reviewed several chapters. We would also like to thank Scott Allen of Mobil; Ralph Wiley of Amoco; Pete Brabeck, Dwight Eggers, Tony Fowler, Tim Lane, Eric Opstad, and John Ziagos of Sohio; and Alex Becker of the University of California for helpful reviews. They all helped us remove glitches from the text. The responsibility for any remaining errors is, of course, our own.

The book would not have been written without the advice, encouragement, and suggestions of Dick Carlson and Paul Kasameyer of the Lawrence Livermore National Laboratory.

This book began as notes for a course at the University of California at Berkeley. The work was supported in part by Sohio Petroleum and the Lawrence Berkeley Laboratory, and it was performed under the auspices of the U.S. Department of Energy by Lawrence Livermore National Laboratory under Contract W-7405-Eng-48.

WELL LOGGING FOR PHYSICAL PROPERTIES

PART 1

Introduction

1

The Well Logging Technology

Well logging is an audacious enterprise. Complex packages of electronics and transducers, referred to as *sondes*, are lowered to a depth of as much as several kilometers in a hole in the ground. The mechanical shock imparted to a sonde is severe under the best conditions. Temperatures and pressures can become high enough to degrade the performance of cables, connectors, and sondes. All this is undertaken to measure the physical properties of rocks and fluids only a small proportion of which will ever be useful. The incentive to overcome the problems must be quite high to have reached the sophisticated level embodied by today's measurement technology.

1-1 ROLE OF LOGGING IN SUBSURFACE EXPLORATION

Information is acquired from a hole in several ways. The nature of the subsurface fluids can be investigated by analyzing the drilling fluid itself. This procedure, called *mud logging*, is especially diagnostic of the presence of gas or oil. Rock chips are recovered from the drill fluid returned to the surface, or even better, cylindrical specimens of rock are *cored* as the drilling progresses. Well logging provides a set of continuous and unbiased physical measurements of rock properties. It is less expensive and quicker than coring; also, in unconsolidated rock, core may be unobtainable, so that logs provide the only continuous data set. An archetypical group, or *suite*, of logs shown in Fig. 1-1 illustrates

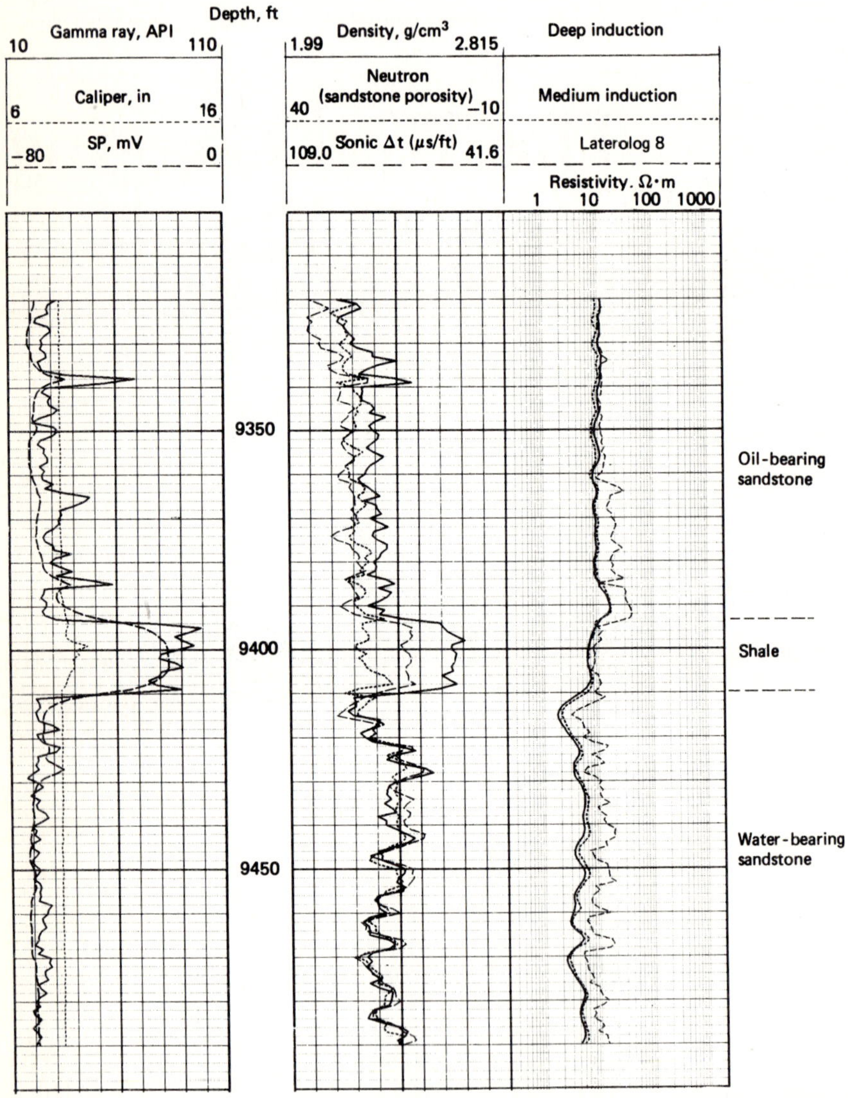

FIGURE 1-1 Suite of well logs from an oil reservoir. (Courtesy Sohio Petroleum Co.)

why logs have become so useful in oil and gas exploration and development. The logs were acquired in a sand-shale sequence which forms an oil reservoir. On the left are γ-ray and SP logs, which, by their deflection to the right, indicate the presence of shale beds. The central three traces are scaled in units of porosity (Chap. 10); each is derived from a different physical measurement. On the right are three electrical resistivity logs. In the water-saturated zone the resistivity is low; it fluctuates between 2 and 5 $\Omega \cdot$m. Above the oil-water interface, resistivity increases as the pore space becomes filled with more oil and less water until it reaches a consistent value of about 15 $\Omega \cdot$m. It is the capability of logs to give information about these three parameters—lithology (in the left track), pore volume (in the center track), and fluid content (in the right track)—which has encouraged the ever-increasing use of logs in oil and gas exploration.

These diagnostic responses were recognized in the late 1920s by C. and M. Schlumberger and their associates, who first developed and employed resistivity measurements and quickly added the self-potential measurement to their logging suite. Their methods caught on, and the techniques were widely adopted (Allaud and Martin, 1977). Early development of logging sondes emphasized the electrical methods; and because of its unique sensitivity to fluid properties, the resistivity measurement remains the mainstay of most logging suites. By the mid-1950s, the resistivity technology was mature, the dependence of resistivity upon porosity and saturation had been established, borehole correction charts were available, the underlying principles and the pitfalls had been explained in the open literature, and the focused and pad-type sondes (Sec. 5-4) had been introduced.

The use of logging increased as the realization grew that not only were qualitative indicators of rock properties available but quantitative results also could be obtained. The value of quantitative logging measurements can be appreciated by inspecting an expression which might be used to estimate the oil or gas content of a reservoir. The total hydrocarbon volume is

$$V_h = TAf_{sd}\,\phi\,(1 - S_w) \tag{1-1}$$

where T is the total vertical thickness of the sand-shale sequence, A is the areal extent of the sequence, f_{sd} is the fraction of oil-bearing sand in the sequence, ϕ is porosity, and S_w is the fraction of the pore space saturated with water. Of these five quantities, all but the area A are routinely obtained from well logs.

Radiation and acoustic methods began to be developed around 1940 and to be accepted by the industry during the 1950s (Johnson, 1962; Segesman, 1980). Detection of γ-rays, both natural and induced, and neutrons was used for measuring radioactive mineral content (Sec. 6-3), density (Sec. 6-4), and hydrogen content (Sec. 6-6). The velocity of acoustic compressional waves was found to be porosity-dependent (Sec. 7-2). By the early 1960s, the three so-called porosity sondes (Chap. 10) were in use, although the electrical methods were still greatly emphasized. Lynch (1962), Pirson (1963), and Wyllie (1963) discuss the formation evaluation methods and the operation of sondes in use at that time.

Logging technology has been developed primarily for oil and gas work, but logging methods have been applied in holes drilled for many other purposes. Table 1-1 indicates the properties which have been obtained from logs in holes drilled for economic and scientific purposes. Logging for uranium (Sec. 15-3) and coal (Sec. 9-3) is often as successful technically as is logging for oil and gas because it is relatively easy to identify and evaluate uranium and coal. Metal-mining applications continue to rely on the acquisition of core because borehole assay methods (Sec. 15-3) have not been developed to the extent that they are reliable enough for engineering purposes. The same is probably true for civil engineering (geotechnical) investigations: holes are generally shallow and core is required for mechanical testing. The drilling of water wells is another application in which logging could benefit an industry more than it has heretofore.

Logging occupies a special niche in the measurement of properties of the earth. Its methods and techniques might be regarded with skepticism by the laboratory experimentalist who is used to controlling the

TABLE 1-1 Various applications of well logs.

Category	Properties from logs	Typical hole diameters, cm	Typical hole depths
Oil and gas	Lithology, porosity, fluid type, saturation	15–30	1–10 km
Uranium and coal	Uranium grade, ash content in coal, bed thickness	11	20–2000 m
Mineral exploration	Lithology, mineralogy, elemental assay	5–9	200–2000 m
Geothermal	Lithology, temperature, porosity	Varies	20–2000 m
Hydrology	Water quality, porosity	Varies	10–1000 m
Geotechnical	Rock quality, fractures	2.5–7.5	10–1000 m
Scientific	Geology, heat flow, crustal studies	Varies	1–10 km

pressure, temperature, and fluid saturation of a rock sample and who can subject the sample to mineralogical and chemical analysis in order to relate its mineral and textural character to the measured property. The log analyst enjoys no such advantage. He must sort out the conflicting contributions from the wellbore environment, and only occasionally can compare his measurements with measured core properties. Working at the other end of the measurement spectrum is the geophysicist, who must accept a set of measurements made on the earth's surface and from them interpret the geological structure. He must in effect, either with or without the aid of computed models, perform an "inversion" on the data to arrive at a plausible model of the earth which produced the observed response. He would be overjoyed to have the relative advantage of the log analyst, who by the standards of surface prospecting, is making a direct measurement.*

It is the geological heterogeneities which are of interest in logging, for without them there would be no sense in making measurements. We shall visualize geological heterogeneities as occurring on the three scales illustrated in Fig. 1-2, and discuss them in the following three subsections. The first scale is substantially greater than that of the *radius of investigation* of most logging sondes. On this scale are features, such as formation boundaries, unconformities, and faults, which can be mapped over distances of tens to thousands of meters. The intermediate scale is that of the radius of investigation of the logging sonde. Included are all phenomena which impact the response of the sonde: the wellbore, the invaded zone, and geological contacts with spacing of the order of one meter and less. Heterogeneities also occur on a third scale, much smaller than that of the measuring array. These are the variations in rock composition itself and in the minerals which make up the rock. Mixing rules are used so that the fractional volumes and physical properties of the minerals and fluids can be extracted from measurement of the aggregate physical response. The devising and use of adequate mixing rules is a prime activity of log analysis.

1-2 GEOLOGICAL INTERPRETATION

Logs are commonly used to interpret the geology of an area. They are used instead of core primarily because they are much less costly and also because they are available in continuous format and are of rela-

*The use of masculine pronouns in such impersonal passages is due to the absence in English of any acceptable alternative, and not to ignorance of the contributions made by the authors' numerous female colleagues.

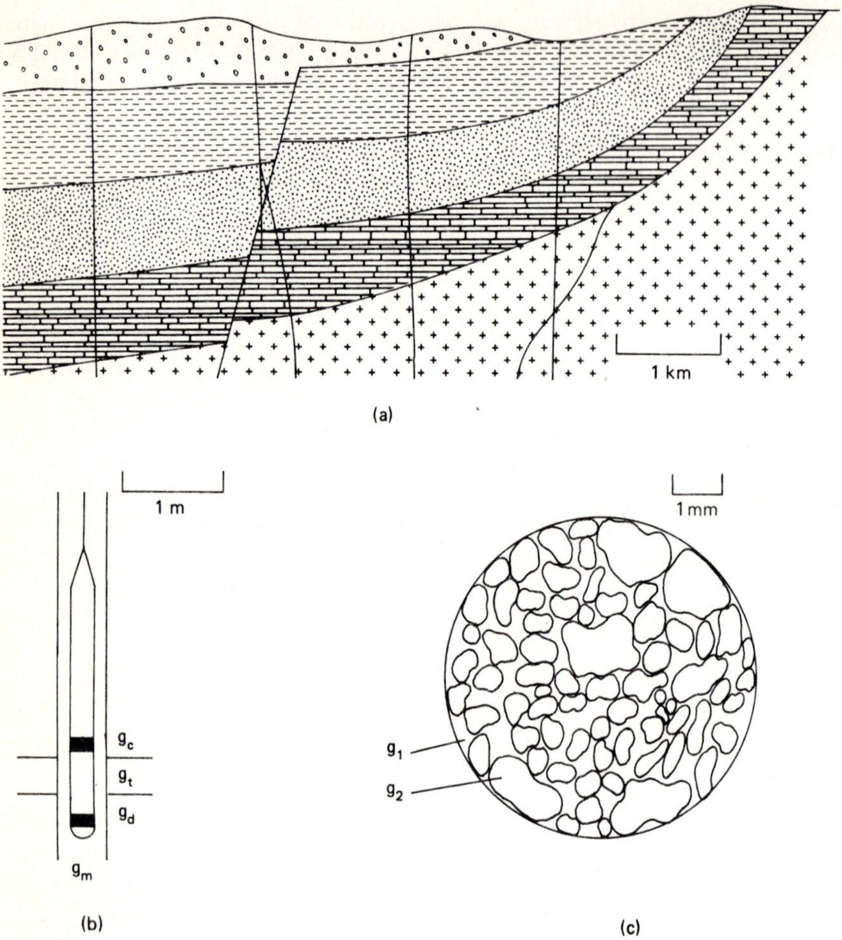

FIGURE 1-2 Three physical scales of investigation of concern in well logging applications.

tively uniform quality from well to well. Once log measurements are correctly calibrated in terms of some physical property, the interpreter working on the regional or sedimentary basin scale no longer needs to be concerned with the details of data acquisition. He is concerned mainly with correlating the logs from one well to the next along a cross section of the subsurface, such as the one shown in Fig. 1-2*a*. By using a knowledge of the way logs respond in particular lithologies, the geologist can often produce an excellent geological interpretation based largely on pattern recognition and extrapolation. The results are presented as a series of cross sections and maps. Maps can be produced of the top or bottom of a given formation, as isopachs of formation thickness, and in many other ways.

The data for this type of analysis are the logs themselves, corrected for depth and presented in terms of true vertical depth with respect to sea level or some other datum. In Chap. 17 we will treat the problem of surveying in wells and converting depths measured in a nonvertical well to true vertical depths. And in Chap. 18 we will examine some aspects of a logging sonde which is considered to be the "geologists log"—the dipmeter. With the dipmeter, the geologist can resolve the strike and dip of sedimentary features. These properties are of great use in understanding the original depositional environment.

We have implied that the interpretation process is simply one of posting logs and extrapolating formation features from one well to the next. This is basically true, but the geologist now has available a powerful set of statistical methods with which to speed the interpretation process. With the aid of digital computers, pattern recognition and correlation schemes can be used to supplement and extend the intuition of the individual interpreter. Such statistical techniques are available for detailed work on a dipmeter log as well as for the interpretation of a sedimentary basin. Although these techniques are a most important application of well logs, the basic problems of geological interpretation and consideration of statistical methods are beyond the scope of this book and will not be discussed further. We now turn to the two topics which will recur again and again throughout this book: sonde response and mixing rules.

1-3 SONDE RESPONSE

Heterogeneities on the scale of the measuring array cause the results of measurements to differ from their values in a homogeneous medium. Suppose we want to determine the true property g_t of a formation (Fig. 1-2*b*). The transducer will give an *apparent response* g_a which will depend upon the mud property g_m, the properties g_c and g_d of the nearby formations, the hole diameter, and the bed thickness. The sonde response g_a must be corrected to remove the contributions of g_m, g_c, and g_d before the desired rock property g_t is obtained.

There are basically two approaches that can be taken to eliminate these effects, and modern logging practice combines the two. The first, direct approach is to configure the sonde so that the influence of rock beyond a certain distance or boundary is much reduced. Excellent examples are the pad-type sondes which have been implemented for particular versions of electrical resistivity, neutron-neutron, and γ-γ

measurements. The sonde is pressed against the wall so that only the mud cake separates the measurement array from the rock. Transducer separation is of the order of centimeters, so the depth of penetration is limited, but the physical properties of rock near the wellbore are measured almost directly, distorted only by the mud cake.

Other examples of the design of a sonde to minimize the effect of heterogeneities are the focused resistivity (Sec. 5-4) and the multicoil induction (Sec. 5-5) sondes. These sondes effectively reduce the vertical zone of influence (by suppressing the contribution of g_c and g_d in Fig. 1-2*b*). They do so by producing cancellation signals from auxiliary electrodes or coils placed above and below the central sensors. A third example is the use of both down-going and up-going signal transmission in acoustic sondes (Sec. 7-3) to help eliminate the effect of borehole rugosity. All these improvements come at the expense of increasing the complexity of the measurement system.

The second approach uses what are sometimes called environmental corrections. Developing these corrections, at least formally, requires solving the partial differential equation which applies to a particular logging measurement. For SP and resistivity it is Poisson's equation,

$$\nabla^2\Psi + S = 0 \tag{1-2}$$

For temperature and for neutron detectors it is the diffusion equation,

$$\nabla^2\Psi - \frac{1}{g}\frac{\partial\Psi}{\partial t} + S = 0 \tag{1-3}$$

And for dielectric and acoustic methods, the wave equation applies:

$$\nabla^2\Psi - k^2\Psi + S = 0 \tag{1-4}$$

where Ψ represents a potential or a flux, g represents the physical properties of the medium, S is the source term, and k is a constant (perhaps including g). Full description of the problem requires that the partial differential equation be written in cylindrical coordinates and that appropriate boundary conditions be stipulated. In some cases the solution can be obtained as an analytical expression. An outstanding example was the solution of the potential problem for two- and three-electrode resistivity sondes. The results were presented as a set of correction curves for converting apparent resistivity to formation resistivity. Parameters were given as dimensionless ratios such as true resistivity to mud resistivity and array spacing to hole diameter.

Analytical solutions cannot be found for many cases of interest, however, and workers turned to the use of two-dimensional electrical networks to simulate geometries in which the electrical properties varied with both radius and depth. These results were likewise presented in graphical format scaled with dimensionless ratios to reduce the large number of variables to a manageable level.

With the advent of digital computing machinery, calculation of the response of sondes in quite complicated geometries is now possible. It is even possible now to simulate a particular sonde design, rather than merely obtain a solution for a simplified version of it. For example, electrodes can be represented as finite bodies rather than as points or lines, or a sonde can be modeled either centralized or decentralized in the wellbore. Consequently, numerical modeling is at the stage where it can not only provide correction factors for an existing sonde configuration which has evolved by trial and error in the field but also be of great use in the design of new sondes.

With such computational power at hand, it might be expected that each logging measurement would be "inverted" to obtain the measurement of true formation properties. The analysis of logging data, however, has evolved quite differently. The trend since the inception of well logging has been to express the measurement directly in terms of the rock property of interest. Much effort in sonde design and log presentation has been expended to provide directly the apparent value of a rock property and relieve the user of the necessity of considering the actual transducer response. Ordinarily, when one wishes to measure some property of a material (in the laboratory, for example), one develops an expression for the value of the desired property in terms of the values of several measured parameters and, by inserting the results of the measurements, calculates the value of the property. In the logging industry, on the other hand, it is common practice to obtain the value of the desired property from a simple expression involving only one measured variable and then apply "corrections" to account for the others. (The correction terms are often designed to avoid changing the simple form of the expression.) Thus the true rock property is derived from the apparent one by applying corrections.

The understanding and use of correction charts is a vital part of log analysis. Such charts are supplied in manuals and papers published by the well logging service companies; Helander (1983) describes many of them. Some will be shown in this book, but our aim is to present the principles and concepts rather than to attempt completeness.

We have stated that correction factors for electrical sondes have been obtained by solving the relevant equation in one way or another. For nuclear sondes, the solution of the appropriate differential equation has not produced the success enjoyed with the electrical methods. In part this is due to the complication caused by sonde housings and detection efficiency, which varies among radiation detectors. These factors can be dealt with experimentally by placing the sonde in a model wellbore. The basic problem, however, is the presence of more variables than measurements; many rock properties and borehole conditions affect γ and neutron transport. Some models have been set up as industry standards, so that the log response is calibrated in terms of the physical properties of the specific model. The responses of γ-ray, γ-γ, and neutron sondes are presented in terms of concentrations of radioactive minerals, rock density, and porosity, respectively, in the model, rather than as counts per second. Correction factors for the radiation sondes are generally obtained experimentally rather than computationally. These arrangements have proved satisfactory for the most part. They do produce some surprises in log analysis nomenclature, as we shall see in Chaps. 6 and 10. For example, the response of a neutron sonde is presented as an apparent porosity which is correct for one rock type but incorrect for another.

There are a few measurement methods for which small corrections to a measured apparent log value, even if available, are not adequate. For example, the radius of investigation of a gravimeter is so great that the log interpreter is often faced with a problem more akin to that produced by a surface geophysical survey. The sonde response cannot be reduced to the density of adjacent beds because a much larger volume of rock influences the measurement. The possibilities must be reduced by geological constraints. A similar situation holds for long-spaced sondes developed to increase the radius of investigation. Examples are long-spaced resistivity arrays used to explore laterally for salt domes and induction sondes useful in mineral exploration.

1-4 MIXING RULES

Let us now assume that a good log has been run, that environmental effects have been "corrected," and that we have data which represent some physical property g of the *bulk rock*, that is, of the rock including the fluids filling the pore space. We now wish to use these data to extract some information about the constituents of the rock. In terms of Fig. 1-2*c* we have measured g and wish to learn something of the distribution of g_1 and g_2. Doing so requires that we use some mixing rule which describes how the properties g_1 and g_2 combine to produce g.

Volume fractions. We must be careful how composition is specified. Most mixing rules used in logging are based on volumetric fractions, and we shall adhere to this practice. Let us consider some volume of rock V. (Sometimes the term "bulk rock" and the symbol V_b are used to emphasize that fluids in the pore space are included in the volume. In this book it will be understood that V includes whatever fluid fills the pore space.) Each constituent with property g_i occupies volume V_i in the rock. Its fractional volume is

$$f_i = \frac{V_i}{V} = \frac{V_i}{\sum_i V_i} \tag{1-5}$$

where

$$\sum_i f_i = 1 \tag{1-6}$$

The porosity, which is the fractional volume occupied by the pore space, is denoted by the special symbol ϕ.

Two-component systems. Korvin (1982) postulated eight physically plausible conditions on the functional relations among g and g_1, g_2, V_1, and V_2. If all eight conditions are satisfied, then the general functional form must be

$$g^\zeta = f_1 g_1^\zeta + f_2 g_2^\zeta \tag{1-7}$$

We will consider a system in which the pore space has property g_1 and fractional volume $f_1 = \phi$. The remainder of the rock volume is occupied by a mineral with property g_2 and fractional volume $f_2 = 1 - \phi$, so Eq. (1-7) becomes

$$g^\zeta = \phi g_1^\zeta + (1 - \phi) g_2^\zeta \tag{1-8}$$

If the exponent $\zeta = +1$, Eq. (1-8) gives the weighted arithmetic mean; if $\zeta = -1$, it gives the weighted harmonic mean. In the limit, as ζ tends to zero,

$$g = g_1^\phi g_2^{1-\phi} \tag{1-9}$$

Equation (1-9) is commonly referred to as the geometric mean rule. Equations (1-8) and (1-9) are graphed in Fig. 1-3 for $g_1 = 1$ and $g_2 = 10$ by using values of $\zeta = 0, +\frac{1}{2}$, and ± 1. Because of the logarithmic scale, the geometric mean rule plots as a straight line. The curves for $\zeta = 1$ and $\zeta = -1$ are symmetric with respect to the geometric mean.

It happens that many of the two-component rules used for log analysis are of the general form of Eq. (1-8), with ζ taking a value between 0 and 1. The rules which specify that quantities combine in direct pro-

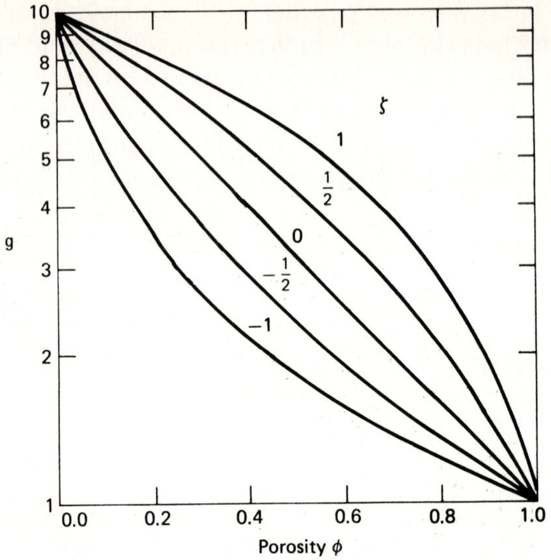

FIGURE 1-3 Mixing rules for a two-component system comprised of a matrix with a physical property value of 10 and a pore space value of 1.

portion to their volumes ($\zeta = 1$) are typified by the equation for density:

$$\rho = \phi\rho_w + (1 - \phi)\rho_{ma} \tag{1-10}$$

where subscripts w and ma refer to fluid and matrix (mineral). The expression for the absorption cross section for neutrons takes on the same form, as does an empirical expression for the reciprocal of the compressional wave speed in some rock types.

It has been found that electrical properties combine with a value of ζ of about ½. The conductivity C and dielectric constant ε at some frequency ω combine as

$$(C - i\,\omega\,\varepsilon)^\zeta = \phi(C_w - i\,\omega\,\varepsilon_w)^\zeta + (1 - \phi)(C_{ma} - i\,\omega\,\varepsilon_{ma})^\zeta \tag{1-11}$$

At low frequencies $\omega\varepsilon$ can be ignored, and since C_{ma} is much less than C_w in most rocks, Eq. (1-11) simplifies to

$$C^\zeta = \phi C_w^\zeta \tag{1-12}$$

It is customary to use m in place of $1/\zeta$. Equation (1-12) is then written as

$$\frac{C}{C_w} = \phi^m \tag{1-13}$$

and is referred to as Archie's equation. Values of the exponent m range from 1.5 to 2.5 (ζ between 0.66 and 0.4). At high frequencies, the terms C in Eq. (1-11) are much less than $\omega\varepsilon$, and we are left with

$$\varepsilon^{1/m} = \phi\varepsilon_w^{1/m} + (1 - \phi)\varepsilon_{ma}^{1/m} \tag{1-14}$$

which has been proposed as a mixing rule for the dielectric constant.

The geometric mean rule, Eq. (1-9), applies for $\zeta = 0$. It has been employed for thermal and magnetic properties. Writing it in logarithmic form for the thermal conductivity k,

$$\log k = \phi \log k_1 + (1 - \phi) \log k_2 \tag{1-15}$$

confirms its simple additive nature, which is shown in Fig. 1-3.

There is no proof that the components of real systems must comply with the functional conditions stipulated by Korvin. But the fact that Eq. (1-8) works for many systems lends credence to Korvin's ideas.

Three or more components. A mixing rule which works for a two-component system is readily extended to a system with three or more components. If the properties mix linearly ($\zeta = 1$), then two or more logs can be used to solve a set of linear equations. For example, if we measure two properties g and h of a three-component system, we have

$$\begin{aligned} g &= f_1 g_1 + f_2 g_2 + f_3 g_3 \\ h &= f_1 h_1 + f_2 h_2 + f_3 h_3 \\ 1 &= f_1 + f_2 + f_3 \end{aligned} \tag{1-16}$$

with the third equation provided by the volumetric constraint. Such sets of equations are easily solved for the volumetric concentrations f_i. In analyzing for hydrocarbon content, the pore space is divided into fractions containing water and hydrocarbon. This special case of saturation with two or more fluids is of such practical importance that it is discussed at length in Chap. 11.

Cross-plots. If one of the response equations from a logging suite is nonlinear ($\zeta \neq 1$), cross-plots can be used to solve for f_i. Suppose we again measure two properties g and h, which can be written as

$$\begin{aligned} g &= G(f_1, f_2; g_1, g_2, g_3) \\ h &= H(f_1, f_2; h_1, h_2, h_3) \end{aligned} \tag{1-17}$$

where the volumetric constraint has been used to eliminate f_3. We next eliminate f_2 by substitution to give

$$g = G'(f_1; g_1, g_2, g_3; h_1, h_2, h_3) \tag{1-18}$$

A plot of g versus h with f_1 as a parameter is called a cross-plot; in Chaps. 9 and 10 on lithology and porosity determinations we give examples. After well log data are plotted on a cross-plot, slopes and intercepts are used to estimate unknown coefficients. Cross-plots are especially useful in highlighting anomalous points due to unusual lithological factors or bad data.

Some of the two-component mixing rules, such as that for density, are almost a matter of definition. A few have been derived from basic principles for simple microscopic geometries. But most are no more than empirical rules based on laboratory core analyses or well log data. Rules for three and more components are (sometimes wishful) extensions of the two-component empirical rules combined with volumetric considerations. At present, most mixing rules must be viewed as no more than handy expressions which are effective in a specific geological environment.

Core measurements. Mixing rules must be checked by core measurements to confirm their validity and to evaluate coefficients. This is particularly vital when logs from a new area are acquired and it is not yet known what rocks are present, what minerals they contain, and what their effect on log readings might be. We regret that a description of the methods of core analysis is beyond the scope of this book. The reader is referred to descriptions by Lynch (1962), Amyx et al. (1960), and Helander (1983), and to current technical literature available from commercial core laboratories.

Statistical approaches. In many cases there may be no known mixing rule. Simple adjustment of coefficients in Eq. (1-8) is inadequate, and it becomes necessary to find a new mixing rule. Multiple-regression methods are one technique for using log responses, either directly or combined into appropriate functions, as independent variables upon which some other variable (usually obtained from core measurement) depends. If a workable relation can be found where core is available, it may be possible to apply it in nearby holes where no core is available. The derived relation may be based in principle upon the physical response of the logs, but it need not be. Although they are effective, statistical techniques will not be discussed further in this book.

1-5 CONVENTIONS AND APPROACH

Units. We shall use SI units whenever possible. A few units, such as grams per cubic centimeter for density, are metric but not SI units, and they are so universal that we shall use them. In heat flow and in some aspects of electromagnetism, several systems of units are in common use. We give conversion factors and tables of units when appropriate. On figures taken from other publications we have left depth expressed in feet. Otherwise depth is expressed in meters, and it always increases downward.

Symbols. Whenever possible, we represent variables with the symbols adopted in the well logging literature. We have used the symbols list of the Society of Professional Well Log Analysts (SPWLA) to determine the proper symbol in most cases. Because of the wide range of subjects discussed in this book, we have used a few nonstandard symbols to avoid confusion. The symbols used are listed in the Appendix.

Porosity and compositions are often expressed as percents. This gives rise to a problem when discussing uncertainties. For example, for a porosity of 15 percent, whose accuracy is 10 percent, is the porosity range from 5 to 25 percent or from 13.5 to 16.5 percent? We avoid this problem by expressing porosities and compositions as fractions. Then we have a porosity of 0.15 with an accuracy of 10 percent, or 0.15 ± 0.015. Clearly, the range is from 0.135 to 0.165.

As we have noted in Sec. 1-4, we use f, with appropriate subscripts, to represent volume fraction. Two-letter subscripts represent solids, and one-letter subscripts represent fluids. For example, f_{sh} and f_w represent fraction of shale and fraction of water, respectively. The symbol S denotes the fraction of pore space filled with a liquid, again with appropriate subscripting. The water saturation S_w is equal to f_w/ϕ.

At times, especially when values are obtained from core data, mineral fractions are expressed as weight fractions rather than volume fractions. The user must be aware that weight fractions may be expressed as either wet or dry weight fraction and make the conversion to volume fraction accordingly.

Approach. A user of well log data is concerned with (1) the quality of the data, and hence with sonde calibration, (2) the response of a given sonde to the rock and wellbore, because corrections must be applied to account for the wellbore or invasion, or layering, and (3) the mixing rules which apply to the system when he computes an "answer," be it for porosity, saturation, or mineral percentage. Treatment of these three topics is interwoven into the discussion of sonde response and mixing rules throughout Parts 2 and 3 of this book.

To fully utilize logs, the user should be familiar with the physical principles and assumptions of each method. This is no easy task, because many different principles have already been embodied in borehole sondes and new ones will be introduced in the future. Consequently, in Part 2 we describe in depth the physical methods most important for well logging. We describe the interaction of fields and particles with media on both microscopic and macroscopic scales. We also discuss the response of minerals and fluids and the mixing rules which have been devised for each physical method. Obviously, we must be quite selective because each of these topics is the subject of many books and papers. For readers wishing to pursue a particular topic, references are included at the end of each chapter.

It is common practice to use combinations of well logs to solve geological and compositional problems. In Part 3 we discuss applications so that the usefulness of logs in combination will be manifest. It is in these sections that shortcomings in present-day log data become apparent. There are still many areas in which development of new sondes and methods of analysis will be especially welcome.

REFERENCES

Allaud, L. A., and M. H. Martin: *Schlumberger: The History of a Technique,* Wiley, New York, 1977.

Amyx, J. W., D. M. Bass, Jr., and R. L. Whiting: *Petroleum Reservoir Engineering,* McGraw-Hill, New York, 1960.

Helander, D. P.: *Fundamentals of Formation Evaluation,* Oil and Gas Consultants, Tulsa, Okla., 1983.

Johnson, H. M.: "A History of Well Logging," *Geophysics,* vol. 27, no. 4, August 1962, pp. 507–527.

Korvin, G.: "Axiomatic Characterization of the General Mixture Rule," *Geoexploration,* vol. 19, no. 4, June 1982, pp. 267–276.

Lynch, E. J.: *Formation Evaluation,* Harper & Row, New York, 1962.

Pirson, S. J.: *Handbook of Well Log Analysis,* Prentice-Hall, Englewood Cliffs, N.J., 1963.

Segesman, F. F.: "Well-Logging Method," *Geophysics,* vol. 45, no. 11, November 1980, pp. 1667–1684.

Wyllie, M. R. J.: *The Fundamentals of Well Log Interpretation,* Academic, New York, 1963.

2

Wellbore Environment

The purpose of logging is to measure the properties of rocks. Yet drilling a hole causes changes of in situ conditions and introduces new materials. Drilling fluids fill the wellbore and often invade the pore space of the rock. If a well is uneconomic, no other materials are added. But if the well is completed as either a water or an oil well, then cement and steel are added. These materials and changes are sensed by the logging instruments. Without knowing why these materials are used and what they are, the log user can make erroneous interpretations. The properties of rocks also change with both temperature and pressure, and it is helpful if these effects can be anticipated before a log is analyzed.

2-1 TEMPERATURE, STRESS, AND PRESSURE

Temperature. It has been found that temperature increases with depth in the earth's crust. Temperature gradients in the United States range from 8 to 50°C/km (0.44 to 2.7°F/100 ft). Thus in a deep well that reaches 6 km (the deepest U.S. wells on record are about 10 km) the temperature will reach 170°C (338°F). As discussed in Chap. 4, the thermal conductivity of rock and the heat flux determine the temperature gradient in a given area. Higher temperature gradients occur in areas of abnormally high heat flow and where hot circulating geothermal waters are present.

Logging cable and sondes must, of course, be constructed of components that can withstand the highest temperatures encountered in the well. Interpretation of log data may require temperature correction; the case of electrical resistivity is the most important (Chap. 5).

Lithostatic stress and fluid pressure. Lithostatic stress and fluid pressure also increase with depth at a near-linear rate simply because of the weight of the overlying rock and fluid columns. The lithostatic stress gradient is

$$\frac{\partial \sigma_z}{\partial z} = \rho_b g \tag{2-1}$$

where σ_z is the vertical stress component and ρ_b is the rock density. Within the wellbore and also within the rock wherever the fluid column is in equilibrium, the hydrostatic pressure gradient is

$$\frac{\partial P}{\partial z} = \rho_w g$$

A freshwater density ρ_w of 1.0 g/cm^3 causes a vertical pressure gradient of 9.8 kPa/m (0.433 psi/ft). A rock with bulk density ρ_b of 2.31 g/cm^3 gives a lithostatic stress gradient of 22.6 kPa/m (1.0 psi/ft). This gradient, illustrated in Fig. 2-1, is typical of gulf coast sediments (Fertl, 1976).

Because salinity increases with depth, so does brine density, with a corresponding increase in pressure gradient. A value of 10 kPa/m is cited for gulf coast conditions. Lithostatic stress gradients may differ substantially with depth, and from one geological setting to another. Compaction plays a role in the lithostatic gradient. Near the surface, relatively uncompacted sediments will be more porous and less dense, so the gradient will be lower than at depth. Where older, compacted rocks occur near the surface, and in metamorphic or igneous terrains, much higher lithostatic stress gradients occur.

The dots in Fig. 2-1 show measured values of fluid pressure above and below those predicted by the normal hydrostatic gradient. This does not mean that the hydrostatic gradient in the holes is uniformly high or low from the surface. It does mean that there are zones in which fluids are not in equilibrium with fluids in zones above or below, so that locally the gradient is quite different from hydrostatic. Such abnormal pressures can be sustained only if the formation is bounded by nearly

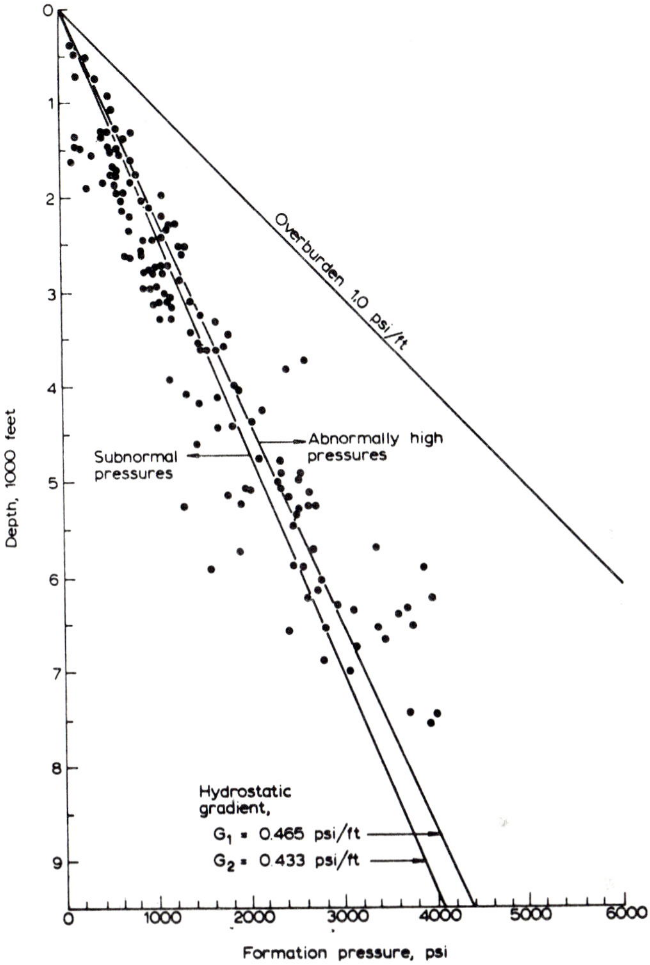

FIGURE 2-1 Lithostatic and hydrostatic pressure gradients (straight lines) and abnormal formation fluid pressures. (After Fertl, 1971. Copyright 1971, SPE-AIME.)

impermeable units such as shales. Underpressured zones may develop if the formation is steeply dipping and has good hydraulic connectivity with a low-pressure regime near the surface. Underpressure can also result from production of fluid from a formation bounded by impermeable units. Erosional processes have also been suggested as causing some underpressured zones (Fertl, 1976).

In overpressured zones (sometimes referred to as geopressured), the fluid partially supports the weight of the overlying rock mass and the load carried by the matrix is reduced correspondingly. Such overpressured zones can originate from the rapid deposition and subsidence of sediments, thus trapping the original fluids within the rock. Tectonic forces can play a role in regions where lateral forces are great enough to compress a rock mass and permeability is so low that the fluids cannot readily escape. Such a condition exists in the coast ranges of California (Berry, 1973).

Effective stress. Effective stress is the difference between the lithostatic stress and fluid pressure. Some physical properties such as acoustic velocity (Chap. 7) depend on effective stress rather than lithostatic or fluid pressure. Because the effective stress is reduced in overpressured zones, the corresponding change in physical properties can be detected with logs and used as an indicator of overpressure conditions (Fertl, 1976).

Lateral stress would ordinarily be no more than one-half the vertical stress, but the existence of tectonic forces tells us that horizontal stresses can sometimes be much greater. Data examined by Jamison and Cook (1980) show that the ratio of horizontal to vertical stress can be as much as 4 at depths ranging from 100 to 500 m. Although rarely done, it is possible to make in situ stress measurements at depths as great as 2000 m (Haimson, 1973).

2-2 DRILLING

Drilling methods vary greatly (Table 2-1). Augers and percussion drills are used for shallow wells in soil and overburden. Rotary methods with diamond core bits are used for most mineral and geotechnical work in which good core recovery is the goal of drilling. The rotary method with tri-cone bits is the standard for modern-day oil and gas drilling. We have drawn from books by Short (1982), Chilingarian and Vorabutr (1981), Campbell and Lehr (1973), Gray et al. (1980), and Moore (1974) to describe the aspects of drilling and drilling fluids which are most important to log acquisition and interpretation.

Percussion drilling. Percussion (cable-tool) drilling systems are commonly used to drill shallow water wells. The drilling instrument (the *bit*) and the pipe to which it is attached (the *tool string*) are suspended on the end of a heavy cable which is repeatedly raised and lowered, allowing the bit to strike the rock while the cable remains in tension. The cable is wound in a manner such that the bit rotates slightly with each succeeding impact, thereby enhancing the breaking action. Another pipe *(casing)* can be driven down as drilling progresses, so that

TABLE 2-1 Bit and core sizes for Diamond Core Drill Manufacturers Association (DCDMA) and metric standards.*

Metric standards		DCDMA standards		
Core diam, mm	Hole diam, mm	Core diam, mm	Hole diam, mm	Casing designation
132	146			
		112.7	145.3	SX
117	131			
		92.1	119.9	PX
102	116			
87	101			
		76.2	98.8	HX
72	86			
62	76			
		54.7	75.3	NX
52	66			
		42	60	BX
42	56			
		30.1	48	AX
32	46			
		21.5	37.3	EX
22	36			
		18.7	29.5	RX
16	28			

*After Weaver (1976).

formations are sealed off and hole collapse is prevented. The rock fragments removed in the drilling process *(cuttings)* are removed from the hole with a bailing device. No fluid return system is required; hence, the formation is not contaminated with drilling fluid *(mud filtrate)* and samples are returned to the surface uncontaminated by mud. Another advantage of bailing is that samples are returned with little loss or diminution in size.

Percussion drilling is generally slower than rotary drilling and less effective as depth increases, but equipment *(rig)* costs are less and the method is effective in unconsolidated formations. Water-bearing strata are easily identified; and because no mud circulation is required, there is minimal contamination of aquifers. Small percussion rigs operate to depths of about 100 m; larger rigs can drive a hole of 0.5 m diameter to about the same depth or a smaller hole to depths of about 600 m.

Rotary methods. Rotary methods are now the most commonly used of all drilling methods. The string of drill pipe is comprised of pipe segments *(joints)*, typically 10 m in length on the larger rigs, which are joined together on the rig floor as drilling progresses. The bit is turned by rotating the entire drill string. Drilling mud flows continuously down the hollow center of the drill string, out across the face of the bit, and up the annulus between the pipe and bore wall to the surface. The mud removes cuttings from the bit face and transports them up the annulus.

Rotary rigs used to drill holes for mineral and geotechnical evaluations obtain rock core continuously as drilling progresses. The bits are hollow cylinders with diamonds set on the leading edge to cut a circular kerf in the rock face. As long as the rock is reasonably competent, the bore cut by this method will be quite smooth, sometimes with wall irregularities *(rugosity)* of no more than a fraction of a millimeter. Core is recovered from a core barrel located immediately above the bit. The core barrel is pulled up the hole with a wireline. Mud circulation is maintained during coring to remove fine cuttings and reduce friction.

Core sampling is the primary goal of mining and geotechnical drilling, so the hole size is no larger than necessary to retrieve the desired core samples. Hole sizes range from 46 to 146 mm (Table 2-1). With such small diameters, logging operations are often restricted by the size of the hole. Small-diameter sondes designed to fit these holes are sometimes referred to as slim hole sondes.

Rotary drilling systems that are used for the larger holes required for oil and gas drilling employ the rolling cutter rock bit, which consists of three (sometimes two) cones mounted on the bit body. As the bit turns, the cones rotate and the teeth on the cones cut the rock by a combination of gouging and scraping in soft formations or by chipping and crushing in hard formations. The rock face is completely reduced to chips and fines which are transported to surface by the action of the circulating mud flow. Jet nozzles provide high mud velocity at the cutting surface, which enhances the cutting action. There are many variants in bit design to provide the most rapid cutting rate in different kinds of rock.

The diameter of oil, gas, and water wells is determined by the diameter required for testing and production equipment. Sometimes a certain minimum diameter is required if the hole is to stay open for any length of time. Tri-cone bits range in diameter from 15 to 45 cm, with holes of about 20 cm diameter most frequently encountered in oil well logging practice. The drilling of very large-diameter holes for shafts and for weapons testing is accomplished by variations of the rolling cutter design. A large number of rolling cutter cones is distributed on the surface of a large-diameter head. Holes as large as 4 m in diameter can be drilled with such bits.

Hole rugosity. Only if drilling practice is good and the rock is hard will the borehole be smooth and of the same diameter as the bit. Often the hole is rugose; that is, the diameter varies rapidly with depth. The diameter will also vary with azimuth at any particular cross section, so rather than being roughly circular in cross section, the hole may be elliptical. Ellipticity can result from the way the bit intersects a weak or faulted zone, or it can arise because the horizontal stress in one direction differs from that in the orthogonal direction. A four-arm caliper sonde (see Sec. 2-4), in which the orthogonal arms act independently, will indicate the elliptical character of a hole. However, the hole may also be highly irregular in cross section, and this cannot be determined from a four-arm caliper. It is known that holes tend to be more irregular in the direction of the long axis of the elliptical opening, where the horizontal stress component is at a minimum.

Rugosity can be caused by intersecting a fault, a fracture, a joint, a pebble, an overpressured zone, or a shale containing clays which expand when contacted by fresh water. Considerable rugosity is common in uncompacted or poorly compacted sediments. In such formations, the hole may enlarge with time as material continues to slough off from the formation and fall into the wellbore. When the hole sloughs or caves, the falling material may form a *bridge* which blocks the hole. Bridging is more likely to occur in holes drilled at a large angle to the vertical than in near-vertical holes. It may be possible (although risky) to work a logging sonde past a bridge, or it may be necessary to run the drill string back in and drill the bridge out. Such conditions increase the likelihood of a sonde becoming stuck in the hole and thereby limit the time after drilling when logging can be performed.

Hole deviation. Most wells are drilled vertically, or at least it is the operator's intention to drill them that way. However, even the straightest wells will wander 1 or 2° off vertical, and 5° inclinations are not uncommon. Wells drilled with low inclination (3° or less) tend to spiral, and wells with angles greater than 5° tend to progress in an arc without spiraling. Deviation occurs most readily where the drill bit traverses a layered sequence of varying drillability. The general tendency is to wander up-dip along the bedding planes when crossing formations of shallow dip and to wander down-dip along the bedding planes where formations are steeply dipping (Short, 1982). The mechanical principles governing hole deviation remain a subject of discussion. It is known, however, that deviation can be controlled to some extent by stiffening the equipment at the bottom of the hole. Small-diameter holes can be expected to deviate more than large ones because the connectors between the segments *(drill collars)* are necessarily more limber.

Deviation in itself is not necessarily detrimental to drilling or logging operations, provided the changes in well course are not severe. A sharp change in well course is referred to as *dogleg*, usually reckoned as the change in deviation angle over 100 ft (30.48 m) (Chap. 17); a dogleg of 1 or 2° per 100 ft is considered manageable. With higher dogleg, friction increases and *keyseats* can develop. As drilling progresses below a section with high curvature (dogleg), the drill pipe and particularly the collars of the pipe work against the side of the hole and abrade the formation until a scalloped indentation called a keyseat is produced. The diameter of the drill pipe is less than that of the collars and bit which comprise the bottom-hole assembly. Consequently, the collars may stick in the keyseat when the drill string is removed from the hole. If the keyseat is not removed, it can cause a logging sonde to stick in the hole. Data obtained in a keyseat section of the well may be misinterpreted if the wrong corrections for hole diameter are applied to the logs.

Sidetracking. During a second drilling operation in the same hole *(reentry)* or during a reaming operation, it sometimes happens that the bit does not follow the original course of the well but instead sidetracks onto a new course. It may be difficult to reenter the original hole, and the driller may decide to continue with the new direction. Sometimes it is possible to cement one of the two holes, sometimes not. In other cases, if there is an abandoned drill string or other "junk" in the hole or if it is not possible to control a leakage of drilling fluid into the formation *(loss of circulation)*, it may be desirable to sidetrack purposely. Whatever the cause, the conditions which have been created can confound a logging run. The sonde may progress down the wrong hole, or it may hang up on a ledge created at the sidetrack. It may respond to the fluid or the cement in the original well, particularly if the two parallel one another closely for some distance.

Directional drilling. There are many reasons to control the course of a well to a specified bottom location which is not vertically below the drilling rig. Multiple wells drilled from an offshore platform, an artificial island, or a drill pad on permafrost are splayed outward and then downward to tap a reservoir at appropriate spacings. Wells may be steered to avoid preexisting wells or geological structures such as faults or salt domes. Or it may be desirable to intersect a well into which there is an uncontrollable leak of formation fluid *(blowout)* at some specified depth with a control well. In these cases it is customary to drill vertically to the depth called the *kick-out point*, where deviation begins. Kickout can commence either by *whipstocking*, a technique in which a long slender wedge is inserted in the well to deflect the bit in the desired direction, or by drilling with a bent section of pipe *(bent sub)* and downhole motor. The bent sub is oriented in the desired direction, and the bit is turned by the downhole motor rather than with the drill string. Once the well is kicked out, directional drilling control is achieved by arranging the weight and centralizing devices in the bottom-hole assembly to increase or decrease the angle as required. Development of doglegs must be monitored and prevented when the doglegs begin to occur. Directional surveys (Chap. 17) are run as needed to check the course of the well. Aside from the higher frequency of mechanical problems which can occur in highly inclined wells, the impact upon logging arises from sampling horizontal beds at an angle, so that edge effects are not as sharp as if the hole were drilled normal to the beds.

Reference elevation. The driller tallies the pipe and casing lengths and measures hole depth from some convenient point on the rig, usually the device that rotates the drill pipe (the *kelly bushing*). Core depths are reckoned from the same point. Usually log depths are also referenced to the kelly bushing, unless the rig has been removed, in which case logging depths must be referenced to a surface collar or flange. Adjustment to a common ground level or sea level elevation must account for any change in reference elevations. Conversion of depths measured along the well course to true vertical depth is discussed in Chap. 17.

2-3 DRILLING FLUID AND MUD CAKE

Drilling mud expedites the drilling operation by cooling and cleaning the bit, reducing friction, and transporting cuttings from the bit to the surface. These mud functions affect logging only insofar as they help produce a good-quality hole. For example, cuttings not removed from the well will settle to the bottom and prevent the logging sonde from reaching the lower portion of the hole. Mud also serves to control pressure downhole, stabilize the wellbore, and form a mud cake which prevents loss of mud from the hole into the formation. To meet these requirements, mud properties are adjusted for different depths and geological conditions. The thickness of the mud cake and the properties and amounts of fluid which invades the formation are in turn controlled by the mud type. The factors which guide the choice of a fluid base and its additives are complex (Gray et al., 1980); we will restrict our discussion to those mud properties which impact the logging program.

Mud properties. Water is most commonly used as the base for drilling fluids, although oil and air also are used. Other materials are added to the fluid base to achieve the desired weight, flow, and caking properties. Solid particles are classed as colloidal (0.005 to 1 μm), silt size (1 to 50 μm), and sand size (50 to 420 μm). The colloidal particles, which are clays and organics, determine the viscosity and other flow properties of the mud. Viscosity is usually controlled by adding bentonite (primarily montmorillonite) or other clays. The silt-size particles, which include barite, determine the fluid density (the "weight"). The sand-size particles, which are made in the cutting process, are regarded as a nuisance because they abrade pumps and pipe.

Salts and organic compounds are added to reduce sloughing of shale, minimize dissolution of rock salt or anhydrite, and control fluid properties at high pressure and temperature. The primary effect of salt on well logs is to lower the electrical resistivity of the mud and mud filtrate. Sometimes the salt is KCl instead of NaCl, and that produces another adverse consequence: the radioactivity of potassium in the wellbore adds an unwanted γ-ray contribution to the natural radioactivity of the formation.

Fluid density. Drilling fluid density is controlled to maintain bottom-hole pressure high enough that formation fluids cannot enter the hole. Near the surface it is possible to drill with water or even air; but as depths increase, mud is required. Bottom-hole pressure is kept higher than formation fluid pressure (often called the pore pressure) but not so high as to reduce the rate of penetration. An example of how mud density must be increased as drilling progresses is shown on the left side of Fig. 2-2. At high mud weights there is increased risk of fracturing the rock and also of sticking the drill pipe. Mud weight naturally increases as small-diameter particles resulting from the drilling process *(drill fines)* are incorporated into the mud stream. If density higher than that caused by the fines is required, barite, which has a specific gravity of 4.25, can be added. A high-density mud in the hole increases the buoyancy of a logging sonde so that sinker bars must be added to get the sonde down the hole. In small-diameter or highly inclined holes, the difficulty becomes even greater. The high density of barite has another side effect: γ-rays are more highly attenuated by barite mud than by water or low-density mud.

Mud cake. As mud infiltrates newly drilled rock surface, the fluid passes into the pore space of the rock and leaves behind a mud cake comprised of the silt- and sand-size particles. The mud cake begins to build once the rock pores are bridged by solid particles. If the particles do not bridge the rock pores, then more solids will enter the formation during the initial "mud spurt" and thereby reduce the permeability of the formation. Hence, it is desirable to build mud cake rapidly and to

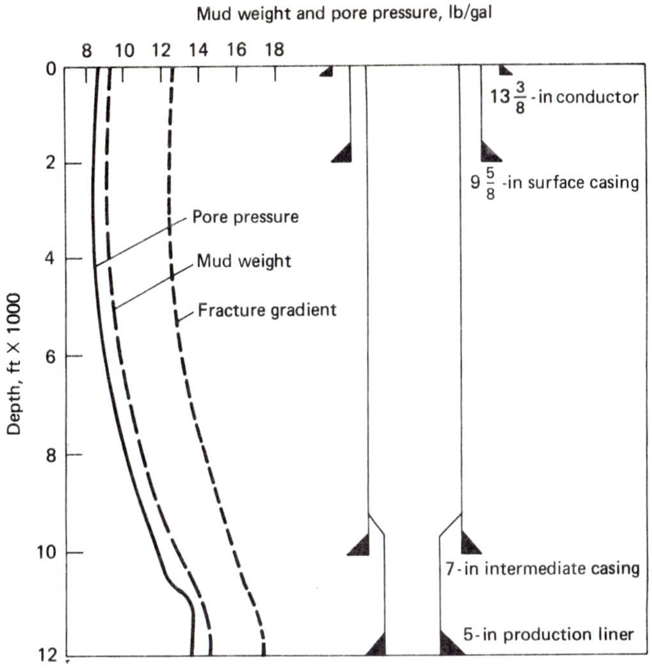

FIGURE 2-2 Typical oil well design. The hydrostatic and mud pressure gradients are given as equivalent densities (lb/gal). Casing stages indicate diameter and depth to casing shoe (solid triangles). (After Short, 1982.)

make it as impermeable as possible. Nevertheless, filtrate does invade the formation and displace the natural *(connate)* movable fluids. The rate at which it does so is controlled by mud cake permeability, which ranges from a fraction of a microdarcy to several hundred microdarcys. If formation permeability is much greater than mud cake permeability, then formation permeability plays no role in determining the volume of filtrate which invades the formation; the volume is determined by the magnitude and duration of the pressure differential acting across the mud cake. For a given filtrate volume, the formation porosity determines the extent of filtrate invasion. Filtrate invades farther into a low-porosity formation than into a high-porosity formation. On the other hand, if formation permeability is comparable to or less than mud cake permeability (as it would be with some shales and igneous rocks), the formation permeability limits the volume of filtrate, and mud cake can be expected to be minimal or nonexistent.

Invasion of a single horizontal and permeable formation is depicted in Fig. 2-3. Mud filtrate has entered the flushed zone and displaced all movable fluids. The uninvaded zone represents rock with its fluid com-

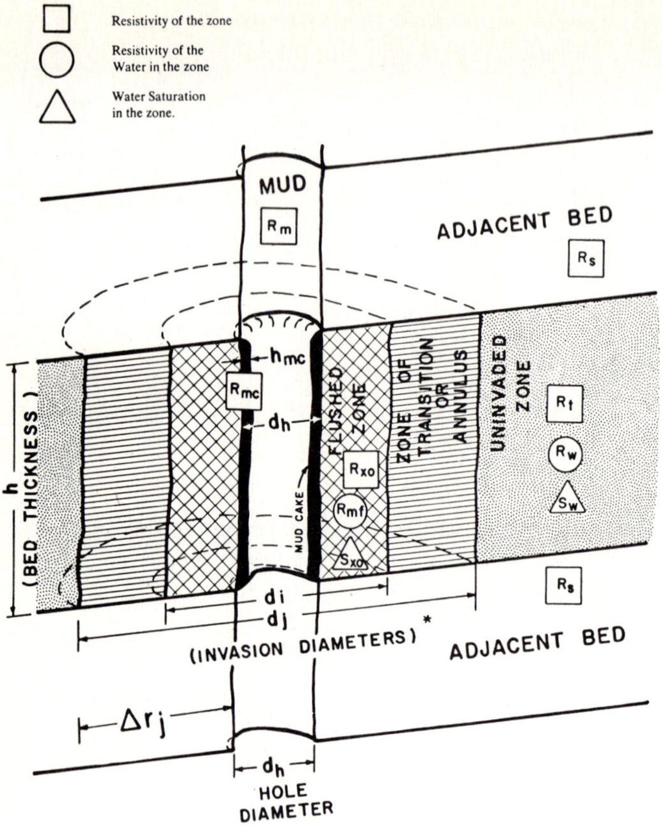

FIGURE 2-3 Schematic of zones invaded by mud filtrate. (Courtesy of Schlumberger, 1979.)

position unchanged. Between the flushed and uninvaded zones is a zone of transition or annulus which contains a mixture of filtrate and reservoir fluids. The subscripts denoting the different invasion zones are now standard nomenclature in the logging industry. R represents the electrical resistivity of rock and fluid (Chap. 5), and water saturations are given by S, both with appropriate subscripts. It can be seen that the adjacent beds are assumed to be uninvaded, since their true resistivity, given by R_s, is the same both near and far from the wellbore.

2-4 CALIPER LOGS

Once a hole has been drilled, it is desirable to know its true size and shape. The instrument used for this purpose is called a caliper log.

Many different types of caliper log are described by Hilchie (1968). There are two major types: independent calipers for supplying detailed information abut hole conditions and calipers attached to other types of sondes to supply hole-size information for correcting data from those sondes. The attached calipers do not require high accuracy, because the effects of their inaccuracy will be second order. The independent calipers, which we shall emphasize, provide greater accuracy.

Most caliper sondes are mechanical; one acoustic type is discussed in Sec. 19-4. The mechanical sonde consists of several arms, which are pushed out against the wall of the hole by springs or hydraulic pressure. Figure 2-4 illustrates a six-arm caliper sonde. A transducer such as a potentiometer or linear differential transformer indicates the angle of the arm, which is converted to the distance from the center of the sonde to the wall.

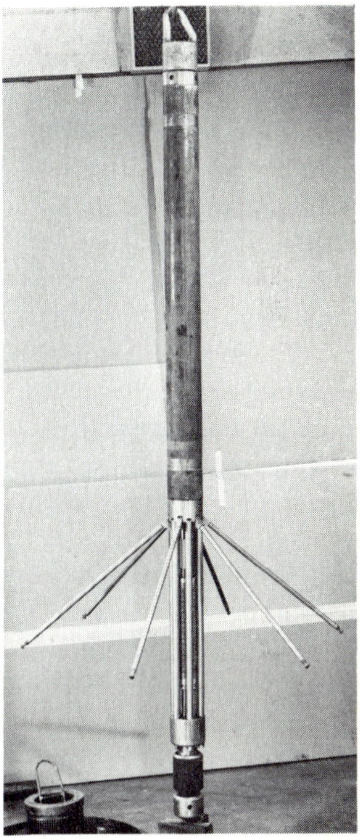

FIGURE 2-4 Six-arm caliper sonde. (Courtesy of Birdwell.)

The vertical accuracy of a caliper measurement depends on the size of the arm tips that touch the wall and on the speed with which the recording device responds to changes in arm position. Hilchie (1968) describes several fairly large contact devices, but they are generally of the attached type; independent calipers tend to have small tips. He also discusses the effect of contact pressure; low pressure permits the contact to float on top of the mud cake, and high pressure will sometimes cut through the mud cake. Low contact pressure is best for an attached caliper, because the diameter of the fluid column is required. High contact pressure is better for an independent caliper if actual hole diameter is required.

Hilchie says "most calipers appear to be designed to log cylindrical bores." By this he means that it is common practice to record twice the average of the extensions of all the arms as the hole diameter. Most independent calipers have three or four arms; some have as many as six. If the hole is circular in cross section, the number of arms does not matter. If, as often happens, the hole is elliptical or even less regular in cross section, a caliper giving only the average of the arm extensions will, of course, not show this. In fact, the area that would be obtained by assuming a circular cross section with a diameter indicated by the average can be very wrong. Hilchie (1968) shows that the center of a three-arm caliper will not be at the center of an elliptical hole, and thus it will not give a good area even if the area is computed from independent recording of the three arms.

Clark (1982) has used television to observe a two-arm sonde in a washed-out hole. The arms lock into the largest diameter of a noncircular hole. This indicates that a four- or six-arm caliper will give a fairly good picture of the situation if the diameters are recorded independently. It should be noted, however, that it is still not certain that the center of the sonde will be in the center of the hole, so one diameter may be less than bit size.

Caliper logs can give lithologic information if they are properly used, because washouts are indicative of formation properties. Weak formations will tend to wash out; strong ones will not. Caliper logs can also give evidence of the presence of fractures. Babcock (1978) has used the four-arm caliper of a dipmeter to identify what appear to be long, vertical fractures. He found increased diameter from one pair of arms at consistent azimuth in many holes. The sonde rotation stopped in these zones of increased diameter, in agreement with Clark's (1982) results.

The azimuths described by Babcock are parallel with the direction of minimum principal stress. Ordinarily, fractures are perpendicular to the direction of minimum principal stress. Therefore, the fractures observed by Babcock are not controlled by the existing state of stress. Thorpe and Springer (1982), on the other hand, find the azimuth of borehole elongation to be perpendicular to the minimum principal stress and therefore to be an indication of the direction of the stress. Evidently both fractures and stress affect hole diameter.

Data from many types of logs (for example, density, neutron, acoustic) are affected by changes in the distance from the sonde to the wall of the hole. When the caliper indicates a rough hole, the user should beware of these effects.

2-5 WELL COMPLETION

Although the majority of logging operations are carried out in uncased *(open)* holes, many are undertaken in wells that have been *completed* to diagnose completion problems or to look for changes in fluid saturation as production progresses. Therefore, the log user needs to have some appreciation of completion methods. Completion is comprised of three stages: the installation of casing, the cementing of casing to the formation, and the perforation of casing to provide access to productive intervals. Water or oil can be pumped from some wells *(produced)* without any casing or screen if the formation is competent enough. Water wells are completed with screen rather than casing. Of course, mineral and geotechnical exploration holes are never completed. They are either plugged with cement or left open until they eventually cave and fill.

Casing. Casing programs vary greatly depending upon the depth of the well, the number of zones to be produced, the amount of formation testing required, and how well the area is known. We shall discuss the four-stage casing installation shown on the right side of Fig. 2-2 to illustrate the reasons for certain aspects of casing design. The first casing to be set is called the conductor. Its purpose is to hold back the unconsolidated formations near the surface, and it may also support some of the weight of subsequent casing strings. The second stage, called the surface casing, varies in depth from about 100 to about 1500 m. It also supports unconsolidated formations, isolates shallow aquifers, and provides support for surface equipment and subsequent casing. The annulus is usually filled with cement to the surface. Next comes the intermediate casing, which serves to isolate subsequent drilling from potential lost circulation zones and sloughing shales and to isolate water and gas flows which would interfere with drilling. The intermediate casing can serve as part of the *production string* if a *production liner* is set. The production casing (or liner, as shown in Fig. 2-2) withstands the maximum shut-in pressure of the producing intervals. If the rock is competent and the formation produces little water, the well can be completed as an open hole, and no production casing is used.

The four-stage example of Fig. 2-2 illustrates completion for a 12,000-ft (3658-m) well. In order to control the well, the mud pressure must be greater than the pore pressure, as shown. However, at 10,000 ft (3048 m) the mud weight exceeds the pressure required to fracture the formation at a depth above 5000 ft (1524 m). Hence, the intermediate casing has been seated just above 10,000 ft (3048 m) to prevent fracturing the formation higher in the hole. It is easy to see that a deeper well will generally require a more complex casing arrangement and that a shallower well can be engineered with less complexity. The greater the complexity of the casing arrangement, the more chances there are for logging sondes to catch on a lip or protuberance.

Casing must have sufficient tensile strength to hold its own weight. It must also withstand external (collapse) pressure and internal (burst) pressure. Specifications for casing used by the water well industry are discussed by Campbell and Lehr (1973). Sizes specified for oil and gas wells by the American Petroleum Institute (1971) range from 4.5 to 20 in (11.4 to 50.8 cm) in diameter. Each size comes in a range of wall thicknesses (specified by weight per unit length, or "weight") with strength properties increasing with weight. For example, 5.5-in (14-cm) casing is available in six weights ranging from 13 to 23 lb/ft (94 to 166 kg/m), with corresponding wall thicknesses ranging from 0.228 to 0.415 in (0.58 to 1.05 cm). Thus in order to know the thickness of steel encountered in a logging measurement, both the size and weight must be given. The threaded couplings (collars) must be as strong as the unthreaded part of the pipe; consequently, their outside diameter is greater than that of the pipe. Couplings on 5.5-in (14-cm) casing have an outside diameter of 6.05 in (15.37 cm) and are 5 to 6 in (12.7 to 15.24 cm) long. Couplings and the bottom of the casing *(casing shoes)* can be detected with the casing collar locator (CCL) on a logging sonde.

Well logs may or may not be run before intermediate casing is set, depending upon the requirements of the operator. In oil and gas wells they are almost always run after the intermediate casing has been set and the production or test interval has been drilled. It is at this stage that timing in log acquisition and interpretation can be critical: the decision either to plug and abandon an exploratory well or to complete and test it may be based on the results of the logging run.

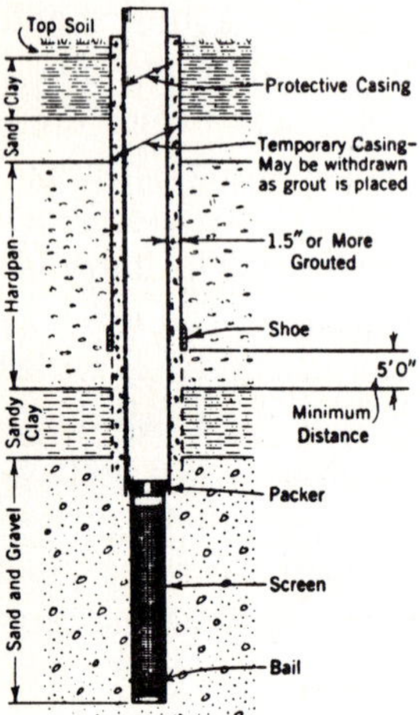

FIGURE 2-5 Typical water well design. (From Campbell and Lehr, 1973.)

Well screens and gravel packs. Water wells are usually completed with a screen in the productive aquifer. Figure 2-5 shows one of 11 different types of water well completion. If the aquifer is unconsolidated, the screen is surrounded with a gravel pack which stabilizes the formation; and if the gravel is properly sized, the pack prevents the finer formation sand from flowing into the screen along with the water. Screens and gravel packs are also used in oil and gas wells in poorly consolidated formations to reduce the amount of sand produced along with the oil. The completion may be similar to that shown in Fig. 2-5, or the screen and gravel pack may be set inside the perforated casing. In some cases, the gravel may be ejected through the perforated casing and into the formation to stabilize the perforations in the formation. The gravel pack affects logs because it reduces the amount of fluid present in the annulus, but it is generally of higher porosity than the formation. Sometimes a well log is run to find the top of the gravel column.

Cement. Cement provides mechanical strength to casing by bonding the casing to the formation. It also prevents fluids from flowing in the annulus between casing and rock and thereby protects productive formations from invasion, minimizes the risk of blowouts, and seals off zones in which drilling fluid flows into the formation *(lost circulation zones)* during drilling. Logs are used to inspect the bond between the cement and casing (Chap. 16) because many production problems are caused by failure of the cement bond.

Cement is composed of calcium aluminates and silicates, some iron, and water (Smith 1976). High pressure and temperature accelerate the setting of cement, so retarding agents are sometimes added to allow enough time to pump the cement down the hole before it sets. Lignins and organics are used as retarders. It may be necessary to reduce the cement density, in which case bentonite may be added in amounts as high as 12 percent. Or if formation pore pressure is high, it may be necessary to add heavier materials such as hematite, barite, or sand. Salt is sometimes added to cement to provide a better bond in salt zones or to prevent sloughing of shales.

The cement slurry is pumped down the inside of the casing, through a valve, and up the annulus to the desired height (Fig. 2-6). A top plug separates the slurry from the displacing fluid. Once the plug seats, the valve closes and sufficient time is allowed to elapse for the cement to set. Then the plug, the cement which has set inside the casing, and the valve are drilled out and the next stage of drilling proceeds. Sometimes the outside of the casing is fitted with centralizers so the cement flow will be distributed more equally around the casing. Cement forms a poor bond with mud cake, so sometimes the casing is fitted with scratchers (not shown) to remove the mud cake as the casing is lowered into the hole.

An inadequate cement bond may necessitate remedial work called *squeeze cementing,* which denotes the application of cement under pressure to a particular point in the well (Rike, 1973). Squeeze cementing may also be done in response to changed conditions in fluid levels as production proceeds. Often the goal is to cement off perforations in the formation. If the formations are clean, then applied pressures are

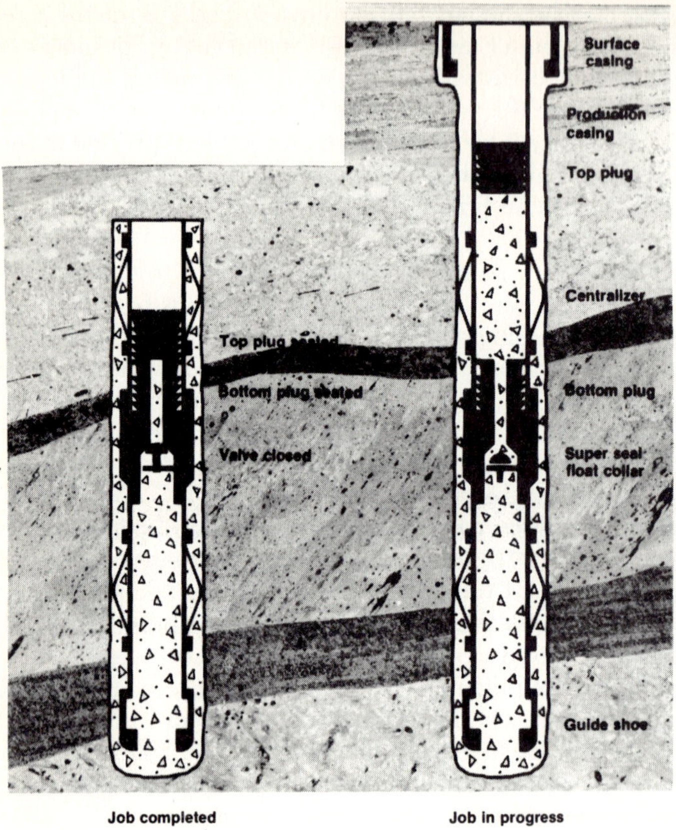

FIGURE 2-6 Emplacement of cement around outside of casing. A top plug separates the cement slurry from the displacing fluid.

low and cement volumes required are small. However, if it is necessary to force mud and debris from the clogged perforations, high pressure is applied, the formation is fractured, cement slurry enters the fracture, and some cement filtrate enters the formation.

2-6 HYDRAULIC FRACTURING

The advent of hydraulic fracturing to improve productivity has been called "one of the major developments in petroleum engineering" (Krueger, 1978). In order to fracture the formation, hydraulic pressure greater than the sum of the minimum compressive stress and the tensile strength of the formation is applied at formation depth. Because the fractures open in a direction normal to the least principal compressive stress, hydraulic fractures thus created tend to be vertical in deep wells and horizontal in shallow wells. Once the fracture is opened, viscous fluid carries sand-sized particles into it, so that it remains propped open once pressure is released. As a result, the effective permeability in the vicinity of the wellbore is notably enhanced. It is sometimes desirable to find the location and extent of fractures created in this manner (Chap. 13).

REFERENCES

American Petroleum Institute: "Specifications for Casing, Tubing, and Drill Pipe," API Standard 5A, 31st ed. April 1971.

Babcock, E. A.: "Measurement of Subsurface Fractures from Dipmeter Logs," *AAPG Bull.*, vol. 62, 1978, p. 1111.

Berry, F. A. F.: "High Fluid Potentials in California Coast Ranges and Their Tectonic Significance," *Bull. Am. Assoc. Petr. Geol.* vol. 57, no. 7, July 1973, pp. 1219–1249.

Campbell, M. D., and J. H. Lehr: *Water Well Technology*, McGraw-Hill, New York, 1973.

Chilingarian, G. V., and P. Vorabutr: *Drilling and Drilling Fluids*, Elsevier, New York, 1981.

Clark, S.: personal communication, 1982.

Fertl, W. H.: "A Look at Abnormally Pressured Formations in the U.S.S.R.," *Soc. Petr. Eng. Paper 3613, 46th Ann. Meeting*, 1971.

————: *Abnormal Formation Pressures*, Elsevier, New York, 1976.

Gray, G. R., H. C. H. Darley, and W. F. Rogers: *Composition and Properties of Oil Well Drilling Fluids*, 4th ed., Gulf Publishing, Houston, Tex., 1980.

Haimson, B. C.: "A Simple Method for Estimating In-Situ Stresses at Great Depths," *ASTM Symp. Field Testing Instrumentation Rock*, 1973.

Hilchie, D. W.: "Caliper Logging—Theory and Practice," *Log Analyst*, vol. 9, no. 1 1968, p. 3.

Jamison, D. B., and N. G. W. Cook: "Note on Measured Values for the State of Stress in the Earth's Crust," *J. Geophys. Res.*, vol., 85, no. B4, 1980, pp. 1833–1838.

Krueger, R. F.: "Advances in Well Completion and Stimulation During JPT's First Quarter Century," Soc Petr. Eng. Reprint Series No. 5a, 1978.

Moore, P. L.: *Drilling Practices Manual*, Petroleum Publishing, Tulsa, Okla., 1974.

Rike, J. L.: "Obtaining Successful Squeeze-Cementing Results," Soc. Petr. Eng. Paper 4608, 1973. Reprinted in Soc. Petr. Eng. Reprint Series No. 5a, 1978.

Schlumberger: *Log Interpretation Charts,* Schlumberger, New York, 1979.

Short, J. A.: *Drilling and Casing Operations,* PennWell Books, Tulsa, Okla., 1982.

Smith, D. K.: *Cementing,* Soc. Petr. Eng., Dallas, Tex., 1976.

Thorpe, R., and J. Springer: "Relationship between Borehole Elongation and In-Situ Stress Orientation at the Nevada Test Site," *Proc. 23d U.S. Rock Mech. Symp.,* Berkeley, Calif., 1982.

Weaver, J. M.: "Exploratory Drilling for Rock Engineering," in Z. T. Bieniawski (ed.): *Exploration for Rock Engineering,* A. A. Balkema, Rotterdam, 1976.

3

Logging Practice

The practice of logging consists of lowering an instrument down a borehole and recording the output of that instrument. In this chapter we discuss the methods for performing that task.

3-1 FIELDING EQUIPMENT

Vehicles. Most logging jobs (on land) require a single truck that carries both the winch and cable, or *drawworks*, and the recording equipment. The cable is usually suspended from the traveling block of the drill rig. If there is no drill rig over the hole, and the instruments are longer than about 3 m, a portable mast is required. Often a mast truck is used; the mast is perhaps 10 m high. Some logging trucks also carry a small mast. When complex on-site computations are to be performed, or if complex logs such as neutron-induced γ-ray (Sec. 6-8) or digitally recorded full-wave acoustic (Sec. 7-3) are to be run, a separate computer van may be required. Offshore, or in remote locations, non-wheeled boxes called cabs or skids are used instead of trucks. Sometimes the logging equipment is semipermanently attached to the drilling platform.

Drawworks. The drawworks consist of a large winch capable of holding 6 km or more of logging cable. There are sheave wheels to guide the cable, and a level-wind (called a *spooler arm*), much like that on a casting reel, to spread the cable evenly on the winch drum. Speed is controlled either mechanically or hydraulically; the hydraulic method seems to offer finer control. A weight or tension indicator tells the operator the tension on the cable.

On the axis of the drum is a device, called a *collector*, which connects the signal wires of the cable, which rotate with the drum, to the nonrotating wires, which lead to the processing and control electronics in the truck. Generally a pair of concentric cylindrical contacts, one rotating and one stationary, called *slip rings*, is used to make the connection for each wire. These slip rings can be a source of noise and other difficulties.

Cable. The logging cable is usually double-armored with two layers of steel wires, wound with opposite lays, outside the conductors. This winding often causes the sonde to rotate slowly as it is pulled up the hole.

Inside the armor are one, four, or seven conductors, depending on the service company and the application. The conductors are commonly copper wires about 0.7 mm in diameter (No. 22). Occasionally cross talk between these conductors can occur, and this can sometimes cause problems in the data. The cable diameter will be in the range of 4 to 15 mm, depending on the number and size of conductors.

If a single conductor is used, the signals are multiplexed in the sonde and then demultiplexed at the surface. Multiplexing is now becoming common even when multiple-conductor cables are used.

A *cable head* connects the cable to the sonde. The head must isolate the inside of the pressure housing of the sonde from the fluid and pressure in the hole. Each service company has its own head design (or designs).

3-2 DEPTH MEASUREMENT

If all depth measurements in a hole were made in exactly the same way, depth accuracy would often be unimportant; relative values would suffice. But in fact, the depth measured by the driller is obtained by counting, and sometimes measuring the length of, segments (or *joints*) of drill pipe, whereas the depth measured by the logger is obtained from the length of cable in the hole. These measurements often disagree. Depths measured by different loggers also disagree. Sometimes one can even get disagreement with different runs by the same logger. Pilkington and Fertl (1975) describe the running of two different cement bond sondes by the same contractor; there was a discrepancy of 3 m at a depth of 1400 m. That is not at all uncommon. It is even possible to get different depths for different runs of the same sonde when the sonde has not been removed from the hole between runs!

3-2a Depth Correlation

The first step in any procedure for comparing different logs in a hole (for example, *cross-plotting*) is to correct the depths. Not only depth accuracy but also sticking is a problem. Often a sonde, especially a pad type like a density or sidewall neutron sonde (Chap. 6) will hang up briefly in a hole, then start up again, hang up again, and so on. The cable stretches while the sonde is stationary, and it can stretch as much as 5 m or more (Cooley, 1974). Cooley pointed out that sticking can be recognized on sonic or resistivity logs because a sticking sonde will give straight lines on the log. Statistical logs like density or neutron (which are more likely to stick) are not as easy to recognize. The problem is compounded when several instruments are combined on one sonde so that the recording point on each instrument is at a different depth. Then the apparent locations of the sticking are different for each record. Even short sticking intervals make crossplots difficult to interpret.

Cooley recommended including a tension or weight curve (derived from the weight indicator) on each log. Then sticking could be recognized by increased tension. He also suggested that the distance between different instruments on the same sonde be recorded on the header so that sticking conditions could be correlated among records.

Zangwill (1982) described a correlation method (Sec. 7-4b) to compensate for both depth inaccuracy and sticking. Instead of shifting depths of an entire section of a log, as is often done to compensate for depth errors, his method is to expand and contract different, small parts of the record to compensate for sticking. Kerzner (1984) has developed an even more sophisticated method.

It is possible to "tie" different logs together by attaching a device, such as a casing collar locator (for a cased well) or a γ-ray sonde to each instrument run. Such a device will locate the same objects at fixed depths in the hole each time and thus supply reference depths.

Casing collar locator. In a cased hole, the places where two segments of casing are joined together *(casing collars)* are thicker than the rest of the casing. They can be located by a device called a casing collar locator (CCL), which senses the change in magnetic field caused by the thicker collars. Often a CCL is added to another logging instrument to provide depth references.

***γ*-Ray.** If more than one contractor is used for logging, it is a common practice to have each run γ-ray logs (Sec. 6-3). Often certain formations will have characteristic *signatures* on a γ-ray log, and peaks or valleys in the γ-ray trace in these formations can be used as depth references. As in the case of the CCL, γ-ray detectors are often attached to other instruments for depth references.

3-2b Depth Measurement

The major difficulty with depth measurement with a wireline is that the cable stretches up to 1 m/km under ordinary circumstances and as much as 2 to 6 m/km with very heavy loads. The length of a piece of cable is a function of the tension on the cable (and, to some extent, the temperature). The total amount of cable in the hole, the weight of the instrument, and the density of the borehole fluid all affect the tension.

Primary depth measurement. The primary depth measurement device usually is a *measure wheel,* whose radius is precisely adjusted to a specified value. The rotation of the wheel is coupled to the depth indicator at the recording device by either mechanical or electronic equipment. The system is designed to convert the number of revolutions of the wheel to the length of cable under stress passing across its rim. The accuracy of the wheel system of measurement is limited, and decreases because of wear, slippage, and mud buildup, so additional controls on depth accuracy are required. The most common of these is the magnetic marker.

Magnetic markers. The common procedure is to magnetize the cable armor at precisely measured regular intervals. This operation is performed at the contractor's base station while the cable is under some specified tension. The accuracy of the placement of the magnetic marks significantly exceeds that of the measure wheel. In service, each time a magnetic marker passes a detector, which is usually mounted on the spooler arm of the truck, some indication (such as a ringing bell) is given to the operator.

When the cable is lowered into the hole, the recorder driven by the measure wheel indicates a certain depth. Near the bottom, when a marker is sensed, the operator knows the exact length of cable out and can correct the depth from the recorder to the true depth indicated by the marker. A chart to correct the length indicated by the marker for the actual tension in the cable is used to provide true depths. Of course, as the cable is reeled in, the tension decreases and this correction must be changed. Often, as each marker passes the detector, a bell rings in the truck and the operator makes a fresh depth correction based on the tension at that time. Ordinarily the procedure of determining the proper correction is required only on the first run of a suite. If the corrections are not satisfactory for subsequent logs, they are tied to the first by the methods described in the preceding subsection.

As the cable ages, a permanent set occurs, and the distance between markers changes. Consequently, the cable must periodically be re-marked. This procedure of marking and correcting is a fecund source of depth error. The user should worry about it.

Calibrated sheave. In some circumstances, such as offshore or remote operations, it is impractical to keep an up-to-date magnetically marked cable on site. In such cases, a calibrated sheave wheel may be substituted. It is machined accurately so that it is in contact with the cable over about half the circumference of the cable and sheave. The calibrated sheave (Brabeck, 1983) is used in place of one of the sheaves over which the cable travels, usually the upper one. The cable is under full tension when it passes over the sheave, so no stretch correction is required. The sheave is attached to a selsyn motor at the axis, and the selsyn drives a digital counter in the truck. By use of offsets in the digital counter, the sheave output is precisely calibrated (at the service company's district office) to reproduce the depth increments of a magnetically marked cable at specified reference tension. Since it is still subject to the vagaries of electromechanical systems, the accuracy of the calibrated sheave is generally not considered as great as that of the regularly marked cable. It nevertheless remains better than that of the measure wheel taken by itself.

3-3 SONDE CHARACTERISTICS

Many terms are used to indicate the instrument that enters the hole: *tool, probe, sonde,* and even *log* are often used. We shall use the term "sonde"; although it strictly applies only to an acoustic instrument, it is in common use to refer to all kinds of instruments.

Borehole conditions can be hostile (Sec. 2-1). High temperatures and high pressures are common, and corrosive fluids are not rare. The diameter of the hole to be logged can vary from less than 3 cm to more than 3 m, depending on the situation.

Up to a point, larger diameters permit the use of sondes that give better data, but very large diameters require ingenuity to provide contact with the borehole wall when it is required (Atkinson and Pointer, 1975). Special problems arise with collimation of radioactivity sondes (Chap. 6) in small-diameter holes. If there is high pressure in the hole, it may be necessary to pass the sonde and cable through a *blowout preventer.* This equipment sometimes limits the diameter of the sonde to less than that which would be satisfactory for the hole itself. It should also be realized that a reasonable amount of clearance between the sonde and the hole is required, since the hole is not perfectly straight.

Ordinary sondes (10 cm or less in diameter) are specified to withstand pressures of about 140 MPa (20,000 psi) and from 125 to 175°C; the parts most sensitive to temperature are the solid-state circuit elements. Some equipment has been built to withstand as much as 315°C for geothermal logging; the most difficult problem in such logging is the temperature capability of the cable and head. All service companies publish temperature, pressure, and diameter specifications.

Because the diameter is limited and much equipment must be installed in the sonde, many sondes are very long (8 m is not unusual) even if a large distance between a source and detector is not required. Some sondes can become quite heavy, but others are buoyant in high-density muds and *sinker bars* are needed to make them go down the hole.

3-4 SIGNAL PROCESSING

The techniques of signal processing are changing so rapidly that we shall present only a brief outline and history of the techniques used in logging. Originally, all logging sondes produced analog signals. With the advent of more sophisticated measurements, such as by radioactive and acoustic logs, it became possible to transmit digital data in the form of count rates, time intervals, and so on uphole. Upon reaching the surface, both digital and analog data were converted and processed to give either digital or analog outputs.

Analog outputs. Until recently, each type of sonde had its own surface *panel* in the truck. This panel contained the electronics to convert the signals from the cable into voltages or currents proportional to the value of the variable to be measured—resistivity, count rate, porosity, or the like. Consequently, each panel embodied the assumptions inherent in the design of the system. For example, a neutron panel might contain, hardwired, the algorithm for conversion from count rate to porosity (Sec. 6-6). There would be switches and knobs to enter constants such as hole size and borehole fluid to be used in the hardwired algorithms. These panels were connected to another patch panel (Fig. 3-1) connected to recording equipment.

Commonly, the currents from the panels drove galvanometers whose indicators were pens on a strip chart recorder. The chart was driven by the depth indicator in some manner. It is interesting to note that almost all subsequent recording methods, and many of the control methods, have been made to look as if the data were being recorded with strip chart recorders and galvanometers.

Some service companies converted to recording on film; the galvanometers drive mirrors like those used in seismic recorders. This method has the advantage of faster response, and problems with ink are eliminated; it has the disadvantage that the user cannot observe the data in real time. Eventually recording was converted from galvanometers to electron beams (like an oscilloscope) for recording on film and to fiber optics cameras for better resolution.

Digital recording. Digital recording has been very useful for the logging industry; it has made possible much more sophisticated processing of individual logs and all sorts of methods of working with combinations of logs. Digitization progressed from using a position-sensitive digitizer tablet and hand-tracing a record from a strip chart recorder through digital recording of the output of the panels (Fig. 3-1). The next step was digitization of the signal at the surface and replacement of the indi-

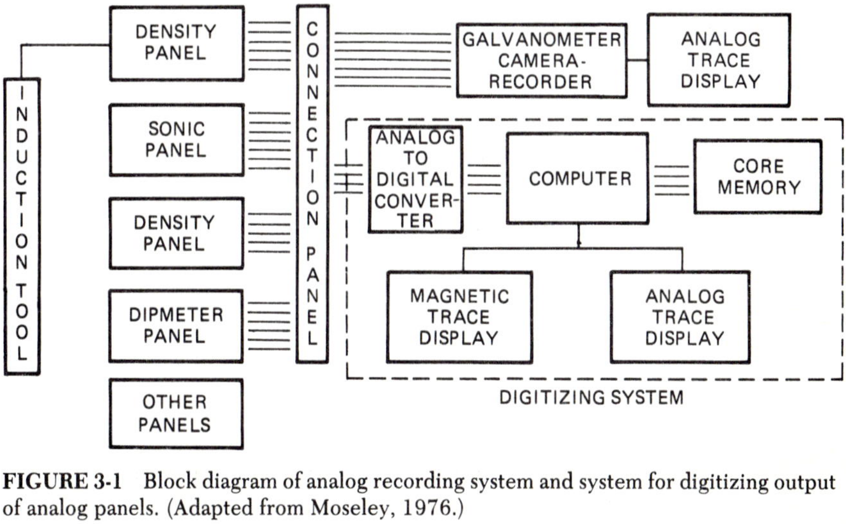

FIGURE 3-1 Block diagram of analog recording system and system for digitizing output of analog panels. (Adapted from Moseley, 1976.)

vidual panels by a computer programmed to reduce data from any sonde (Fig. 3-2).

Currently, digitization, and often a fair amount of digital processing, is done in the sonde itself, and digital signals are sent up the cable. This permits much greater bandwidth and the transmission of many more channels of information. The result is to increase the capability of logging sondes because of the large number of variables that can be recorded. With the advent of microcomputers, one can do pulse height analysis (Sec. 6-2d) downhole and include things like temperature compensation. It also permits more control by the operator. This is a mixed blessing. The user must still be aware of the assumptions embodied in the programs and the changes made by the operator; it may be that more computation places an even greater burden on the user to be sure the quality of the data is preserved.

3-5 LOGGING PROCEDURE

Checkout and calibration. First, of course, the sonde is attached to the cable head and the sonde's operation is checked. The sonde is then calibrated. Often there are three grades of calibration: (1) a very elaborate one for the prototype at the service company's headquarters (the primary calibration), (2) a somewhat less elaborate one at the field office (the shop calibration), usually performed once a month, and (3) a simpler field calibration at the wellsite. Service companies issue books describing their field calibration procedure. Although the explanations may not be complete, the user can follow a given procedure to be sure that all of it is done as described. Sometimes the field calibration is not

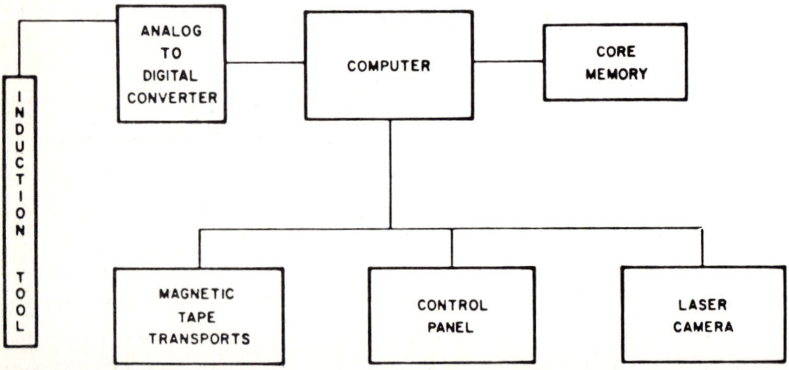

FIGURE 3-2 Block diagram of surface digitization system. (Moseley, 1976.)

a calibration with known values of the variable of interest; instead, it is a checkout to make sure the equipment performs correctly in a setup referenced to the primary or shop calibration. For example, dual-detector density logs (Sec. 6-4c) are calibrated in a jig containing two γ sources so designed that the ratio of count rates in the detectors is the same as would be seen in a material of specified density.

Waller et al. (1975) describe in detail the calibration procedures used by one service company. Stephen (1982) describes ways in which operators fail to apply the procedures properly and gives several examples of situations in which service company procedures have been shown, after years of use, to be significantly erroneous. The user must be aware of both the procedures used and the possibilities of error. Understanding is required.

The calibration is usually done at the surface; in some instances part of it is done downhole. For example, an acoustic log may be checked in a formation or casing of known velocity. Some of the calibration is usually repeated after the log is finished. It is important that the two calibrations agree within tolerance—one of the best checks the user has for a faulty log is a difference between before-log and after-log calibration. The tolerance is sometimes specified in the company's book or even on a printed part of the log (Fig. 3-7). This tolerance is, of course, the absolute limit on the accuracy to which the parameter can be measured. For example, one porosity log has two positions of the calibration jig which represent porosities of 0.11 ± 0.002 and 0.22 ± 0.01. The system cannot possibly measure porosity to greater accuracy.

The user should not hesitate to demand that the operator keep running logs until the before and after calibrations are within tolerance, *without changing panel settings between them.*

Precision versus accuracy. Accuracy is defined as "total error compared to the true value," or the difference between the measured value and the true value. Precision, on the other hand, is the repeatability of a measurement, the difference of individual measurements from the mean of an infinite number of measurements of the same quantity, or perhaps even the number of significant figures to which the measurement can be made.

For example, consider a meter stick that was ruled with an accuracy of 0.01 percent. But it is made of aluminum and it was ruled in a warm room, at 30°C. In that room, it was accurate to 0.01 percent. Now assume it is used to measure the length of an object which is in fact 0.73526 m long and is unaffected by temperature. The meter stick will measure the length to within ±0.00007 m. But if the measurement is made outdoors, on a cold day, at −20°C, it is still *precise* to 0.01 percent, but all readings will be high by 0.125 percent because of the thermal contraction of the meter stick. Consequently, the observed value of the length of the object will be 0.73617 ± 0.00007 m, a value still very precise but no longer as accurate.

Many precise, inaccurate instruments are available in the logging industry. The user must be careful, when reading quoted results in a paper or on being told specifications of an instrument by a vendor, to learn whether he is being told accuracy or precision.

The run. Occasionally a mandrel, or a simple sonde such as a γ-ray, may be run first to be sure the hole is not caved or blocked and is safe for more valuable equipment. Then, one by one, the sondes to be run (the suite of logs) are lowered to the bottom of the interval to be investigated. At the bottom, depth is corrected with the magnetic marker, and the run is started. Sondes are generally lowered rapidly and logging is done coming up, because the cable is then under fairly constant tension. This results in more uniform speed and more accurate depth measurement. The temperature log is an exception to this practice (Sec. 4-3).

It is common practice to combine several instruments (for example, caliper, γ-ray, neutron) in one package and run them all at once. Some sondes (for example, acoustic) are centered in the hole, often by devices such as bowsprings (Fig. 3-3). Others, such as density, are often pressed against the side of the hole (Fig. 3-4) by means such as hydraulic arms.

The speed depends on the type of log being run. Nuclear logs require slow speeds (Secs. 6-1c and 6-4f), and some even need to be stationary in the hole for several minutes, whereas resistivity logs can be run much faster. Sometimes the operator may run fast in some of the hole and slow down in parts he has reason to believe are of particular interest.

Special field problems. If a drill rig is on the hole, time spent in logging can be enormously costly. But there is no point in paying for bad logs. Watching logging can be a boring task. Rigging up and lowering the sonde into the hole is time-consuming. Often sondes require repair, or a new sonde must be fetched from a distant field office. It seems that the most important log in a suite is invariably started at two o'clock on a sleety Sunday morning.

Nevertheless, if it is important to the user that a really good log be obtained, he should be on the truck when it is run. Operators can be sleepy or careless. A study of 67 density and neutron logs (Lang, 1980) showed that in 58 percent of the cases a calibration correction was

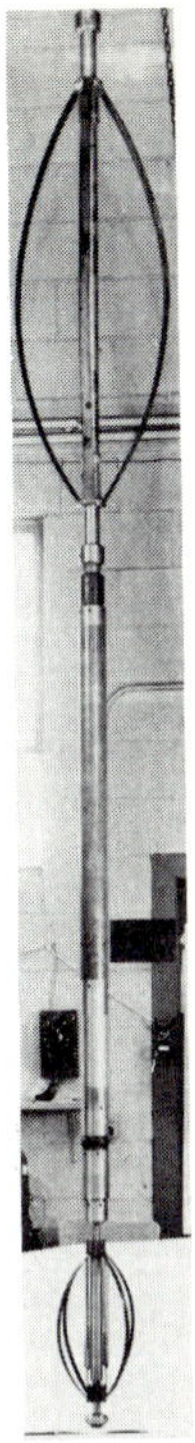

FIGURE 3-3 Bowspring-centralized logging sonde. (Courtesy of Birdwell.)

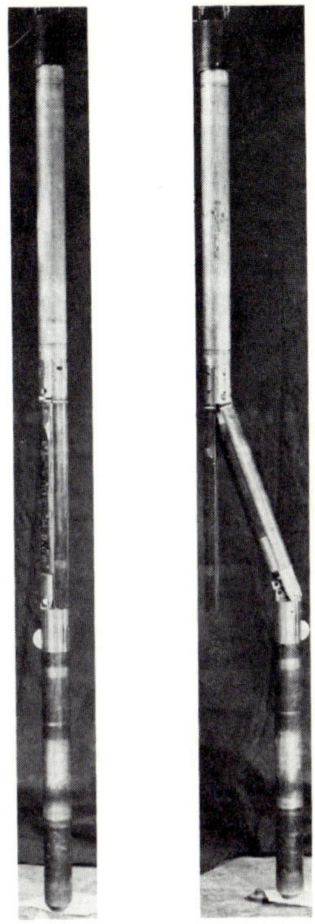

FIGURE 3-4 Sidewall logging sonde: (*a*) closed for going down hole; (*b*) open to press against side of hole. (Courtesy of Birdwell.)

required for at least one log. Farnan and McHattie (1984) used repeatability of logs as a quality criterion, and they showed that there are sometimes large differences between logs that should almost be identical. Other studies (Neinast and Knox, 1973, Knox, 1974) have given similar results. These bad logs often cannot be fixed after the truck leaves the hole, but an alert observer can catch many problems. The operator may adjust settings of some of the controls from time to time. This practice should be regarded with suspicion. It is not impossible for the operator to forget to turn the sonde on until it is halfway up the hole, and then try to pass the result off as a good log! Do not be afraid to ask what seem to be stupid questions and to *demand* that you be shown that the logs are good. You are paying for the log. This book should give you the understanding necessary to tell a good log from a bad one. For example, an operator has been observed to record the density calculated from only one of the two detectors of a compensated density sonde (Sec. 6-4c) and claim this is the proper procedure. You will learn why it is not.

3-6 THE LOG ITSELF

A log is a record of events. The term first came into the field of exploration geophysics as a record of formations encountered while drilling. We may now consider the log to be the piece of paper (or tape or disk memory) containing the geophysical data from the borehole. It consists of three parts: the header, the calibration record, and the data from the run in the hole.

The header. The header contains vital information; the user should insist that it is completely filled out. It usually consists of two pages. The format of the first page is fairly standard among service companies; an example is shown as Fig. 3-5. Items of interest are:

1 Company, well name, field, county, state. These identify the hole.

2 Permanent datum, "log measured from," "drilling measured from." These identify the zero point of the depth measurement. Sometimes depth discrepancies arise from the use of different zeros for log and drill depth.

3 Date, run number. These identify the specific log.

4 Depth—driller, depth—logger. Note that they are not the same and that the logger's depth differs on the two runs. The first time the operator only ran the sonde to the bottom of the 22-in (55.88-cm) part of the hole.

5 Top and bottom of logged interval. One does not always log the entire hole. Here on run 4 the operator went to the bottom of the hole and then just logged up far enough to overlap the bottom of run 3.

6 Casing and bit sizes.

7 Type of fluid, fluid level, density, viscosity, pH. Fluid level is important for most logs, because the logs operate differently in liquid than in air.

8 Resistivity of mud, filtrate, and mud cake at measured and bottom-hole temperatures. Necessary for resistivity logs; not given here for a nuclear log.

9 Time, since circulation affects some fluid properties.

T20 WW

BIRDWELL

Epithermal Neutron Porosity

COMPANY D.O.E.

WELL U20 WATER WELL

FIELD N.T.S.

COUNTY NYE STATE NEVADA

AREA 20
N - 910,569'
E - 569,101' (LO)

SEC. TWP. RGE.

OTHER SERVICES: DF CA6 FDD GR V3-D ES

PERMANENT DATUM GROUND LEVEL ELEV. ※
LOG MEASURED FROM G.L. 0 FT. ABOVE PERM. DATUM
DRILLING MEASURED FROM G.L.

ELEV. K.B. ※
D.F. ※
G.L. ※

DATE	JULY 19, 1982	JULY 19, 1982		
RUN NO.	THREE	FOUR		
DEPTH–DRILLER	3268'	3268'		
DEPTH–LOGGER	2676'	3232'		
BTM. LOG INTER.	2676'	3232'		
TOP LOG INTER.	50'	2650'		
CASING–DRILLER	24" @ 64'	24" @ 64'	@	@
CASING–LOGGER	64'	NOT REACHED		
BIT SIZE	22"	17-1/2"		
TYPE FLUID IN HOLE	MUD	MUD		
LIQUID LEVEL	2034	NOT REACHED		
DENS. | VISC.				
pH | FLUID LOSS	ml	ml	ml	ml
SOURCE OF SAMPLE				
R_m @ MEAS. TEMP.	@ °F	@ °F	@ °F	@ °F
R_{mf} @ MEAS. TEMP.	@ °F	@ °F	@ °F	@ °F
R_{mc} @ MEAS. TEMP.	@ °F	@ °F	@ °F	@ °F
SOURCE R_{mf} | R_{mc}				
R_m @ B.H.T.	@ °F	@ °F	@ °F	@ °F
TIME SINCE CIRC.				
MAX. REC. TEMP.	92° °F	106° °F	°F	°F
LOGGING UNIT LOC.	MERCURY	MERCURY		
RECORDED BY	BARKER/SORG	BARKER/SORG		
WITNESSED BY	HARRINGTON	HARRINGTON		

CORRECTED COPY

SSC–1172-A

FIGURE 3-5 First page of a log header.

10 Maximum recorded temperature. Needed for some temperature corrections.

11 Unit location, recorded by, witness. Valuable in case of problems. Sometimes one operator has different results than another; it is good to know the operator's name.

Note that the log is labeled "corrected copy." This indicates something was wrong the first time it was printed.

The second page of the header, shown as Fig. 3-6, has a different format for each contractor. Most contain the same data in different orders.

12 Remarks. This one describes changes in setting of a panel between runs. It also describes hole conditions that might cause problems. This section can sometimes be the most important part of the header.

13 Model and serial numbers of the equipment used. This is important if things go wrong and one wants to check the equipment or if one is searching for evidence that troublesome equipment was used on several wells.

14 Logging data. These are the settings of all the controls on the panels; they are the assumptions on which the conversion of volts or count rate to porosity is based. There is space for different settings at different depths. It also shows logging speed.

FOLD HERE 3 3 W

Remarks NEUTRON POROSITY CURVE IS SET FOR LIMESTONE MATRIX IN REPEAT SECTION ONLY! CURVE IS SET AT SANDSTONE MATRIX IN LOG. REPEAT SECTION IS 2' DEEP. HOLE WAS DRILLED WITH AIR MIST. REACHED T. D. 8.0 HRS AFTER INITIALLY FILLING HOLE WITH KCL WATER. FLUID LEVEL 816'

EQUIPMENT DATA

	RUN NO. 1			RUN NO. 2			RUN NO. 3		
	CDL	CNLog	GR	CDL	CNLog	GR	CDL	CNLog	GR
Series No	2208	2413	1306						
Serial No	29227	29901	24543						
Diameter	3"	3 5/8	3 5/8						
Elec. No.									
Detect: Model No.	166-161	D6N3	D6G4						
Type	GM	HE3	SCINT.						
Dist. to Source			6'						
Source Model No.	S3E20	S17S20							
Serial No	C-116	30431							
Computer No.	3457	3457	3621						
Serial No	37004	37004	34481						

Logging Data

General				Compensated Densilog				Gamma Ray			
Run No.	Depths From	Depths To	Speed Ft./Min.	T.C. Sec.	Density Scale	Correction Scale	Porosity Scale Data	T.C. Sec.	Sens. Settings	Zero Div. L or R	API G.R. Units/Div.
1	2501	816	20	3			0-30	3	587	0	15

General				Compensated Neutron							Caliper
Run No.	Depths From	Depths To	Speed Ft./Min.	T.C. SS	T.C. LS	Sens. SS	Sens. LS	Porosity Zero	Porosity Units/Div	Porosity Scale	Scale
											6"-16"
1	2484	816	20	3	3	134	134	R-15	2	0-30	

FIGURE 3-6 Second page of a log header.

All of this information is of great importance if any checking is to be done.

Calibration. Somewhere on the log, usually at the bottom, are the records of the calibrations. They contain some standard information about the method and perhaps a code to identify individual parts of calibration traces. Figure 3-7 is a pasteup of several parts of a calibration record from a combination neutron and density log. At the bottom are the code numbers to identify the traces and the values they should have. At the top are scales taken from the log itself to enable you to see the values taken by the trace segments.

Between the codes and the scale are the calibration traces. The panel is set to the conditions specified in the code below, and the position of the galvanometer is recorded for a short distance on the paper. The numbers on the left refer to the codes below. Sometimes the operator neglects to record those numbers; this is unacceptable. Some contractors supply a computer-printed record of the calibration data rather than traces. Since the traces are produced by the computer in any case, this is perfectly satisfactory, and it eliminates many problems. The figure shows the neutron "before survey" and the density "after survey" calibrations. The values agree within tolerance with those specified by the codes below. Note that, for each log, there is only one test in "log" position; it is the position actually used for running the log. All the others are "panel tests," or tests to be sure the panel is operating correctly, or checks of zero and calibration settings. There is no check that the equipment is operating properly over some range of actual data.

Calibrations are not always done properly. One log available as an example for this chapter had no scale for the calibration. It was a combination neutron and density log, and there was no way of telling which calibration applied to which instrument. The log should not have been accepted.

The data traces. Logs are recorded on *tracks* on the records. Figure 3-8 shows several tracks. The different types of data can be distinguished by trace thickness and dash pattern; each is identified on the figure. The "CNL" and "FDC" are neutron and density porosity (described in Chap. 10), respectively. Often the same data are recorded on the same scale, but with different zero settings, to permit greater resolution. If the second scale is needed, a different dash pattern is used. Such a trace is referred to as the *backup* trace. It is designated specifically at 393 ft (119.79 m) on the right side, and the backup for the caliper trace can be seen from 392 to 400 ft (119.48 to 121.9 m) and from 436 ft (132.9 m) to the bottom. Hatching is used to indicate

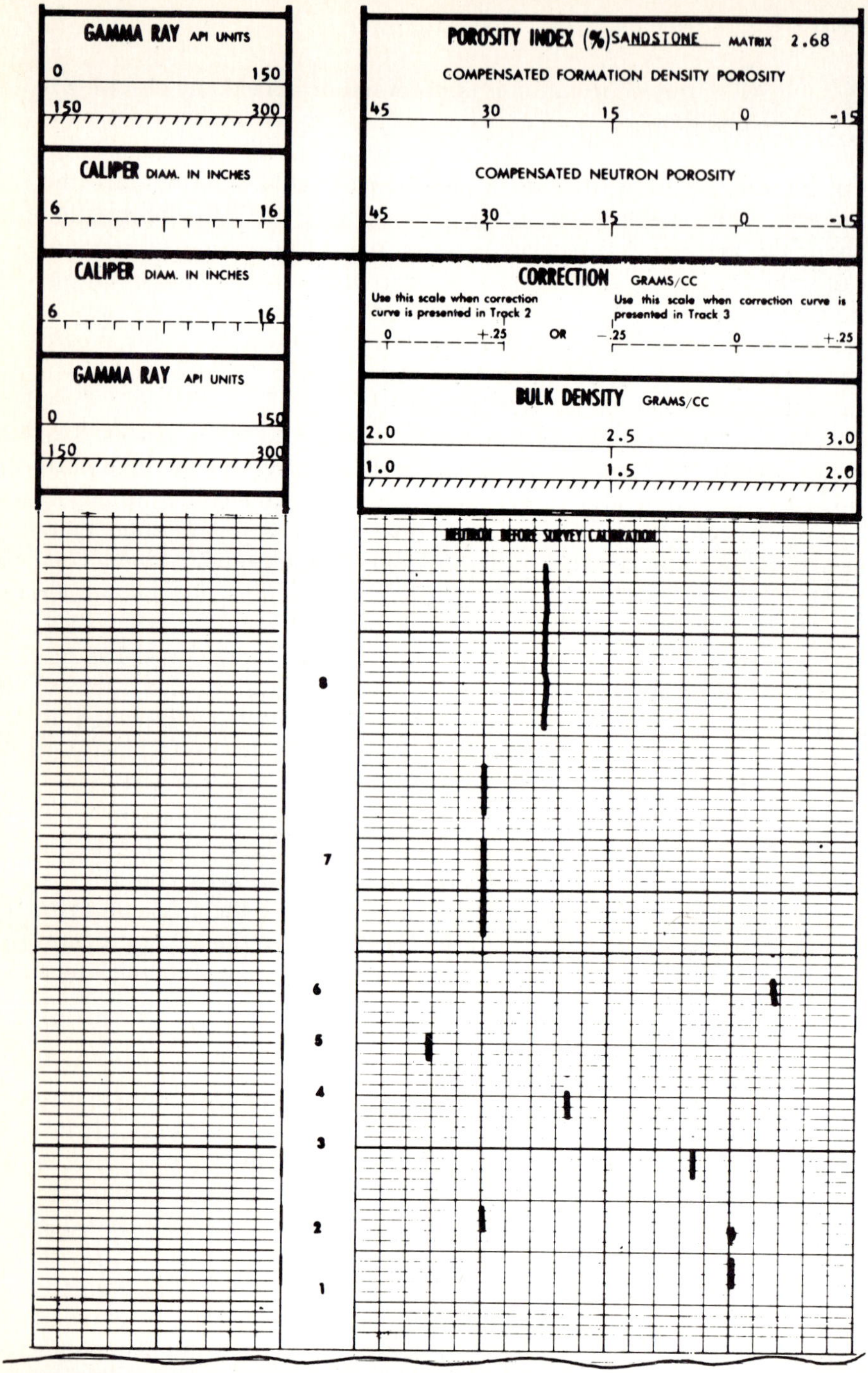

FIGURE 3-7 Pasteup of parts of a calibration record from a combination neutron and density log.

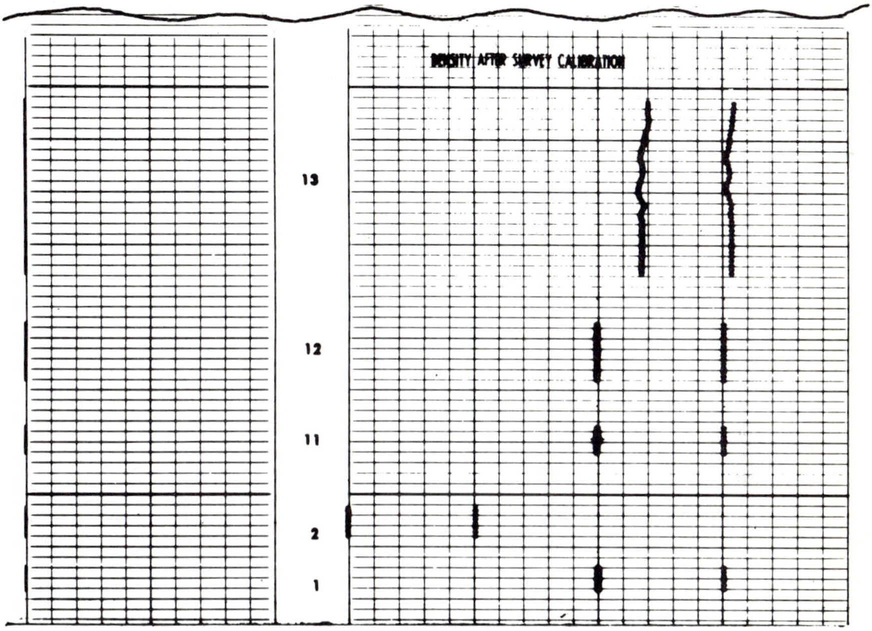

COMPENSATED NEUTRON CALIBRATION CODING

1. MECHANICAL ZERO
2. RECORDED SENSITIVITY (THRU MEMORIZER IF USED

PANEL TEST

	RATIO	OH POROSITY			CH POROSITY	
		LS ☐	SS ☒	DOL ☐	SS ☐	LS ☐
3.	1	1.6	4.9	−0.2	2.4	0.1
4.	2	15.6	19.7	8.1	13.0	9.0
5.	3	30.5	36.0	25.2	29.1	24.1
6.	4	45.4	53.1	47.5	47.4	43.2

7. POROSITY NORMALIZED WITH CNB-A IN PLACE
   7A. TOOL IN NCT-B
8. LOG POSITION WITH CNB-A IN PLACE
   8A. LOG POSITION WITH TOOL IN NCT-B

OH			CH	
LS	SS	DOL	SS	LS
18	22.2	10.4	15.3	11.2

$$\text{RATIO (NORMALIZED)} = \frac{2.17}{\text{RATIO (NCT-B)}} \quad \text{RATIO LOG}$$

FORMATION DENSITY COMPENSATED CALIBRATION CODING

1. MECHANICAL ZERO
2. RECORDER SENSITIVITY

PANEL TEST

FDC LIQUID

	POS	ρ	$\Delta\rho$
3.	# 1	2.92	.00
4.	# 2	2.78	+.14
6.	# 3	2.42	−.10
6.	# 4	2.35	.00
7.	# 5	2.08	.01

8. MECHANICAL ZERO CALIPER
9. 8″ RING
10. 12″ RING
11. TOOL CALIBRATE #1 SET ρ = 2.50
12. TOOL CALIBRATE #2 SET $\Delta\rho$ = .00
13. LOG POSITION ρ = 2.59, $\Delta\rho$ — .015

FIGURE 3-7 *(continued)*

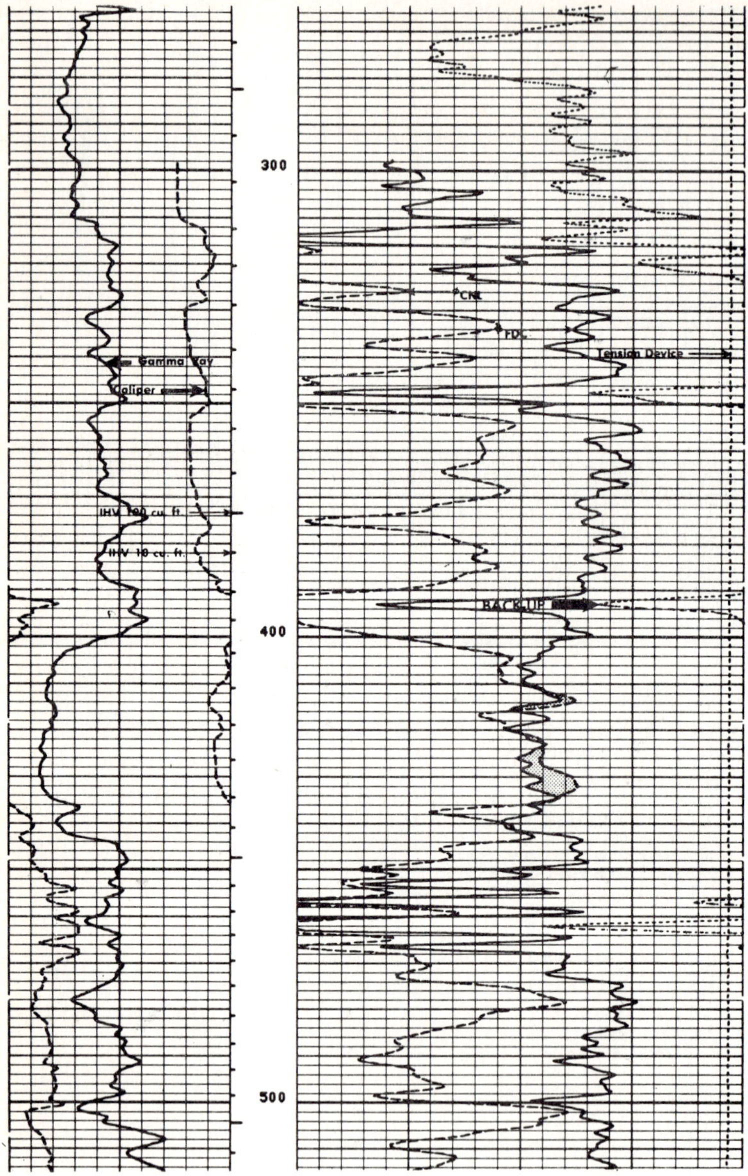

FIGURE 3-8 Typical log traces for a combination neutron and density log, with caliper and γ-ray sections.

the part of the log in which the CNL porosity is less than the FDC porosity. No caliper or FDC data are recorded above 298 ft (90.8 m); presumably this is in casing. Tick marks on the right side of the left scale indicate volume measured by the caliper.

There is usually some sort of time record on the log as well; on this one, time is recorded by gaps in the outer margins of the vertical boundaries of the record. There is a gap every minute. This permits the user to check logging speed [which in Fig. 3-8 is about 30 ft/min (0.15 m/s)]. On some logs time is recorded by tick marks somewhere on the record.

Scale. The scale of the log is, of course, of great importance. The operator sets the zero and span of the galvanometers or the zero and the units of data per unit of width on the record. Setting the galvanometers manually can be tricky, but the computer sets zero and span automatically. Sometimes more than one scale is used as a backup, as described above. Figure 3-9 shows a typical set of scales. There is always a scale at the top of the log; sometimes there are other scales elsewhere.

Often, for good reason, the scale is changed during the run. For example, one may be logging two formations one of which has transit times between 50 and 100 μs/ft (30.48 μs/m) and the next has transit times between 100 and 150 μs/ft (30.48 and 45.73 μs/m). One can obtain better resolution by changing scale when the sonde enters the second formation. But it is *vital* that the change be recorded. Figure 3-9 shows such a record. One of the commonest pitfalls in logging is an unrecorded scale change.

For "porosity" logs such as the acoustic, neutron, or density (Chap. 10) one can display either the physical property being measured, such as density or transit time, or the porosity computed from that property. It is advisable to have both recorded, if possible, in case the assumptions used in calculating porosity are not valid.

Sometimes one sees a log that looks beautiful until it is observed that there is no scale. The log should be rejected.

Generally logs are recorded on two depth scales simultaneously: a small scale (ordinarily 1:240) for a quick look and a larger one (ordinarily 1:600) for detailed study.

Quality control. The operator can change things during the run. You should insist that every change be recorded on the log. The presence of changes in scale or speed, but not always their magnitude, can usu-

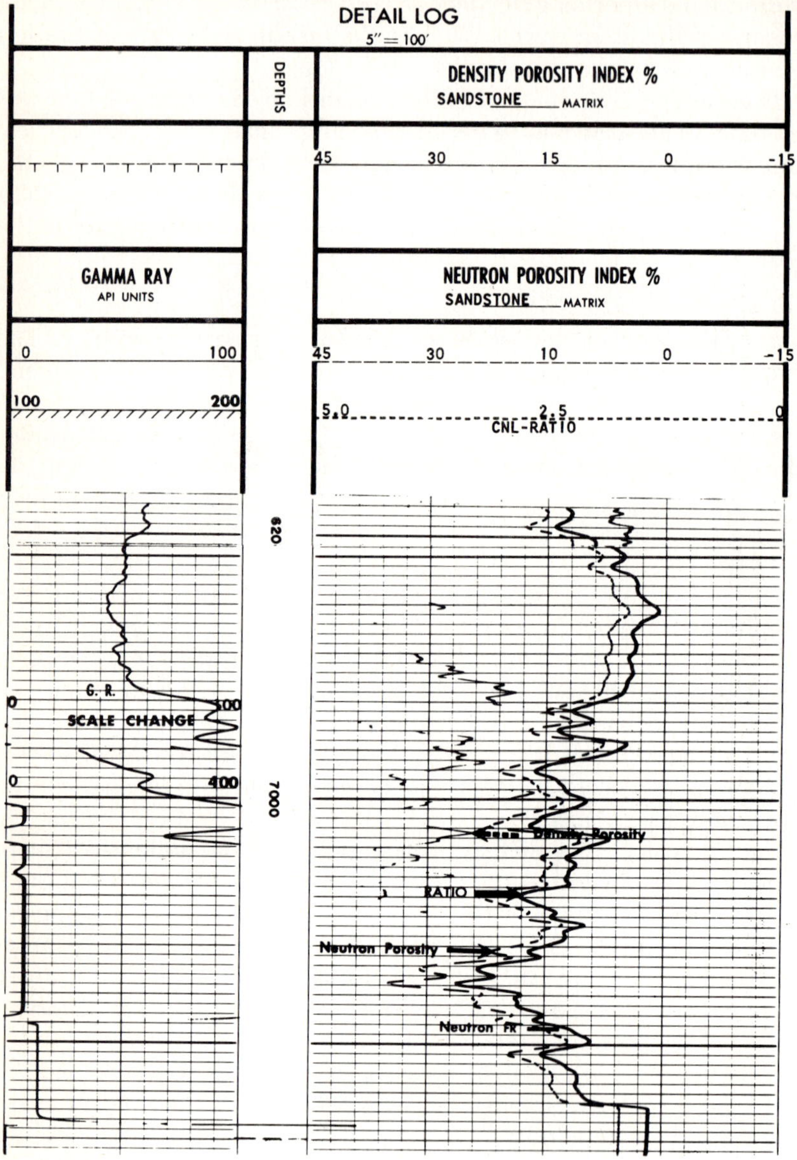

FIGURE 3-9 Typical scales for a combination neutron and density log with a γ-ray section.

ally be inferred by good detective work. Being there is best. Alertness when studying the logs is second best. Make a practice of rejecting logs with unrecorded changes. This should reduce the number of autopsies you must perform on a suite of logs

Field offices do not always implement changes dictated by service company headquarters. For example, Wiley and Zittel (1982) discovered that a certain service company's spectral γ-ray logs were miscalibrated by a factor of about 2. The company had also learned of this situation and, in 1981, had issued instructions to its field offices to change the values of the calibration constants. Yet in mid-1983, at least one of the field offices was still using the old calibration. The problem was discovered by comparing log and sample data.

There is no way to be sure that you will always get good data. Constant suspicion is the best policy.

REFERENCES

Atkinson, A, and R. Pointer: "Logging Services for Large Diameter Boreholes," *Log Analyst* vol. 17, no. 6, 1975, p. 25.

Brabeck, P.: personal communication, 1983.

Cooley, B. B.: "The Delta Tension Curve for Better Log Quality," *Proc. Soc. Prof. Well Log Analysts 15th Logging Symp.*, 1974, paper F.

Farnan, R. C., and C. M. McHattie: "Log Quality Enhancement: A Systematic Assessment of Logging Company Wellsite Practice and Log Quality," *Proc. Soc. Prof. Well Log Analysts 25th Logging Symp.*, 1984, paper L.

Kerzner, M.: "A Solution to the Problem of Automatic Depth Matching," *Proc. Soc. Prof. Well Log Analysts 25th Logging Symp.*, 1984, paper VV.

Knox, C. C.: "Quality Control of Well Logs," *Proc. Soc. Prof. Well Log Analysts 15th Logging Symp.*, 1974, paper A.

Lang, W. H.: "SPWLA Ad Hoc Calibration Committee Report: Porosity Log Calibrations," *Log Analyst*, vol. 21, no. 2, 1980, p. 14.

Moseley, L. M.: "Field Evaluation of Direct Digital Well Logging," *Proc. Soc. Prof. Well Log Analysts 17th Logging Symp.*, 1976, paper NN.

Neinast, G. S., and C. C. Knox: "Normalization of Well Log Data," *Proc. Soc. Prof. Well Log Analysts 14th Logging Symp.*, 1973, paper I.

Pilkington, P. E., and W. H. Fertl: "Field Tests of Cement Bond Logging Tools," *Log Analyst*, vol. 16, no. 4, 1975, p. 13.

Stephen, W. F.: "Participatory Log Quality Control," *Proc. Soc. Prof. Well Log Analysts 23d Logging Symp.*, 1982, paper D.

Waller, W. C., M. E. Cram, and J. E. Hall: "Mechanics of Log Calibration," *Proc. Soc. Prof. Well Log Analysts 16th Logging Symp.*, 1975, paper GG.

Wiley, R., and R. J. Zittel: "Natural Gamma Ray Data of Various Sandstones," *Proc. Soc. Prof. Well Log Analysts 23d Logging Symp.*, 1982, paper K.

Zangwill, J.: "Depth Matching—A Computerized Approach," *Proc. Soc. Prof. Well Log Analysts 23d Logging Symp.*, 1982, paper EE.

Note: An excellent new book that describes the details of log calibration procedures is *Log Quality Control* by Richard M. Bateman, published in 1985 by International Human Resources Development Corporation (IHRDC). The book includes many useful examples of improperly run logs and gives explanations of the causes of the problems.

PART 2

Logging Methods

4

Temperature

Subsurface temperature measurements are used to correct other temperature-sensitive borehole measurements, to explore for high-temperature geothermal resources, and less frequently, to map stratigraphy and to infer the movement of groundwater. For these applications, the temperature of the rock is desired, typically to an accuracy of $\pm 1°C$. However, the measurement itself is usually made in borehole fluid, which is not necessarily in thermal equilibrium with the formation temperature. Hence, it is often necessary to correct for the thermal imbalance between fluid and rock.

It is also possible to make use of the temperature imbalance caused by transient phenomena. When a cool fluid is used for drilling or a hot fluid is injected into or produced from a permeable formation, the heat exchanged between fluid and rock causes temperature imbalances that require days or even months to equilibrate. Consequently, by measuring the resulting thermal effects in the wellbore fluid, temperature logs can be used to map zones of fluid loss or entry in open wells or to gain information on the nature and quantity of convected fluid. In cased wells, temperature logs can detect the movement of fluid behind casing or confirm the location of cement by detecting the heat produced during the curing stage.

We have already seen in Chap. 2 that temperature in the earth's crust increases with depth. The temperature gradient is equal to the local vertical heat flux divided by the thermal conductivity of rock. Hence, from accurate measurement of temperature gradient and knowledge of rock thermal conductivity, the earth's heat flux can be determined. To avoid transient effects, temperature sensors are implanted and monitored for some time period, or else a sensor is lowered into a borehole after the borehole has been allowed to equilibrate. Desired accuracy is of the order of $\pm 0.01°C$. Because boreholes used for heat flow determinations are often shallow, the effects of surface temperature, topography, and groundwater movement must be taken into account. More discussion and references on crustal heat flow are given by Lachenbruch and Sass (1977) and Kappelmeyer and Haenel (1974). In the following treatment we will emphasize the use of temperature measurements for applications other than heat flow determinations.

4-1 HEAT FLOW EQUATIONS

The change in internal energy ΔU (joules per cubic meter) caused by a temperature change ΔT can be expressed by

$$\Delta U = \rho C \, \Delta T \tag{4-1}$$

where ρ is the density of the rock-fluid system (kg/m^3), C is heat capacity per unit mass [J/(kg)(K)], and temperature ΔT will be given in °C rather than K because we will be discussing temperature changes.

Heat flux Q is the amount of thermal energy crossing an area per unit time. In an isotropic medium,

$$Q = -k \, \nabla T \tag{4-2}$$

where k is thermal conductivity [W/(m)(°C)]. (These units are compact and technically correct, but it is helpful to remember that the units for k are heat flux divided by temperature gradient, or J/(m^2)(s)/(°C)/(m), which then condenses to W/(m)(°C). For engineering purposes it is convenient to specify the temperature difference ΔT across a boundary or a material instead of the gradient ∇T. The heat transfer coefficient h is defined by

$$Q = h \, \Delta T \tag{4-3}$$

and has dimensions W/(m^2)(°C). It is useful in specifying the heat exchange characteristics of a cased well.

Within a unit volume in the material, the rate of change of internal energy U must equal the energy generation rate S [J/(s)(m^3)] minus the energy losses given by the divergence of Q,

$$\frac{\partial U}{\partial t} = -\nabla \cdot Q + S \tag{4-4}$$

Combining Eqs. (4-1), (4-2), and (4-4) gives the equation of heat conduction,

$$\rho C \frac{\partial T}{\partial t} = \nabla \cdot (k \, \nabla T) + S \tag{4-5}$$

and if the material is homogeneous and isotropic so that k is constant everywhere,

$$\frac{\partial T}{\partial t} = \frac{k}{\rho C} \nabla^2 T + \frac{S}{\rho C} \tag{4-6}$$

The term $k/\rho C$ is called the thermal diffusivity κ. In wellbore heat conduction problems we assume cylindrical symmetry, so that Eq. (4-6) can be written as

$$\frac{\partial T}{\partial t} = \kappa \left(\frac{\partial^2 T}{\partial z^2} + \frac{1}{r}\frac{\partial T}{\partial r} + \frac{\partial^2 T}{\partial r^2} \right) + \frac{S}{\rho C} \tag{4-7}$$

Conversions. Thermal quantities are commonly expressed in either English, cgs, or SI units. Conversion among systems will remain a burdensome nuisance until SI is fully adopted, but the transition will not come easily. When Peppermint Patty, a cartoon character drawn by C. Schultz, is asked in school to convert between the Fahrenheit and Celsius scales, she decides that it will probably be easier to put on a sweater! As an alternative, we recommend the use of Table 4-1.

4-2 THERMAL PROPERTIES OF ROCKS

4-2a Conductivity

If steady-state heat flow conditions prevail, then $Q = -k \, \nabla T$, and k is the only material parameter of interest. Thermal conductivity of rocks is determined by conductivities of the constituent minerals and fluids, by the porosity, and by heat transfer coefficients between dissimilar minerals within the rock. Values for silicate minerals and water at room temperature are given in Fig. 4-1. Mean values decrease from 7.7 for quartz to 1.9 W/(m)(°C) for the feldspars. Horai (1971) points out that thermal conductivity of the silicates is controlled by the structure of the

TABLE 4-1 Factors for conversion of thermal parameters from British and cgs units to SI units. For example, a geothermal gradient of 2°F/100 ft converts to 36.46°C/km.

Quantity	British or cgs units	Multiplier	SI units
T	°F	$5(T-32)/9$	°C
dT/dz	°F/100 ft	18.23	°C/km
k	Btu/(h)(ft)(°F)	1.730	W/(m)(°C)
k	cal/(s)(cm)(°C)	418.4	W/(m)(°C)
Q	Btu/(ft^2)(h)	3.152	W/m^2
Q	cal/(cm^2)(s)	41,840	W/m^2
C	Btu/(lb)(°F)	4184	J/(kg)(°C)
C	cal/(g)(°C)	4184	J/(kg)(°C)
ρ	lb$_m$/ft^3	16.02	kg/m^3
ρ	g/cm^3	1000	kg/m^3
κ	ft^2/h	0.00002581	m^2/s
κ	cm^2/s	0.0001	m^2/s

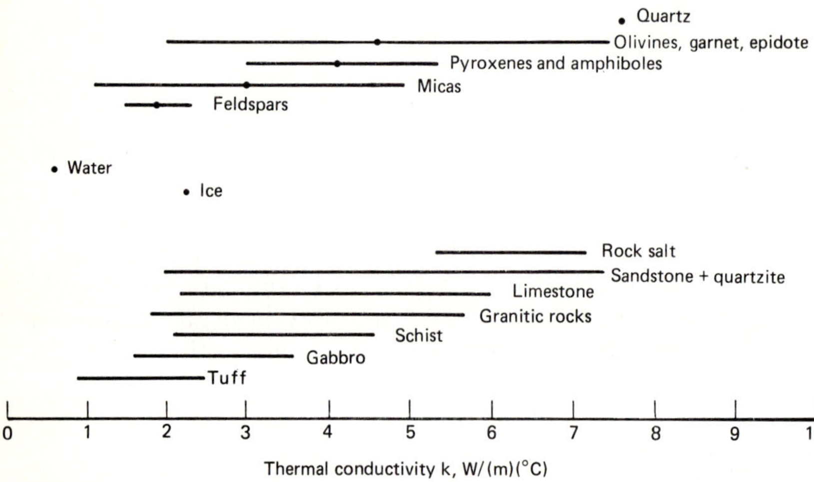

FIGURE 4-1 Thermal conductivity k for minerals and rocks. Values measured at room temperature except ice (0°C). (Mean and standard deviation values for minerals from Horai, 1971; approximate ranges for rocks adapted from Roy et al., 1981; other values from Clark, 1966.)

silicon-oxygen framework. More oxygen atoms are shared in the framework silicates (feldspars) than in the sheet silicates (micas) or the chain silicates (pyroxenes). We see that k is less in the more complicated crystal structures.

The thermal conductivity of water is considerably lower than that of most minerals, and that of a gas is less than water. Consequently, as water-saturated pore space increases, total rock k will drop. And if the pore space is gas-saturated rather than water-saturated, rock k will be further reduced. These expectations are confirmed by the data of Table 4-2. As porosity increases from 3 to 59 percent in sandstones, k drops from 7.4 to 2.0 W/(m)(°C) when the rock is saturated with water and from 6.5 to 0.5 when it is saturated with air. The dependence upon saturant conductivity is also shown by the data on vacuum and *n*-heptane saturations. At high porosities, we see that the effect of changing from water to air in a porous sandstone will decrease k by a factor of 2 to 4. On the other hand, k of ice is greater than k of water (Fig. 4-1), so k of permafrost will be greater than that of unfrozen rock. King (1979) reports that k of a Berea sandstone sample increased 15 percent upon freezing.

Saturated sandstones can be considered as two-component (silicate-water) systems, one component having high k and the other low k. Many relations have been proposed to compute conductivity of a rock from the conductivities of the rock's constituents. One of the simpler such relations [Eq. (1-15)] computes the geometric mean of the constituent properties:

$$\log k = \phi \log k_1 + (1 - \phi) \log k_2 \tag{4-8}$$

where k_1 and k_2 are thermal conductivities of the pore space and quartz and ϕ is porosity. This relation gives a good fit to the data of Table 4-2 (Woodside and Messmer, 1961). However, other empirical relations fit the data just as well (Beck, 1976), and there is no consensus as to the most appropriate method to use. Robertson and Peck (1974) and Roy et al. (1981) consider the merits of mixing relations which have been derived from a study of thermal conductivity data.

TABLE 4-2 Thermal conductivity k of porous, quartz-rich sandstones with various interstitial fluids.*

Sandstone	Porosity	Thermal conductivity, W/(m)(°C)			
		Vacuo	Air	*n*-Heptane	Water
Berkeley sandstone	0.3	2.90	6.49	7.11	7.41
St. Peters sandstone	0.11	2.49	3.56	5.34	6.36
Tensleep sandstone	0.155	2.62	3.04	4.37	5.86
Berea sandstone	0.22	1.68	2.39	3.74	4.48
Teapot sandstone	0.29	1.09	1.54	2.65	4.05
Tripolite	0.59	0.22	0.53	0.879	2.03
Saturant	(1.0)	0.00	0.026	0.128	0.628

*After Clark (1966), based on original data by Woodside and Messmer (1961).

It is clear that variations in conductivity of rocks can be attributed to compositional and porosity variations. Representative ranges of values for specific rock types are shown in Fig. 4-1. Although the ranges overlap considerably, systematic differences among the several rock types are apparent. Values of k for sandstone and quartzite are high, owing to the high silicate content. The consistently low values for gabbro are caused by high mafic mineral content. Limestones, granitic rocks, and schist have median values close to 3 W/(m)(°C), which can be taken as a typical value for low-porosity, consolidated rock if no other is available.

4-2b Diffusivity

Thermal diffusivity κ (square meters per second) characterizes a material by the ratio of its ability to conduct heat to its heat capacity. Hence, a high value of κ signifies that thermal transients will dissipate much more quickly than in a material of low κ. Solutions to time-varying temperature problems are a function of the dimensionless group $\kappa t/L^2$, where L is a characteristic length of the physical system.

Values of $\kappa = k/\rho C$ can be calculated from measured values of k and from inferred or measured values of ρC. Roy et al. (1981) report that most minerals and low-porosity rocks have ρC values within 20 percent of 2.3×10^6 W·s/(m³)(°C). Values of ρC for porous rocks can be obtained by weighting the ρC values of matrix and fluid according to their volume fractions. Because values for ρC, except for air, are so tightly grouped, one would not expect the range of values for diffusivity to be much different on a percentage basis than the ranges for k pictured in Fig. 4-1.

Numerical values for κ are given in Table 4-3. The diffusivity value for a typical rock is 1.3×10^{-6}, based on a k of 3 W/(m)(°C), ρ of 2.55 g/cm³, and C of 900 J/(kg)(°C). Ranges in a given rock can be estimated from the ranges shown in Fig. 4-1 for k. For comparison, k and κ are

TABLE 4-3 Representative thermal properties of five materials at room temperature.

Material	k, W/(m)(°C)	ρ, kg/m³	C, J/(kg)(°C)	ρC, J/(m³)(°C)	κ, m²/s
Air	0.0263	1.29	1004	1.295×10^3	2.03×10^{-5}
Water	0.619	1000	4184	4.184×10^6	1.48×10^{-7}
Cement	0.950	1750	879	1.53×10^6	6.21×10^{-7}
Rock	3.0	2550	902	2.3×10^6	1.30×10^{-6}
Steel	45.0	7770	502	3.90×10^6	1.15×10^{-5}

also shown for other materials which influence the thermal state in wellbores. Values of k and κ for cement are somewhat lower than typical values for rock. Steel performs as an excellent thermal conductor, whereas a low k illustrates the insulating capacity of air. Water is quite slow to respond to change in thermal conditions: its κ is about one-tenth that of rock.

4-2c Temperature Dependence

The thermal conductivity of crystalline solids decreases with increasing temperature. As a result, the k of brine-saturated, high-quartz rocks does likewise (Somerton, 1982). The temperature dependence appears to be less for porous sands, particularly if the sands are partially water-saturated. Empirical equations for standstone-fluid systems permit extrapolation of room temperature values of k to higher temperatures without requiring any other petrophysical parameters. Somerton also points out that large amounts of heat will be transported if the saturating fluid boils, convects, and then condenses at a point where temperatures are less than boiling. Under such conditions, apparent thermal conductivity can increase dramatically.

Graphs of k and κ as a function of temperature for a variety of rocks and minerals are presented by Roy et al. (1981). If no empirical relations are available for a material of interest, it is recommended that thermal properties at elevated temperatures be estimated by paralleling experimental data in a known material, starting from an appropriate room temperature value.

4-3 TEMPERATURE MEASUREMENT

Borehole fluid temperature is measured by devices in which a temperature change produces a visible alteration in the physical state of a material (for example, expansion or melting), produces a change in electrical properties (for example, a resistance which is temperature-dependent), or induces a voltage. Most familiar is the mercury-in-glass thermometer which is used in logging to obtain a maximum temperature. Because maximum readings are almost invariably encountered at the bottom of the hole, values obtained in this manner are referred to as bottom-hole temperatures (BHT). Sensors used for continuous logging are ordinarily either resistance thermometers or thermistors. The following description of sensor properties is adapted from Doebelin (1975).

Resistance thermometers, sometimes called resistance temperature devices (RTD), are made from metals such as platinum, nickel, and copper. Of these, platinum is preferred because of its stability and wide operating range; platinum RTDs have been adopted as an international standard for temperature measurement. Variation of resistance with temperature can be expressed as a polynomial

$$\mathcal{R} = \mathcal{R}_0(1 + a_1T + a_2T^2 + a_3T^3 + \cdots + a_nT^n) \qquad (4\text{-}9)$$

Platinum RTDs are quite linear, so the term $\mathcal{R}_0(1 + a_1T)$ is adequate for many purposes. For example, platinum is linear to within ± 0.4 percent over the range -75 to $+150°$C. Metallic elements may be formed into shapes convenient for borehole probes, and they are commonly used for continuous logging.

Thermistors are resistance temperature-measuring devices made of a semiconducting material such as manganese, nickel, or cobalt. Their resistance-temperature curve is nonlinear, being of the form

$$\mathcal{R} = \mathcal{R}_0 \exp b\left(\frac{1}{T} - \frac{1}{T_0}\right) \qquad (4\text{-}10)$$

where T is in kelvins. Measurement sensitivity is obtained by differentiating Eq. (4-10):

$$\frac{1}{\mathcal{R}}\frac{d\mathcal{R}}{dT} = -\frac{b}{T^2} \qquad (4\text{-}11)$$

which shows that the sensitivity coefficient is negative rather than positive. A typical value for b is 4000, so at 20°C (298 K) the sensitivity is $-4.5\%/°$C and at 125°C it is $-2.5\%/°$C. Resolution is quite good, but is temperature-dependent. To improve linearity, manufacturers combine one or more thermistors with resistors. Glass-coated bead thermistors are frequently used for borehole measurements, particularly for heat flow determinations where an inexpensive thermistor string can be left in a borehole for long time periods (Judge, 1973). The merits of various empirical formulas used to relate $\mathcal{R}$ and T are discussed by Lee (1977).

Thermocouples are thermoelectric devices made by forming junctions of two different metals. One junction is placed at some stable reference temperature and the other at the unknown temperature. A voltage nearly proportional to the temperature difference is generated by this junction. The types of material used to form the junction determine the sensitivity. Common metals used are platinum/rhodium (6 μV/°C), Chromel/Alumel (40 μV/°C), and copper/constantan (60 μV/°C). The reference temperature, usually chosen to be 0°C, can be an ice water bath or an arbitrary temperature corrected to 0°C equivalent by a voltage source, but it must be accurately known. The requirement for a reference temperature complicates the use of thermocouples in borehole probes.

A device coming into use is the integrated circuit temperature sensor. It incorporates a transistor in which the base-emitter junction voltage is proportional to T. The additional circuitry of the integrated package provides constant current sourcing and compensates for the nonlinear characteristic of the junction. Available devices function either as temperature-controlled voltage sources with nominal outputs of 1 mV/°C or as temperature-controlled current sources with outputs of 1 mA/°C. These devices are relatively new and have not seen much application in borehole sondes.

Temperature sensors are typically mounted on the bottom tip of a sonde to minimize the temperature or flow perturbation induced by the body of the sonde. For the same reason, continuous logs are obtained with the sonde descending into the borehole, and it is generally desirable that temperature be obtained first in a sequence of logging runs. Sensors are calibrated in a controlled oven by using an RTD standard to measure oven temperature. Checks can be done in the field by making side-by-side comparisons with a mercury-in-glass thermometer.

4-4 THE GEOTHERMAL GRADIENT

Steady-state heat flow in the earth's crust can be treated by considering a half-space of uniform conductivity k with constant heat flux Q directed upwards from great depth. The surface temperature T_0 is constant. The source term and the derivatives with respect to time and radial direction in Eq. (4-7) are zero, leaving

$$\frac{\partial^2 T}{\partial z^2} = 0 \tag{4-12}$$

for which the appropriate solution is

$$T(z) = T_0 + Gz \tag{4-13}$$

where $G = Q/k$ is called the geothermal gradient. Values for Q lie in the range between 40 to 80 mW/m^2 in areas of normal heat flow. Using 3 W/(m)(°C) as a reasonable value for k gives 13 to 26°C/km for G, as is indeed the case for sedimentary sequences.

The geothermal gradient is clearly reflected in the bottom-hole temperature data of Fig. 4-2. The scatter of the data at a given depth (Carvalho and Vacquier, 1977) is due to the effects of drilling. The average gradient is 15.6 ± 3.2°C/km; core and log data provide a mean

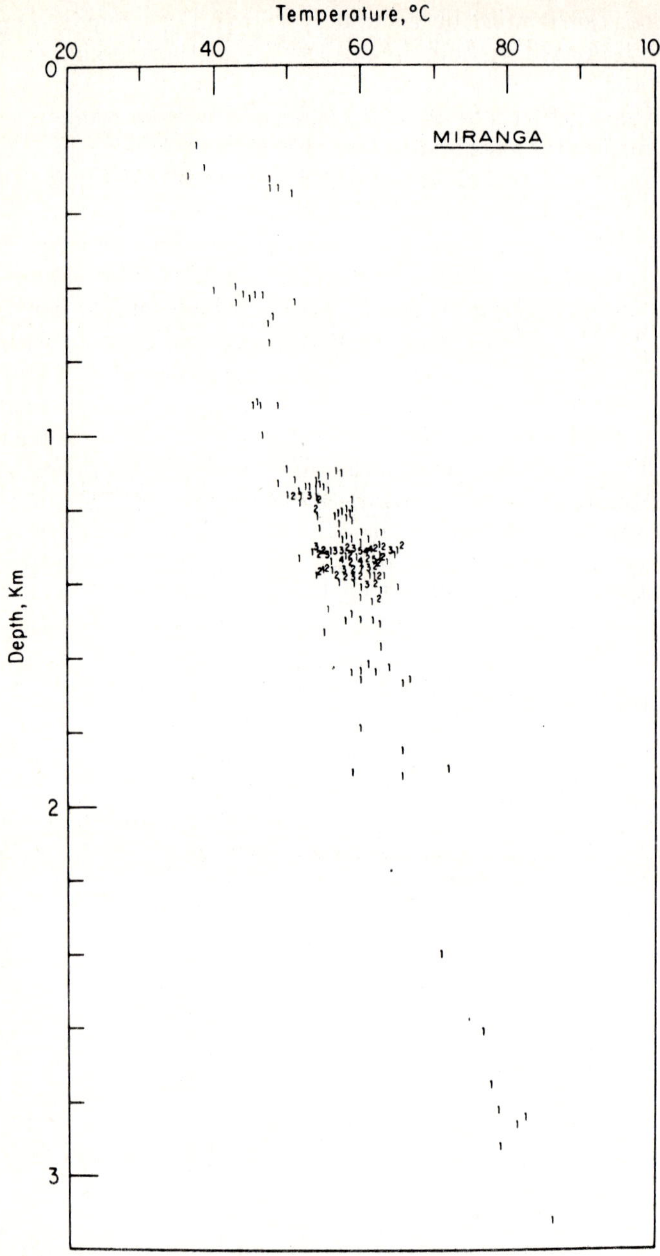

FIGURE 4-2 Bottom-hole temperatures as a function of depth in Miranga field, Brazil. (From Carvalho and Vacquier, 1977.)

k of 2.44 W/(m)(°C). By Eq. (4-2), the computed heat flow is 38 mW/m^2. This value is slightly below the 40 to 80 mW/m^2 range cited above, probably because the measured bottom-hole temperatures had not equilibrated with the true rock temperatures.

We obtained that average gradient by assuming $\Delta T/\Delta z$ to be constant over the entire depth of a well. But a continuous temperature log will reflect any departures from a uniform thermal conductivity distribution. In fact, at one time borehole temperature measurements were made to explore for salt domes, since k for rock salt is greater than for most rocks (Fig. 4-1). Another important case arises where the earth is comprised of a layered sequence of N beds each of thickness Δz_i and conductivity k_i. We then have

$$T(z) = T_0 + Q \sum_{1}^{N} \frac{\Delta z_i}{k_i} \tag{4-14}$$

In this case the temperature profile consists of connected line segments of constant characteristic slope in each bed. An example will be discussed in Sec. 4-5b.

The geothermal gradient is also distorted near the earth's surface. Periodic variations in surface temperature propagate into the subsurface, with temperature amplitudes attenuating as $\exp\,(-2\pi z/\sqrt{4\pi\kappa t_0})$, where t_0 is the period of the variation. Daily temperature variations decay to less than 1 percent within 0.5 m of surface, and annual variations are negligible at a depth of about 20 m. Hence, these effects are unimportant for all wellbore measurements except heat flow determinations. Temperature perturbations due to long-term climatic changes such as glacial loading can propagate to great depths, but the effect on the geothermal gradient will not be great. Climatic changes can be represented by a series of step-like changes in temperature; solutions are given by Carslaw and Jaeger (1959).

4-5 THERMAL TRANSIENTS DUE TO WELLBORE FLUID MOVEMENT

Drilling a hole upsets the thermal equilibrium in and around the hole because drilling fluids will cool or warm the rock, particularly where fluid infiltrates a porous formation. Because temperature surveys are made in the wellbore fluid, these questions arise: How well does the measured temperature reflect the formation temperature? How long a time must pass before equilibrium is attained? As suggested in Sec. 4-4, the disequilibrium effect is a likely cause for the scatter in temperature shown in Fig. 4-2. Temperature logs are also often run in cased wells to diagnose flow behavior during injection or production of fluids. To what extent does the temperature profile reflect the volumes of fluids injected or produced? After injection of cool fluid into a warm formation, how long will a measurable temperature anomaly persist?

An answer to these questions requires solution of the problems of heat exchange between the rock formations and a fluid moving down the wellbore and into the formation. We next present four cases which illustrate the principles involved. As shown schematically in Fig. 4-3, they are: injection to some depth far below the zone of observation in a homogeneous earth (*a*) and in a layered earth (*b*), injection into a single permeable zone (*c*), and the case of fluid circulating through drill pipe with complete return to surface (*d*).

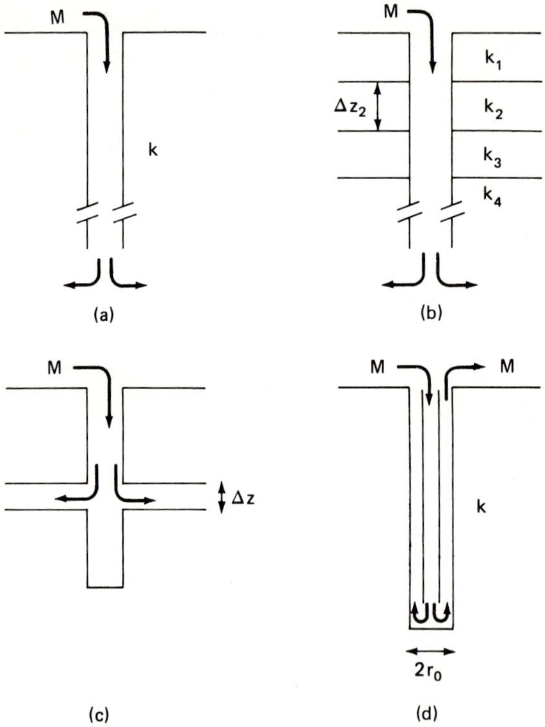

FIGURE 4-3 Four cases of wellbore fluid injection and circulation. Fluid at temperature T_i is injected at volumetric flow rate M. Thermal conductivity structure k and layer thicknesses Δz vary from case to case. Well radius is r_0. (*a*) Deep injection; (*b*) deep injection in layered earth; (*c*) injection into permeable zone; (*d*) circulation of drilling fluid.

4-5a Injection of Fluid, Homogeneous Earth

Ramey (1962) provided a solution for the temperature profile produced by continuous fluid injection at a volumetric rate M into a deep formation of thermal conductivity k. This case represents the field situations in which water is injected for disposal or to waterflood a petroleum reservoir. The fluid injection temperature $T_i(t)$ can vary with time if desired. The model assumes that heat is transferred radially from the wellbore to the formation. Fluid exodus from the well occurs at a depth great enough that it has no impact upon the observed temperature profile. Ramey's solution is

$$T(z, t) = T_0 + Gz - GL(t) + \left[T_i(t) + GL(t) - T_0 \right] \exp\left[-\frac{z}{L(t)} \right] \qquad (4\text{-}15)$$

where $T_0 + Gz$ is the geothermal temperature gradient, and

$$L(t) = \frac{M\rho C}{2\pi r_0 h} + \frac{M\rho C}{2\pi k} f(t) \qquad (4\text{-}16)$$

is a time-varying effective length. The dimensionless function $f(t)$ depends upon the boundary conditions assumed at the rock-wellbore interface; h is the heat transfer coefficient of the tubing or casing; and r_0 is the inside radius of the tubing or casing. Temperature-depth curves based on Eq. (4-15) are shown in Fig. 4-4 for three injection rates and three injection time periods. The leftmost curves, representing high injection rates, show nearly constant temperatures extending to almost 1000 m below surface. The high downward velocities do not permit the injection fluid to be heated by the formation until it reaches considerable depth.

At lower injection rates, the curves quickly equilibrate to the geothermal gradient, but at temperatures which are lower than geothermal. Where this asymptotic behavior has been reached, the effect of cool fluid injection is to translate the geothermal temperature profile to a depth greater than its actual depth in the formation. The asymptotic behavior results when the exponential term in Eq. (4-15) is negligible, that is, when $z \gg L(t)$. For most cases of interest, $L(t)$ (in meters) can be approximated by (Witterholt and Tixier, 1972)

$$L(t) \simeq \frac{M\rho C}{2\pi k} f(t) \qquad (4\text{-}17)$$

Using numerical values of $\rho C = 4.18 \times 10^6$ J/(m^3)(°C) for water and $k = 2.43$ W/(m)(°C) for sedimentary rock gives us

$$L(t) = 3.2Mf(t) \qquad (4\text{-}18)$$

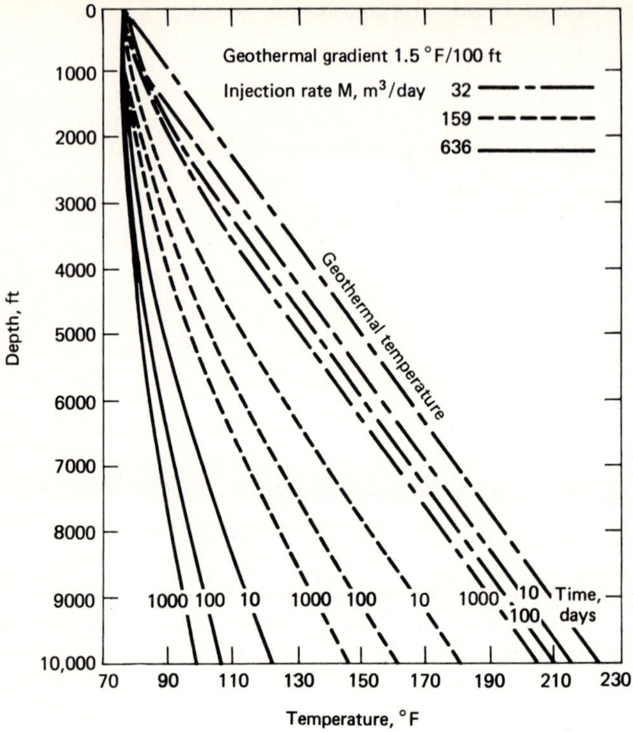

FIGURE 4-4 Temperature profiles due to fluid injection into a homogeneous earth. (Based on solution by Ramey, 1962. Figure from Witterholt and Tixier, 1972. Copyright, 1972, SPE-AIME.)

where $f(t)$ has values 0.6 to 1.0 at 0.1 day, 1.6 to 1.9 at 1 day, 2.8 to 3.0 at 10 days, and 4.0 at 100 days. The $f(t)$ values vary over a limited range for the lower values of injection time t to accommodate the range in possible boundary conditions at the well-rock interface. The temperature profiles will asymptotically approach the geothermal gradient at depths z several times greater than L calculated by Eq. (4-18). At these depths, the difference between the geothermal and observed temperatures will be GL, and hence it will be proportional to injection rate M.

4-5b Injection of Fluid in a Layered Earth

We next examine the case of injection past a series of layers of different thermal conductivities (Fig. 4-3*b*). Fluid injected down the wellbore will lose heat at different rates to different layers because thermal diffusivity $\kappa(z) = k(z)/\rho C$ is a function of depth z. For temperature T' in the formation, the diffusion equation (4-7) must now be written

$$\frac{\partial T'}{\partial t} = \frac{\partial}{\partial z}\left[\kappa(z)\,\frac{\partial T'}{\partial z}\right] + \kappa(z)\left(\frac{1}{r}\frac{\partial T'}{\partial r} + \frac{\partial^2 T'}{\partial r^2}\right) \qquad (4\text{-}19)$$

The rate of heat exchange per unit depth between wellbore and formation is given by $h(T - T')$. Consideration of the heat balance within a small wellbore volume gives the equation for temperature T in the wellbore,

$$M\rho C\,\frac{\partial T}{\partial z} + \rho\pi C r_0^2\,\frac{\partial T}{\partial t} + 2\pi r_0 h(T - T') = 0 \qquad (4\text{-}20)$$

The coupled equations (4-19) and (4-20) include the time derivatives of temperature T' and T. This permits us to consider thermal recovery following cessation of injection during the important time period often referred to as the shut-in phase.

Thermal profiles have been computed by Hoang and Somerton (1981). As in Fig. 4-3*a*, fluid exodus occurs far enough below the zone of observation that it does not affect the temperature profile. Before injection begins, rock and fluid are at a temperature governed by Eq. (4-13) for heat flow in a layered earth. Throughout injection and recovery, the formation temperature at the surface is a constant T_0; during injection, fluid temperature is T_i. Vertical conduction of heat in the fluid is assumed to be negligible compared with radial conduction. Changes of fluid density due to heating are accounted for by letting the downward fluid velocity be inversely proportional to density.

By using numerical methods to solve Eqs. (4-19) and (4-20), the fluid temperature $T(z, t)$ is determined as injection and recovery proceed. For the case shown in Fig. 4-5, hot water at $T_i = 150°C$ is injected for 24 h at a rate $M\rho = 40{,}000$ kg/h. Four temperature profiles are computed at the times shown following injection. Wellbore temperatures have nearly returned to geothermal 72 h after injection has ceased. However, the earlier profiles reflect both the geothermal gradient and the earth's conductivity structure. Reversals in the temperature gradient exist at the lower edge of the transitions from $k_2 > k_1$ and $k_3 > k_2$ because zones of high thermal conductivity return more rapidly to geothermal temperatures than do zones of low thermal conductivity. If only temperature profiles at early shut-in times were available, such as the 4-h curve in Fig. 4-5, it would be difficult to distinguish these peaks and troughs from zones of fluid loss discussed in the next subsection.

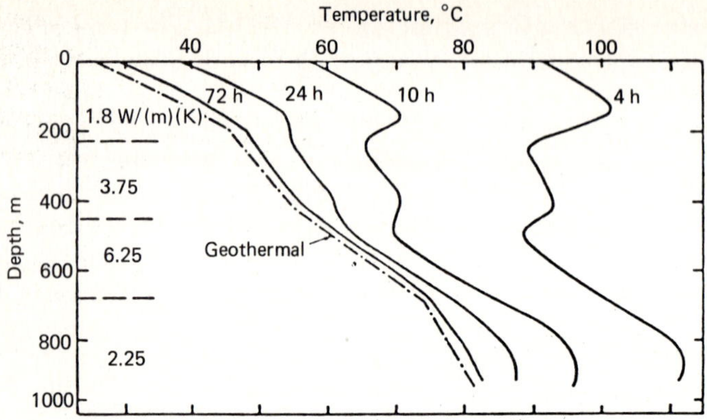

FIGURE 4-5 Temperature profiles during shut-in following injection into a layered earth. Values of k shown on left. (From Hoang and Somerton, 1981.)

4-5c Injection into a Permeable Zone

As a cool fluid is injected into a hot permeable zone (Fig. 4-3*c*), a cool front progresses steadily outward from the wellbore. Its character and rate of progress depend upon fluid temperature at the injection depth, the formation temperature, and the injection flow rate per unit depth. Temperature within the permeable zone obeys the diffusion equation (4-7) with the addition of a final term to account for convective heating

$$\frac{\partial T}{\partial t} = \kappa \left(\frac{\partial^2 T}{\partial z^2} + \frac{1}{r} \frac{\partial T}{\partial r} + \frac{\partial^2 T}{\partial r^2} \right) + \frac{S}{\rho C} - \frac{M}{2\pi r_0 \, \Delta z} \frac{\partial T}{\partial r} \tag{4-21}$$

The factor $M/\Delta z$ is the volumetric flow rate per meter thickness of permeable formation; the $1/2\pi r_0$ term accounts for cylindrical spreading of the flow front. Vertical and radial heat conduction are again accounted for by the first term on the right-hand side of the equation. Additional relations stipulate initial and boundary conditions (Smith and Steffenson, 1970); the full set of equations is solved using numerical methods.

A comparison between a computed solution and a field case is shown in Fig. 4-6. Cool water was injected for 5 months at a rate of 240 m^3/day into a zone 34 m thick. The temperature log was obtained 48 h after shut-in. The fluid temperature has increased from the final injection temperature back toward the formation (geothermal) temperature of approximately 36°C. The smallest temperature recovery occurs at the injection zone, where the formation was cooled the most. Higher radial temperature gradients were established in the open hole interval

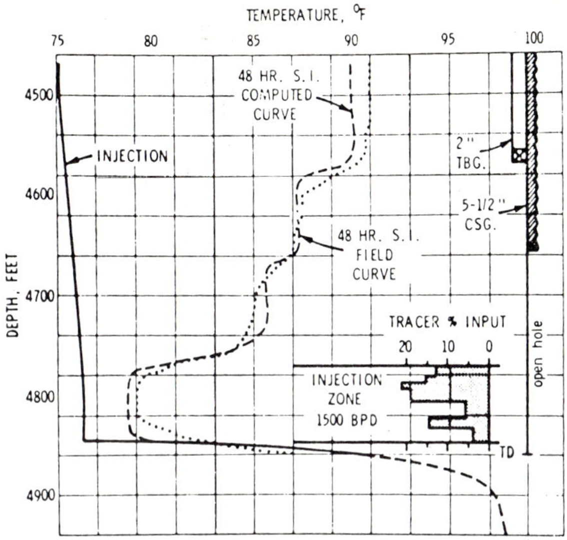

FIGURE 4-6 Computed and measured shut-in temperature profiles due to injection of a cool fluid into a warm permeable zone. Dashed curve below 4900 ft indicates the undisturbed (geothermal) temperature in the formation. (From Smith and Steffensen, 1970. Copyright 1970, SPE–AIME.)

immediately above the injection zone; therefore, this zone warms more rapidly than the injection zone. And even more rapid warming occurs in the upper intervals where the fluid passes through casing and tubing. The authors attribute this to the lower thermal diffusivity of cement (and also of water surrounding the tubing) which surrounds the casing (see Table 4-3).

The temperature of the injected water increases within the formation as the water undergoes a pressure drop from wellbore pressure to formation pressure. This is referred to as the Joule-Thomson effect (Steffenson and Smith, 1973); in Eq. (4-21) it is represented by the distributed source term S. Water warms approximately 1°C per 4 MPa pressure drop. If there were no Joule-Thomson effect, the 48-h shut-in temperature at the injection zone in Fig. 4-6 would have remained equal to the injection temperature, according to computations by Steffenson and Smith. Thus the result of Joule-Thomson heating in water injection wells is to reduce the magnitude of the temperature anomaly. If the injection temperature is quite different from geothermal, then the anomaly remains recognizable. However, if the temperature contrast between fluid and formation is not great, then Joule-Thomson heating can cancel the effect of convective cooling or even produce a temperature in the wellbore fluid which is greater than geothermal.

4-5d Circulation of Fluid During Drilling

Jaeger (1961) considered a steady-state simulation of the drilling process in which drilling fluid at initial temperature T_i is injected at volumetric flow rate M down the drill pipe (Fig. 4-3*d*). The fluid exchanges heat with the drill pipe as it moves downward, cools the bit as it flows from the drill pipe, and exchanges heat with both drill pipe and bore wall as it returns up the annulus between pipe and bore wall to the surface. In actual drilling circumstances, the temperature of fluid inside the pipe $T_p(z)$ will increase with depth, while inside the annulus, the temperature of fluid $T_a(z)$ will decrease as the fluid returns to surface. At surface, $T_p(0) = T_i$; and at bottom depth z_1, $T_p(z_1) = T_a(z_1)$. Formation temperature is prescribed by the geothermal gradient $T_0 + Gz$.

The rate of heat exchange per unit length between pipe and annulus fluids is given by $h(T_p - T_a)$, where h is the heat transfer coefficient of the drill pipe. Consequently, the equation for heat exchange through the drill pipe over the depth range $0 < z < z_1$ is

$$M\rho C \frac{dT_p}{dz} + 2\pi r_0 h(T_p - T_a) = 0 \tag{4-22}$$

where C is fluid heat capacity. The fluid in the annulus exchanges heat through both drill pipe and bore wall,

$$M\rho C \frac{dT_a}{dz} + 2\pi r_0 h(T_p - T_a) - 2\pi r_0 h'[T_a - (T_0 + Gz)] = 0 \tag{4-23}$$

Jaeger (1961) considers the evaluation of h' in detail. The analytical solutions obtained for $T_p(z)$ and $T_a(z)$ are similar to Eq. (4-15), involving linear dependence on z due to the geothermal gradient plus exponential dependence on z reflecting the heat exchange processes. Coefficients for the exponential terms are expressed as combinations of the dimensionless ratios hz_1/MC and $h'z_1/MC$.

Calculated results for three cases are shown in Fig. 4-7. Curves *E* and *N* represent holes drilled with EX (38 mm) and NX (76 mm) diamond bits. Curve *R* corresponds to a 250-mm hole. Injection rate M is much greater in the large-diameter hole than in the small-diameter hole.

We see that temperature T_p inside the small-diameter drill rods (curves *E* and *N*) is the same as the geothermal temperature over much of the depth, departing only near the top and bottom. However, in the rotary-drilled hole (curve *R*), fluid temperature remains constant and

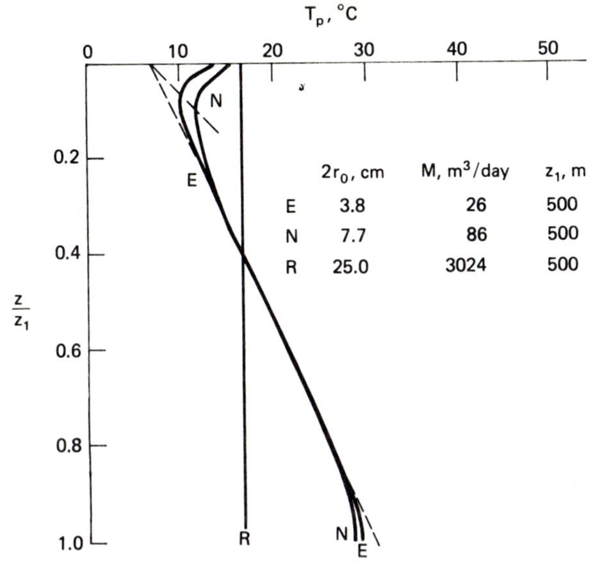

FIGURE 4-7 Temperature T_p of circulating fluid inside drilling pipe for three cases with different borehole diameters $2r_0$, flow rates M, and total z_1. Dashed line is geothermal temperature. Curves E and N are for diamond bits; curve R is for a rotary drill. (Adapted from Jaeger 1961.)

equal to the injection temperature T_i. The injection rates used in the rotary hole example are so high that fluid makes the round trip with no significant gain or loss of heat. In deeper holes or with lower circulation rates, the fluid temperature will more closely approach the formation temperature. Wyllie (1963) states that, after drilling a rotary hole, good agreement between fluid and formation temperatures will generally be found at depths between 0.4 to 0.6 of total depth. This statement is in agreement with the results shown in Fig. 4-7. The relation between fluid and formation temperatures at the bottom of a large-diameter hole is discussed further in the next section.

4-6 RELATION OF BOTTOM–HOLE TEMPERATURE TO FORMATION TEMPERATURE

In many wells, a single bottom-hole temperature (BHT) is recorded by using a maximum-reading thermometer. The BHT is recorded after drilling and circulation operations have stopped and (unavoidably) before thermal equilibrium between formation and wellbore fluid is reached. Determination of true formation temperature rests on analytic solutions to idealizations of the cooling caused by drilling fluid.

Lachenbruch and Brewer (1959) assumed that the drilling process lasts for a time t_0. During this time, heat is removed from the rock at a constant rate q (W/s per vertical meter). A solution is found by integrating the instantaneous response of a linear heat source over the time duration t_0:

$$T(t) - T_f = \frac{q}{4\pi k} \ln \frac{t}{t - t_0} \tag{4-24}$$

where T_f is the formation temperature and time t is reckoned from commencement of drilling at the depth of interest. Equation (4-24) is applied by plotting a sequence of measurements $T(t_1)$, $T(t_2)$, . . . as a function of ln $(t/t - t_0)$. The formation temperature T_f is then found by extrapolation to very long times $t \gg t_0$. The method does not require that q and k be known, but it does require that t_0 be known. It also requires that q be constant during drilling and circulation, although there is usually no way to determine whether that is the case. Moreover, several temperature measurements must be made over long enough time to define the recovery curve.

A second method of finding T_f from BHT assumes that, at the time drilling and circulation cease, a step-function temperature distribution exists such that the temperature is T_m out to a radius a and T_f for r/a (Barelli and Palama, 1981; Leblanc, et al., 1981). Radius a can correspond to the borehole radius, in which case T_m is mud temperature and T_f is formation temperature, or it can correspond to an estimate of the size of the thermal front. Thermal diffusivity is assumed to be the same value κ in both formation and wellbore. The recovery of fluid temperature $T(t)$ on the borehole axis to T_f is

$$\frac{T(t) - T_m}{T_f - T_m} = \exp\left(-\frac{a^2}{4\kappa t}\right) \tag{4-25}$$

where time t commences when drilling and circulation cease. If a series of temperatures $T(t_1)$, $T(t_2)$, . . . are available, the unknown parameters T_m, a, and κ can be estimated by curve fitting. On the other hand, if an assessment of a and κ can be made, Eq. (4-25) can be used to calculate $T(t)$. T_f can then be obtained by comparing the calculated $T(t)$ with measured values.

In this discussion we assume that the fluid column is in static equilibrium. But if the temperature at the bottom of a column of fluid is much greater than at the top, the column is unstable and convective overturn occurs. This condition of instability does in fact exist in large-diameter wells and in some small-diameter boreholes (Gretener, 1967). However, field observations have shown that the resulting temperature oscillations are of the order of ±0.01°C and that individual convective cells extend for a vertical length of no more than a few borehole diameters (Beck, 1976; Gretener, 1967). Hence, convective overturn can be ignored for most logging purposes.

REFERENCES

Barelli, A., and A. Palama: "A New Method for Evaluating Formation Equilibrium Temperature in Holes during Drilling," *Geothermics*, vol. 10, no. 2, 1981, pp. 95–102.

Beck, A. E.: "An Improved Method of Computing the Thermal Conductivity of Fluid-Filled Sedimentary Rocks," *Geophysics*, vol. 41, no. 1, 1976, pp. 133–144.

———: "The Use of Thermal Resistivity Logs in Stratigraphic Correlation," *Geophysics*, vol. 41, no. 2, 1976, pp. 300–309.

Carslaw, H. S., and J. C. Jaeger: *Conduction of Heat in Solids*, Oxford, New York, 1959.

Carvalho, H., and V. Vacquier: "Method for Determining Terrestrial Heat Flow in Oil Fields," *Geophysics*, vol. 42, no. 3, 1977, pp. 584–593.

Clark, S. P., Jr. "Thermal Conductivity," in S. P. Clark, Jr. (ed.), *Handbook of Physical Constants*, Geol. Soc. of Am. Memoir 97, New York, 1966.

Doebelin, E. O.: *Measurement Systems, Application and Design*, McGraw-Hill, New York, 1975.

Gretener, P. E.: "On the Thermal Instability of Large Diameter Wells—An Observational Report," *Geophysics*, vol. 43, no. 4, 1967, pp. 727–738.

Hoang, V. T., and W. H. Somerton: "Effect of Variable Thermal Conductivity of the Formations on the Fluid Temperature Distribution in the Wellbore," *Trans. 22d Ann. Symp. SPWLA*, 1981, pp. L2–L24.

Horai, K.: "Thermal Conductivity of Rock-Forming Minerals," *J. Geophys. Res.*, vol. 76, no. 5, 1971, pp. 1278–1308.

Jaeger, J. C.: "The Effect of the Drilling Fluid on Temperatures Measured in Bore Holes," *J. Geophys. Res.*, vol. 66, no. 2, 1961, pp. 563–569.

Judge, A. S.: "Thermal Regime of the Mackenzie Valley: Observations of the Natural State," Environmental-Social Committee, Northern Pipelines Task Force on Northern Oil Development Report 73-38, 177 pp., Information Canada Catalog no. R72-11973, 1973.

Kappelmeyer, O., and R. Haenel: "*Geothermics with Special Reference to Application*," Geopublication Associates, Berlin, 1974.

King, M. S.: "Thermal Conductivity Measurements on Saturated Rocks at Permafrost Temperatures," *Can. J. Earth Sci.*, vol. 16, 1979, pp. 73–79.

Lachenbruch, A. H., and M. C. Brewer: "Dissipation of the Temperature Effect of Drilling a Well in Arctic Alaska," U.S. Geol. Survey Bull. 1083-C, 1959.

——— and J. H. Sass: "Heat Flow in the United States and Thermal Regime of the Crust," in J. G. Heacock (ed.), *The Earth's Crust: Its Nature and Physical Properties*, Am. Geophys. Union Monograph 20, Washington, D.C., 1977.

Leblanc, Y., L. J. Pascoe, and F. W. Jones: "The Temperature Stabilization of a Borehole," *Geophysics*, vol. 46, no. 9, 1981, pp. 1301–1303.

Lee, T. C.: "On Shallow-Hole Temperature Measurements," *Geophysics*, vol. 42, no. 3, 1977, pp. 572–583.

Ramey, H. J., Jr. "Wellbore Heat Transmission," *Trans. Soc. Petr. Eng. AIME*, vol. 225, April 1962, pp. 427–435.

Robertson, E. C., and D. L. Peck: "Thermal Conductivity of Vesicular Basalt from Hawaii," *J. Geophys. Res.*, vol. 79, no. 32, 1974, pp. 4875–4888.

Roy, R. F., A. E. Beck, and Y. S. Touloukian: "Thermophysical Properties of Rocks," in Y. S. Touloukian and C. Y. Ho (eds.), *Physical Properties of Rocks and Minerals*, McGraw-Hill, New York, 1981, chap. 12.

Smith, R. C., and R. J. Steffensen: "Computer Study of Factors Affecting Temperature Profiles in Water Injection Wells," *J. Petr. Tech.*, November 1970, pp. 1447–1458.

Somerton, W. H.: "Porous Rock-Fluid Systems at Elevated Temperatures and Pressures," in T. N. Narasimhan (ed.), *Recent Trends in Hydrogeology*, Geol. Soc. Am. Special Paper 189, 1982.

Steffensen, R. J., and R. C. Smith: "The Importance of Joule-Thomson Heating (or Cooling) in Temperature Log Interpretation," Soc. Petr. Eng. Paper 4636, 1973.

Witterholt, E .J., and M. P. Tixier: "Temperature Logging in Injection Wells," Soc. Petr. Eng. Paper 4022, 1972.

Woodside, W., and J. H. Messmer: "Thermal Conductivity of Porous Media. II. Consolidated Rocks," *J. Appl. Phys.*, vol. 43, no. 9, 1961, pp. 1699–1706.

Wyllie, M. R. J.: *The Fundamentals of Well Log Interpretation*, Academic, New York, 1963.

5

Electrical and Magnetic Methods

Electrical logging methods, and the electrical resistivity method in particular, were first pursued intensively when it was learned that individual strata could be identified and correlated by such measurements and, even better, that oil-producing zones could be distinguished from water-saturated zones. Later it was found that in many cases the amount of oil in place could be quantified from electrical logs. The electrical resistivity of rocks in the upper portion of the earth's crust varies with:

1. Water content. Natural waters are much more conductive than most rock-forming minerals.
2. Salinity. Natural waters are conductive in proportion to the concentration of ionized salts in the water.
3. Temperature. A rock at a depth of 1 km can be twice as conductive as the same rock at the surface.
4. Geologic strike of the formation. Many sedimentary and metamorphic rocks are anisotropic; they have a lower resistivity along the bedding plane than perpendicular to it.

5 Clays and conductive minerals. Clays can augment the ionic conduction of pore water. Electronic conduction in sulfide and oxide minerals can dominate if such minerals are abundant enough.

Most of these factors were recognized in the early days of well logging (Schlumberger and Schlumberger, 1929). Understanding of the physical and chemical nature of these effects has evolved steadily over the past 50 years. Sections 5-2 and 5-3 summarize current knowledge of fluid and rock properties.

The first resistivity logging sondes had combinations of current and voltage electrodes spliced into a cable or mounted on a mandrel. Methods of correcting the resulting readings for the distortion produced by wellbore and invasion zones were developed, but several outstanding problems remained and inspired the development of more sophisticated systems:

1 Bed definition. Oftentimes, unfocused electrode arrays either could not delineate thin beds or else gave readings quite different from the true formation resistivity. Focused sondes solved this problem.

2 Lateral definition. Conventional logs measured the combined effect of the invaded and uninvaded zones. Separation of these contributions required an independent measurement of the resistivity of the invaded zone, which was accomplished by small arrays mounted on pads pressed against the bore wall.

3 Resistive borehole conditions. Because electrode-type sondes require conductive fluid, their use was precluded above the mud column and in wells where oil-based muds were used. With the introduction of induction sondes, electrical measurements can be made under virtually all open-hole conditions.

A historical account of these developments, made predominantly by Schlumberger Ltd., is given by Allaud and Martin (1977). Our discussion of the electrode and induction sondes in Secs. 5-4 and 5-5 emphasizes the fundamental characteristics of those sondes. Other electrical logging methods are discussed in Secs. 5-6 to 5-8. The relations of the various methods can be understood by first examining the fundamental equations and constitutive laws governing electromagnetic wave propagation and current flow in a homogeneous medium.

5-1 ELECTROMAGNETIC PARAMETERS AND EQUATIONS

5-1a Electrical Quantities

Our purpose here is to define the terms and establish the fundamental equations to be used in Secs. 5-2 to 5-8. We have relied upon two texts on electromagnetism: one by Grant and Phillips (1975), which uses SI units, and one by Purcell (1965), which uses cgs units. We define an electric field **E** as the force **F** imposed on a static electric charge q

$$\mathbf{E} = \frac{\mathbf{F}}{q} \tag{5-1}$$

The units of **E** are volts per meter (V/m). The field can be expressed as the gradient of a potential Ψ,

$$\mathbf{E} = -\nabla\Psi \tag{5-2}$$

and we define a potential difference V between two points in a field **E** as the work done moving a unit charge between those two points in the field

$$V = -\int_{P_1}^{P_2} \mathbf{E} \cdot d\mathbf{s} \tag{5-3}$$

In many materials the current density **J**, given in amperes per square meter (A/m^2), is directly proportional to **E**:

$$\mathbf{J} = C\mathbf{E} = \frac{\mathbf{E}}{R} \tag{5-4}$$

where C is the conductivity of the material, expressed in siemens per meter (S/m). Note that S replaces the older mho, or reciprocal ohm. The resistivity R of a material is $1/C$ and is given in units of ohm-meters. The symbols R and C have been adopted in the well logging literature to represent resistivity and conductivity. We shall use them throughout this book, although texts on electromagnetism customarily use ρ and σ.

Circuit elements. Place a material (for example, a rock sample) of uniform cross section A and length ℓ between two electrodes (also of cross section A) and impose a constant potential difference V across the sample. Then from Eq. (5-3), $V = E\ell$. Current density is taken to be uniform across the sample, so the current $I = JA$. According to Ohm's law,

$$I = \frac{V}{\mathcal{R}} \tag{5-5}$$

The resistance $\mathcal{R}$ of the sample is given in ohms. By substituting $V = E\ell$ and $I = EA/R$ we get

$$\mathcal{R} = \frac{R\ell}{A} \tag{5-6}$$

and thereby relate R, a property of bulk material, to its circuit element representation $\mathcal{R}$.

If an inductance $\mathcal{L}$ is added to the circuit in series with $\mathcal{R}$ (all materials have some inductance) and the impressed voltage varies with time,

$$V = I\mathcal{R} + \mathcal{L}\frac{dI}{dt} \tag{5-7}$$

Let ω be the angular frequency, equal to 2π times the cyclic frequency (Hz). If the voltage is varied sinusoidally, as $V = V_0 \cos \omega t$,

$$I = I_0 \cos (\omega t - \varphi) \tag{5-8}$$

where φ is called the *phase* of the current. Equations (5-7) and (5-8) can be solved to give

$$I = \frac{V_0 \cos [\omega t - \tan^{-1}(\omega \mathcal{L}/\mathcal{R})]}{(\mathcal{R}^2 + \omega^2 \mathcal{L}^2)^{1/2}} \tag{5-9}$$

where $\omega\mathcal{L}$ has the dimensions of resistance and is called the inductive reactance.

Since I peaks later than V, the current of an inductive circuit is said to lag the voltage by an amount φ, where

$$\tan \varphi = \frac{\omega \mathcal{L}}{\mathcal{R}}$$

There is a convenient mathematical artifice often used to deal with currents that are out of phase with voltages. For any φ, we can write the complex function with a real part $\cos \varphi$ and an imaginary part $\sin \varphi$,

$$\exp (-i\varphi) = \cos \varphi - i \sin \varphi$$

where $i^2 = -1$. Then the current of Eq. (5-8) is the real part of

$$I_0 \exp [-i(\omega t - \varphi)] = I_0 \cos (\omega t - \varphi) - iI_0 \sin (\omega t - \varphi) \tag{5-10}$$

The complex term can also be written

$$I_0 \exp\left[-i(\omega t - \varphi)\right] = I_0 \exp(-i\omega t) \exp(i\varphi) \qquad (5\text{-}11)$$

Consequently, the time dependence of a sinusoidal system is contained in the term $\exp(-i\omega t)$. It can be deleted from calculations of the spatial variations of the system and later added by multiplying by $\exp(-i\omega t)$. We can now represent the impedance Z of a combination of a resistance and inductance by $Z = \mathcal{R} - i\omega\mathcal{L}$. Capacitance can be treated in a similar manner, yielding an impedance for a series resistance and capacitance of $Z = \mathcal{R} - i/\omega\mathcal{C}$. Then Eq. (5-9) can be written

$$I = \frac{V}{Z} \qquad (5\text{-}12)$$

Electric polarization. In a "good dielectric" there are no free charges, but there are bound charges which can be displaced with the application of an electric field. The material is then said to be polarized. Then the current is $\mathbf{J} = -i\omega\varepsilon\mathbf{E}$, called the displacement current. The *dielectric permittivity* ε is a measure of the polarizability of a material, and the $-i$ term signifies that the displacement current $\mathbf{J}$ lags the electric field $\mathbf{E}$ by 90° in phase.

Our arrangement of electrodes and rock sample can be considered to be a capacitor. The capacitance of two parallel plates of area A separated by a distance ℓ is $\varepsilon A/\ell$. The impedance of a capacitor is

$$Z = \frac{-i}{\omega\mathcal{C}} = \frac{-i\ell}{\omega\varepsilon A}$$

Since capacitance is given in farads, the units of ε must be farads per meter. The value of ε in vacuum, designated ε_0, is 8.854×10^{-12} F/m (8.854 pF/m). If a polarizable material is inserted between the two plates, the capacitance increases in proportion to $\varepsilon/\varepsilon_0$. If we now allow the voltage across the capacitor to vary as $\exp(-i\omega t)$, there will be a flow of alternating current of density $\mathbf{J} = -i\omega\varepsilon\mathbf{E}$ between the plates.

The above description is correct only for ideal dielectrics. It is inadequate for real earth materials because some free charge is almost always present to provide a conduction current. Moreover, the dielectric component is not constant with frequency but has relaxation peaks or resonances at frequencies characteristic of molecular, atomic, and electronic oscillations. At these frequencies, displacement currents cannot maintain the 90° phase relation with an applied electric field. To treat these phenomena, C and ε are written as complex quantities,

$$\begin{aligned} C(\omega) &= C_r(\omega) + iC_i(\omega) \\ \varepsilon(\omega) &= \varepsilon_r(\omega) + i\varepsilon_i(\omega) \end{aligned} \qquad (5\text{-}13)$$

where the subscripts r and i refer to the real and imaginary parts of the quantity. The total current density is

$$\mathbf{J} = [C_r + \omega\varepsilon_i + i(C_i - \omega\varepsilon_r)]\mathbf{E} \tag{5-14}$$

where we have dropped the explicit dependence on ω. The four parameters C_r, C_i, ε_r, and ε_i are not independent, but we shall not discuss their relations here.

Although the theoretical expressions are more easily expressed as conductivity C, logging results are usually expressed as resistivity R. Since $R = 1/C$,

$$R_r = \frac{C_r}{C_r^2 + C_i^2} \qquad \text{and} \qquad R_i = \frac{-C_i}{C_r^2 + C_i^2} \tag{5-15}$$

In most logging applications C and R are considered to be independent of frequency. At low frequencies the reactance current is usually small in most rocks, and R_i and C_i can be neglected.

5-1b Magnetic Quantities

Three distinct vectors, **B**, **H**, and **M**, are used to describe magnetic fields. Of these three, the magnetic field vector **B** is fundamental. If a charge q is moving with velocity **v**, the total force acting on it is

$$\mathbf{F} = q\mathbf{E} + q\mathbf{v} \times \mathbf{B} \tag{5-16}$$

The $\mathbf{v} \times \mathbf{B}$ force is orthogonal to both **v** and **B**. The vector **B** can be regarded as the sum of **H** and **M**. If the magnetic field is caused by electric currents, it is described by the field intensity **H**; whereas the field caused by polarization within magnetic bodies is described by the magnetic polarization **M**. The symbols **B** and **H** are almost universally accepted, but some authors use **I** or **J** in place of **M**.

Most of the literature regarding the earth's magnetism is expressed in cgs units. The relations of magnetic quantities expressed in cgs units to the equivalent quantities in SI units are especially confusing because **B** and **H** have different physical dimensions in SI yet are dimensionally the same in cgs. These differences stem from the way force relates to charge and current in the two systems. In SI, the dimension of **H** is amperes per meter (A/m) and **B** has the dimension newtons per ampere-meter (N/A·m), as can be seen from Eq. (5-16). In cgs units, both **B** and **H** have dimensions of force per unit charge, given as dynes per esu. These and other quantities are summarized in Table 5-1.

In SI, the unit of **B** is called the tesla (T), and in cgs, it is called the gauss (G). A field of 1 G converts to 10^{-4} T. A more practical unit is the

TABLE 5-1 Terminology for SI and cgs magnetic units.

	Magnetic field vectors			Constitutive parameters		Circuit parameter
Symbol	**B**	**H**	**M**	μ	χ	$\mathcal{L}$
Name	Induction or flux density	Intensity or field strength	Polarization or magnetization	Permeability or inductive capacity	Susceptibility	Inductance
SI-named unit	tesla (T)	None	None	None	None	henry
SI dimensions	N/A·m	A/m	A/m	N/A^2	Dimensionless	$N \cdot m/A^2$
cgs-named unit	gauss	oersted	None	None	None	None
cgs dimensions	dynes/esu	dynes/esu	dynes/esu	Dimensionless	Dimensionless	statvolts/esu

γ, defined as 10^{-5} G. The earth's field, nominally 0.6 G, converts to 60,000 γ, or 60,000 nanoteslas (nT).

To clarify the conversions between the two systems, we shall develop the relations among the magnetic field vectors in parallel, with SI equations given first and cgs equations in brackets. In a region occupied by matter which is magnetically polarizable and isotropic but has no permanent magnetic moment, **B** and **H** are related by the permeability μ,

$$\mathbf{B} = \mu\mathbf{H} \qquad [\mathbf{B} = \mu\mathbf{H}] \tag{5-17}$$

In free space $\mathbf{B} = \mu_0\mathbf{H}$ in SI, where μ_0 has the value $4\pi \times 10^{-7}$ H/m; whereas in cgs units, $\mathbf{B} = \mathbf{H}$ in free space so that **B** and **H** are identical numerically as well as dimensionally. Note that μ has the dimension newtons per ampere squared in SI but is dimensionless in cgs. Only inside magnetic matter are **B** and **H** numerically different in cgs. For that reason alone they have been assigned different names, **B** being expressed in gauss and **H** in oersteds. At low field strengths, the polarization **M** is linearly related to the field intensity **H** through the magnetic susceptibility χ, which is defined by

$$\mathbf{M} = \chi\mathbf{H} \qquad [\mathbf{M} = \chi\mathbf{H}] \tag{5-18}$$

The relations among **B**, **H**, and **M** are defined differently in the two systems:

$$\begin{aligned} \mathbf{B} &= \mu_0(\mathbf{H} + \mathbf{M}) \qquad [\mathbf{B} = \mathbf{H} + 4\pi\mathbf{M}] \\ \mathbf{B} &= \mu_0(1 + \chi)\mathbf{H} \qquad [\mathbf{B} = (1 + 4\pi\chi)\mathbf{H}] \end{aligned} \tag{5-19}$$

Consequently, the relations between permeability μ and susceptibility χ must be

$$\frac{\mu}{\mu_0} = 1 + \chi \qquad [\mu = 1 + 4\pi\chi] \tag{5-20}$$

Both expressions (5-20) are dimensionless, and μ/μ_0 in SI is numerically equal to μ in cgs. Consequently, a susceptibility value expressed in cgs must be divided by 4π to convert to SI units, even though χ is dimensionless in both systems.

5-1c Maxwell's Equations

The magnetic and electric field vectors **B** and **E** are related by Maxwell's equations. In a homogeneous isotropic medium free of accumulation of electric charge,

$$\nabla \times \mathbf{B} = \mu\varepsilon\frac{\partial\mathbf{E}}{\partial t} + \mu C\mathbf{E} \tag{5-21}$$

$$\nabla \times \mathbf{E} = \frac{-\partial \mathbf{B}}{\partial t} \tag{5-22}$$

$$\nabla \cdot \mathbf{E} = 0 \tag{5-23}$$

$$\nabla \cdot \mathbf{B} = 0 \tag{5-24}$$

From Eq. (5-21) with a constant electric field,

$$\nabla \times \mathbf{B} = \mu C\mathbf{E} = \mu \mathbf{J} \tag{5-25}$$

and from this it can be shown that, for any element of a conductor of length dl, the contribution to the field at radius $\mathbf{r}$ is

$$d\mathbf{B} = \frac{\mu_0 I}{4\pi} \frac{dl \times \mathbf{r}}{|r^3|} \tag{5-26}$$

Equation (5-26) is known as the Biot-Savart law.

From Eq. (5-22), $\nabla \times \mathbf{E} = -\partial \mathbf{B}/\partial t$, we see that a changing magnetic field gives rise to an electric field. From Eq. (5-3) and a result from vector calculus, the potential difference V can be shown to be

$$V = \frac{d}{dt} \int \int \mathbf{B} \cdot d\mathbf{A} \tag{5-27}$$

Equation (5-27) gives the voltage measured at the terminals of a coil of area A which is cut by a changing magnetic field. The polarity of V is opposite to the electromotive force established in the circuit by the magnetic field. Equation (5-27) is known as Faraday's law of induction.

5-1d Electromagnetic Waves

If we set $C = 0$ for the present, Equations (5-21) and (5-22) can be combined to get the electromagnetic wave equations

$$\nabla^2 \mathbf{B} - \mu\varepsilon \frac{\partial^2 \mathbf{B}}{\partial t^2} = 0 \tag{5-28}$$

$$\nabla^2 \mathbf{E} - \mu\varepsilon \frac{\partial^2 \mathbf{E}}{\partial t^2} = 0 \tag{5-29}$$

Each component of **B** and **E** in Eqs. (5-28) and (5-29) is of the form

$$\nabla^2 \psi(x, y, z, t) = \mu\varepsilon \frac{\partial^2 \psi}{\partial t^2} \tag{5-30}$$

where ψ is any component (x, y, or z) of **E** or **B**. By introducing a sinusoidal frequency dependence $\exp(-i\omega t)$,

$$\psi(x, y, z, t) = \psi_0(x, y, z) \exp(-i\omega t) \tag{5-31}$$

The second time derivative becomes

$$\frac{\partial^2 \psi}{\partial t^2} = -\omega^2 \psi \tag{5-32}$$

We can then express Eq. (5-30) as

$$\nabla^2 \psi_0(x, y, z) + k^2 \psi_0(x, y, z) = 0 \tag{5-33}$$

where $k^2 = \mu\varepsilon\omega^2$ is called the propagation factor. Then for a plane wave propagating in the z direction, the solution is

$$\psi(z, t) = \psi_0 \exp i(kz - \omega t) \tag{5-34}$$

The phase velocity of such a plane wave is ω/k, or $(\mu\varepsilon)^{-1/2}$.

This discussion does not include conductivity. When C is nonzero, we replace $\mu\varepsilon\omega^2$ with $\mu\varepsilon\omega^2 + i\mu\omega C$ (Stratton, 1941). When ε and C are complex, the propagation factor becomes

$$k^2 = \mu\varepsilon_r\omega^2 - \mu\omega C_i + i\mu\varepsilon_i\omega^2 + i\mu\omega C_r \tag{5-35}$$

Following Stratton (1941), we square the expression $k = k_r + ik_i$ and equate it to Eq. (5-35) to obtain

$$\begin{aligned} k_r^2 - k_i^2 &= \mu\varepsilon_r\omega^2 - \mu\omega C_i \\ 2k_r k_i &= \mu\omega C_r + \mu\omega^2\varepsilon_i \end{aligned} \tag{5-36}$$

from which the real and imaginary parts of k are found to be

$$k_r^2 = \frac{\mu\varepsilon_r\omega^2 - \mu\omega C_i}{2}\left\{\left[1 + \frac{(\omega\varepsilon_i + C_r)^2}{(\omega\varepsilon_r + C_i)^2}\right]^{1/2} + 1\right\} \tag{5-37a}$$

and

$$k_i^2 = \frac{\mu\varepsilon_r\omega^2 - \mu\omega C_i}{2}\left\{\left[1 + \frac{(\omega\varepsilon_i + C_r)^2}{(\omega\varepsilon_r + C_i)^2}\right]^{1/2} - 1\right\} \tag{5-37b}$$

Equation (5-37) can be simplified in accordance with the relative values of the conductivity and permittivity components for each problem we shall consider.

5-2 ELECTRICAL RESISTIVITY OF WATER AND MINERALS

5-2a Conduction in Electrolytes

Electric current in aqueous solutions is carried by the ions of dissociated salts migrating through the fluid in response to an applied electric field. The current density $\mathbf{J}_i$ (amperes per square centimeter) due to a single ionic species with n_i ions per cubic centimeter is

$$\mathbf{J}_i = n_i z_i e \mathbf{v}_i \tag{5-38}$$

where z_i (electrons per ion) is the valence, e (coulombs per electron) is the electronic charge, and $\mathbf{v}_i$ (centimeters per second) is the drift velocity of the ion. The ionic mobility ν is the ratio of velocity $\mathbf{v}$ to electric field $\mathbf{E}$ (volts per centimeter). The fractional current borne by a species becomes

$$\mathbf{J}_i = n_i z_i e \nu_i \mathbf{E} \tag{5-39}$$

It is customary to express ionic density n_i in terms of concentration c_i (moles per cubic centimeter) by using $n_i = c_i \mathcal{F}/e$, where $\mathcal{F}$ is Faraday's constant (96,500 C/mol). Then Eq. (5-39) becomes

$$\mathbf{J}_i = c_i \mathcal{F} z_i \nu_i \mathbf{E} \tag{5-40}$$

From the definition of conductivity by Eq. (5-4), we can write

$$C_i = c_i \mathcal{F} z_i \nu_i \tag{5-41}$$

This simple expression shows that ionic conductivity C_i is the product of ionic charge per volume of solution (given by $c_i \mathcal{F} z_i$) and the ion velocity per unit voltage gradient (given by mobility ν_i). In dilute solutions we expect that the contribution of each species is independent of the others, so

$$C = \sum C_i = \sum c_i \mathcal{F} z_i \nu_i \tag{5-42}$$

For a simple salt, Eq. (5-42) can be written as

$$C = \mathcal{F}(C_+ z_+ \nu_+ + C_- z_- \nu_-) \tag{5-43}$$

However, this represents an upper limit to the conductivity, because a salt will not necessarily dissolve completely (MacInnes, 1961).

Ions of different size have different mobilities, although the relation between size and mobility is complicated by the tendency of charged ions to solvate, that is, to attract and carry a sheath of water molecules. Consequently, some ions carry more current than others even if their concentrations and valences are the same. The fraction of current carried by an ionic species is called the transport number

$$t_i = \frac{\mathbf{J}_i}{\mathbf{J}} = \frac{C_i}{C} = \frac{c_i \mathcal{F} z_i \nu_i}{\sum c_i \mathcal{F} z_i \nu_i} \tag{5-44}$$

In a system with one salt whose ions are univalent, the transport numbers for cations and anions can be written in terms of mobility ratios:

$$t_+ = \frac{\nu_+}{\nu_+ + \nu_-} \qquad (5\text{-}45)$$

$$t_- = \frac{\nu_-}{\nu_+ + \nu_-}$$

Resistivity and electrolyte composition. At low salt concentrations where ion-ion interactions can be neglected, we might expect that Eq. (5-42) would lead to a simple working relation between concentration and conductivity. If the average mobilities of different species do not differ much,

$$C \sim \sum z_i c_i \text{ (mol/cm}^3\text{)} \qquad (5\text{-}46)$$

or in practical units where ΣA_i is the molecular weight

$$C \sim \frac{\sum z_i c_i \text{ (mg/L)}}{\sum A_i} \qquad (5\text{-}47)$$

If the weight concentration c_i is written in chemical equivalents, Eq. (5-47) becomes

$$C \sim \sum c_i \text{ (meq/L)} \qquad (5\text{-}48)$$

where meq designates milliequivalents.

Resistivity of natural waters. As shown in Fig. 5-1 the resistivity of natural waters varies over five decades, rainwater being the most resistive and saturated brines the least resistive. The resistivity of pure water without ions other than H^+ and OH^- is $2.8 \times 10^5\ \Omega\cdot$m at 17.6°C (Dorsey, 1940). All natural waters are less resistive than pure water due to the presence of dissolved ions.

The six common ionic species in surface waters are Na^+, Ca^+, Mg^{2+}, Cl^-, HCO_3^-, and SO_4^{2-}. The balance among the ionic constituents in groundwater is determined by the ionic concentration of precipitation and subsequent chemical reactions within particular geological environments (Freeze and Cherry, 1979). Concentrations in regions underlain by igneous rocks tend to be somewhat less than in sedimentary areas, as the resistivity distributions in Fig. 5-1 indicate.

Salt concentration in today's oceans is about 35,000 mg/L; the resistivity range of seawater is indicated in Fig. 5-1. The pore waters in marine sediments become increasingly saline with depth: total dissolved solids in sandstone waters increase at a rate of 60,000 to

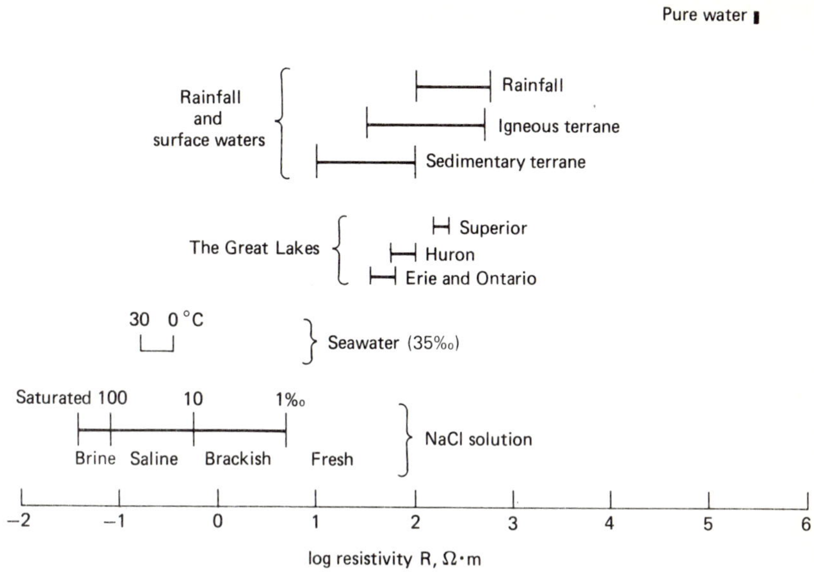

FIGURE 5-1 Range of resistivity values of pure and natural waters. (Data sources: pure water, Dorsey, 1940; rainfall, Sarma and Rao, 1972; Great Lakes, Doherty, 1963; NaCl, Schlumberger, 1972; all other information, Freeze and Cherry, 1979.)

150,000 mg/L per 1000 m (Dickey, 1969), although this trend can reverse where abnormal fluid pressures occur (Schmidt, 1973). At the depths of many hydrocarbon reservoirs, NaCl is the predominant salt, as reflected by the water chemistry distributions in Fig. 5-2. For this reason, water chemistry concentrations are usually referenced to NaCl solutions for logging purposes.

Equation 5-48 indicates that the electrical conductivity of water can be predicted if we know the water chemistry. A test of Eq. (5-48) is shown in Fig. 5-3 for a variety of natural waters. To a first approximation,

$$C\ (\text{S/m}) \sim 0.01 c_i\ (\text{meq/L})$$

or

$$R\ (\Omega\cdot\text{m}) \sim \frac{100}{c_i\ (\text{meq/L})} \tag{5-49}$$

which holds quite accurately for very low concentrations ($R > 100\ \Omega\cdot$m) and becomes less accurate as concentration increases. Logan (1961) gives empirical expressions relating C and c_i which can be used to improve the correlation over various concentration ranges and for different dominant ions.

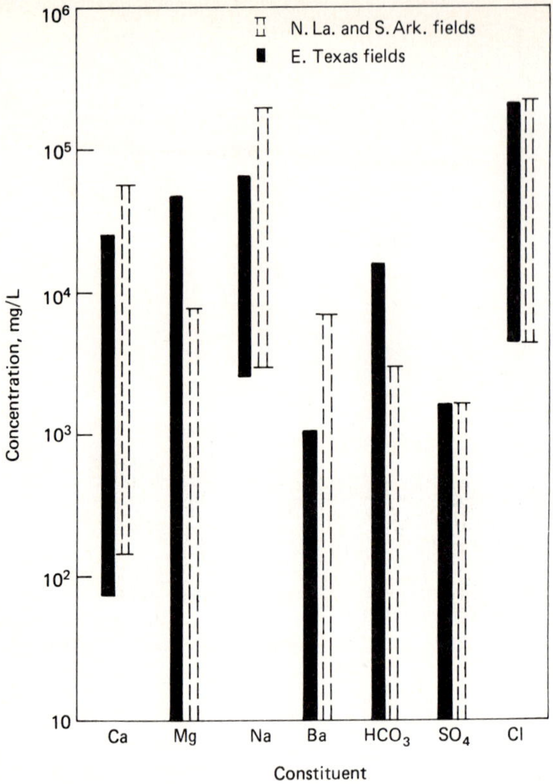

FIGURE 5-2 Concentrations of ionic species in subsurface waters from oil-producing formations. (After Talash and Crawford, 1965. Copyright 1965, SPE–AIME.)

At high brine concentrations the effects of individual ionic species must be taken into account. Talash and Crawford (1965) used brine data having concentration ranges shown in Fig. 5-2 to derive coefficients for an expression relating the logarithms of R to c_i. The results of straightforward regression analyses are usually limited to the range of data, however, and application to concentrations outside that range requires yet another analysis.

Several authors have derived quasi-empirical relations which convert the concentrations of individual species to an equivalent NaCl concentration. The concentration of each ion, in ppm, is multiplied by a factor unique to the species, and the results are summed. Because the multiplier factors vary with concentration, they are presented graphically in Fig. 5-4.

The chart in Fig. 5-4 can be used to determine the solution resis-

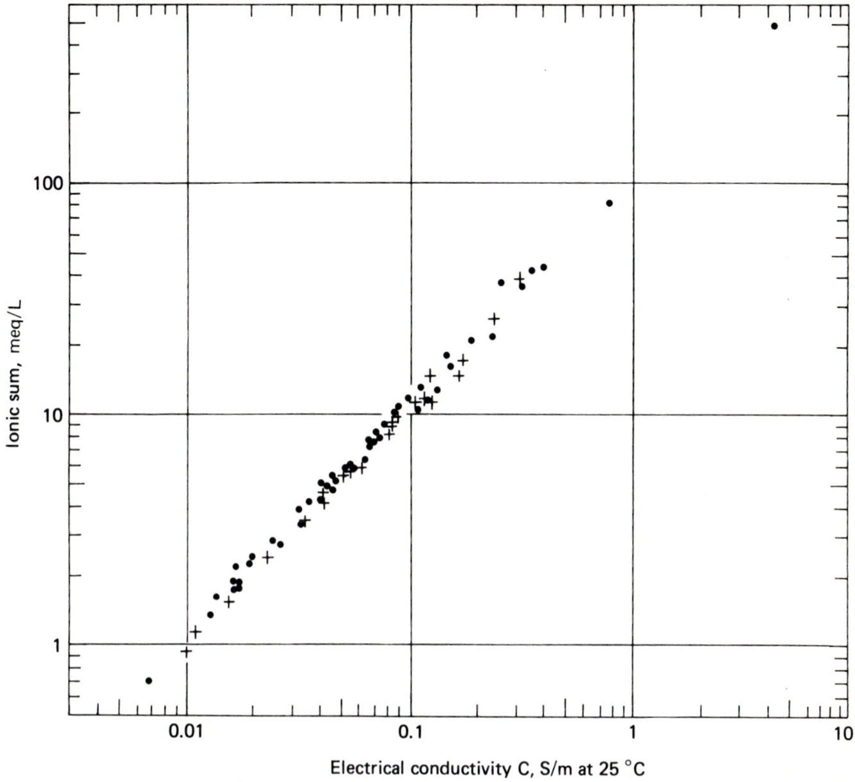

FIGURE 5-3 Ionic concentration and electrical conductivity of surface and groundwaters. The ionic sum is half the sum of all anions and cations, in equivalents per million. (After Logan, 1961.)

tivity R_w when the water chemistry is known or can be used to check the consistency between chemical analysis and an R_w measurement.

Effect of viscosity on resistivity. Stokes' law states that the drag force retarding a sphere moving in a liquid at velocity v is $6\pi\eta rv$, where η is liquid viscosity and r is the sphere radius. If an electrical field $\mathbf{E}$ is present, the electrical driving force $z_i e\mathbf{E}$ is balanced by the drag, so that

$$\nu_i = \frac{\mathbf{v}_i}{\mathbf{E}} = \frac{z_i e}{6\pi\eta r_i} \tag{5-50}$$

Using Eq. (5-41) for ionic conductivity gives us

$$\eta C_i = \frac{c_i \mathfrak{F} z_i^2 e}{6\pi r_i} \tag{5-51}$$

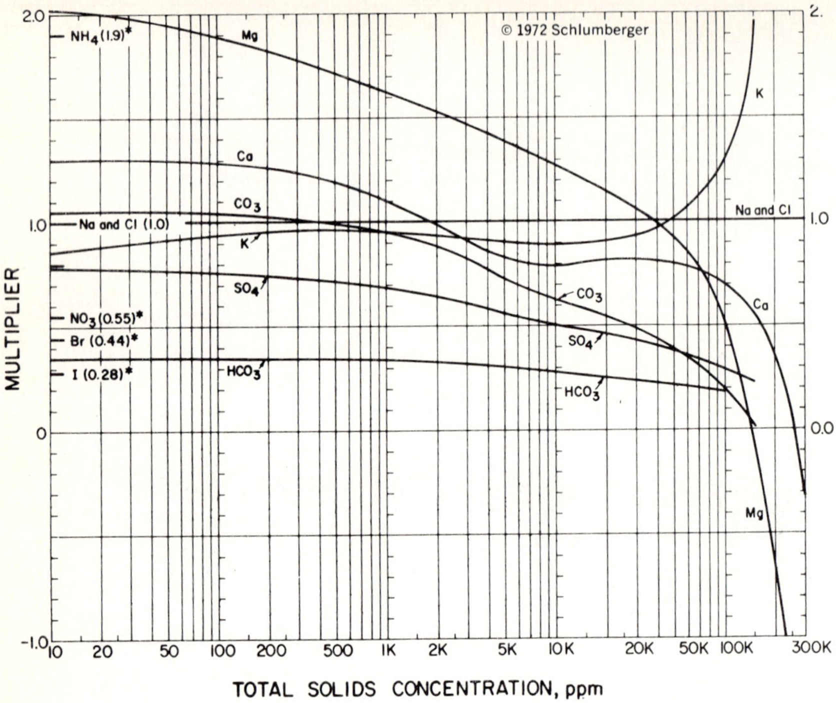

FIGURE 5-4 Multiplier coefficients for conversion of a multiple-species electrolyte to an equivalent NaCl solution. (Courtesy of Schlumberger, 1972.)

which shows that the viscosity-conductivity product is a constant at a given concentration. Because fluid resistivity R is proportional to the reciprocal sum of C_i, it can be expected to be proportional to the viscosity.

The viscosity of water varies with temperature. The temperature dependence of viscosity determines the temperature dependence of resistivity. Temperature effects are important, and corrections must be made for well logging applications. Arps (1953) used measured data for NaCl solutions to obtain the empirical correction equation,

$$R_w(T_2) = R_w(T_1)\,\frac{T_1 + 6.77}{T_2 + 6.77} \tag{5-52}$$

where T is given in degrees Fahrenheit. Correction for temperature is commonly done with a chart such as the one shown in Fig. 5-5, which incorporates the effect of concentration as well as temperature.

At low temperatures, conductivity increases with pressure. Bradshaw and Schleicher (1965) show that the conductivity of seawater at

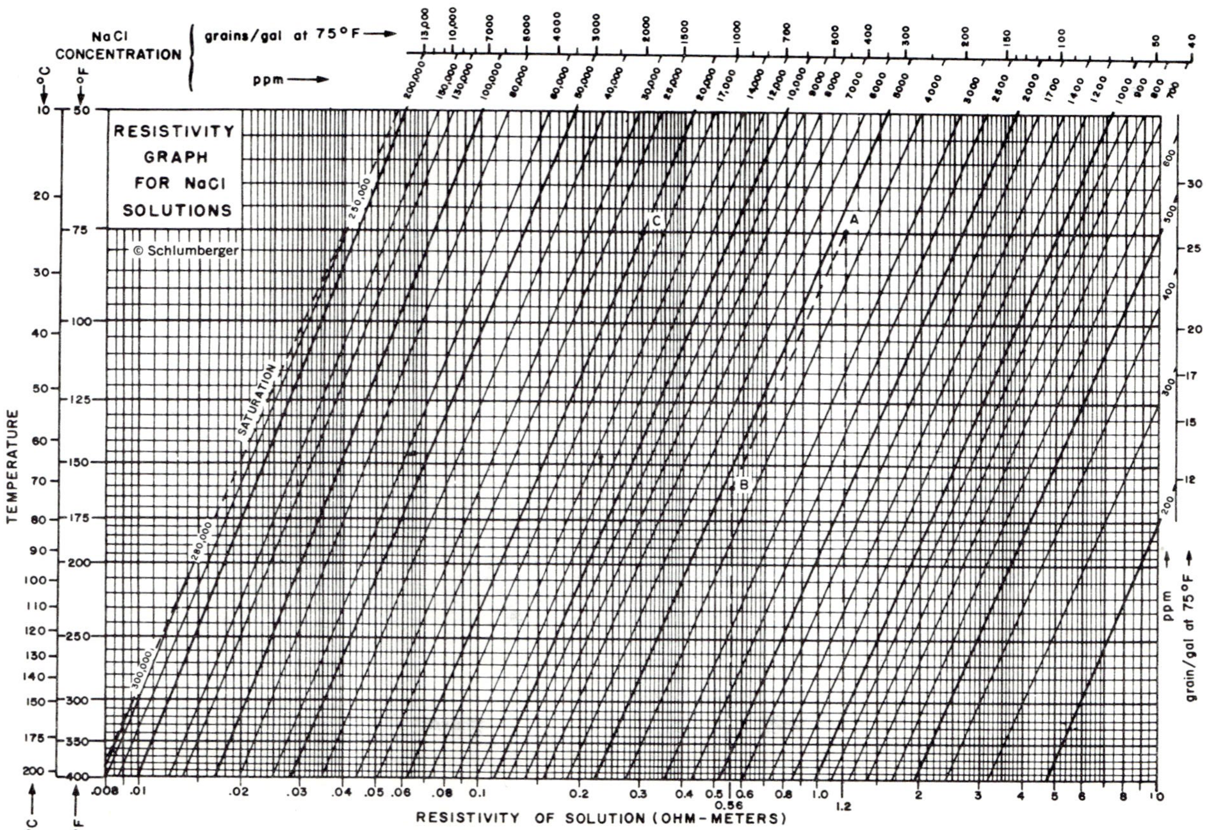

FIGURE 5-5 Dependence of resistivity of NaCl solution on temperature and salinity. (Courtesy of Schlumberger, 1972.)

100 MPa (equivalent to 10 km of hydrostatic head) is higher than at atmospheric pressure by approximately 10 percent at 5°C and 5 percent at 25°C. The effect of pressure on conductivity continues to diminish as temperature increases up to 300°C; above 300°C conductivity is highly dependent upon pressure (Gancy, 1972). Since most logging measurements are made at temperatures between 25 and 300°C, the effect of pressure on brine conductivity can usually be ignored.

5-2b Resistivity of Minerals

In contrast to ionic conduction in fluid-filled pore space, electric current in the rock matrix is carried by electrons. In most rocks electrical conduction in the matrix is negligible because silicates and carbonates are classed as insulators, having resistivities which are very high ($>10^7$ $\Omega \cdot m$). By comparison, conductive minerals are comparatively rare in the earth's crust, being restricted to the sulfides, some oxides, and native elements.

Compilations of resistivity data for conductive minerals are given by Shuey (1975) and Parkhomenko (1967). Representative values shown in Fig. 5-6 indicate that, except for the native elements, there is a broad range of values for any given mineral. These wide variations are attributed to chemical impurities, structural defects, and crystalline structures which produce electrical anisotropy. The lack of predictability of resistivity even in small selected samples is one reason why

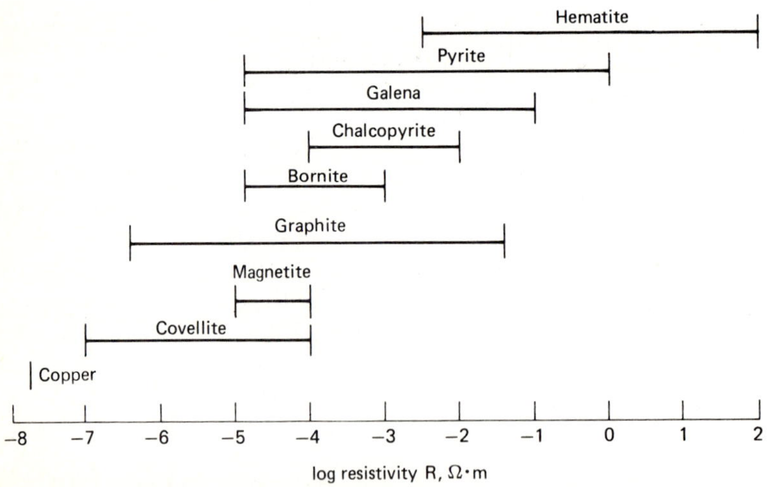

FIGURE 5-6 Range of resistivity values of some conductive minerals. (Data taken from compilations by Shuey, 1975.)

little attention has been devoted to the subject, despite the immense economic importance of the conducting minerals. A second reason is that the resistivity of metallic ores is even less predictable than that of individual minerals because of accidents of texture and porosity.

5-2c Interfacial Phenomena

Within a rock, ionically conducting fluid is in contact with solid mineral surfaces. At these interfaces either of two phenomena occurs. If the mineral is conductive, charge transfer across the fluid-solid interface requires that an electrochemical reaction take place. At low ($<1\ \mu A/cm^2$) current densities, the applied voltage is proportional to the current, and transient effects are controlled by the diffusion of reactive species in the electrolyte (Angoran and Madden, 1977). At high current densities the relation between voltage and current is nonlinear and is dependent upon the polarization history of the electrode as well as solution chemistry and mineral composition (Klein and Shuey, 1978).

If the mineral grain has a net structural charge imbalance, excess ions congregate in the solution near the mineral surface. Clay particles are commonly negatively charged, and in solution they attract an excess of salt cations. This creates what is often referred to as the *double layer* comprised of the fixed negative charge on the clay particle and the mobile positive charge in a thin solution layer adjacent to the grain. The enhanced cation concentration results in an increase in electrical conductivity in a direction parallel with the grain surface.

5-3 ELECTRICAL RESISTIVITY OF ROCKS

We have seen that the electrical resistivity of most rocks is controlled by the pore fluids because the resistivity of silicate minerals is extremely high compared with that of the fluids. The relation between rock and pore fluid resistivity has been studied in the laboratory by determining the porosity of a core sample, saturating the pore space with a fluid of known resistivity, and measuring the resulting resistivity of the sample. Because of statistical variations, a large number of samples from the same formation is required to produce meaningful results. A few experimenters have constructed artificial samples so that the shape, size, and chemistry of the constituents are well known. Such samples are then measured in the same way as natural samples. From these studies, reliable relations among porosity, pore fluid resistivity, and rock resistivity have been found, but theoretical understanding of these empirical findings is less than satisfactory.

5-3a Archie's Equation

The most influential study of rock resistivity was Archie's (1942) examination of sandstone cores from the gulf coast region. Archie first established that R_o, the resistivity of a core sample fully saturated with brine, is proportional to the brine resistivity.

$$R_o = FR_w \tag{5-53}$$

where F is the *formation factor*. If the core porosity is between 0.1 and 0.4 and the brine concentration is between 20,000 to 100,000 mg NaCl per liter of water, then F is constant. The restricted range of brine concentration is of particular importance, as will be discussed below.

Furthermore, by determining F and porosity ϕ on suites of samples from several formations, Archie established the relation

$$F = \frac{R_o}{R_w} = \phi^{-m} \tag{5-54}$$

An example of this dependence is shown in Fig. 5-7*a*. Equation (5-54), often referred to as Archie's equation, is a reasonable fit to the data, although individual samples deviate considerably from the fit.

The exponent m was found to range between 1.8 and 2.0 in consolidated sandstones and to decline to values around 1.3 in clean unconsolidated sands. Because of its dependence on the state of consolidation, m is often called the *cementation exponent.*

Modifications to Archie's equation. Winsauer et al. (1952), after studying 40 sandstone cores from a wide range of locales and geological age, introduced two empirical formulas:

$$F = (\phi - \phi_0)^{-m} = (\phi - 0.06)^{-1.48} \tag{5-55}$$

and

$$F = a\phi^{-m} = 0.62\phi^{-2.15} \tag{5-56}$$

The numerical values of the coefficients represent the best fit to Winsauer's data base of 40 samples. Equation (5-56) is often cited because the extra degree of freedom provided by the second parameter improves the fit to a given data set. From the definition of F, we expect $F = 1$ at $\phi = 1$, but neither of the above equations satisfies this constraint. For practical purposes, this is not a serious deficiency as long as the relation between porosity and resistivity is sufficiently accurate over the range of interest.

Unconsolidated samples. Jackson et al. (1978) studied the $F - \phi$ relation in artificial samples and natural marine sands. Eight artificial sam-

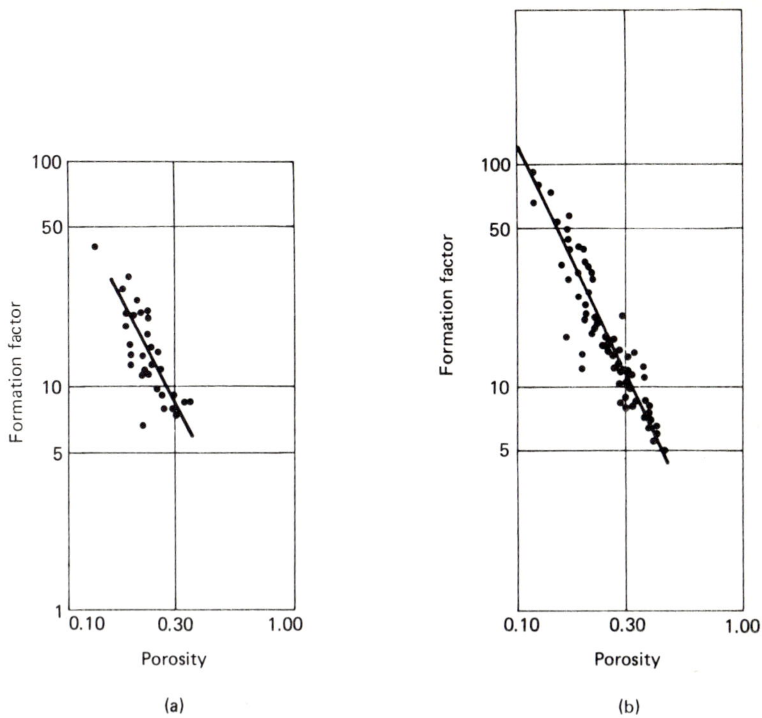

FIGURE 5-7 Relation of formation factor to porosity. Samples are consolidated sandstones from the gulf coast (*a*) and a low-permeability sandstone from Louisiana (*b*). (After Archie, 1942. Copyright 1942, SPE–AIME.)

ples of unconsolidated quartz sand were fit by m values of 1.41 to 1.45, even though the mean size and spread of grain sizes varied considerably among the samples. These results were similar to values of m from eight natural marine sands, which ranged from 1.39 to 1.58. The authors concluded that grain size is not a factor in the value of m for a mix of spherical grains, and hence a spread of grain size will not affect m as long as the shape of all particles is the same.

Results from a range of particle shapes, compared with Archie's equation for three values of m are shown in Figure 5-8. Particle shapes in the five samples grade from spherical grains to plate-like shell fragments. When shells are added to the sample, m increases with increasing shell fraction as illustrated by the progressive steepening of the slopes in Fig. 5-8. For unconsolidated samples the exponent m appears to be entirely dependent upon particle shape.

Crystalline rocks. Porosity in many igneous and metamorphic rocks is quite low, often less than 0.01. At low confining stresses, electrical

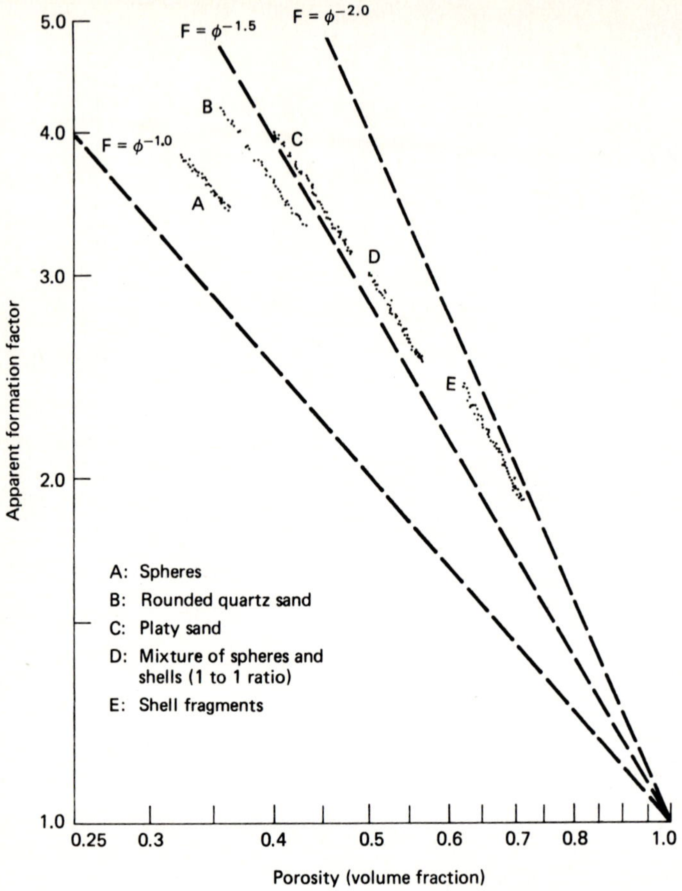

FIGURE 5-8 Formation factor of unconsolidated artificial samples. Key indicates trend of decreasing sphericity from *A* to *E*. Solid lines show Archie's equation for three values of *m*. (After Jackson et al., 1978.)

resistivity in such rocks is controlled by fractures and cracks. Because fractures are difficult to sample, their properties often require study in situ, but the behavior of microcracks and pores can be studied by using laboratory samples. Brace et al. (1965) showed that resistivity increases sharply as confining stress is increased, often increasing threefold with the initial application of 20 to 50 MPa confining stress. This initial increase is attributed to the closing of cracks, an effect which continues up to stresses of 200 MPa. At still higher stresses, resistivity increases at a much reduced rate as pores compress. In this high-stress regime, cracks have closed, and the relation between *F* and pore porosity is as shown in Fig. 5-9.

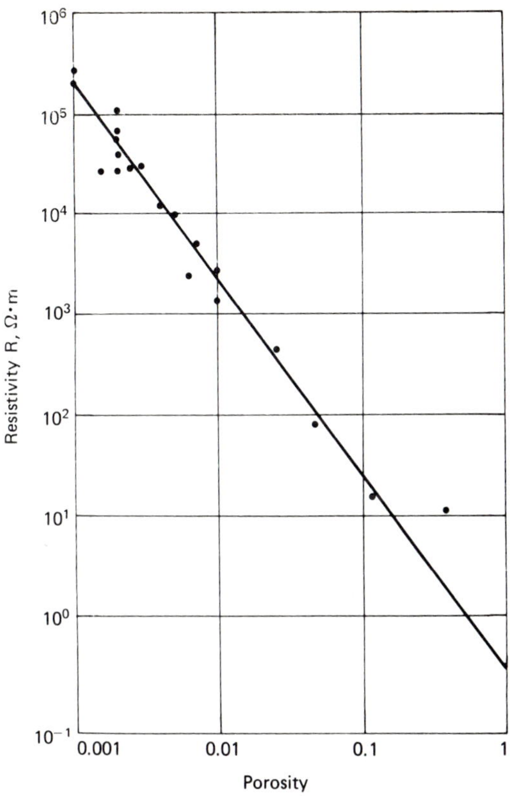

FIGURE 5-9 Resistivity of low-porosity igneous and metamorphic rock samples. Measurements taken at 400-MPa effective confining stress on 22 samples saturated with 0.25 Ω·m NaCl solution. (After Brace and Orange, 1968.)

The data in Fig. 5-9 conform to Archie's equation with $m = 2$ for a rather disparate sampling of rock types: granites, tuffs, gabbros, dunites, limestones, and marbles. The effective confining stress (defined in Sec. 2-1) of 400 MPa corresponds to a depth of about 25 km, but a similar trend would be expected wherever stresses are great enough to close cracks or where cracks are unimportant in the conduction process. However, at shallow depths, substantial deviations from a predictable dependence of F upon ϕ can be expected to occur wherever unhealed fractures exist.

Theoretical understanding of Archie's equation. Over the years numerous investigators have sought without much success for a theory to explain the experimental results of Archie and his successors. Pho-

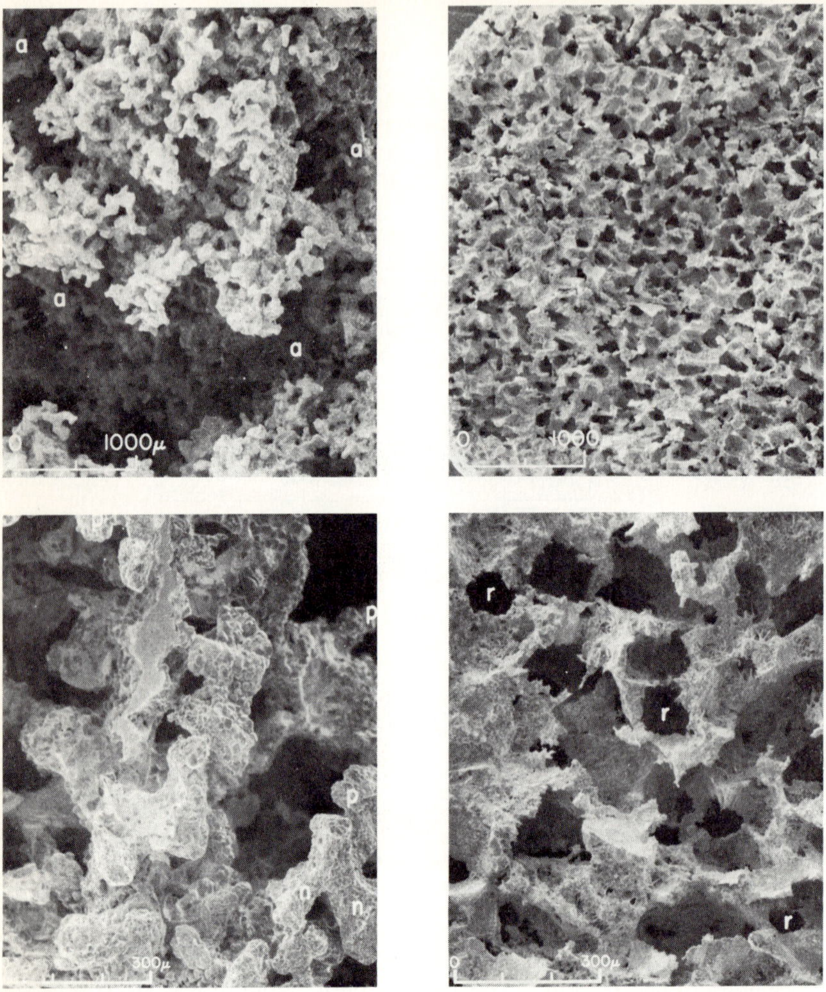

FIGURE 5-10 Cast of pore space in Berea sandstone. Woods metal has been introduced into the pores at two saturations. (After Swanson, 1979. Copyright 1979, SPE–AIME.)

tographs taken with a scanning electron microscope (Fig. 5-10) illustrate the problem of characterizing pore space. To produce these photographs, Woods metal (an alloy which melts at low temperature) was injected into the pore space of a sample of Berea sandstone (Swanson, 1979). The sandstone was then dissolved in hydrofluoric acid to produce a metal cast of the pore space, which was photographed with the scanning electron microscope. The photographs reveal the extraordinary complexity of the pores and their interconnections. It appears difficult to construct a workable theory beginning at this level of observational detail.

One would be reluctant to predict a simple relation between porosity and resistivity after inspection of Fig. 5-10. The apparent simplicity of Archie's equation becomes even more surprising considering the wide range of rock types over which it appears to be valid. The nature of pore space in detrital sediments varies greatly with sorting and mineral constituents, yet according to Archie's equation, one or, at most, two empirical parameters suffice to characterize the effects of grain shape and distribution. Apparently this holds true even to the very low (<0.01) porosities in unfractured crystalline rocks, where the character of the pore space is quite different from that in sediments. Given the number of rock characteristics to be taken into account, it is amazing that we need determine only one empirical variable, m, in order to relate rock resistivity to pore space!

Effect of partial saturation. If part of the pore space is occupied by nonconducting fluid such as oil or gas, the water saturation S_w is less than 1 and the rock resistivity increases. The resistivity index I is the ratio of the "true" resistivity R_t of the partially brine-saturated rock to its value R_o when completely water-saturated. Experimental results show that a power law relation

$$I = \frac{R_t}{R_o} = S_w^{-n} \tag{5-57}$$

is a useful approximation over certain ranges of S_w. The saturation exponent n varies with rock type and with the disposition of the brine and the nonconducting fluid within the pore space. It is usually determined by measurements on core samples, and it typically has a value of 2 ± 0.5. If water adheres to grain surfaces in preference to oil, the rock is said to be water-wet as opposed to oil-wet. Water exists as the continuous phase in water-wet rocks and provides good continuity for electrical conduction. When rocks are artificially made oil-wet in laboratory conditions, oil becomes the continuous phase, water occurs as isolated droplets, and resistivity becomes much higher. Consequently, the value of n can be much greater than 2 in oil-wet rocks. However, there is evidence that, in real reservoirs, such extreme conditions of oil wetness do not occur (Swanson, 1980), and the subject remains a matter of investigation. In practice, if laboratory data are not available for a particular formation, a value of $n = 2$ is used.

Equations (5-54) and (5-57) are used to relate ϕ and S_w to resistivity logs and are frequently cited in the logging literature. Examples of their application are given in Chaps. 10 and 11 of this book.

5-3b Exceptions to Archie's Equation

It should be noted that Eqs. (5-54) and (5-57) are really two-component mixing rules (Sec. 1-4) in which the second component, which represents the insulating rock grains plus any nonconducting fluid, does not appear explicitly. One must expect that these rules will not hold when conducting minerals are present in the rock or when fluid conduction is augmented by surface conduction. Surface conduction leads to a net increase in the effective fluid conductivity and can occur in both clay-bearing and clay-free rocks. Effects are more pronounced when clay is present, because many clay particles possess a net negative charge which is compensated by an excess number of cations in solution close to the clay surfaces. When clay is not present, some surface conduction occurs because silicate minerals, although electrically neutral, nevertheless do attract a mobile layer of excess cations in solution adjacent to grain surfaces.

These effects are demonstrated by the curves shown in Fig. 5-11 for three sandstone samples, one of which contains high, one moderate, and one very little clay. A change in F reflects the contribution of surface conduction. At any given brine concentration, surface conduction in the clay-free sample c is a much smaller fraction of total conduction than in sample a or b. Consequently, in the clay-free sample F remains constant until R_w exceeds 1 to 2 Ω·m, whereas in the clay-bearing samples F declines at considerably lower values of R_w.

The clay-free case typified by sample c is of interest in water well

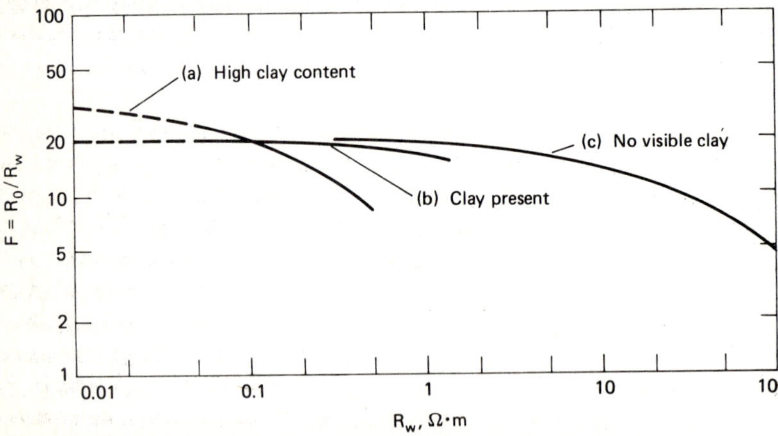

FIGURE 5-11 Effect of surface conduction on formation factor in three sandstones. Samples a and b, from Waxman and Smits (1968), have porosities of 0.253 and 0.187. Sample c, from Evers and Iyer (1975), has a porosity of 0.183.

exploration (Alger, 1966) because good-quality water in a permeable near-surface aquifer is characterized by high resistivity. However, there currently exists no well-accepted model which will permit construction of a curve of type *c* from other rock parameters, although the topic has received some attention (Rink and Schopper, 1974; Evers and Iyer, 1975). On the other hand, clay-rich sandstones have been well studied; a model established by Waxman and Smits (1968) will be discussed next. An alternative approach to the problem has been proposed by Clavier et al. (1977).

Surface conduction in clay-bearing rocks. Figure 5-12 shows observed C_o in terms of C_w for clay-bearing sands. A model to explain these data, due to Waxman and Smits (1968), expresses $C_o = 1/R_o$, where R_o is the conductivity of the rock sample, in terms of the conductivity of the pore fluid C_w and the exchangeable cations C_e. C_e represents the conductivity of excess mobile ions held in solution near the surface of the clay particles. It is first assumed that the pore fluid and exchangeable cations reside in two independent rock systems of formation factor F_1 and F_2, which are in parallel electrically.

$$C_o = \frac{C_w}{F_1} + \frac{C_e}{F_2} \tag{5-58}$$

This mixing rule assumes that it is the tortuous conductive pathways in a porous rock which are in parallel and additive, rather than the component conductivities themselves. This assumption is easily forgotten because it is masked by a second assumption, $F_1 = F_2 = F^*$, which implies that the tortuous conduction paths are statistically simi-

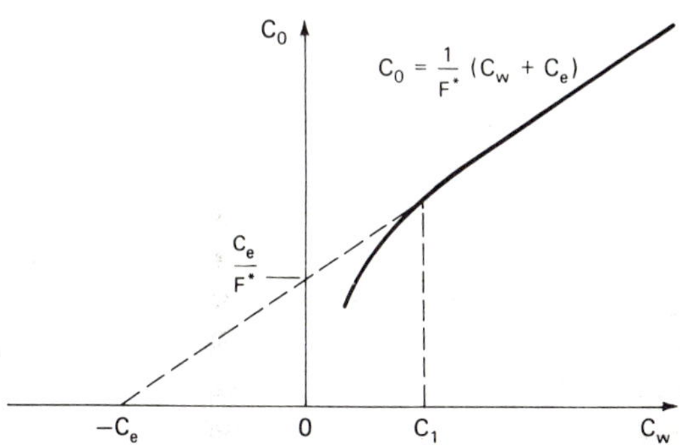

FIGURE 5-12 Model for conduction in a shaly sand. The heavy line represents the trend of typical experimental data. (After Waxman and Smits, 1968.)

lar for both conduction modes, so that

$$C_o = \frac{C_w + C_e}{F^*} \tag{5-59}$$

where F^* reduces to the usual F in a clean sandstone. This equation is represented by the straight line in Fig. 5-12. It represents quite well the data shown by the heavy line for C_w greater than some experimental value C_1. The deviation below C_1 is attributed to loss of mobility of the exchange ions.

The conductivity C_e is now expressed in terms of the concentration of exchangeable ions by recalling Eq. (5-41), which expresses conductivity in terms of ionic concentration,

$$C_e = c_e \mathcal{F} z_e \nu_e \tag{5-60}$$

The valence-concentration product $z_e c_e$ is expressed as Q_v, given in chemical equivalents per liter:

$$C_e = \frac{\mathcal{F} \nu Q_v}{1000} \tag{5-61}$$

where the factor 1000 converts from liters to cubic centimeters. Waxman and Smits introduced a parameter B to fit the experimental deviations found at low C_w. Their final expression for a fully saturated shaly rock represented by the solid line in Fig. 5-12 is

$$C_o = \frac{C_w + BQ_v}{F^*} \tag{5-62}$$

where

$$B = \frac{\mathcal{F} \nu [1 - a \exp(-C_w/b)]}{1000}$$

Evaluation of the constants from experimental data (Waxman and Thomas, 1974) gives

$$B = 3.83[(1 - 0.83 \exp(-0.5/R_w)] \tag{5-63}$$

Equation (5-62) can also be written in terms of the resistivity

$$R_o = \frac{F^* R_w}{1 + BQ_v R_w} \tag{5-64}$$

The term $R_w/(1 + R_w B Q_v)$ can be considered an effective water resistivity R_w'. Its dependence on Q_v is shown in Fig. 5-13. For reference, Q_v determinations on samples a and b in Fig. 5-11 were 1.530 and 0.239 eq/L, respectively.

Q_v must be established for the formation by direct measurement. It

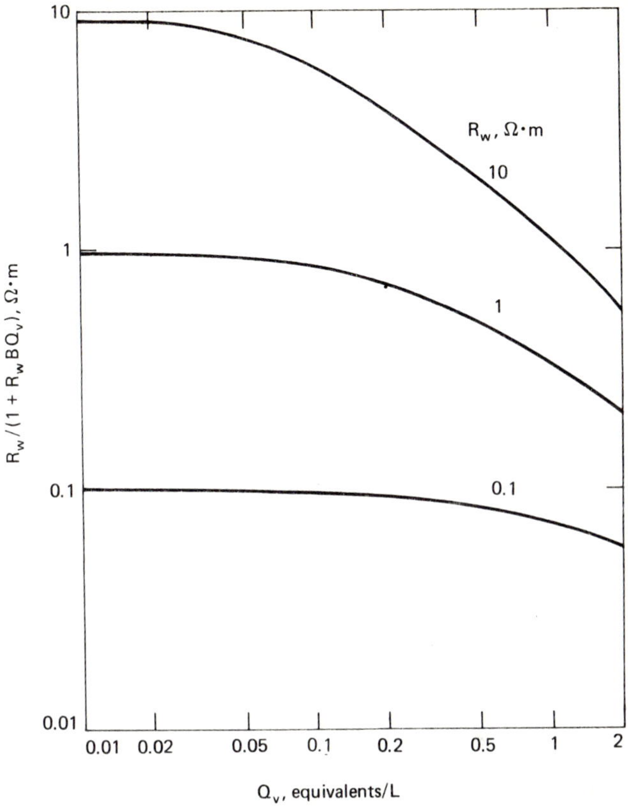

FIGURE 5-13 Effective resistivity of pore fluid incorporating surface conduction in a fully saturated rock, using the Waxman-Smits formulation.

has been found that positive ions of some species will readily leave a clay site and be exchanged by ions of certain other species. This property of a clay is called its cation exchange capacity (CEC). An accepted laboratory method for measuring CEC involves the controlled substitution of ammonium or barium ions for the exchangeable ions in the sample. The CEC determined on crushed samples in this manner is given as milliequivalents per 100 g of rock, which is converted to Q_v by

$$Q_v = \frac{(1 - \phi)\rho_{ma}\,\text{CEC}}{100\phi} \tag{5-65}$$

where ρ_{ma} is the grain density. With Q_v determined, it is possible to determine the water saturation in oil-bearing shaly reservoir rocks. The method will be discussed in Sec. 11-1.

Conductive minerals. It was shown in Fig. 5-6 that the resistivity of many sulfides and oxides is low, much less than that of silicate minerals. The effect of conducting minerals in rock depends greatly on abundance. Figure 5-14 shows data collected in open-pit copper mines in the southwestern United States. Clay alteration products were present, but they play a minor role in the gross conductivity.

At sulfide contents of less than 2 percent by weight rock resistivity is not well correlated with sulfide abundance. It is unlikely that the sulfides form continuous conduction pathways at these low concentrations. However, rock resistivity begins to decline as the sulfide content exceeds 2 percent by weight, and it becomes low and erratic at 10 percent by weight. This behavior is attributed to the presence of sulfide veins at higher concentrations, which provide more and more continuous electronic conducting paths within the rock. Eventually electronic conduction in the sulfides dominates ionic conduction in the pore space.

These two different conduction modes have their counterparts in surface prospecting systems developed for mineral exploration. High-sulfide ores are generally quite conducting and are sought with inductive systems designed to induce and detect large-scale current flow in the conductive body. Borehole induction methods are discussed in Sec. 5-5. Low-sulfide ores, on the other hand, cannot be detected by their resistivity properties, but they can be detected with induced polarization.

5-3c Induced Polarization

When conductive minerals are present, a rock-fluid system can store electric energy for short time periods. As a consequence, an applied steady current will induce a polarization voltage which decays slowly with time after the cessation of current. This phenomenon is responsible for the observed decrease in resistivity with increasing frequency (Fig. 5-15). The decrease is due to a reactive component within the rock which contributes an impedance small in comparison with the total resistance. The presence of a reactive component requires that complex resistivity $R = R_r + iR_i$ or complex conductivity $C = C_r + iC_i$ be used to characterize the electrical properties. The two are related by Eq. (5-15). An applied sinusoidal current of frequency ω will induce a voltage slightly out of phase with the current. The observed phase shift φ is

$$\tan \varphi \simeq -\frac{R_i}{R_r} = \frac{C_i}{C_r} \tag{5-66}$$

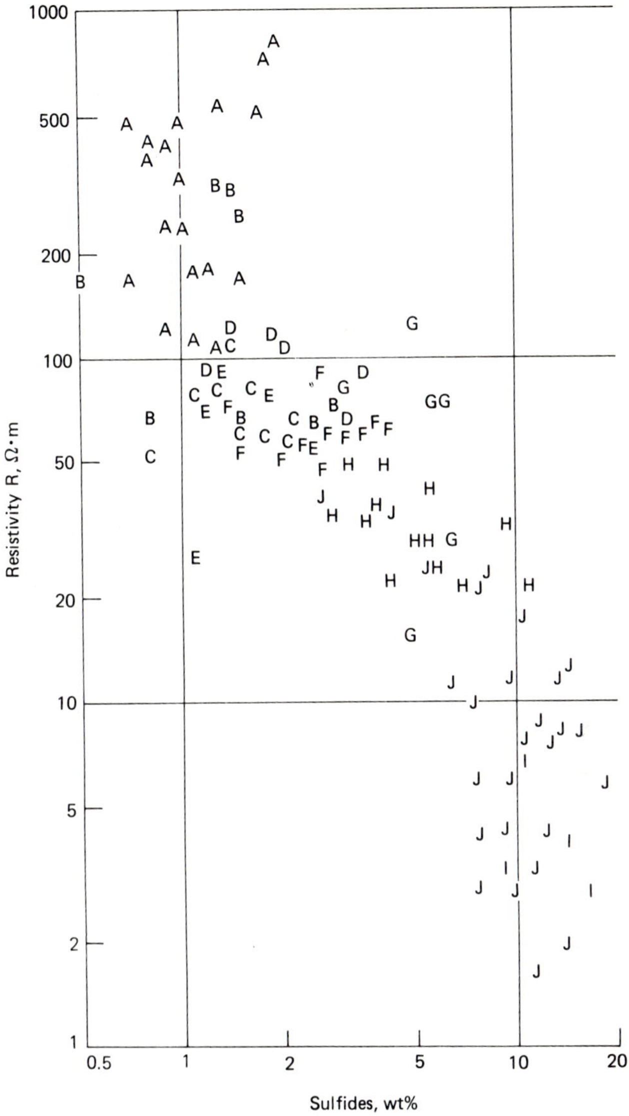

FIGURE 5-14 Resistivity of sulfide-bearing rock, measured in open-pit porphyry copper mines. Each data point represents several cubic meters of rock. (After Nelson and Van Voorhis, 1982.)

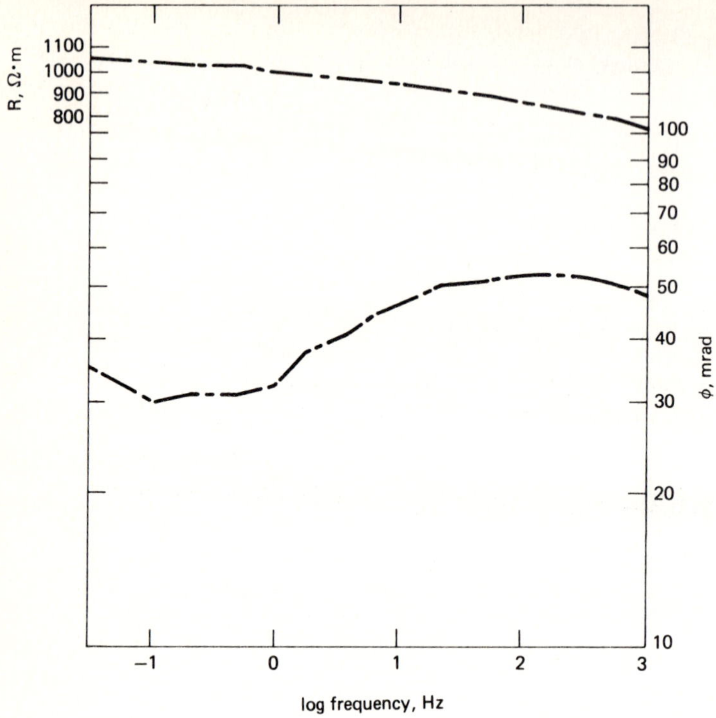

FIGURE 5-15 Resistivity and phase spectra of a quartzite sample from Bingham, Utah. Sample contains 0.01 volume fraction pyrite and an unknown amount of kaolinite; it is saturated with 10.7 Ω·m water.

Because the phase shifts are small, typically less than 0.1 rad (5.7°), $\varphi \simeq \tan \varphi$. The phase curve shown in Fig. 5-15 ranges between 30 and 55 mrad, which indicates that the R_i component is no more than 5.5 percent of R_r. Where φ is higher, the slope of the resistivity curve becomes more negative, that is, resistivity declines more rapidly where the reactive component is larger.

Polarization is attributed to the presence of interfaces between zones of electronic and ionic conduction (electrode polarization) and to the presence of zones of unequal ionic transport properties (membrane polarization). Sulfide minerals and graphite are sources of electrode polarization; clays and zeolites are sources of membrane polarization. The observed magnitudes differ: sulfide minerals can produce polarization phase angles as large as 120 mrad, whereas phase angles attributed to clay minerals seldom exceed 15.

The induced polarization response of a rock couples the complexities of pore space variations with the nature and geometry of the individual polarization sources. No general mixing rule for induced polarization has yet been established. Only a few of the experimental findings are summarized here.

1. Phase angle increases with the volume fraction of conductive mineral present as long as the grains are disseminated throughout the rock matrix. At high concentrations grains are not isolated, and sulfide minerals occur as veins. Consequently, phase angle does not continue to increase at high concentrations. It has been observed empirically that C_i is then a better indicator of sulfide content than φ (Nelson and Van Voorhis, 1983).
2. If the size of both sulfide and nonsulfide grains is uniform, a well-defined polarization maximum will be observed at some frequency. An example is shown in Fig. 5-16. But uniform grain size and properties are the exception rather than the rule in most rocks. Variations from grain to grain within the rock matrix produce a continuum of time constants and smear the observed phase spectra so that maxima are obscured or obliterated. Figure 5-15 shows an example of a partially smoothed spectrum.
3. Partial saturation can alter the spectrum and increase the polarization response. Figure 5-16 shows results from a sand column loaded with sized pyrite grains. When saturated with brine, the uniform sand-pyrite system produced a peaked phase spectrum. The system was drained until only water retained by capillary action remained at $S_w = 0.13$. Resistivity of the column increased fortyfold, and the phase spectral shape changed from peaked to flat. Desaturation undoubtedly altered the contact area between fluid and conductive particles quite markedly and produced a distribution of time constants and smoothed the phase spectrum.

The induced polarization properties of rocks are important to logging in two ways. A polarization measurement can detect the presence of conductive minerals even when concentrations are less than 1 percent. It promises to provide quantitative information as well. As a nuisance factor, polarization will produce erroneous resistivity information. Resistivity measurements made with a logging sonde operating at one frequency will differ from measurements made at another frequency unless they are corrected for the effect of polarization.

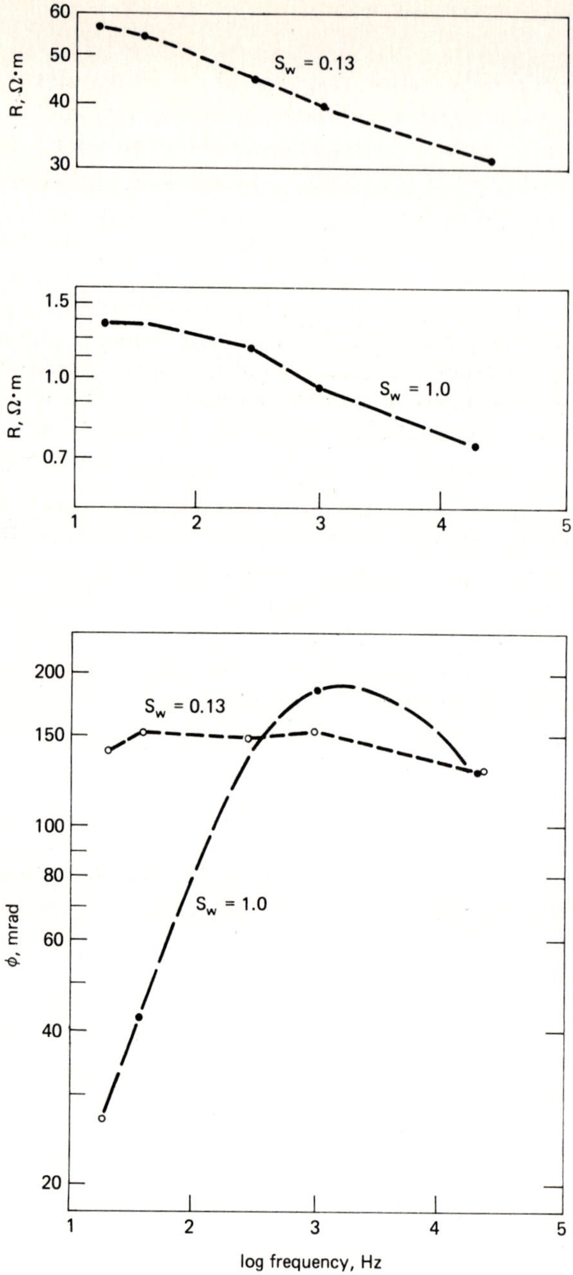

FIGURE 5-16 Resistivity and phase spectra of a sieved mixture of quartz sand and pyrite saturated with 0.3 Ω·m NaCl solution. Pyrite grain diameter is approximately 1.0 mm; pyrite occupies 16.5 percent by volume of sample. Porosity is 0.42; $m = 1.53$; $n = 1.85$. (Data from Clavier et al., 1976.)

5-4 GALVANIC RESISTIVITY

Galvanic resistivity sondes consist of one or more metallic electrodes through which current is transmitted into the wellbore fluid and thence into the formation. One or more potential electrodes measure the resulting voltage differences established by the current flow. Although the galvanic methods are considered to be direct current, that is, current of very low frequency, in practice the signals alternate at some frequency between 10 and 1000 Hz to avoid problems of electrode polarization and pickup of spontaneous polarization voltages.

Different electrode configurations have been devised over the years to measure formation resistivity more accurately, increase bed resolution, and increase the depth of investigation. Older systems measure the voltage at one or two points in response to a constant current input. More modern systems are adaptive in the sense that the current is varied in order to maintain a zero voltage gradient at specific locations on the sonde. All systems can be understood, at least in principle, by superposing the patterns of current flow and voltage drop from individual electrodes. Moreover, all system outputs are referenced to the resistivity measured in some standard condition, usually that obtained with the sonde positioned in a medium of uniform and isotropic resistivity.

5-4a Homogeneous Media

The frequencies used in galvanic systems are low enough that displacement currents can be neglected. Recalling Eq. (5-4), the current density is

$$\mathbf{J} = \frac{\mathbf{E}}{R} = -\frac{\nabla \Psi}{R} \tag{5-67}$$

Except at a source or a sink, the current density must be divergence-free; that is, the current leaving a volume must equal the current entering. If we consider a single current source of I amperes at the center of a spherical coordinate system,

$$\nabla \cdot \mathbf{J} = I\delta(0) \tag{5-68}$$

where $\delta(0) = 1$ at $r = 0$ and is zero elsewhere. Combining Eqs. (5-67) and (5-68) gives us

$$\nabla \cdot \left(\frac{1}{R} \nabla \Psi\right) = \nabla \left(\frac{1}{R}\right) \cdot \nabla \Psi + \frac{1}{R} \nabla^2 \Psi = -I\delta(0) \tag{5-69}$$

If we assume the medium is homogeneous, resistivity R is constant everywhere and the first term vanishes. Away from the source, Eq. (5-69) reduces to Laplace's equation,

$$\nabla^2\Psi = 0 \tag{5-70}$$

All galvanic resistivity problems require the solution of Laplace's equation for the appropriate source and boundary conditions.

The solution for a single current source at $r = 0$ (Fig. 5-18) can be found by integrating **E** to get the potential

$$\Psi(r) = -\int_{\infty}^{r} \mathbf{E}\, d\mathbf{r} = -\int_{\infty}^{r} R\mathbf{J}\, d\mathbf{r} \tag{5-71}$$

The current density **J** at any radius **r** is the source strength divided by the area of the sphere, $I/4\pi r^2$. The other condition necessary for complete determination of Ψ is obtained by defining $\Psi = 0$ at $r = \infty$. The result of evaluating Eq. (5-71) is then

$$\Psi(r) = -\int_{\infty}^{r} \frac{RI}{4\pi r^2}\, dr = \frac{RI}{4\pi r} \tag{5-72}$$

This solution obeys Laplace's equation for $r > 0$. Figure 5-17 indicates the spatial behavior of Ψ and emphasizes that a voltage V is determined by taking the difference of the potential Ψ at two different points. For two points distant r_M and r_N from the current electrode,

$$V_{MN} = \frac{RI}{4\pi}\left(\frac{1}{r_M} - \frac{1}{r_N}\right) \tag{5-73}$$

If the current electrode is a sphere of radius r_e rather than a point electrode, the voltage on its surface is found by letting $r_M = r_e$ and $r_N = \infty$.

$$V(r_e) = \frac{RI}{4\pi r_e} \tag{5-74}$$

so that the impedance V/I of a single spherical electrode in a homogeneous material is $R/4\pi r_e$. If the electrode is not spherical, the radius r_e is replaced by an appropriate length factor. In any case, electrode impedance varies inversely with electrode size.

Resistance logs. The simplest galvanic logging measurement can be made by grounding an electrode at the surface, suspending a second electrode in the borehole fluid, and measuring the resistance of the resulting loop formed by the wire, electrodes, and ground (Fig. 5-18*a*). Voltage and current are measured with the same electrodes, and logs

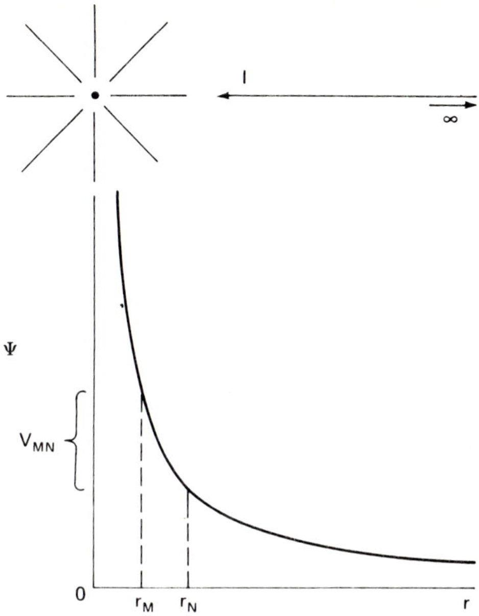

FIGURE 5-17 Potential Ψ and voltage *V* in presence of a single-point electrode in a uniform isotropic medium.

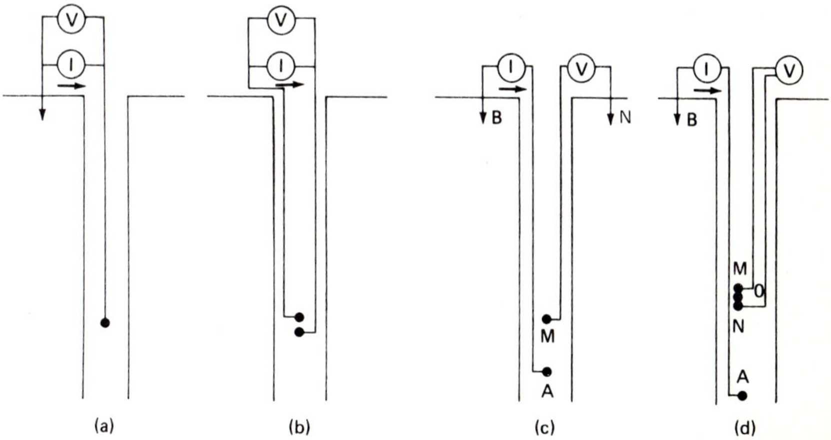

FIGURE 5-18 Electrode arrangements of four common arrays with current source *I* and voltage measurement *V*. (*a*) Single-point resistance; (*b*) differential resistance; (*c*) normal array; (*d*) lateral array.

obtained in this fashion are called single-point resistance logs. The voltage drop across the surface electrode can be made much less than that of the subsurface electrode, or it can be eliminated by measuring its impedance, so that the measured V of Fig. 5-18*a* represents $V(r_e)$ of Eq. (5-74). Consequently, from Eq. (5-74) the earth resistivity R can then be determined from V/I. Unfortunately, in practice the mud close to the electrode dominates the reading, as shown by an approximate solution given by Keller and Frischknecht (1966), so that V is often more sensitive to mud resistivity and to changes in borehole diameter than to formation resistivity. Consequently, single-point resistance logs give no more than a qualitative indication of the formation resistivity.

If both electrodes of the circuit are placed on the sonde, as shown in Fig. 5-18*b*, the resulting log is referred to as a *differential resistance log*. Now, most of the current flows in the borehole fluid, so that the resistance of the circuit is controlled almost entirely by the geometry and resistivity of the mud path between the two electrodes. Although the vertical resolution of the single-point and differential-resistance sondes is excellent, it is not feasible to extract quantitative resistivity data from either of them. To measure formation resistivity, the voltage-measuring circuit and electrodes must be separated from the current circuit.

Electrode arrays. Multiple-electrode arrays are used to extend the effective depth of investigation and obtain a more representative value of true formation resistivity R_t. Because potentials add, Eq. (5-73) shows that the voltage V_{MN} between points M and N in a homogeneous medium can be found by combining the potentials generated by the individual current sources. A four-electrode system with a $+I$ source at point A and a $-I$ source at point B will produce a voltage

$$V_{MN} = \frac{RI}{4\pi}\left[\left(\frac{1}{r_{AM}} - \frac{1}{r_{AN}}\right) - \left(\frac{1}{r_{BM}} - \frac{1}{r_{BN}}\right)\right] \tag{5-75}$$

where r_{AM} is the distance between point electrodes A and M. The resistivity $R = GV/I$, where

$$G = 4\pi\left(\frac{1}{r_{AM}} - \frac{1}{r_{AN}} - \frac{1}{r_{BM}} + \frac{1}{r_{BN}}\right)^{-1} \tag{5-76}$$

is the *geometric* or *array factor*. It is dependent only upon the array geometry and has the dimension of length.

The formulation simplifies if current electrode B and voltage electrode N are placed on the surface, leaving only two electrodes on the sonde as depicted in Fig. 5-18*c*. This is called the *normal array*. In this case only the $1/r_{AM}$ term in Eq. (5-76) is significant, $G \simeq 4\pi r_{AM}$, and the measured resistivity $R = 4\pi r_{AM}(V/I)$. Normal arrays were introduced early in electrical logging by the Schlumbergers (Allaud and Martin, 1977). The 16-inch normal array remains available commercially today in combination with other electrical logs.

The *lateral array* depicted in Fig. 5-18*d* consists of a current pole at *A* and a voltage dipole at *MN*. Because $(r_{AM}, r_{AN}) \ll (r_{BM}, r_{BN})$, Eq. (5-76) shows that the geometric factor for the lateral array is

$$G = 4\pi \left(\frac{1}{r_{AM}} - \frac{1}{r_{AN}} \right)^{-1} \tag{5-77}$$

The spacing r_{AO} between the center of the voltage dipole (point *O*) and current electrode at point *A* is usually much greater than the separation r_{MN} between *M* and *N*, so that

$$G \simeq \frac{4\pi r_{AO}^2}{r_{MN}} \tag{5-78}$$

Because the separation between the potential electrodes is small, a measurement from a lateral array is very close to the potential gradient, that is, it approximates the electric field at point *O* quite closely. It was introduced early to supplement the normal array and delineate bed boundaries. With care it could be used to ascertain the true formation resistivity R_t (Lynch, 1962). However, because of its asymmetrical design, its response when passing through a bedded formation is asymmetric. It also produces spurious effects and blind zones (see discussion of Fig. 5-23). For these reasons it has fallen into disuse, having been replaced by the focused and induction sondes.

5-4b Inhomogeneous Media

Thus far we have considered only homogeneous media. The actual resistivity distribution in a layered rock sequence penetrated by a wellbore is shown in Fig. 2-3. To understand sonde response in a wellbore, we must look at the effect that spatial variations in resistivity have upon the measurement.

Apparent resistivity. Suppose an array is placed in a medium with varying properties; that is, the resistivity changes from place to place and may even be anisotropic. For any current *I*, the potential electrodes at *M* and *N* produce some voltage *V* which can be expressed as

$$R_a = \frac{GV}{I} \tag{5-79}$$

The resulting quantity R_a is called the apparent resistivity, but its exact value is determined by the pattern of current flow set up in this particular geological environment and by the position of the potential electrodes therein. We can, however, at least conceptually, return the sonde to a uniform isotropic earth of resistivity R_a and obtain the same reading. Because of this conceptual convenience, results from logging sondes are given as R_a, or else as a parameter very closely akin to it.

In most real heterogeneous situations, the value R_a is affected by several zones of differing resistivity, including the mud resistivity R_m. The goal of resistivity interpretation is to derive from R_a the desired formation resistivity R_t, whereas the goal of sonde design is to minimize the impact on R_a of all but the desired zone. We next examine the behavior of R_a for the simplest geometrical situations.

Single electrode in presence of plane interface. We examine the effect of current flow upon R_a by considering the potential field established by a current electrode at point A of medium 1 with resistivity R_1. A horizontal boundary separates the first medium from a second with resistivity R_2. As shown in Fig. 5-19, the ordinary spherical outflow of current from A is distorted. It diverges away from medium 2 if $R_2 > R_1$ and converges toward it if $R_2 < R_1$.

The potential can be obtained exactly by using the method of images (Dakhnov, 1962; Logn, 1954), whereby an image source I' is placed at point A' to simulate the potential perturbation in medium 1

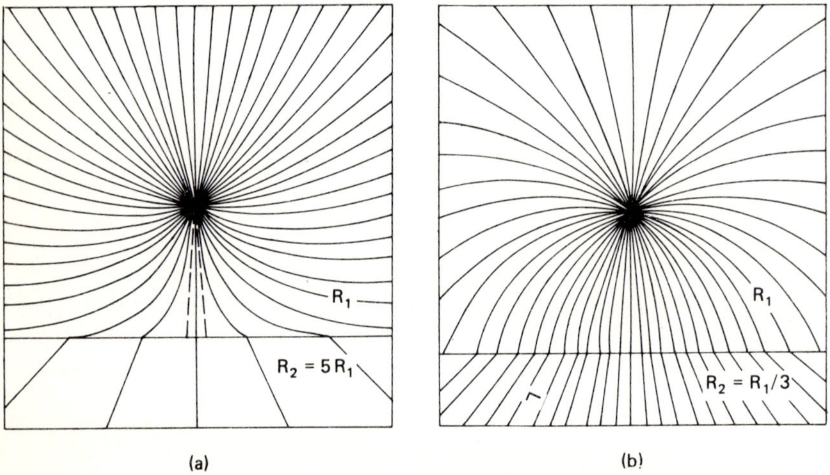

FIGURE 5-19 Patterns of current flow in presence of plane boundary separating media of different resistivity R. (*a*) $R_2 = > R_1$; (*b*) $R_2 < R_1$. (After Guyod, 1944.)

caused by the presence of medium 2 (Fig. 5-20). Likewise, to treat the perturbed potential in medium 2, the true current I emanating from point A is replaced by a current I''.

The potentials in media 1 and 2 can be written in terms of the currents and distances defined in Fig. 5-21.

$$\Psi_1 = \frac{R_1 I}{4\pi r} + \frac{R_1' I'}{4\pi r'} \qquad \Psi_2 = \frac{R_2 I''}{4\pi r} \tag{5-80}$$

Defined in this way, Ψ_1 and Ψ_2 satisfy Eq. (5-70) in their respective media. The magnitudes of the image currents are determined by applying boundary conditions on the plane $z = z_0$. Continuity of current components normal to the interface and continuity of potential $\Psi_1 = \Psi_2$ at the interface are used to evaluate Eq. (5-80),

$$I' = R_{21} I \qquad I'' = (1 - R_{21}) I \tag{5-81}$$

where R_{21} is the reflection coefficient

$$R_{21} = \frac{R_2 - R_1}{R_2 + R_1} \tag{5-82}$$

which determines the ratios of the image currents to the real current. Its properties are illustrated in Fig. 5-21. With the ratio R_2/R_1 plotted logarithmically, the reflection coefficient is antisymmetric about $R_2 = R_1$. As R_2 becomes greater than R_1, R_{21} changes sign. Also, the image current I' changes from a sink to a source and thereby deflects current away from the boundary instead of attracting it. When $R_2 \gg R_1$, $I' = I$, and no current from I crosses the boundary. Because no current then penetrates medium 2, the image I'' and the potential Ψ_2 decline to zero. Using Eqs. (5-80) through (5-82) in combination with the definition of apparent resistivity R_a yields

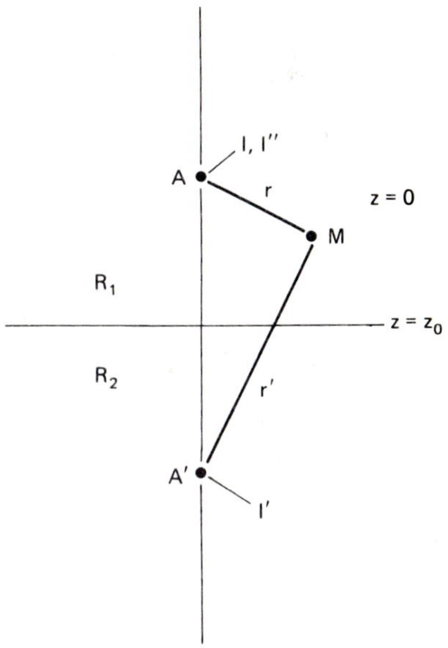

FIGURE 5-20 Point electrode A and image A' in presence of plane interface.

$$R_a = R_1 \left(1 - R_{21} \frac{z/z_0}{2 - z/z_0}\right) \qquad z \leq z_0 \tag{5-83}$$

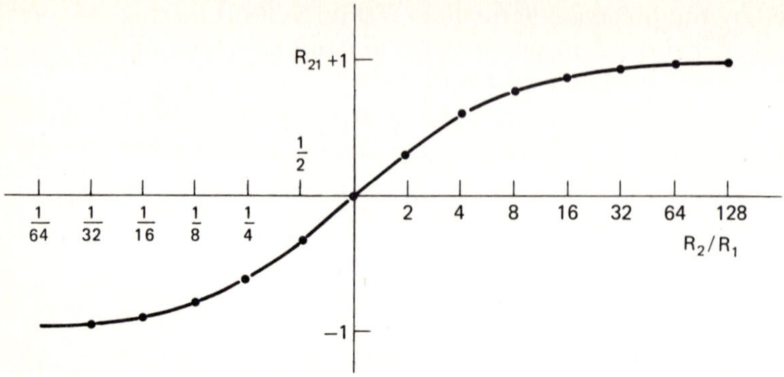

FIGURE 5-21 Dependence of reflection coefficient R_{21} on resistivity ratio.

$$R_a = \frac{2R_1R_2}{R_1 + R_2} = R_1(1 + R_{21}) \qquad z \geq z_0 \tag{5-84}$$

for the case in which the potential is sensed along a vertical line extending through the current source. From Eq. (5-84) and Fig. 5-20 we see that the apparent resistivity of a normal array straddling an interface will be a constant value! It ranges from zero, when $R_2 \ll R_1$, to $2R_1$, when $R_2 \gg R_1$. This characteristic of a normal array limits the definition to thin resistive beds, as we shall see again in connection with Figs. 5-22 and 5-23. The fundamental problem is that current emitted by a single electrode in a conductive medium will not penetrate a resistive medium.

Results for a normal array of fixed spacing L crossing the interface are shown in Fig. 5-22. The equations given in Fig. 5-22 for three positions of the array are extensions of Eqs. (5-84). Values are plotted with respect to the midpoint between A and M, so that values of R_a are constant as the array straddles the interface and reach a maximum value of 2 for high R_2. With both electrodes in a single half-space, R_a values are representative of the half-space values when the midpoint is $3L$ or more distant from the interface.

Beds of finite thickness. The case of a layered bed can be treated by using the method of images or electrical networks to examine the behavior of multiple-electrode sondes passing through a layered rock sequence (Guyod, 1944). Figure 5-23 summarizes the salient features of the single-point, normal, and lateral sondes passing through beds either more resistive or more conductive than the surrounding formations. The transitions across bed boundaries lack the crisp cusps and

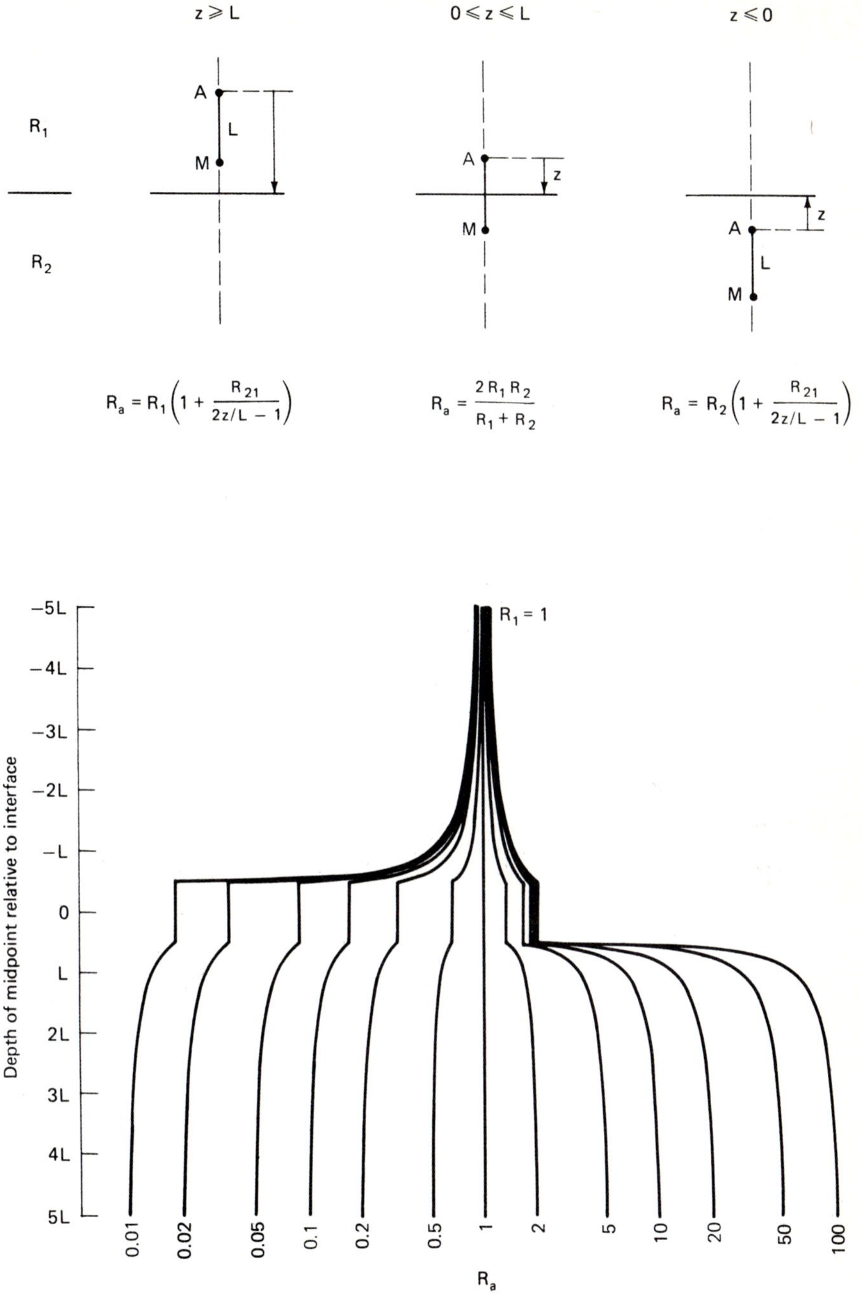

FIGURE 5-22 Normal array of separation L between current A and voltage M electrodes traversing a plane interface between two media of differing resistivity. Graph shows apparent resistivity R_a as function of depth.

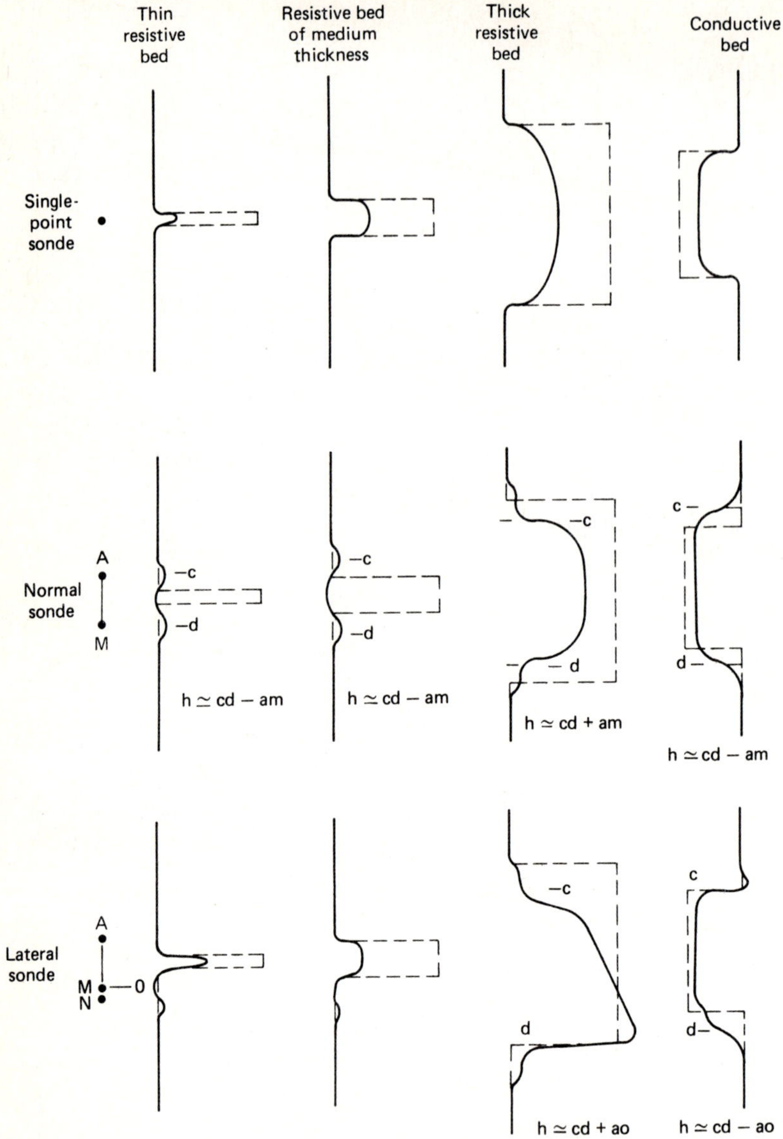

FIGURE 5-23 Apparent resistivity (solid curves) obtained with three arrays passing through horizontal layers of thickness h. Dashed line shows true resistivity structure. Bed boundaries picked from logs (points c and d) sometimes differ from actual locations. (After Guyod, 1944.)

linear segments of Fig. 5-22 because the solution accounts for the presence of the wellbore.

The single-point sonde produces the simplest response. Deflections are in the same sense as the formation; bed boundaries are sharply delineated; and there are no ghosts or distortions. However, as explained above, the single-point R_a value is so dominated by R_m that wellbore irregularities usually make it impossible to determine R_t accurately.

The anomaly produced by the normal sonde passing through a thick resistive bed is thinner than the bed by a length equal to the sonde length. This behavior, which is due to the deflection of current by the resistive bed, is in accord with Fig. 5-22, as is the corresponding behavior in the thick conductive bed in which the anomaly is actually broader than the bed. The main failing of the normal sonde occurs when a resistive bed is thinner than the spacing *AM*. Because the potential electrode on the far side of the bed is screened from current flow, R_a decreases rather than increases, and lobes are produced above and below the bed. Consequently, normal logs are blind to thin resistive beds.

The lateral array is more effective at delineating thin resistive beds, although R_a remains considerably less than R_t. Note that the lobes below the thin beds occur as the current electrode approaches and enters the bed. The asymmetrical character of the array is reflected in the asymmetrical response in thick resistive beds, the distance between *c* and the upper bed boundary being equal to the spacing *AO*. Bed definition is excellent on the lower edge of the thick resistive bed and on the upper edge of the thick conductive bed. In both cases R_a is comparable with R_t.

Anisotropy. As mentioned in Sec. 5-3, many rocks possess anisotropic resistivity properties, commonly with a value R_h along the bedding plane which is less than a value R_v normal to the bedding. When the bedding is horizontal and a vertical borehole is aligned with the R_v principal axis, Laplace's equation can be written (Kunz and Moran, 1958)

$$\frac{1}{r}\frac{\partial}{\partial r}\left(\frac{r}{R_h}\frac{\partial \Psi}{\partial r}\right) + \frac{\partial}{\partial z}\left(\frac{1}{R_v}\frac{\partial \Psi}{\partial z}\right) = 0 \tag{5-85}$$

where *r* and *z* are radial and vertical coordinates in a cylindrical system. Introducing the coefficient of anisotropy, $\lambda^2 = R_v/R_h$, the solution for the potential is

$$\Psi = \frac{R_h I}{4\pi(r^2/\lambda^2 + z^2)^{1/2}} \tag{5-86}$$

Surfaces of constant potential take the form of ellipsoids in anisotropic media rather than the spheres formed when $R_h = R_v$. An equivalent view is that the anisotropic case can be transformed to an isotropic case by stretching the radial coordinates by the factor $1/\lambda$. If the potential measurement is made along the z axis, as is usual for the borehole situation, then, since $r = 0$, Eq. (5-86) shows that R_a will be equal to R_h and be unaffected by the value R_v. This surprising consequence is known as the paradox of anisotropy. From this result, we expect that R_a values obtained in such a circumstance will be comparable with values obtained from cores cut with their long axes lying within the bedding plane.

Some anisotropic geometries can produce quite surprising effects as shown theoretically by Moran and Gianzero (1979), who treat the situation of two anisotropic half-spaces separated by a plane interface. Anisotropy is usually ignored in logging practice, however, because it is difficult to recognize, and there are no adequate methods to deal routinely with its effects.

Wellbore effects. When $R_m \ll R_t$, current flow is channeled within the wellbore rather than spreading into the rock as it would if the resistivity contrast were minimal. As a result, R_a can differ substantially from R_t. Much effort has been expended on compensating for or eliminating the effect of R_m on R_a. For a normal array, the potential can be expressed as a solution somewhat analogous to the expression for a plane interface [Eq. (5-80)]. However, the image current term is replaced by the function H, an integral involving products of Bessel functions which requires numerical evaluation.

$$\Psi = \frac{R_m I}{4\pi}\left[\frac{1}{(r^2 + z_2)^{1/2}} + H(r, z, a, R_m, R_t)\right] \qquad (5\text{-}87)$$

Evaluation of the integral has been carried out (Schlumberger, 1950) and is discussed further by Gianzero and Rau (1977) for the case of a sonde not centered in the wellbore. Calculated R_a curves for a normal sonde centered in a wellbore of mud resistivity R_m, surrounded by a medium of resistivity R_t, are shown in Fig. 5-24 in a format which has come to be known as *departure curves.* Some observations made by Lynch (1962) bear repeating here. For electrode spacings AM much less than the well diameter, R_a approaches R_m, as expected. As the spacing AM increases, R_a increases to the true formation value, which is shown by the dashed line n–n'. With continuing increase in spacing, however, the R_a/R_m curves continue to rise and overshoot to a maximum (or minimum) designated by the dashed line m–m' before reaching the true value R_t. The greater the contrast between R_t and R_m, the greater

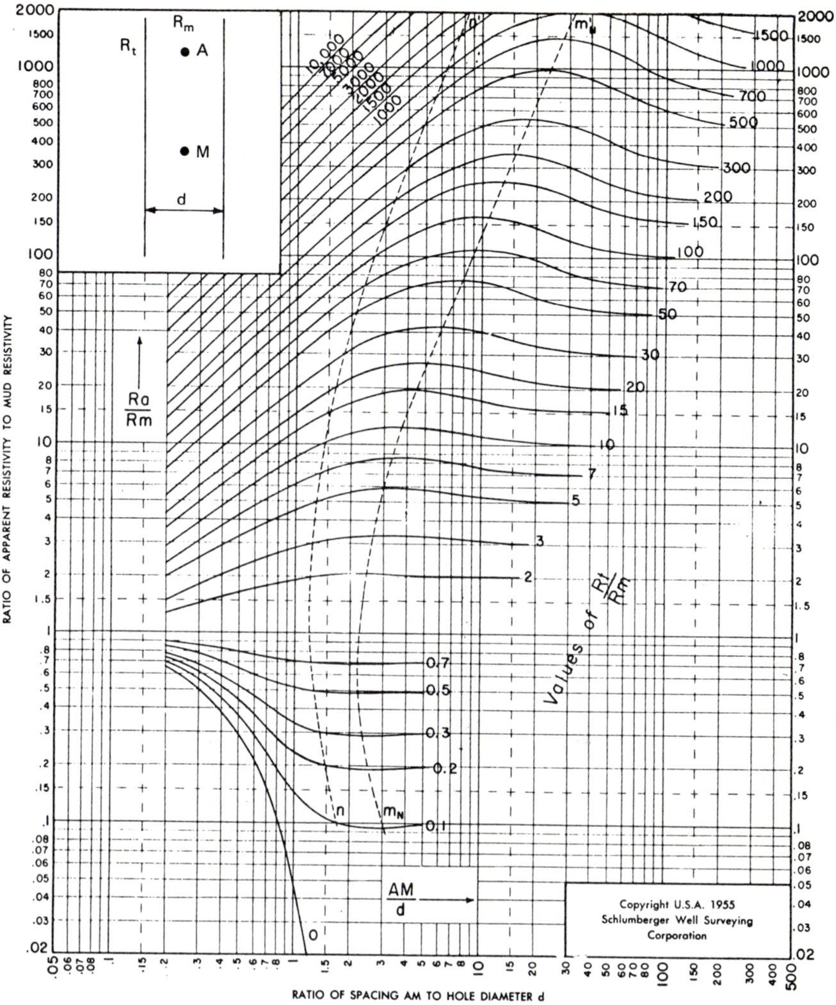

FIGURE 5-24 Departure curves for normal array. (Courtesy of Schlumberger, 1955.)

the separation required to reach the true asymptotic value. This presents a practical design problem, because large separations are incompatible with good vertical resolution of thin beds.

A workable compromise is reached by selecting an *AM* spacing for the normal array which will operate close to the line *n*–*n*′ in many situations of interest. For moderate contrasts (R_t between 0.1 and 50 R_m), the spacing is 1.5 to 2 times the wellbore diameter. That dictates a spacing of 25 to 40 cm for a 20-cm bore. Over the years, 40 cm (16 in) has been retained as the standard spacing for the short normal array.

Invasion. The existence of an invaded zone of resistivity R_{xo} (refer to Fig. 2-3 for nomenclature) extending a diameter D_i into the formation is treated by an elaboration of the formulas developed for the no-invasion case (Schlumberger, 1950). Ratios of R_{xo}/R_m and R_t/R_{xo} are incorporated into the term designated H in Eq. (5-87). Results are plotted in a suite of departure curves similar in form to Fig. 5-24, and their characteristics are discussed by Lynch (1962). Such curves are most useful when data from normal arrays of several spacings are available. Newer methods of handling the invaded zone rely on direct measurement with pad-type sondes, and they are discussed below.

Determination of R_t. We have seen that electrical measurements are most conveniently expressed as R_a and that the quantity sought is R_t. The approach for obtaining R_t has been touched upon in the preceding paragraphs: R_a is determined for specific cases of interest and plotted as a function of depth (Fig. 5-23) or as a function of the resistivity and length parameters (Fig. 5-24). R_t is then determined (we might say the inverse problem is solved) by curve matching or by applying a correction factor to the data.

Calculation of R_a for a given resistivity structure can be done analytically only for simple cases. Solutions for realistic cases which combine the wellbore, a finite sonde diameter, and layered or invaded formations are obtained by using model tanks, circuit network analogs, and numerical techniques. A sonde can be studied in a model tank at either reduced or full scale. Resistor networks can simulate situations of almost any complexity. Each resistor incorporates the resistivity and geometry of an annular ring, so that a two-dimensional resistor grid can represent any situation with cylindrical symmetry (Guyod, 1955; Dakhnov, 1962). Resistor networks have now been replaced by numerical methods implemented on digital computers.

A primary objective of sonde design is the reduction of effects from beds other than the one of interest. The focused logs and the pad-type sondes are designed to minimize the effect of R_m and to increase vertical resolution. Consequently, R_a corresponds much more closely to R_t than for the unfocused sondes, and the corrections required are much smaller.

5-4c More Complicated Electrode Arrays

Guard and focused logs. Guard and focused logs are designed to define the boundaries of thin beds, particularly resistive ones, and measure their resistivity (Doll, 1951). To do so, auxiliary electrodes are placed above and below a central current electrode (Fig. 5-25*a*). The

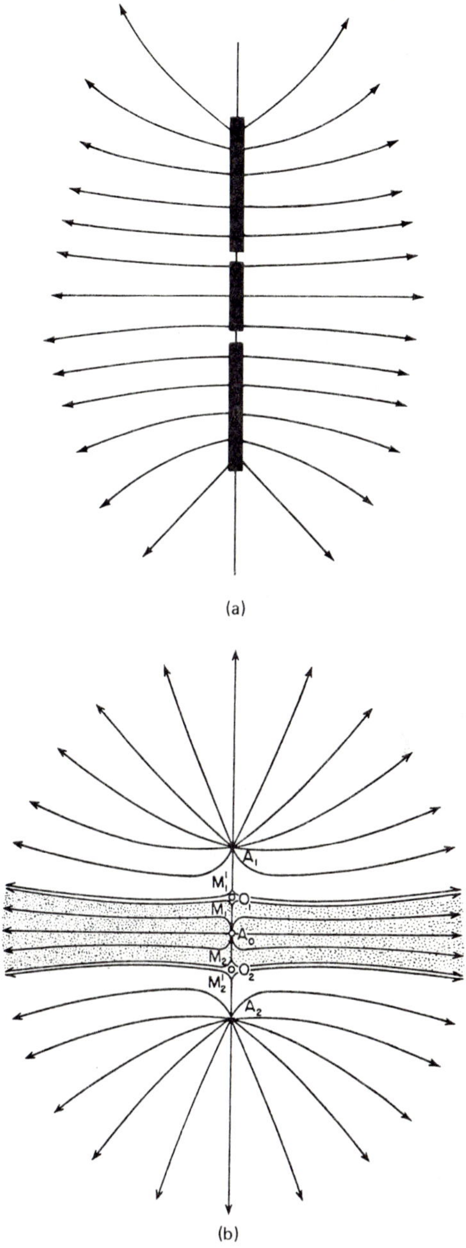

FIGURE 5-25 Schematic of current flow lines emanating from (*a*) a guard sonde and (*b*) a focused sonde. (After Doll, 1951. Copyright 1951, SPE–AIME.)

guard sonde uses a long metallic electrode to maintain a constant potential surface and thereby discourage current flow along the wellbore. A controlled current is transmitted from the auxiliary electrodes to maintain the two guard electrodes at the same potential as the central electrode. Apparent resistivity is obtained by monitoring the potential of the central electrode. The geometric factor (Keller and Frischknecht, 1966) is

$$G = \frac{2\pi\ell[(L/d)^2 - 1]^{1/2}}{(L/d)\ln(L/d) + [(L/d)^2 - 1]^{1/2}} \tag{5-88}$$

where L is the total length of the array, ℓ is the length of the central electrode, and d is the hole diameter. The guard sonde can be viewed as a single-point sonde with current flow constrained by the guards.

Focused sondes are similar in principle to the guard sonde (Fig. 5-25*b*): current from the auxiliary electrodes A_1 and A_2 is varied to maintain the vertical voltage gradient at points O_1 and O_2 equal to zero and thereby reduce to zero the vertical current flow in the mud column at those two points. The potential for the arrangement introduced by Doll (1951) can be obtained by using Eq. (5-72) and summing the contribution of the three current electrodes:

$$\Psi(O_1) = \frac{RI_0}{4\pi L_0} + \frac{RI_1}{4\pi L_1} + \frac{RI_2}{4\pi L_2} \tag{5-89}$$

where I_0, I_1, and I_2 are the currents, all of the same polarity, transmitted at electrodes A_0, A_1, and A_2 which are located at lengths L_0, L_1, and L_2 from position O_1.

Because the sonde is symmetrical about the point A_0, in a homogeneous medium the auxiliary currents must be equal, $I_1 = I_2 = I'$. If V and I_0 are the measured quantities, then apparent resistivity is defined as $R_a = GV/I_0$, where, from Eq. (5-89),

$$G = \frac{4\pi}{1/L_0 + I'/I_0(1/L_1 + 1/L_2)} \tag{5-90}$$

This type of electrode arrangement is called an adaptive array because the auxiliary current(s) I' changes as required to maintain a zero voltage gradient at points O_1 and O_2. We now face a dilemma in evaluating G from Eq. (5-90). Do we use the value of auxiliary current(s) I' which is applied at the point of measurement, or do we use the value which is applied in a homogeneous medium in order to comply with the concept of apparent resistivity? This question is peculiar to adaptive arrays. Since the focused resistivity log is such an adaptive array, it would appear from Eq. (5-90) that the array factor G should vary with the applied current I', but this leads to the following problem: Suppose that a measurement V/I_0 is made in some heterogeneous system and then, while maintaining those same currents I' and I_0, the sonde is transported to a homogeneous medium which produces the same result V/I_0. According to convention, the resistivity of the homogeneous medium would then be R_a for the heterogeneous system. However, in holding I' fixed we have violated a design premise of the sonde. The sonde is now out of balance and the voltage gradient at O_1 and O_2 is no longer zero. To avoid violating the zero gradient condition, the auxiliary current I' must change between the homogeneous and heterogeneous situations. For this reason the value of I' obtained for a given sonde in a homogeneous medium is used to evaluate G for that sonde (Moran, 1976), and this is the convention adopted for commercial logging sondes. Roy (1975, 1982) argues the case for defining R_a in terms of the actual I' used at the point of measurement.

To find I/I_0, set $d\Psi/dz = 0$ at position O_1 in Eq. (5-89). Then

$$\frac{I'}{I_0} = \frac{1/L_0^2}{1/L_1^2 - 1/L_2^2} \tag{5-91}$$

which shows that in a homogeneous medium the ratio I'/I_0 is independent of R and is a constant for any specific sonde. Dimensions of a focused sonde called the Laterolog 7 are $L = 2.03$ m between A_1 and A_2 and 0.81 m between O_1 and O_2, so that $L_0 = 0.2L$, $L_1 = 0.3L$, and $L_2 = 0.7L$. By Eq. (5-91), $I/I_0 = 2.76$; by Eq. (5-90), $G = 1.41$ for the Laterolog 7.

Apparent resistivity measured by a focused sonde must be corrected for the effects of mud and the "shoulder" beds above and below the target layer (Schlumberger, 1972). Applied first is a correction factor for mud and hole size (Fig. 5-26*a*) which seldom exceeds +20 percent. The effect of the shoulder bed can be larger, depending upon the contrast. Curves such as those in Fig. 5-26*b* are based on solutions obtained with a resistor network model.

Pad-type sondes. In terms of their basic principles, arrays mounted on pads pressed against the bore wall operate no differently than arrays mounted on mandrels. Both unfocused and focused arrays are placed on pads, as shown in Fig. 5-27. The combination of *microinverse* (utilizing electrodes A, M_1, M_2) and *micronormal* (utilizing electrodes A and M_2) arrays, designated the *microlog* by Doll (1950), provided a tool effective at delineating the presence of mud cake and hence useful for inferring permeable zones. The pattern of current flow through the mud cake and into the formation is suggested in Fig. 5-27*a*. Charts for

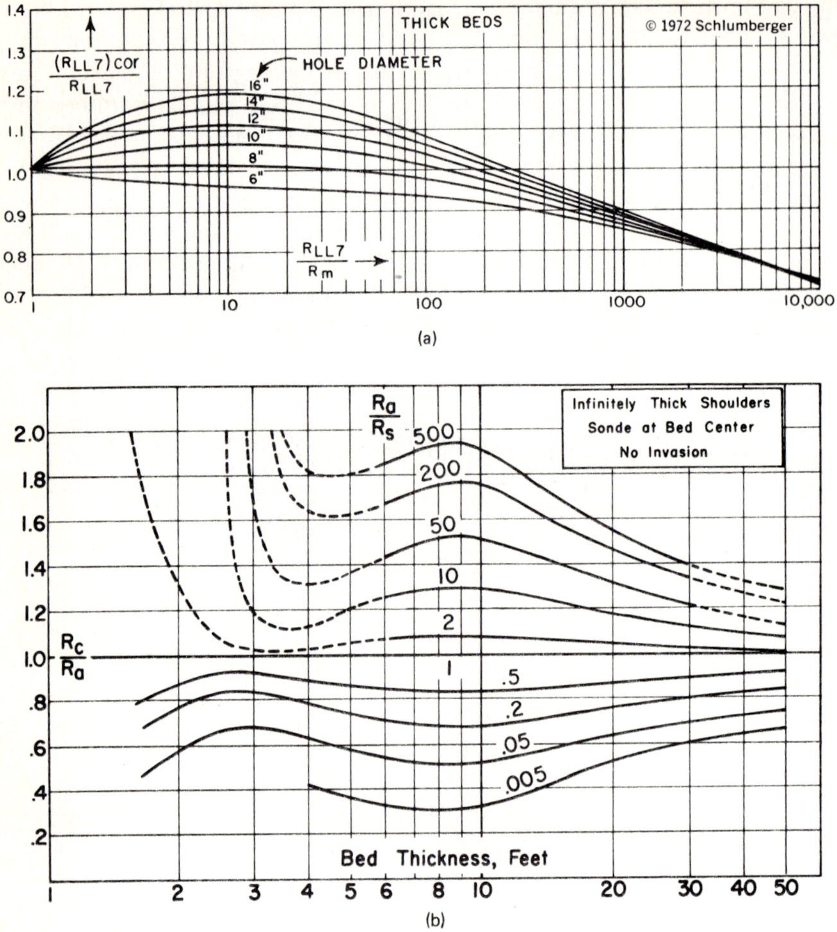

FIGURE 5-26 Correction charts for (*a*) hole size and (*b*) shoulder beds for a Laterolog 7. (Courtesy of Schlumberger, 1972.)

the estimation of mud cake thickness and R_{xo} from micrologs are available (Schlumberger, 1972); however, current distortion is such that the estimate of R_{xo} is of marginal quality.

The *microlaterolog*, shown in Fig. 5-27*b*, is an adaptive array patterned after the mandrel-type laterolog (Doll, 1952). Focusing greatly reduces the loss of current through the mud cake and thereby minimizes the dependence of the measurement upon mud cake properties so that a more accurate measure of R_{xo} can be made. Other versions of pad-type focused sondes have been introduced (Schlumberger, 1972) to further minimize the effect of thick mud cake.

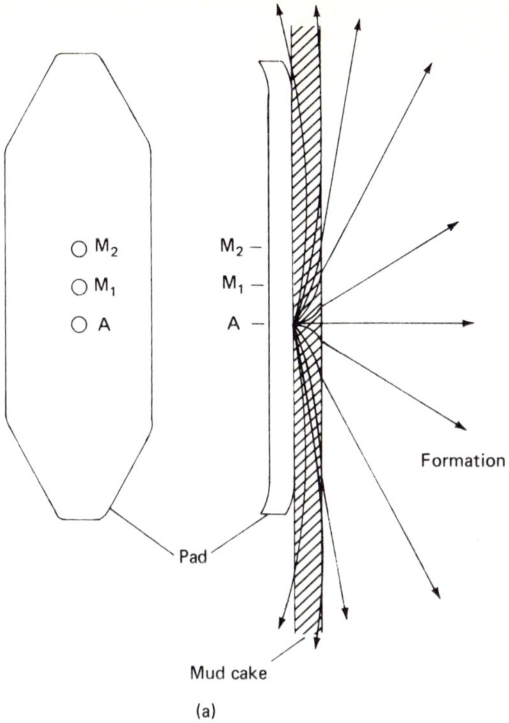

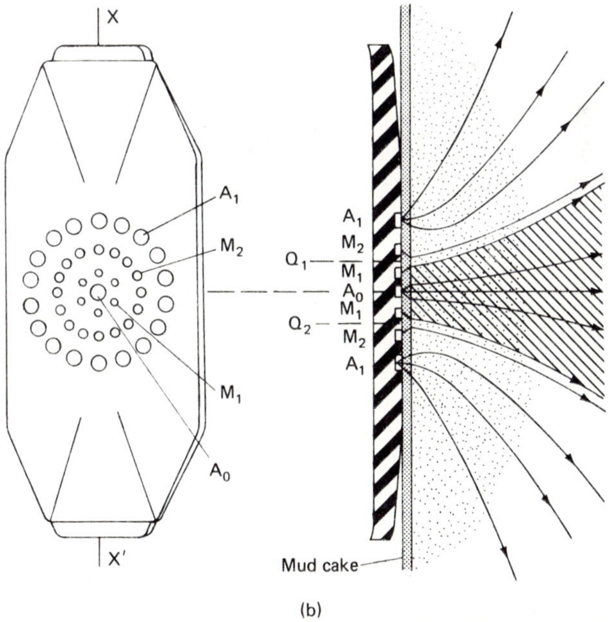

FIGURE 5-27 Electrode layout and schematic of current flow lines for (*a*) microlog and (*b*) microlaterolog. (After Doll, 1950, 1953. Copyright 1950, 1953, SPE–AIME.)

5-5 INDUCTION

Up to this point we have described methods of resistivity measurement which pass electric current into the earth by means of galvanic contact. The inductive method, on the other hand, utilizes a transmitter coil to set up an alternating magnetic field which induces current flow in the earth (Fig. 5-28). The induced currents generate a secondary magnetic field which is sensed by a receiver coil. The character and magnitude of the secondary field are governed by the frequency of the transmitted current and the distribution and magnitude of electrical properties in the vicinity of the borehole.

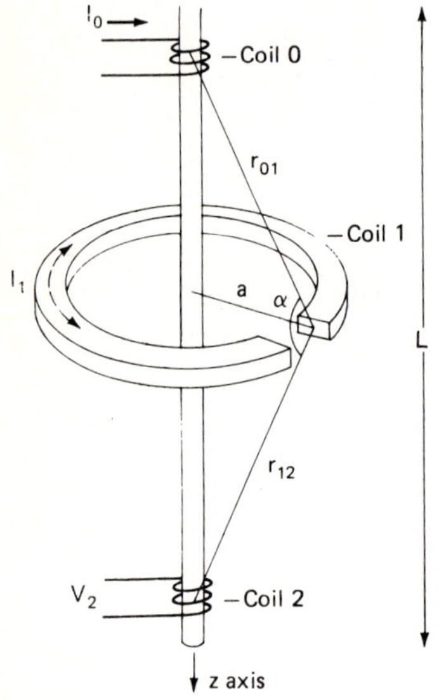

FIGURE 5-28 A two-coil induction sonde in presence of an elemental conducting loop.

Because galvanic contact is unnecessary, induction sondes can be operated in wells in which fluid is either not present or is highly resistive. In fact, the development of the induction method was partially motivated by the introduction of nonconductive, oil-based muds. For evaluation of true resistivity beyond the invaded zone, it follows that induction methods are most advantageous when R_m and R_{xo} are high. On the other hand, the method loses sensitivity at high R_t because the induced currents are weak.

Inductive methods have been developed and widely applied for mineral exploration, where highly conductive targets are sought, as well as for resistivity logging. The terminology, sonde design, and interpretive methods for determining R_t in sediments differ from those adopted for locating metallic ores. However, the relation between the two approaches can be readily grasped by developing the concept of induction number.

5-5a Induction Sonde Concepts

Propagation and skin depth. The inductive methods are described by treating the transmitter coil, receiver coil, and conductive earth as a set of transformer coils linked by their mutual inductances. However, it is useful to recognize at the outset that the electromagnetic energy coupling the loops undergoes attenuation and phase shift as it propagates. As shown in Sec. 5-1, an electromagnetic plane wave is characterized by a term exp $[i(kx - \omega t)]$, where

$$k^2 = \mu\varepsilon\omega^2 + i\mu\omega C \tag{5-92}$$

If $C \gg \omega\varepsilon$, the second term dominates; that is, conduction currents are more important than displacement currents. Figure 5-29 shows the ranges of resistivity and frequency in which these two current flow mechanisms become important. Because many induction sondes operate at a frequency of 20 kHz, the first term need be considered only for the most resistive environments, so that

$$k^2 \simeq i\mu\omega C \tag{5-93}$$

and the complex-valued propagation factor is

$$k \simeq (1 + i)\left(\frac{\mu\omega C}{2}\right)^{1/2} \tag{5-94}$$

If we define *skin depth* as

$$\delta = \left(\frac{2}{\mu\omega C}\right)^{1/2} \tag{5-95}$$

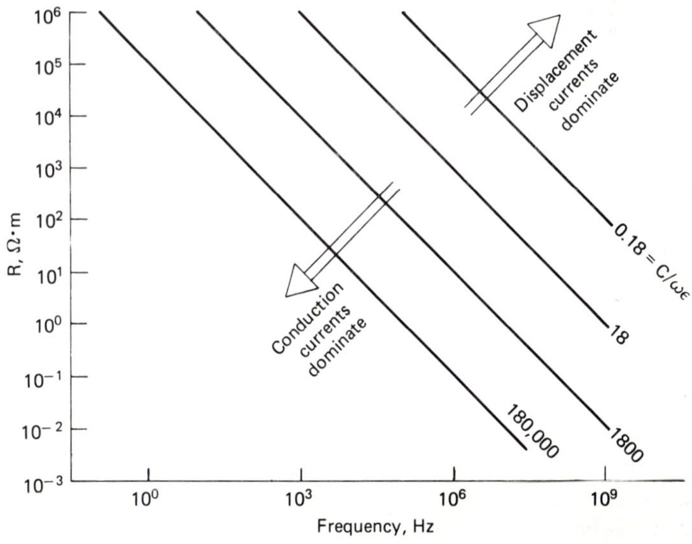

FIGURE 5-29 Ratio $C/\omega\varepsilon$ as a function of resistivity and frequency, illustrating relative importance of conduction and displacement currents.

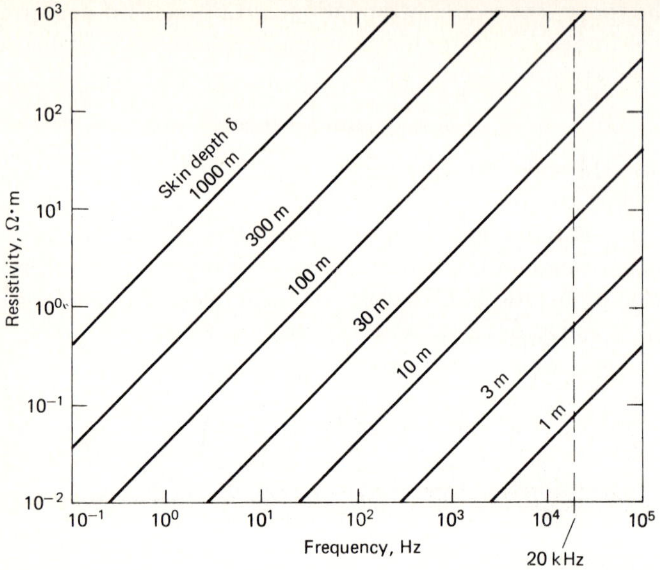

FIGURE 5-30 Skin depth as a function of resistivity and frequency.

then $k = (1 + i)/\delta$, and from Eq. (5-34) any electric or magnetic component of a propagating wave is described by

$$\psi \simeq \exp\,(-x/\delta)\,\exp\,[i(x/\delta - \omega t)] \tag{5-96}$$

which shows that the ratio of distance x to skin depth δ determines both the spatial decay rate and the phase shift. The skin depth is, therefore, a measure of distance to which an electromagnetic wave will penetrate. Physically this arises because the propagating wave loses energy to the currents set up in the medium. The more conductive the medium, the larger the currents and the shorter the distance over which the disturbance can propagate. Skin depth is shown as a function of resistivity and frequency in Fig. 5-30. The figure shows that an induction sonde with an operating frequency of 20 kHz will transmit wave energy with skin depths greater than the typical 1-m coil separation of a sonde unless the resistivity drops below 0.1 Ω·m.

Inductive coupling. We can describe the operation of an induction sonde by assuming that the skin depth is much greater than the coil separation so propagation effects can be ignored at first. Consider the sonde comprised of two small coaxial loops shown in Fig. 5-28. The currents produced in the earth will be circular and concentric with the z axis, so we can model the continuum as filled with elemental circular coils. For the time being, we consider just one such loop, designated coil 1. We follow Grant and West (1965) to obtain an expression for the signal generated at a receiver (coil 2) that is due to currents induced in the earth (coil 1) from a magnetic field generated by a transmitter (coil 0). The vertical magnetic field produced by current I_0 flowing through a small source coil of area A_0 and number of turns N_0 can be shown from Eq. (5-26) to be

$$B_z = \frac{\mu N_0 A_0 I_0}{4\pi} \frac{1}{r^3} \left(\frac{3z^2}{r^2} - 1 \right) \tag{5-97}$$

where r is the distance from the source coil to any point in the surrounding medium. By Faraday's law of induction [Eq. (5-27)] the voltage induced in the surrounding medium at coil 1 must be

$$V_{10} = \frac{d}{dt} \iint B_z \, dA \tag{5-98}$$

which, after insertion of Eq. (5-97) and integration, becomes

$$V_{10} = \frac{\mu N_0 A_0}{2} \frac{a^2}{r_{01}^3} \frac{dI_0}{dt} \tag{5-99}$$

We can represent the electrical properties of the elementary loop (coil 1) by discrete circuit elements: its resistance $\mathcal{R}$, in ohms, and its inductance $\mathcal{L}$, in henrys. The impressed voltage V_{10} causes a current I_1 to flow in the loop. As we found for Eq. (5-12), the current I_1 is the quotient of V_{10} and the loop impedance $\mathcal{R} - i\omega\mathcal{L}$, so from Eq. (5-99)

$$I_1 = \frac{\mu N_0 A_0}{2} \frac{a^2}{r_{01}^3} \frac{dI_0}{dt} \frac{1}{\mathcal{R} - i\omega\mathcal{L}} \tag{5-100}$$

From this expression the vertical magnetic field at coil 2 can be found by again using Eq. (5-26),

$$B_z = \frac{\mu a^2}{2r_{12}^3} I_1 \tag{5-101}$$

Applying Faraday's law once again,

$$V_{21} = \frac{d}{dt} (B_z N_2 A_2) \tag{5-102}$$

Substituting Eqs. (5-100) and (5-101),

$$V_{21} = \frac{\mu^2 N_0 A_0 N_2 A_2}{4(\mathcal{R} - i\omega\mathcal{L})} \frac{a^4}{r_{01}^3 r_{12}^3} \frac{d^2 I_0}{dt^2} \tag{5-103}$$

which is the expression for the voltage induced in a receiver coil that is due to a single loop of current in the earth. For comparison, a single application of Faraday's law to Eq. (5-97) gives the voltage in coil 2 that is due solely to the current in the source coil:

$$V_{20} = \frac{\mu N_0 A_0 N_2 A_2}{2\pi L^3} \frac{dI_0}{dt} \tag{5-104}$$

where L is the spacing between coils. V_{20}, sometimes called the primary voltage, is the voltage measured if the two coils are suspended in air away from any conducting bodies.

The current source I_0 is taken to be of form $I_0 \exp(-i\omega t)$, so the first and second time derivatives can be written $-i\omega I_0$ and $-\omega^2 I_0$, respectively, making the time dependence implicit. Then Eqs. (5-103) and (5-104) become

$$V_{21} = -\frac{\mu^2\omega^2 N_0 A_0 N_2 A_2 I_0}{4} \frac{a^4}{r_{01}^3 r_{12}^3} \frac{1}{\mathcal{R}} \frac{1 + i\omega\mathcal{L}/\mathcal{R}}{1 + (\omega\mathcal{L}/\mathcal{R})^2} \tag{5-105}$$

$$V_{20} = -\frac{i\mu\omega N_0 A_0 N_2 A_2 I_0}{2\pi L^3} \tag{5-106}$$

Induction number. The desired signal V_{21} can be normalized with respect to the primary voltage, resulting in

$$\frac{V_{21}}{V_{20}} = \frac{\pi}{2} \frac{\mu}{\mathcal{L}} \frac{L^3 a^4}{r_{01}^3 r_{12}^3} \frac{(\omega\mathcal{L}/\mathcal{R})^2 - i(\omega\mathcal{L}/\mathcal{R})}{1 + (\omega\mathcal{L}/\mathcal{R})^2} \tag{5-107}$$

The ratio $\mathcal{L}/\mu$ is proportional to a, the radius of loop 1. We define the induction number $\Theta = \omega\mathcal{L}/\mathcal{R}$, the ratio of reactance to resistance of loop 1. Now Eq. (5-107) can be written

$$\frac{V_{21}}{V_{20}} = b \frac{L^3 a^3}{r_{01}^3 r_{12}^3} \frac{\Theta^2 - i\Theta}{1 + \Theta^2} \tag{5-108}$$

where b is a constant. Equation (5-108) shows that the normalized signal is the product of two dimensionless factors. The first factor, called the *geometric factor*, incorporates only the distances between coils and loop. Grant and West (1965) show that the geometric factor in Eq. (5-108) is equivalent to a combination of the coupling coefficients of mutual induction among the coils.

The second term of Eq. (5-108), called the *induction factor*, depends only upon the properties of the elemental loop as given by the induction number Θ. As frequency or earth conductivity increases, so does Θ. This can be shown more clearly by expressing the induction number of a one-turn elementary loop as

$$\Theta = \frac{\omega\mathcal{L}}{\mathcal{R}} = b\,\frac{\mu\omega a}{R(2\pi a/\Delta z\,\Delta x)} \tag{5-109}$$

where $\Delta z\,\Delta x$ is the cross-sectional area of the loop. By writing R as its reciprocal C and incorporating all constants into b, we get

$$\Theta = b\mu\omega C\,\Delta z\,\Delta x \tag{5-110}$$

Equation (5-110) shows that Θ increases linearly with frequency and with the conductivity and area of an elemental loop. For more complex situations, the induction number is of the form

$$\Theta \simeq \frac{x_1 x_2}{\delta^2} \tag{5-111}$$

where x_1 and x_2 are characteristic lengths of a conductive zone in the earth and δ is skin depth as defined above.

The induction factor in Eq. (5-108) is complex-valued, so the received voltage V_{21} consists of components which are in phase [proportional to $\Theta^2/(1 + \Theta^2)$] and out of phase [proportional to $\Theta/(1 + \Theta^2)$] with V_{20}. The behavior of the two components is shown in Fig. 5-31. At large values of Θ, the in-phase component dominates, showing that V_{21} must be in phase with V_{20}. Referring to Eq. (5-110), we see

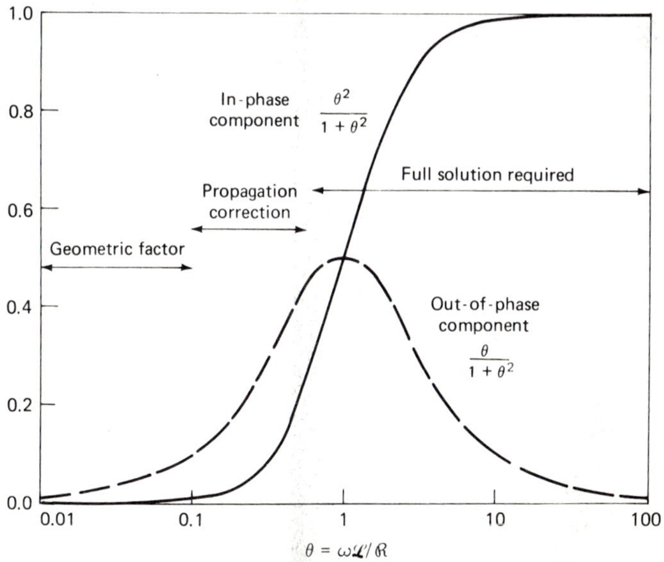

FIGURE 5-31 Induction factor showing regimes in which the geometric factor and propagation corrections are valid.

that this will occur when the earth is conductive or when high frequencies are used. As induction number decreases, the two components become comparable. Hence, the induction number governs the phase shift of the voltage contributed by the earth relative to the free-space voltage.

Although only a few real situations can be represented by an expression as simple as Eqs. (5-108) and (5-109), the induction number concept can be used to consider regimes in which theoretical treatments of varying difficulties are required for induction logging problems. These regimes are denoted by the horizontal arrows in Fig. 5-31. At very low induction number, where the effective penetration of the sonde is much less than a skin depth (earth conductivity is low), electromagnetic propagation can be ignored altogether and the static field development of the preceding paragraphs is adequate. These results will be expressed in terms of Doll's (1949*a*) "geometric factor" in the following subsection. As conductivity increases, the straightforward geometric approach requires a "propagation correction" because interaction among the elemental loops in the earth cannot be ignored as attenuation and phase shift of the propagating wave become important. These "propagation effects" are treated as an expansion of the full solution of the electromagnetic problem with appropriate boundary conditions. As the situation becomes even more complex, because of either high conductivities or complex geometries which cannot be treated with analytical solutions, numerical solutions and scale-modeling techniques are required.

Geometric factor. In the low-frequency limit, the terms in $\omega\mathcal{L}/\mathcal{R}$ on the right-hand side of Eq. (5-105) can be ignored. Loop resistance $\mathcal{R}$ is equal to $2\pi a/(C\,dz\,dr)$. We are then able to write Eq. (5-105) as

$$V_{21} = -\frac{\mu^2\omega^2 N_0 A_0 N_2 A_2 I_0}{4\pi L}\,\frac{La^3}{2r_{01}^3 r_{12}^3}\,C\,dz\,dr \tag{5-112}$$

which is equivalent to the expression given by Doll (1949*a*). The first term, referred to as K, contains only sonde parameters. The second term is the geometric factor G, where $L/2$ has been inserted to normalize G so that

$$\int_0^\infty \int_{-\infty}^\infty G\,dz\,dr = 1 \tag{5-113}$$

G can also be expressed in term of the angle α subtended by the transmitter and receiver coils from some point of interest in the medium (Fig. 5-28).

$$G = \frac{La^3}{2r_{01}^3 r_{12}^3} = \frac{\sin^3 \alpha}{2L^2} \tag{5-114}$$

Consequently, G has the same value for all loops in the medium from which the coils subtend the same angle α, and all such loops of unit cross section contribute the same signal to the receiver coil. Figure 5-32 shows cross sections of eight such loops lying on two circles which intersect the coils, for which $\sin^3 \alpha = 0.65$. The closer proximity of the 120° loop is compensated by the greater horizontal area of loops on the 60° arcs. Maximum signal is contributed by loops on a 90° arc lying intermediate to the two shown in Fig. 5-32.

This concept of the geometric factor was exploited by Doll to subdivide the earth surrounding a two-coil sonde into 10 sectors of equivalent geometric factor and hence of equivalent signal contribution (Fig.

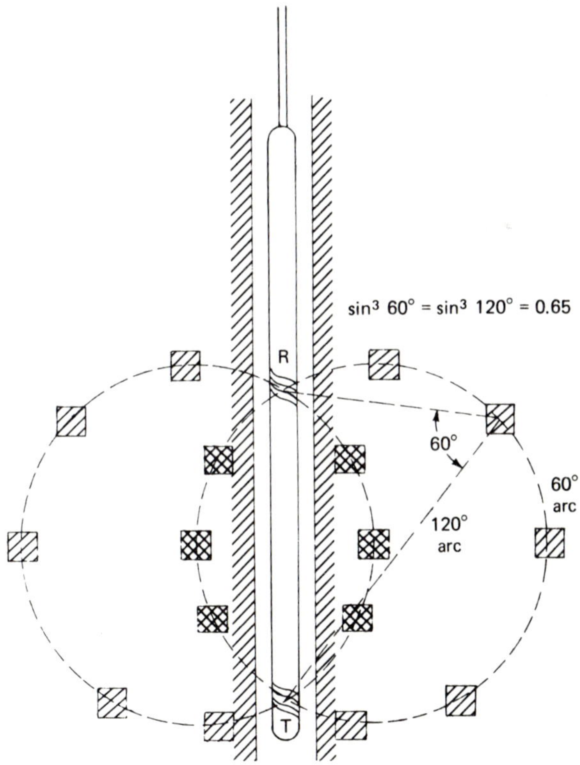

FIGURE 5-32 Locus of unit-cross-section loops of equal geometrical factor. (After Doll, 1949*a*; copyright 1979, SPE–AIME.)

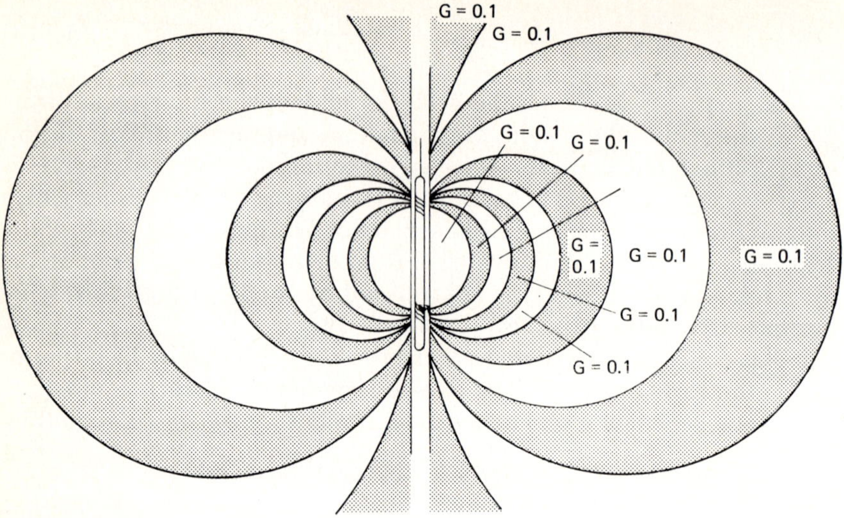

FIGURE 5-33 Cylindrical sectors of equal geometrical factor around a two-coil sonde Each sector contributes 0.1 of total signal. (After Doll, 1949*a*; copyright 1949, SPE–AIME.)

5-33). By summing contiguous sectors outward from the coil axis, the volume of earth contributing, say, 70 percent, of the signal can be quickly ascertained. If the signals contributed by all elementary loops are summed,

$$V'_{21} = \int_0^\infty \int_{-\infty}^\infty KGC\, dz\, dr \tag{5-115}$$

If the sonde is operating in a medium of uniform conductivity C, the voltage $V'_{21} = KC$. Then, by definition, the apparent conductivity $C_a = V'_{21}/K$.

If the medium consists of units of cylindrical symmetry and of different conductivities $C_m, C_n, \ldots,$ as is the case for a well penetrating horizontal layers, the voltage is

$$V'_{21} = K(\iint GC_m\, dr\, dz + \iint GC_n\, dr\, dz + \cdots) \tag{5-116}$$

or

$$V'_{21} = K \sum_i G'_i C_i \tag{5-117}$$

where G_i' is the geometric factor integrated over the ith area. Apparent conductivity can be computed as the sum of $G_i'C_i$ for horizontal beds, either invaded or noninvaded. In practice, however, induction sondes employ multiple coil pairs to increase both vertical and horizontal resolution. The response of multiple-coil sondes requires computational and calibration techniques more sophisticated than those indicated by Eqs. (5-112) through (5-117). In addition, V must be corrected for electromagnetic effects.

Full solution for homogeneous medium. Proper treatment of the two-coil sonde in a homogeneous medium requires solution of Maxwell's equations by means of vector potentials (Duesterhoeft, 1961; Moran and Kunz, 1962). The solution obtained for the total (and complex-valued) voltage in the receiver coil is

$$V_2 = \frac{2i\mu\omega N_0 A_0 N_2 A_2 I_0}{4\pi} (1 - ikL) \frac{e^{ikL}}{L^3} \tag{5-118}$$

Series solutions are obtained by expanding in powers of $kL = (1 + i) L/\delta$. In discussing Eq. (5-108) we used V_{20} as the phase reference for V_{21}, but now we shall adopt I_0 as the phase reference for V_2. Separating components which are in phase (^{r}) and out of phase (^{i}) with I_0:

$$V_2^r = -KC\left(1 - \frac{2L}{3\delta} + \cdots\right) \tag{5-119}$$

$$V_2^i = V_{20} - KC\frac{2L}{3\delta} + \cdots \tag{5-120}$$

In conformity with the geometric factor development, apparent conductivity is defined as $-V_2^r/K$,

$$C_a = C\left(1 - \frac{2L}{3\delta} + \cdots\right) \tag{5-121}$$

showing that the observed C_a will be less than the true homogeneous earth conductivity C because of the contribution of terms L/δ. The value C would be equal to C_a if the geometric theory were strictly valid, that is, if either the frequency or conductivity were low enough that $L/\delta \simeq 0$. In general, however, some correction will be required. Figure 5-34 illustrates the behavior of the ratio C_a/C with increasing L/δ. The correction is 7 percent at $L/\delta = 0.1$ and approximately 60 percent at $L/\delta = 1$. The correction is variously referred to as a propagation factor or *skin effect* correction, although the latter term is a misnomer because perturbations to the current flow in the formation increase progres-

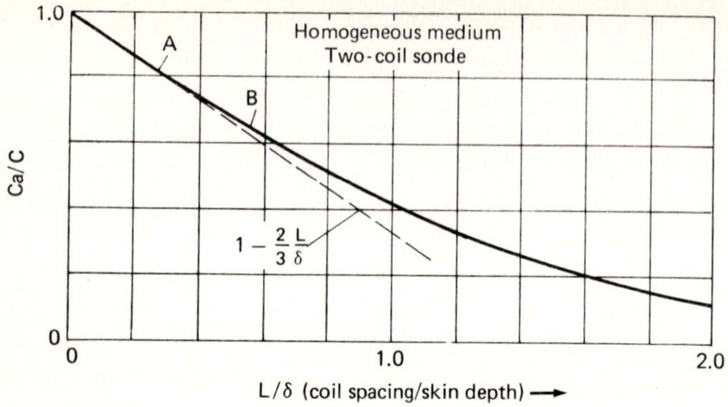

FIGURE 5-34 Effect of electromagnetic propagation upon apparent conductivity. Point *A* (*B*) corresponds to a 1-m coil spacing in a 1.0 (0.25) Ω·m formation. (After Moran and Kunz, 1962.)

sively with distance from the transmitter rather than being greatest at the wellbore. Hence, the correction is fairly insensitive to variations in resistivity of the invaded zone. Commercial logging systems generally apply a *propagation correction* before recording the log.

Thadani and Hall (1981) and Gianzero and Anderson (1982) have developed complete solutions for the geometric factor in simple geometries. They have also computed the radial and vertical investigation characteristics of two-coil and multicoil sondes, for both the resistive and reactive signal components.

Multicoil sondes. At small L/δ, V_{20} can be orders of magnitude greater than the desired signal V_2^r. It is common practice to null out the primary field V_{20} by coil arrangement. In addition, supplementary coils are used to cancel the contribution of the field above and below the main coils and to reduce the near-borehole field and thereby allow more "penetration" into the formation. Cancellation is optimized by arranging the spacing and turns of the individual coils, but it can be optimized in only one given electrical geometry. It is usually done in a homogeneous medium simulated by a fluid-filled tank. Moran and Kunz (1962) indicate how the concepts of effective coil spacing and conductivity can be extended from the two-coil to the multicoil sonde.

5-5b Evaluation of Formation Parameters

Both induction and galvanic sondes can be designed to obtain different depths of investigation into the formation. When apparent resistivity (conductivity) is measured by a combination of methods, the data can be used to infer the resistivity profile as a function of radius. If three independent resistivity measurements are available, then R_{xo}, d_i, and R_t (refer to Fig. 2-4 for definitions) can be determined. The method assumes an abrupt transition between the invaded and uninvaded zones. The integrated geometric factor for each measurement is defined as

$$G'(d_i) = \int_0^{d_i/2} \int_{-\infty}^{\infty} G \, dz \, dr \qquad (5\text{-}122)$$

It is common practice to use a sonde which incorporates a focused galvanic array to measure shallow properties in combination with medium-depth and deep-focused induction systems. Their integrated geometric factors are shown in Fig. 5-35. According to this chart, at a diameter of 140 in, the shallow focused log has 95 percent of its contributed response, the medium induction log has 80 percent, and the deep induction has only 53 percent.

Let the three integrated geometric factors and apparent resistivity values be designated by the subscripts 1, 2, and 3, respectively. If adjacent formations do not influence the measurement, and if propagation

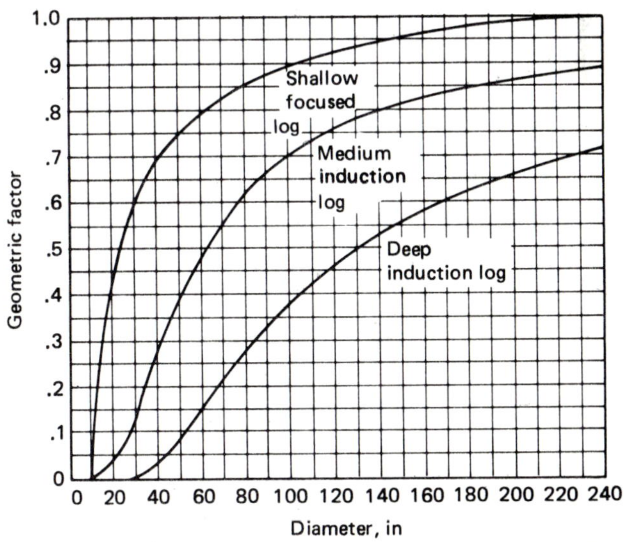

FIGURE 5-35 Integrated geometric factors for a sonde incorporating dual induction and focused resistivity measurements. (From Dresser Atlas, 1974.)

effects have been corrected, the response can be written as

$$
\begin{aligned}
R_1 &= G_1' R_{xo} + (1 - G_1') R_t \\
\frac{1}{R_2} &= \frac{G_2'}{R_{xo}} + \frac{1 - G_2'}{R_t} \\
\frac{1}{R_3} &= \frac{G_3'}{R_{xo}} + \frac{1 - G_3'}{R_t}
\end{aligned}
\tag{5-123}
$$

Conductivities (reciprocal resistivities) add if measured with an induction sonde [Eq. (5-117)], whereas resistivities add if measured with a galvanic system. This can be understood by considering the current patterns set up by the two systems. Current around an induction sonde flows concentrically around the wellbore, whereas current from a galvanic system flows radially.

The set of nonlinear equations (5-123) can be solved to construct charts (often called *tornado charts*) for obtaining R_t/R_3, R_{xo}/R_t, and d_i (Fig. 5-36). The charts are entered by using the ratios of the measured resistivities on the abscissa and ordinate. By evaluating the ratios, unique values for R_{xo}, R_t, and d_i can be obtained. Obviously, such charts are valid only if conditions are such that the geometric factors can be used.

5-5c Scale Modeling of Highly Conductive Bodies

Highly conductive geologic units cannot be treated by using the simple propagation correction because electromagnetic effects become dominant as induction number increases (see Fig. 5-31). In such cases a full solution such as that given by Eq. (5-118) for the homogeneous medium is required. However, conductive ore bodies are of finite extent and of geometry not readily amenable to analytic solution. Grant and West (1965) present some of the simple models for which analytic solutions can be found. Others have pursued numerical solutions to the governing differential equations or to formulations based on integral equations.

A third approach to treating finite conductive bodies is scale modeling, whereby the field situation is replicated on the laboratory scale. The scaling law simply requires that induction number be preserved:

$$
(\mu\omega C x_1 x_2)_m = (\mu\omega C x_1 x_2)_f \tag{5-124}
$$

where subscripts m and f designate the model and field parameters. To simulate conductive ore bodies, Drinkrow and Duffin (1978) constructed scale models of graphite slabs suspended in brine (Fig. 5-37). One model represented a field slab 50 m thick and of 5 Ω·m resistivity;

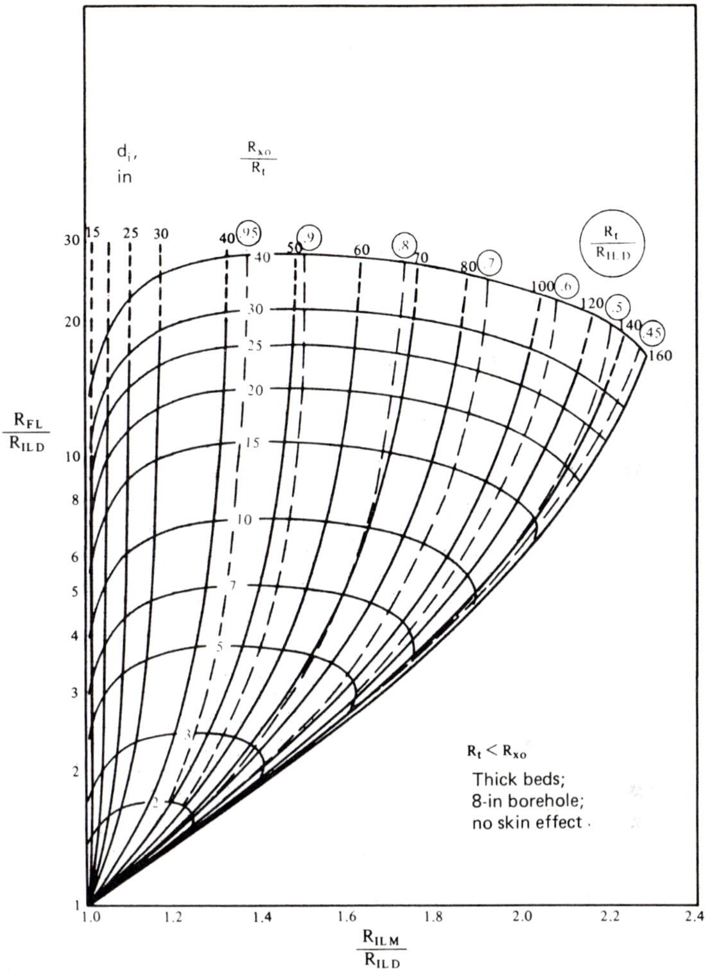

FIGURE 5-36 Interpretation chart for R_t, R_{xo}, and d_i from a dual induction sonde. (From Dresser Atlas, 1974).

a second smaller slab was scaled to a 6-m thickness and 0.006 Ω·m. The brine represented a host rock resistivity of 1000 Ω·m. A pair of coaxial coils passing through a hole in the center of the slab simulated an induction sonde with $L = 100$ m.

Results shown in Fig. 5-38 are plotted as z/L, with z taken as the distance of the probe center point from slab midplane. The receiver signal, analogous to V_{21} of Eq. (5-105), is given as a percent of the free-space primary field [V_{20} of Eq. (5-106)]. The anomalies produced by the two slabs are quite different. That is particularly true of the in-phase

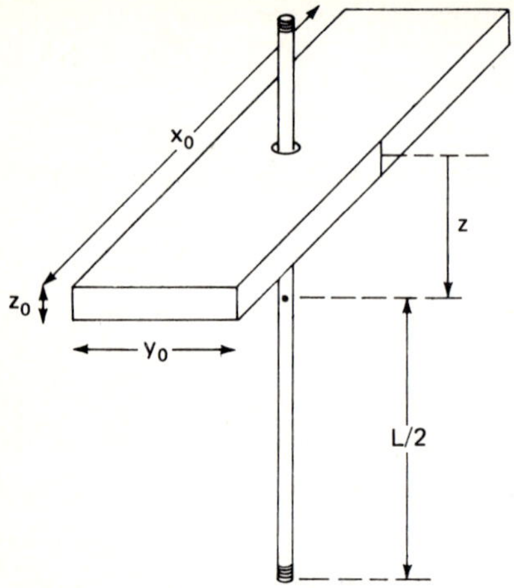

	Slab 1	Slab 2
x_0, m	100	300
y_0, m	30	230
z_0, m	6	50
R, Ω·m	0.006	5
δ, m	2.25	65

FIGURE 5-37 Coaxial coil system penetrating conductive slab modeled by Drinkrow and Duffin (1978). Values given are for field scale. Other values are coil separation L of 100 m, frequency 300 Hz, and host rock resistivity 1000 Ω·m.

component, which is bicuspid in shape for slab 1 and trough-like for slab 2.

The anomalies produced can be understood by visualizing the current loops induced in the slabs. If the slabs were highly conductive planes of great extent, the induced currents would "mirror" the dipole source and completely screen the field so that zero field would be sensed at the receiver. Complete screening occurs here only at the tips of the bicuspid anomaly, where one of the two coils is inside slab 1. Screening should occur, because the slab thickness is equivalent to several skin depths, as given in Fig. 5-37. However, as a coil moves away from the slab, the image current pattern grows larger and becomes distorted because of the finite size of the slab. This "imperfection" in the image current reduces the secondary field when the coils straddle the slab, and thereby accounts for the bicuspid anomaly.

Slab 2, being considerably larger than slab 1, produces a current pattern relatively uninfluenced by its boundaries, and hence a smooth trough-like anomaly. Because the slab thickness is less than a skin depth, cancellation is not possible even when a coil is inside the slab.

The dependence of voltage upon induction number, given in Eq. (5-108) and portrayed in Fig. 5-31, is reflected in the magnitudes of the in-phase and out-of-phase components of the two slabs. Slab 1, with a very high in-phase component, is characterized by a very high induction number commensurate with its low resistivity. Slab 2, by contrast,

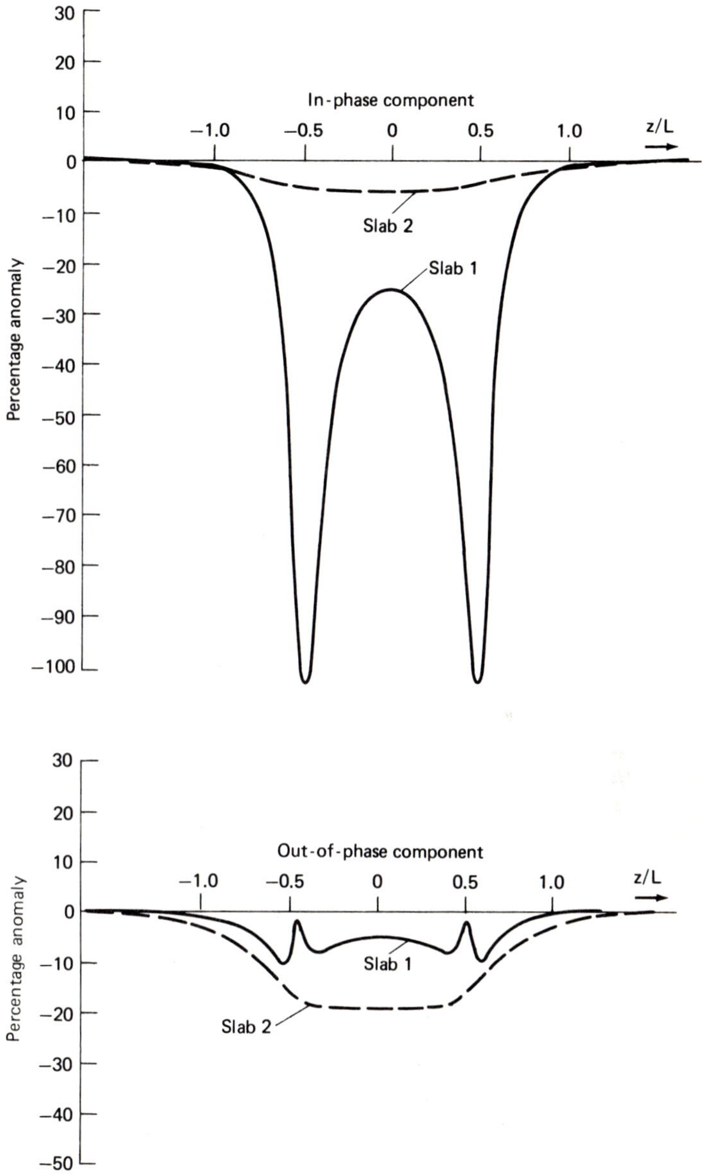

FIGURE 5-38 In-phase and out-of-phase components along a line through the center of two conductive slabs. (After Drinkrow and Duffin, 1978.)

produces an out-of-phase component greater than its in-phase one; and according to Fig. 5-31 its induction number must be less than 1.

Scale-modeling techniques are also useful for studying effects arising from the presence of more than one conductive body and from the interaction of resistive host rock with finite conductive bodies. Scale models can be used to examine geometric effects such as dip and plunge, offset between borehole and conductive body, and coil spacing.

5-6 SPONTANEOUS POTENTIAL

If an electrode is placed in a well, a potential difference can be observed between it and another electrode on the surface. The downhole electrode is metallic, often lead, and it should be separated from other metallic conductors sufficiently that no spurious voltages are created. No current is introduced into the formation; the measurement is completely passive. Because no excitation is used, the potential has come to be known as *spontaneous potential* (SP), a term which is used interchangeably with *self-potential* and secondary potential.

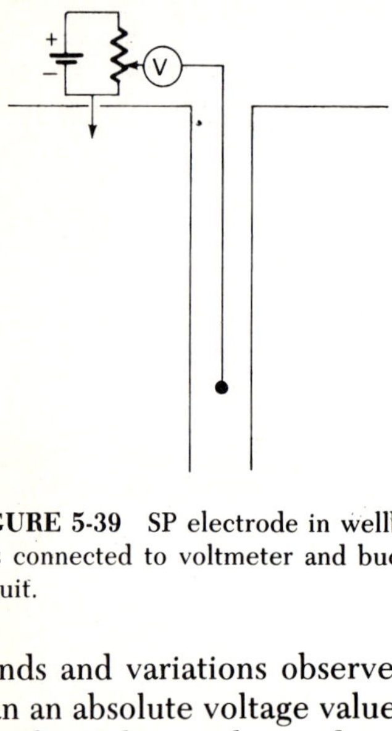

FIGURE 5-39 SP electrode in wellbore; it is connected to voltmeter and bucking circuit.

The battery and variable resistor shown in Fig. 5-39 are used to counter any constant voltage offsets introduced by the reference electrode. The SP measurement is a relative one in the sense that trends and variations observed within the well are of interest rather than an absolute voltage value.

The SP log can be used to distinguish between shales and other formations which are porous and permeable, provided the salt concentration of the formation fluid differs from that of the mud. The SP log is also used to correlate beds within a field. In addition, it has been found that the SP log can be used to determine the value of R_w if the value of R_{mf} is known.

5-6a Causes of SP

The simplicity of the method belies the complexity of the mechanisms which generate the SP log. The single electrode responds to potential gradients produced by the flow of current in the wellbore; that is, the observed variations in voltage are nothing more than ohmic $I\mathcal{R}$ potential drops. The mud resistivity and well diameter determine the resistance to current flow within the wellbore.

Some of the sources of current flow are external to the earth's surface. Industrial sources of electric power are usually readily recognized and avoided. In the high northern latitudes, telluric currents can have a significant vertical component in some locales (Dakhnov, 1962). The effect can be readily noted because the SP log then tracks the resistivity log.

Within the earth, electrical potentials arise from the transport of charged ionic species, chemical reactions, and the flow of fluids. Hence, the electrical phenomena are due to forces which are basically nonelectrical. The coupling between these forces and the electrical potentials which they generate can be isolated and studied in the laboratory, but in well logging practice it is sometimes difficult to attribute the cause of SP to a particular source. Of the three mechanisms mentioned above, concentration imbalances cause ionic transport and generate diffusion potentials which are generally recognized as the dominant contributor to SP in sedimentary sequences. The other two mechanisms, chemical (reduction-oxidation) reactions and fluid flow (electrokinetic) effects, are less commonly observed in conventional practice.

Diffusion potentials. The diffusion of ions under the influence of a concentration gradient is recognized as the prime source of SP. Ions tend to migrate from zones of high to low concentration. Because the transport properties of cations and anions differ, migration along the chemical gradient proceeds more rapidly for one than the other, with consequent separation of charge and the establishment of a voltage to counter it.

The equation relating ionic transport characteristics to voltage (MacInnes, 1961) is

$$V = -\frac{RT}{\mathcal{F}} \sum_i \int_{a'_i}^{a''_i} \frac{t_i}{z_i}\, d \ln a_i \qquad (5\text{-}125)$$

where a_i is the *activity* of the ith ionic species. Because of interionic attraction, the properties of real solutions differ from those of ideal solutions, so the effective concentration, called activity, is less than the actual concentration. In very dilute solutions, activity and concentration are equal. Other terms, including the transport numbers t_i, have been discussed in Sec. 5-2. The integral extends from the activity in solution′ (to be denoted as a') to solution″. All cations and anions in solution are represented by the summation under the integral.

Two cases pertinent to wellbore SP can be treated with Eq. (5-125).

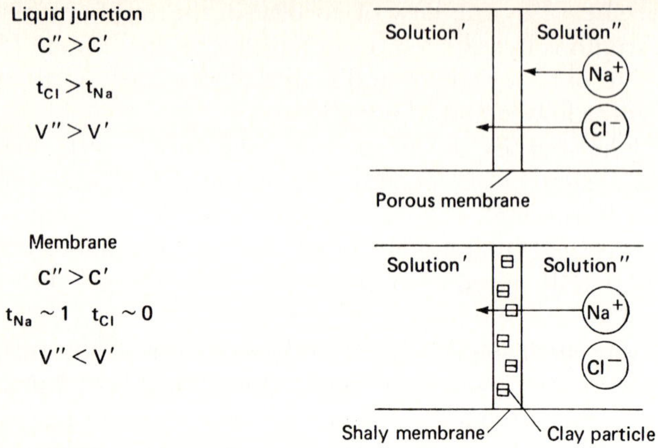

FIGURE 5-40 Behavior of Na and Cl ions in presence of liquid and membrane junctions.

The first, called a liquid junction, is shown in the upper half of Fig. 5-40. A thin porous membrane serves to separate the two solutions while allowing migration of ions from the high-concentration solution″ to solution′. Since NaCl is the only salt in solution, Eq. (5-125) becomes

$$V_l = -\frac{RT}{\mathcal{F}}\left[\frac{t_{Na}}{z_{Na}}\ln\,(a''_{Na}/a'_{Na}) + \frac{t_{Cl}}{z_{Cl}}\ln\,(a''_{Cl}/a'_{Cl})\right] \qquad (5\text{-}126)$$

where the primes specify the solution. Because $t_{Na} + t_{Cl} = 1$, $z_{Na} = +1$, and $z_{Cl} = -1$, if we assume $a_{Na} = a_{Cl} = a$, we have the result

$$V_l = -\frac{RT}{\mathcal{F}}(2t_{Na} - 1)\ln\left(\frac{a''}{a'}\right) \qquad (5\text{-}127)$$

Evaluation of the voltage developed across an ideal membrane follows in a manner similar to that for the liquid junction. An ideal membrane, indicated in the lower half of Fig. 5-40, contains fixed charges which block the passage of Cl ions while permitting Na ions to pass from solution″ to solution′. The fixed negative charges represent clay particles in shale. Evaluation of Eq. (5-125) for the membrane leads to

$$V_m = \frac{RTt_{Na}}{\mathcal{F}z_{Na}}\ln\left(\frac{a'_{Na}}{a''_{Na}}\right) \qquad (5\text{-}128)$$

For this case, $t_{Na} = +1$ and $z_{Na} = +1$, so that

$$V_m = \frac{RT}{\mathcal{F}}\ln\left(\frac{a''}{a'}\right) \qquad (5\text{-}129)$$

As shown in Fig. 5-41, the situation existing in the wellbore can be represented by a liquid junction between the saline pore waters in sandstone and the low-salinity drilling mud and a membrane junction formed by the shale interposed between the same two fluids. To represent the total voltage developed by these two sources, Eqs. (5-127) and (5-129) are summed:

$$V_l + V_m = 2\frac{RT}{\mathcal{F}}(1 - t_{\text{Na}}) \ln\left(\frac{a''}{a'}\right) \tag{5-130}$$

The working equivalent of this expression is obtained by using $t_{\text{Na}} = 0.4$, $\ln x = 2.30 \log_{10} x$, $R = 8.315$, and $\mathcal{F} = 96{,}500$, so that

$$V_l + V_m = K \log_{10}\left(\frac{a''}{a'}\right) \tag{5-131}$$

where $K = 64.9 + 0.238T$ (°C)
$K = 60.7 + 0.132T$ (°F)

so that $K = 71$ at 25°C (77°F). Voltages V_m and V_l are expressed in millivolts.

Thus far it has been assumed that only NaCl solutions are present

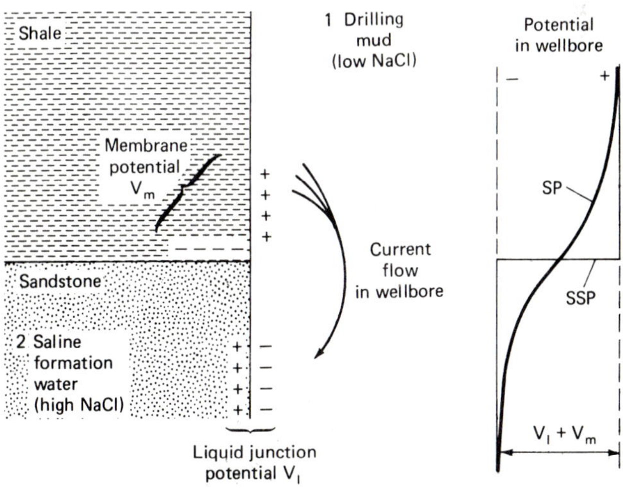

FIGURE 5-41 Electrochemical mechanisms and current flow around a conducting ore body. (After Sato and Mooney, 1960.)

and that the membrane is ideal; that is, it allows complete passage of Na ions and complete blockage of Cl ions.

Reduction-oxidation processes. Metallic ore bodies often give rise to SP anomalies on the surface with maxima on the order of −500 mV centered above the ore. The origin of these SP expressions is attributed to a system of circulating current, depicted in Fig. 5.42, which is driven by electrochemical reactions in the weathering zone above the ore (and above the water table) and in the zone below the water table (Sato and Mooney, 1960). The ore body serves as part of the three-dimensional circuit; it transports electrons from depth to the zone of weathering. An SP log obtained in a borehole penetrating the entire system of circulating current would detect a maximum potential difference equal to the source potential established by the chemical reactions.

Oxidation of ore actually reduces the magnitude of SP; hence, the largest effects are observed above graphite bodies that serve as electrodes and are chemically inert yet conductive. If the reactions in solution involve ions contributed by the ore body, such as pyrite, chalcocite, or galena, the driving potential difference is lower. An important aspect of this model is that the ore body serves primarily as a conductive pathway for electron flow. Potential gradients associated with such systems could be of the order of 500 mV/100 m, and they are likely to

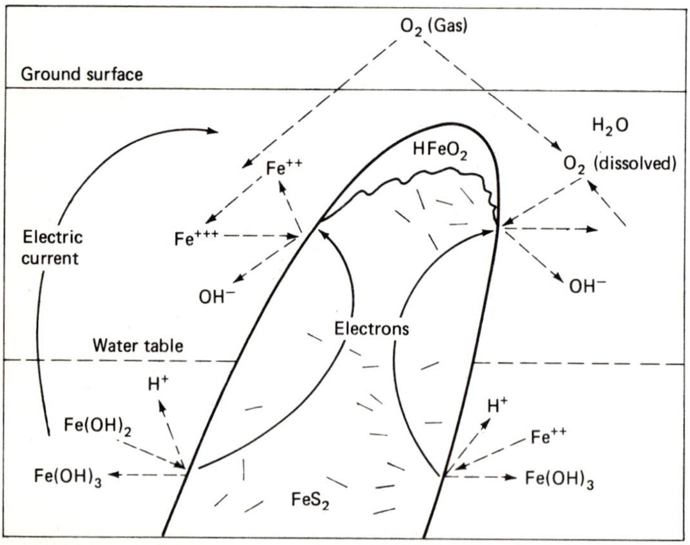

FIGURE 5-42 Representation of diffusion potentials in a wellbore with liquid and membrane junctions.

be restricted to shallow depths. Incidentally, one would expect an analogous situation to hold for a partially cased well in which the casing serves to connect an oxidizing environment with a reducing one. The effect of the circulating current would be observed in an open hole for some distance below the bottom of casing.

Some evidence for the existence of similar, but much larger, oxidation-reduction systems around petroleum reservoirs has been presented by Pirson (1977). Small amounts of hydrocarbons in rocks above the reservoir create a reduced environment with an oxidized zone lying outside the reduced "chimney." The resulting pattern of current flow is such that the SP log becomes progressively more positive with depth above the petroleum-bearing zone and more negative with depth in the oxidized regions above the water-bearing zones. Gradients are of the order of 5 to 10 mV per 100 m. The model must be regarded as speculative at this point because there is no apparent mechanism for electron transfer. Pirson (1977) also points out the difficulties of removing trends that are due to temperature gradients and the relative instability of some of the newer SP electrodes.

Streaming potential. Because of the fixed negative charge on mineral grain boundaries, water in pore spaces contains an excess of cations, as discussed in Sec. 5-3. When a differential fluid pressure is applied across a rock sample, fluid flows in accordance with Darcy's law. As fluid moves through the pore space, transporting mobile charge, the charge separation causes a voltage. This voltage is referred to as the *streaming potential* or *electrokinetic effect.*

Measurements of streaming potential on laboratory samples of sandstones and shales have shown that the voltage increases linearly with applied pressure. Shales typically have streaming potentials of 10 to 15 mV per 1000 psi (Hill and Anderson, 1958). In permeable sandstones, most of the streaming potential develops across the mud cake, where most of the pressure drop occurs. Measurements of the potential developed by mud filtrate passing through mud cake give streaming potentials ranging from 0 to 100 mV per 1000 psi. The values vary greatly with the type of mud.

In the wellbore, streaming potentials are probably always present to some extent because the wellbore pressure is greater than formation pressure. They are not usually detectable on the SP log, however, because only the difference between the streaming potential developed across the mud-caked permeable zones and that across shale beds will be observed. The difference is usually but not always small relative to the diffusion potentials. Corrections to the SP log can be made if the streaming potential characteristics of the mud and the pressure distribution in the permeable zone are known, but that is not usually the case. Another method is to measure SP with the well under different mud pressures and note the relative differences produced in SP opposite sand and shale as a result (Gondouin and Scala, 1958). Neither method has proved easy to implement in practice. Therefore, since the correction is small, streaming potential is ignored under most circumstances (Wyllie, 1963; Schlumberger, 1972).

5-6b Static SP

We now turn to the application of SP observed in wells penetrating sedimentary sequences. Present theory rests on the diffusion potentials developed across liquid and membrane junctions, as described in Eqs. (5-125) through (5-131). Let us recast Eq. (5-131) by replacing the activity in the concentrated solution a'' with that of the formation water a_w, which is usually a brine with an activity higher than that of mud filtrate. Solution activity a' is replaced by a_{mf}, so that Eq. (5-131) becomes

$$V_l + V_m = -K(T) \log \left(\frac{a_w}{a_{mf}} \right) \tag{5-132}$$

A negative sign has been adopted at this point to comply with the usual sign convention for SP. A final approximation is made by assuming that the activity ratio in Eq. (5-132) is equivalent to the concentration ratio and that the latter is inversely equivalent to the resistivity ratio. Equation (5-132) is then written

$$V_l + V_m = \text{SSP} \simeq -K(T) \log \left(\frac{R_{mf}}{R_w} \right) \tag{5-133}$$

The SSP, or *static SP*, designates the total voltage developed by the diffusion of ions across the liquid and membrane junctions. As shown in Fig. 5-41, it represents a voltage $V_l + V_m$ which would be observed as a step-like discontinuity if current flow were completely blocked, say, by a thin insulating wafer placed in the wellbore at the shale-sandstone elevation. Then the mud above and below the wafer would take on the potential of the adjacent formation. In an actual situation, however, the voltage $V_l + V_m$ causes current to flow in the wellbore and the resulting ohmic drop gives rise to the observed SP. The relation between SP and SSP is determined by the geometry of the formation and wellbore resistivities, and it will be discussed further below.

Equation (5-133) shows that SSP results from the contrast between the resistivity (actually, the ionic activity) of the mud and the formation fluid. In other words, SSP results from the presence of drilling fluid in the well and was not present before the well was drilled. Further, Eq. (5-133) shows that SSP (1) diminishes as the resistivity contrast decreases and (2) reverses in sign if the mud is more conductive than the formation fluid.

Equation (5-133) also shows that SP deflections can be used to determine R_w, and SP logs have been used extensively for this purpose. Once the SP reading is corrected to SSP, all that is required is knowledge of R_{mf} and $K(T)$. However, corrections are needed because Eq. (5-133) was derived by first assuming that both the mud filtrate and formation waters have the properties of an ideal NaCl solution. It was also assumed that the fluid's resistivity is inversely proportional to its ionic activity. The first assumption is least valid for low brine concentrations, and the second becomes progressively invalid as NaCl concentration increases.

To handle conditions which violate these assumptions, Gondouin et al. (1957) introduced the concept of equivalent resistivity, which is defined (rather than assumed) to be inversely proportional to activity. More precisely, it is defined as $R_{we} = A/a_w$, where A is a constant so chosen that $R_w = R_{we}$ in dilute solutions. When the equivalent resistivity concept is applied to both the mud filtrate and formation water, Eq. (5-133) becomes

$$\text{SSP} = -K(T) \log \left(\frac{R_{mfe}}{R_{we}} \right) \tag{5-134}$$

The equivalent values are useful only if they can be related to true values of R_w and R_{mf}. The relation is shown in Fig. 5-43. In pure NaCl solutions of resistivity greater than 0.3 $\Omega \cdot$m, $R_{we} = R_w$ as shown by the solid line in Fig. 5-43, but when other salts are present or brine concentrations are high and therefore R_w is low, the equivalent value differs from the true value.

The presence of cations other than Na is most important in fresh waters. Experimental work by Gondouin et al. (1957) on comparing R_{we} from the SP log with R_w measured on recovered formation waters established the dashed lines in Fig. 5-43. The 25°C dashed line diverges from the 25°C NaCl line because of the presence of ions other than Na. These empirical dashed lines are supported by a relation which accounts for the presence of two common divalent ions, Ca^{2+} and Mg^{2+}. For resistive muds, however, the freshwater correction cannot be used and other corrections are recommended (Schlumberger, 1972). Schlumberger (1972) also points out that the coefficient $K(T)$ will differ from Eq. (5-131) for solutions which are not predominantly NaCl. This

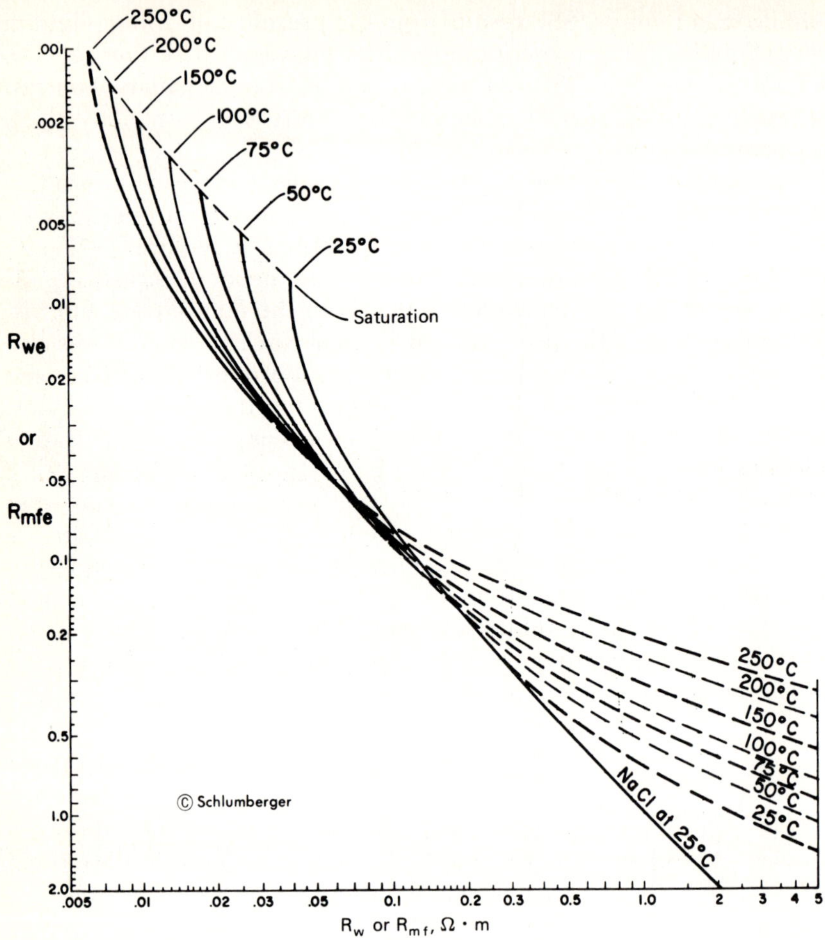

FIGURE 5-43 Relations between equivalent and true fluid resistivities. Solid curves are based on NaCl solutions; dashed curves are based on the solution of different salts representing an "average" fresh water. (Courtesy of Schlumberger, 1972.)

is understandable, since $K(T)$ incorporates the transport properties of Na^+ and Cl^-.

At high NaCl concentrations, ionic interactions play an increasingly important role in the conduction process and true resistivity is no longer inversely proportional to activity as it is at low concentrations. The resistivity of NaCl brines at elevated temperatures was determined experimentally and checked with theoretical values (Gondouin et al., 1957). The resulting data establish the solid curves shown in Fig. 5-43 for low R_w.

Experience has shown that the modifications to Eq. (5-133) which are embodied in Eq. (5-134) and Fig. 5-43 constitute a workable method for the determination of R_w from SSP values, but only if the measurement is made properly. If the SP electrode is not isolated from other metallic elements and from stray currents, the measurement will not be valid. Moreover, it must be remembered that Fig. 5-43 embodies assumptions that the fluids involved are predominantly NaCl solutions. Schlumberger (1972) urges that empirical K values and R_w-R_{we} relations be established for areas where other salts are found in significant concentration.

5-6c Character of SP Logs

Patterns of current flow set up in the formations and wellbore are governed by the location of the diffusion potential sources, the geometry of the resistive units, and the resistivity contrasts of the units. Since SP is proportional to wellbore resistivity and magnitude of the current within the wellbore, any distortion of the current flow pattern will be reflected in the SP. Doll (1949*b*) first described the nature of SP distortions resulting from variations in geometry and resistivity contrast. Correction factors to convert SP to SSP values have been evaluated by using a large resistor network simulator (Segesman, 1962).

Many of the features of SP logs can be understood with the aid of a voltage divider network (Fig. 5-44) which serves as a crude analog of the situation depicted in Fig. 5-41. Each resistor value represents the resistivity and geometry of a section of the rock-wellbore situation. A valid representation requires a distribution of resistors and voltage sources, so the lumped circuit of Fig. 5-44 serves only as a heuristic model. The two voltage sources cause current to flow through resistors representing the upper shale bed, the mud in the wellbore, the invaded zone, and the uninvaded formation. A bypass resistor $\mathcal{R}_{by}$ provides a path for current flow directly from the invaded formation to the shale bed. It does not exist when there is no invaded zone.

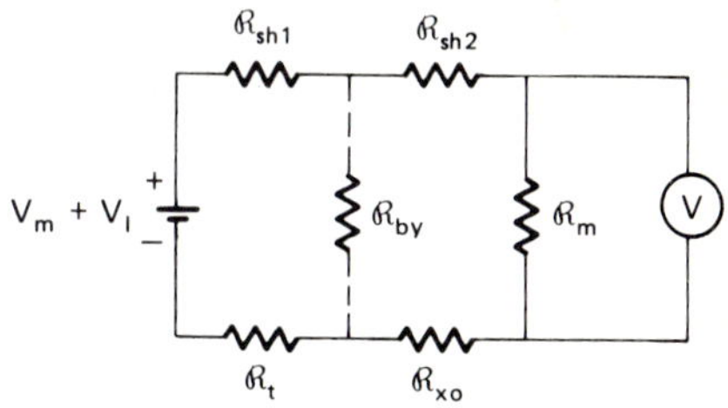

FIGURE 5-44 Voltage divider representation of current flow for SP. Subscripts for resistor elements represent shale bed (*sh*), mud (*m*), invaded zone (*xo*), uninvaded zone (*t*), and bypass (*by*).

The SP recorded across $\mathcal{R}_m$ will equal SSP $= V_l + V_m$ only if all the return current flows through the wellbore and if the formation and shale resistances are small enough compared with $\mathcal{R}_m$ that the voltage drops in the rock are negligible compared with that in the wellbore. Here we must caution that the resistance $\mathcal{R}_m$ can be larger than $\mathcal{R}_t$, even though the resistivity R_m is less than R_t. This occurs because the cross-sectional area available for current flow is much smaller in the wellbore than in the rock. With this in mind, the simple model of Fig. 5-44 can be used to explain the influence of several factors. For convenience the term SP will refer to the maximum SP deflection, expressed as a fraction of SSP.

1 Bed thickness. SP recorded opposite a thin bed is less than opposite a thick bed with the same resistivity because current flowing in a thin bed encounters a higher resistance $\mathcal{R}_t$.

2 Resistive beds. A high formation resistivity will diminish SP, again because $\mathcal{R}_t$ is increased.

3 Hole diameter. As hole diameter increases, the resistance $\mathcal{R}_m$ decreases and hence SP declines.

These points are illustrated in Fig. 5-45 for the no-invasion case. The results are summarized in terms of bed thickness as a multiple of wellbore diameter, and of R_t/R_m. Figure 5-45 shows that if the bed thickness exceeds $8d$ and R_t is less than $10R_m$, the observed SP maximum will be 90 percent or more of SSP.

The voltage divider circuit can also be used to explain the effect of

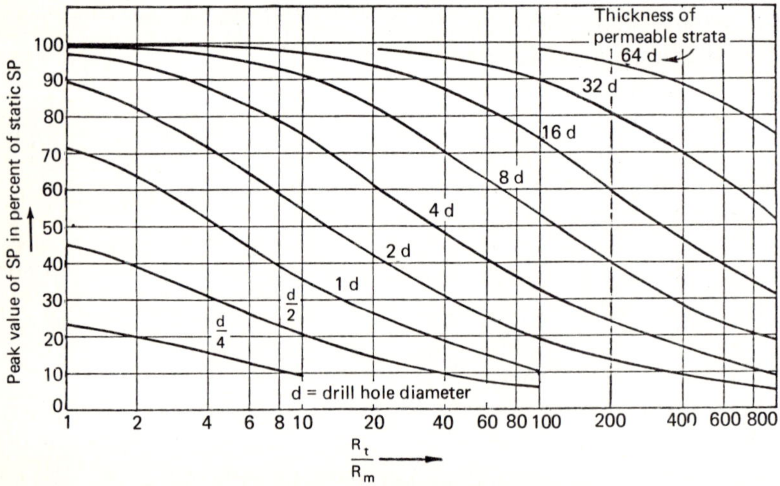

FIGURE 5-45 SP maxima as a function of bed thickness and resistivity contrast. (After Doll, 1949*b*. Copyright 1949, SPE–AIME.)

increased depth of invasion. As invasion depth increases, the effective location of $V_m + V_l$ moves further back into the formation. Resistors $\mathcal{R}_{xo}$ and $\mathcal{R}_{s2}$ increase and $\mathcal{R}_{by}$ decreases, so that more current flows through the bypass element $\mathcal{R}_{by}$ rather than through $\mathcal{R}_m$. Hence, the effect of invasion is to decrease SP, and the correction required for many cases of interest can be quite substantial as shown by Segesman (1962).

5-7 DIELECTRIC METHODS

We have seen how electrical resistivity is measured with galvanic resistivity methods and induction sondes. These sondes operate at frequencies low enough that the dielectric properties of rock can be ignored or corrected because they are small compared to conduction processes. At high frequencies and high resistivities, dielectric properties (that is, displacement currents) begin to dominate the propagation characteristics of electromagnetic waves as shown in Fig. 5-29. The presence of water governs the sonde response in this high-frequency regime just as it does at low frequencies, because the dielectric permittivity of water is significantly greater than that of most earth materials.

The amount of water present determines the logging application at frequencies of 100 kHz and higher. In materials with low water content, such as salt and granite, heterogeneities such as fractures or xenoliths often contain enough water to reflect or attenuate electromagnetic waves. High-frequency probes have been employed in ranging (radar) and transmission modes to map discontinuities in the rock mass. In porous rock, on the other hand, water attenuates electromagnetic energy so rapidly that the depth of investigation is limited. In this situation a dielectric measurement can be used to infer water-saturated porosity because the dielectric permittivity of rock increases directly with water content. A sonde which exploits this property is described later in this section.

In Sec. 5-1 we pointed out that complex representations of C and ε are required to characterize fully the conduction of electric current in rock. The total current density [Eq. (5-14)] is

$$\mathbf{J} = [C_r + \omega\varepsilon_i + i(C_i - \omega\varepsilon_r)]\mathbf{E} \tag{5-14}$$

Rather than retain all four conductivity terms, most workers choose one term for in-phase and another for out-of-phase conduction. At frequencies of 1 GHz and above, the choice is usually

$$\mathbf{J} = (\omega\varepsilon_i - i\omega\varepsilon_r)\mathbf{E} \tag{5-135}$$

and at frequencies of 1 to 100 MHz, most workers use

$$\mathbf{J} = (C_r - i\omega\varepsilon_r)\mathbf{E} \tag{5-136}$$

The choices are not clear-cut. An electrical measurement over a limited frequency range cannot distinguish between the conduction and displacement current contributions to each of these two components. Consequently, a response due to conduction may be erroneously attributed to polarization. Fuller and Ward (1970) discuss this problem at length. They point out that, if Eq. (5-135) is used, the phenomena properly described by $\varepsilon_i + C_r/\omega$ are lumped together as ε_i. As a result, unrealistically high values of ε_i are obtained at the low-frequency end of a measurement spectrum.

For practical use, the ratio of dielectric permittivity in a material to that in vacuum is defined as

$$\kappa = \frac{\varepsilon}{\varepsilon_0} \tag{5-137}$$

It also has real and imaginary components κ_r and κ_i. The ratio κ_i/κ_r is often referred to as the dielectric loss tangent. κ itself is called the relative dielectric permittivity or the *dielectric constant.* In fact, κ is not constant at all, but varies with frequency and temperature.

5-7a Dielectric Properties

Water. The two hydrogen protons of a water molecule extend away from one another at an angle of 105° and lie about one angstrom distant from the oxygen nucleus. A net positive charge is associated with each hydrogen proton and a net negative charge with the oxygen nucleus. It is this charge separation which gives water its unusually high dielectric permittivity: κ_r is close to 80 at 20°C and is constant with frequency up to 1 GHz (Fig. 5-46). Above 1 GHz the dielectric spectrum becomes rather complex, as shown by the molecular relaxation peak appearing at 20 GHz.

If the internal (thermal) energy is increased, individual molecules are less likely to align with the applied electric field. As a result, κ decreases with increasing temperature as shown in Fig. 5-47. Clearly the temperature dependence is quite important for the well logging environment, and corrections must be made. Pressure changes the value of κ only slightly, so that pressure can be adequately accounted for by assuming typical depth profiles for temperature and pressure and combining the effect of pressure with that of temperature (Calvert et al., 1977). Alternatively, measured values of κ have been fitted to products of polynomials in temperature and density (Helgeson and Kirk-

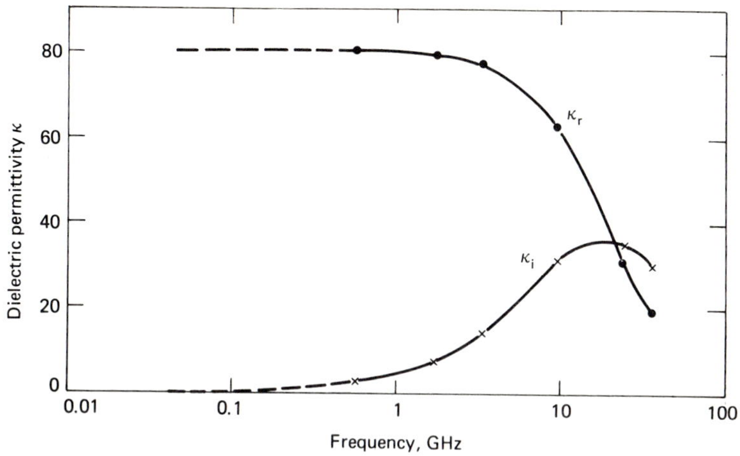

FIGURE 5-46 Components of relative dielectric permittivity for water at 20°C based on data compiled by Hasted (1972).

ham, 1974) and to polynomials in temperature and pressure (Olhoeft, 1981). Formulas and coefficients given by these authors can be used to calculate κ over the range of interest for logging.

When a salt is added to pure water, the disassociated ions hydrolyze and attract a sheath of water molecules. κ decreases because water molecules are both displaced by salt and immobilized in the hydration sheath. Olhoeft (1981) gives an empirical equation for the effect of salt concentration c (moles per liter)

$$\kappa = \kappa(T) - 13.00c + 1.065c^2 - 0.03006c^3 \qquad (5\text{-}138)$$

where $\kappa(T)$ is the temperature-corrected value. This expression shows that a solution of 1 M NaCl (58,000 ppm) in water reduces the value of κ from 78.3 to 66.3, a reduction of approximately 15 percent.

Minerals and dry rocks. Most dry earth materials have values of κ_r less than 10. Data compiled by Keller (1982) on minerals (Table 5-2) show that, among the silicates, quartz has the lowest κ_r. All the other silicates, plagioclase, muscovite, and biotite, have a higher density than quartz and, since there is correspondingly more polarization charge per unit volume, a higher κ_r. Consequently, κ_r of dry rocks will depend on mineralogy. Measurements by Olhoeft (1981) on dry sandstones give values of 4.7 for an arkosic sandstone and 6.0 for a ferrigenous sandstone. Limestone and dolomites measured by Olhoeft ranged from 6.1 to 8.7, in agreement with the values for calcite and dolomite in Table 5-2. Because they have such a varied mineralogy, shales can be expected to

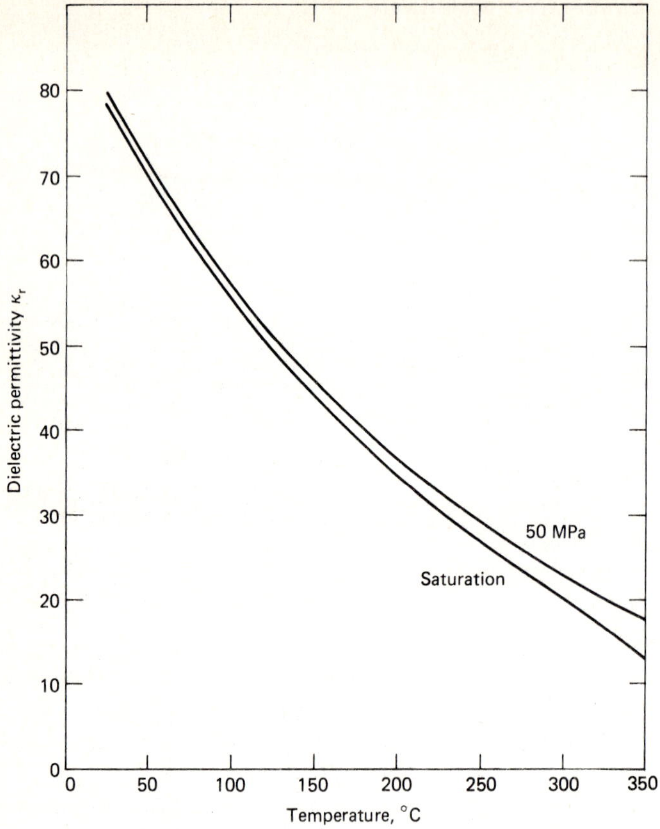

FIGURE 5-47 Relative dielectric permittivity of water as a function of temperature at two pressures based on polynomial fit by Helgeson and Kirkham (1974). Curve marked "saturation" indicates equilibrium pressure of water and steam.

TABLE 5-2 Measured relative dielectric permittivity of minerals. Compiled by Keller (1982).

Class	Mineral	Composition	κ_r
Halides	Halite	$NaCl$	5.7–6.2
	Sylvite	KCl	4.4–6.2
Carbonates	Calcite	$CaCO_3$	7.8–8.5
	Dolomite	$CaMg(CO_3)_2$	6.8–8.0
Sulfates	Barite	$BaSO_4$	7.0–12.2
	Gypsum	$CaSO_4 \cdot 2H_2O$	5.0–11.5
Silicates	Biotite	$K(Mg, Fe)_3AlSi_3O_{10}(OH)_2$	6.2–9.3
	Muscovite	$KAl_3Si_3O_{10}(OH)_2$	6.2–8.0
	Plagioclase	$NaAlSi_3O_8$-$CaAl_2Si_2O_8$	5.4–7.2
	Quartz	SiO_2	4.2–5.0

have a wide range of κ_r. Values of κ_r do not vary much with frequency for frequencies less than 1 MHz. The κ_i component is typically 1 to 5 percent of κ_r.

Water-saturated rocks. Measurements of κ versus frequency shown in Fig. 5-48 are representative of spectra obtained on sandstones saturated with salt water. The high slopes and high values at low frequencies are actually due to conductivity, not permittivity. Dielectric measurements are usually reported as κ_r and κ_i with no separation of conductive and displacement components. Hence, according to Eq. (5-14), κ_r includes the term C_i/ω and κ_i includes C_r/ω. These terms undoubtedly account for the high values obtained for both components at low frequencies and for the $1/\omega$ slope of the κ_i curve. Not until the frequency exceeds 200 MHz does the κ_r component approach a value uninfluenced by conduction. At these very high frequencies, the effects of conduction are minimal.

We would, of course, like to be able to combine the spectra for water (Fig. 5-46) with the dielectric properties of unsaturated rock to predict the resulting properties of saturated rock (Fig. 5-48). However, there does not yet exist a generally accepted mixing model for dielectric properties comparable to that provided by Archie's equations for resistivity. The effect of the conduction mechanism is one problem;

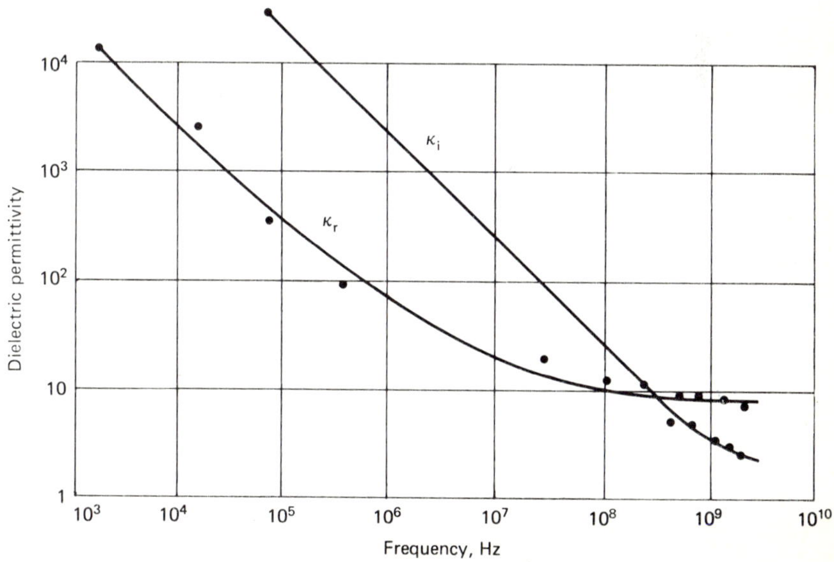

FIGURE 5-48 Relative dielectric permittivity of sandstone (of 0.145 porosity) saturated with brine. (From Poley et al., 1978.)

another is the lack of data at high frequencies. It is perhaps illustrative to mention some of the mixing models for dielectric properties which have been proposed:

1 Olhoeft (1981) notes the general utility of the geometric mean law. In addition, his data lead him to suggest that

$$\kappa \simeq 2^{\rho} \kappa_w^{\phi} \tag{5-139}$$

where κ_w is the value of κ for water and ρ and ϕ are the bulk density and porosity, is a good first-order approximation to κ_r for a water-saturated rock.

2 Sen et al. (1981) used self-consistent medium theory to develop an expression for effective dielectric permittivity of a model comprised of rock grains coated with a layer of water containing smaller grains similarly coated. This nested model produced excellent agreement with laboratory measurements on saturated fused glass beads. In the low-frequency limit the model reduces to a form of Archie's equation.

3 In Sec. 1-4 we introduced a generalized mixing rule which for a good dielectric (negligible conductivity) is

$$\varepsilon^{\zeta} = (1 - \phi)\varepsilon_{ma}^{\zeta} + \phi\varepsilon_w^{\zeta} \tag{5-140}$$

Meador and Cox (1975) expanded this formulation to include the case in which both water and hydrocarbon fill the pore space.

$$\varepsilon^{\zeta} = (1 - \phi)\varepsilon_{ma}^{\zeta} + S_w\phi\varepsilon_w^{\zeta} + (1 - S_w)\phi\varepsilon_h^{\zeta} \tag{5-141}$$

where S_w is the water saturation and ε_h is the dielectric permittivity of hydrocarbon. They indicate that ζ is not a constant but varies between 0 and 2, depending on the rock type.

4 Calvert et al. (1977) introduced a relation which follows from Eq. (1-14) by setting $m = 2$:

$$\varepsilon = (1 - \phi)^2\varepsilon_{ma} + \phi^2\varepsilon_w \tag{5-142}$$

In a lossless medium, $\varepsilon^{1/2}$ can be replaced by the inverse phase velocity, or travel time Δt. Hence, for a rock composed of matrix and water-filled pore space, the total lossless travel time is

$$\Delta t_0 = (1 - \phi)\,\Delta t_{ma} + \phi\,\Delta t_w \tag{5-143}$$

(In Chap. 7 we will encounter a similar relation for acoustic properties called the *time-average formula.*) In applying Eq. (5-143) to dielectric logs, measured travel times are corrected for losses after measuring attenuation. Equation (5-143) is then solved for porosity ϕ by using known or derived values of Δt_w and Δt_{ma}.

At this stage no generally accepted model for dielectric properties has been developed. Current practice is to develop empirical laws for materials and frequencies of interest.

5-7b Plane-Wave Propagation

Several aspects of determining κ with a borehole sonde can be understood by a discussion of the propagation characteristics of plane waves, even though the fields generated in the borehole are not planar. Equations (5-35) to (5-37) can be simplified in accordance with the relative values of the conductivity and permittivity components. The approximate form for induction problems has already been discussed in Sec. 5-5. For this discussion, we assume that $C_r \gg \omega\varepsilon_i$ and $C_i \ll \omega\varepsilon_r$, so Eq. (5-35) becomes

$$k^2 = \mu\varepsilon_r\omega^2 + i\mu\omega C_r \tag{5-144}$$

and Eqs. (5-36) become

$$k_r^2 - k_i^2 = \mu\varepsilon_r\omega^2 \tag{5-145}$$

$$2k_r k_i = \mu\omega C_r \tag{5-146}$$

The components of k [Eqs. (5-37)] then become

$$\begin{aligned} k_r^2 &= \frac{\mu\varepsilon_r\omega^2}{2}\left\{\left[1 + \left(\frac{C_r}{\omega\varepsilon_r}\right)^2\right]^{1/2} + 1\right\} \\ k_i^2 &= \frac{\mu\varepsilon_r\omega^2}{2}\left\{\left[1 + \left(\frac{C_r}{\omega\varepsilon_r}\right)^2\right]^{1/2} - 1\right\} \end{aligned} \tag{5-147}$$

In some applications, either the operating frequency will be sufficiently high or the rock resistive enough that $C_r \ll \varepsilon_r\omega$, and then

$$\begin{aligned} k_r^2 &\simeq \omega^2\mu\varepsilon_r\left(1 + \frac{C_r}{8\varepsilon_r^2\omega^2}\right)^2 \\ k_i^2 &\simeq \frac{C_r^2\mu}{4\varepsilon_r} \end{aligned} \tag{5-148}$$

Attenuation can be expressed in terms of the skin depth $\delta = 1/k_i$. As discussed in Sec. 5-5, δ determines the penetration depth of an electromagnetic wave. From Eq. (5-148),

$$\delta \simeq \frac{2}{C_r}\left(\frac{\varepsilon_r}{\mu}\right)^{1/2} = \frac{\sqrt{\kappa_r}}{188.3C_r} \tag{5-149}$$

Hence, penetration of a propagating wave is dependent upon frequency only implicitly through its dependence on ε and C.

5-7c Dielectric Sondes

To measure ε with a borehole sonde, an electromagnetic wave must be established by a transmitter and the propagation characteristics (k) of the wave sensed by one or more receivers. Two types of sondes have been developed; one uses slot antennas operating at microwave frequencies and the other uses coils.

Slot antennas. A sonde which transmits and receives electromagnetic pulses of frequency 1.1 GHz is constructed with the antenna package mounted on the side of a mandrel which is held against the bore wall by a decentralizing arm. The arrangement of transmitter and receivers is similar to that of the compensated acoustic sonde (Fig. 7-14). As described by Calvert et al. (1977) and Freedman and Vogiatzis (1979), the sonde measures phase difference $\Delta\varphi$ and attenuation ln (E_1/E_2) between two receivers during transmission from one of the two transmitters. [In practice, the phase and attenuation results are averaged from sequential readings obtained from each of the two transmitters. This technique is a refinement which compensates for borehole irregularities (see Fig. 7-14) but does not alter the basic operation of the sonde.]

The phase and amplitude measurements are used to determine the components of the propagation factor, k_r and k_i. The k_r component is determined from phase shift $\Delta\varphi$ by

$$k_r = \frac{\Delta\varphi}{L} \tag{5-150}$$

where L is the separation between the two receivers. The k_i component is obtained from the amplitude attenuation

$$k_i = \frac{1}{L}\ln\left(\frac{E_1}{E_2}\right) - k_s \tag{5-151}$$

The additional term k_s accounts for spreading losses which occur because the sonde does not generate a true plane wave.

To convert k_r and k_i to ε_r, it is possible to use Eq. (5-145) to get

$$\varepsilon_r = \frac{k_r^2 - k_i^2}{\mu\omega^2} \tag{5-152}$$

which can be used in a mixing expression such as Eq. (1-14) or (5-142). Calvert et al., however, find it convenient to express Eq. (5-152) in terms of reciprocal phase velocity $\Delta t_0 = (\mu\varepsilon_r)^{1/2}$, and $\Delta t_l = k_r/\omega$, so that

$$\Delta t_0^2 = \Delta t_l^2 - \frac{k_i^2}{\omega^2} = \frac{k_r^2 - k_i^2}{\omega^2} \tag{5-153}$$

Consequently, the log-based measurements are expressed as nanoseconds per meter rather than relative dielectric constant. The lossless inverse velocity t_0 is used in Eq. (5-143) to compute an apparent porosity.

Freedman and Vogiatzis (1979) point out that k_r determined from Eq. (5-150) is a true value for the formation only for the case of plane-wave propagation. To simulate the wellbore case, they consider the field generated by an electric dipole on a layered model representing the wellbore fluid, the mud cake, and the formation. Their numerical results indicate that measured values of t_ℓ are too high by about 5 percent, depending on electrical parameters and mud cake thickness. They indicate how corrections might be applied to actual well log data.

Sonde with coils. Huchital et al. (1981) describe an experimental sonde which is constructed similarly to an induction sonde (transmitter and receiver coils coaxial with the z axis), but operates in a much higher frequency range of 10 to 30 MHz. At these frequencies, both conduction C_r and dielectric polarization ε_r play a role in the propagation of electromagnetic waves, so that k is defined by Eq. (5-144).

The voltage induced in the receiver coil [Eq. (5-118)] is

$$V_2 = \frac{2i\mu\omega N_0 A_0 N_2 A_2 I_0}{4\pi}(1 - ikL)\frac{e^{ikL}}{L^3} \tag{5-118}$$

If the ratio of the voltage at two receivers spaced L_1 and L_2 from the transmitter is measured,

$$\frac{V_1}{V_2} = \exp\left[ik(L_1 - L_2)\right]\frac{L_2^3}{L_1^3}\frac{1 - ik_1L_1}{1 - ik_2L_2} \tag{5-154}$$

By measuring phase difference and amplitude attenuation, the components k_r and k_i (or, equivalently, ε_r and C_r) can be determined. Huchital states that the depth of investigation is greater for the attenuation measurement than for the phase measurement. To compensate, they use two pairs of receiver coils, which measure phase on the pair farther from the transmitter and amplitude on the closer pair.

Equations (5-118) and (5-154) apply only if both coils are in a homogeneous medium. The actual sonde and wellbore combination can be considered as a series of concentric cylinders each with constant properties ε_r and C_r (Chew, 1982; Yu et al., 1982). For a given sonde geometry, operating frequency, and mud cake properties, the amplitude and phase response can be computed as a function of rock ε_r and C_r. This type of sonde is the newest method of measuring conductivity and dielectric constant.

5-8 MAGNETIC PROPERTIES AND FIELDS

The magnetic field observed in an uncased hole consists of the sum of the normal ambient field of the earth and superimposed fields caused by induced and remanant magnetism. Induced magnetism occurs when a material with magnetic susceptibility is placed in a magnetic field. The induced field has a strength proportional to the susceptibility and the strength of the applied field. The induced field is of the same sign as the applied field, so the total field is increased. Remanant magnetism is that observed in the absence of an applied field. When hot volcanic material cools below its Curie point, the induced field is "frozen" into the formation. The amplitude and direction of the remanant field are proportional to those of the ambient field at the time of the cooling. The direction can be *normal,* that is, in the same direction as today's ambient field, or *reversed,* in the opposite direction.

The earth's magnetic field has long been used to survey borehole declination, the horizontal angle between magnetic north and hole direction (Chap. 17). The earth's field can be used reliably for this purpose in open holes as long as the induced and remanant components are negligible. Because the earth's main dipole is not coincident with its rotational axis, magnetic north generally differs from geographic north, so that magnetic declination must be corrected to true north. The properties of the earth's field are described in many texts; Jacobs (1967) provides a compact summary. The magnitude of the earth's field is about 60,000 γ's, which is equivalent to 60,000 nanotesla (nT).

The earth's magnetic field displays a highly complex structure when mapped with sufficient resolution at the surface. The complex patterns are caused by crustal rocks having sufficient iron content to possess induced or remanant magnetization of their own. The local magnetic fields add vectorially to the earth's main field, and the total perturbed field is mapped and interpreted for local geological structure (Grant and West, 1965). By contrast with surface mapping using ground and airborne magnetometers, there has been little implementation of borehole magnetometers for subsurface exploration. Because of this lack of application, we give only a short description of magnetic data acquisition and interpretation.

Within a borehole it is possible to measure the magnetic properties of rocks directly. Because magnetic susceptibility is controlled by the volume fraction of iron-bearing minerals, usually magnetite, its measurement is of practical use when it can be correlated with iron ore grade or when the presence of magnetite is indicative of geochemical (reduction-oxidation) conditions.

5-8a Magnetic Properties of Minerals and Rocks

As already noted, rock magnetism **M** is the sum of its remanent and induced components $\mathbf{M}_n$ and $\mathbf{M}_i$. The Koenigsberger ratio Q is introduced as a measure of the relative contribution of these two components,

$$Q = \frac{\mathbf{M}_n}{\mathbf{M}_i} = \frac{\mathbf{M}_n}{\chi \mathbf{H}} \tag{5-155}$$

Data cited by Strangway (1967) show that Q can range as high as 100 in basalt flows, can exceed 1 in mafic sills and dikes, and ranges from 0 to 1 in acidic intrusives. Both remanence and susceptibility can vary widely within a rock unit. Remanent polarizations are induced at the time of formation or subsequently during cooling, and hence they are directed along the earth's field lines existent at that time. This property of remanent polarization is utilized in the study of paleomagnetism (McElhinny, 1973).

The magnitude of the induced component is expressed in terms of rock susceptibility $\chi = 1 - \mu$ (cgs) rather than permeability μ, because permeability values are very close to 1.0 for most rocks. Data compiled from a number of sources are shown in Table 5-3. Although dimensionless, values of susceptibility in cgs units are often stated as cgs/cm^3 for a practical reason. The polarization M can be considered as magnetic

TABLE 5-3 Range of magnetic susceptibility (cgs/cm^3) in major rock types. From Lindsley et al. (1966); compilation of data credited to P. M. Hurley.

		Percent of samples with susceptibility			
Rock type	Number of samples	Less than 10^{-4}	Between 10^{-4} and 10^{-3}	Between 10^{-3} and 4×10^{-3}	Greater than 4×10^{-3}
Mafic effusive rocks	97	5	29	47	19
Mafic plutonic rocks	53	24	27	28	21
Granites and allied rocks	74	60	23	16	1
Gneisses, schists, slates	45	71	22	7	0
Sedimentary rocks	48	73	19	4	4

moment per unit volume, and an experimenter determining the magnetic moment of a small sample has the choice of expressing it per unit volume or per unit mass. The nomenclature cgs/cm^3 specifies the volumetric choice.

Most sedimentary rocks can be expected to have values of χ less than 10^{-4} cgs, whereas values for mafic extrusives will generally exceed that value. However, a susceptibility measurement cannot be used to identify a rock, because there is complete overlap of χ values among rock types.

Carmichael (1982) and McElhinny (1973) discuss the different types of magnetic minerals. Rock susceptibility is controlled by a very limited number of iron-rich minerals which are classified as ferrimagnetic: these are pyrrhotite, an iron sulfide, and members of the iron–titanium oxide system. Magnetite is the most common and the most magnetic mineral of the iron–titanium oxide series. The correlation between rock susceptibility and magnetite content is shown in Fig. 5-49. Data collected from a specific site can usually be fit fairly well by a power law

$$\chi = af^b$$

where f is volume fraction of magnetite and the exponent b is a number ranging between 1.0 and 1.4 (Grant and West, 1965). Establishment of a mixing law relating rock χ to mineral χ and V is complicated by the demagnetization effect which occurs because the magnetic field set up by each individual grain opposes the inducing field; that is, grain-to-grain effects cannot be ignored where magnetic phenomena are concerned. Expressions incorporating a demagnetizing factor also provide good fits to specific data sets, but they still rely on empirical parameters to accommodate grain geometry (Strangway, 1967).

5-8b Susceptibility Measurements

Susceptibility logs are acquired by measuring the change in inductance of a multiturn coil. The principle is similar to that used for the measurement of laboratory samples in which an air-core coil occupies one arm of an induction bridge. Insertion of a core sample into the coil changes the coil's self-inductance and unbalances the bridge, thereby producing a signal which is proportional to the susceptibility of the sample. In the borehole sonde, the coil-sample arrangement is "turned inside-out" so that self-inductance is altered by the susceptibility of the surrounding rock.

Coils can be so wound that the magnetic field is aligned either along the borehole or transversely to it. Coils wrapped around a high-perme-

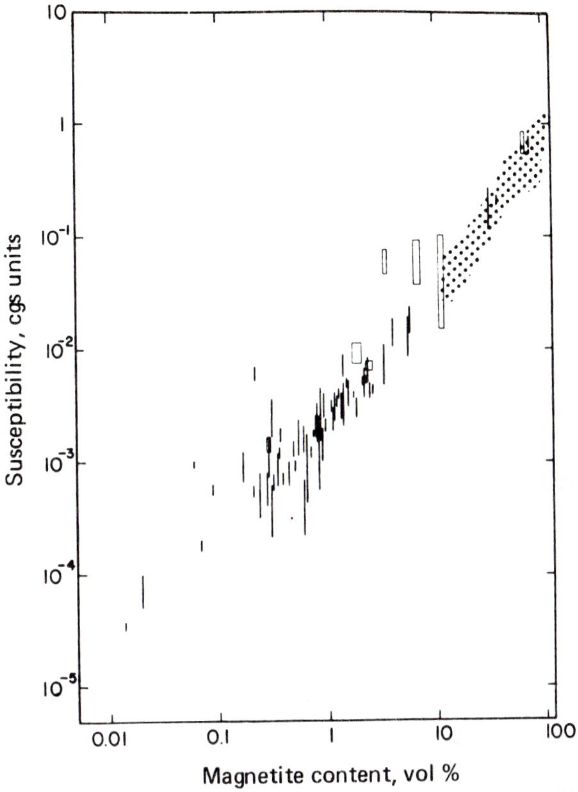

FIGURE 5-49 Relation between susceptibility and magnetite content of a variety of rocks and ores compiled by Lindsley et al. (1966). Vertical lines show range due to anisotropy; box widths show variation among samples.

ability core establish a dipolar field with its axis aligned along the borehole as shown in Fig. 5-50*a*. Broding et al. (1952) related the coil signal to the electromagnetic field energy. The coil is represented by a series resistance $\mathcal{R}$ and inductance $\mathcal{L}$ driven by a constant current I_0. With the sonde suspended in air, the output is nulled by adjusting the bridge. As the coil is moved from air to borehole, $\mathcal{R}$ and $\mathcal{L}$ change because of the interaction of the electromagnetic field with the surrounding rock. The energy supplied to the coil changes by

$$\Delta U \simeq \tfrac{1}{2} I_0^2 (\Delta \mathcal{R} + i\omega \Delta \mathcal{L}) \tag{5-156}$$

The complex notation indicates that the resistive and inductive components are 90° out of phase with one another. These changes in

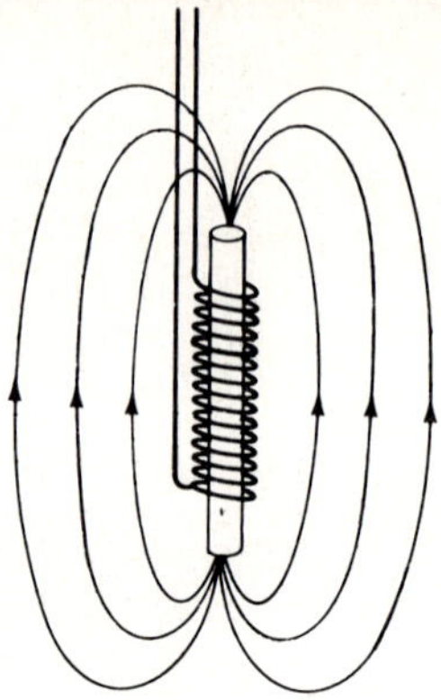

(a)

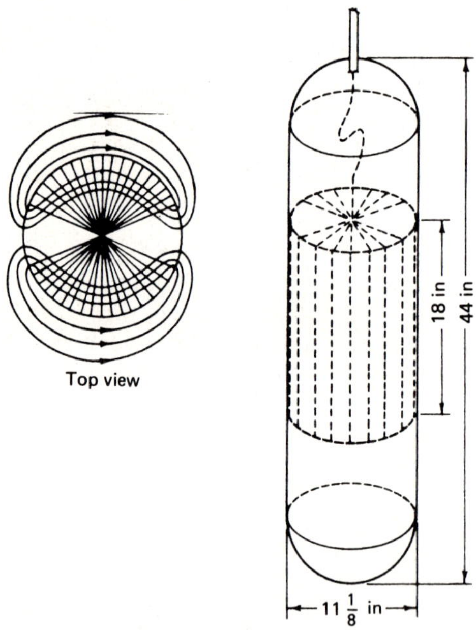

(b)

FIGURE 5-50 Susceptibility sondes. The coil wound on a high-permeability core (*a*) is the type developed by Broding et al. (1952). Rectangular coils (*b*) are arrayed with coil plane vertical. (After Zablocki, 1974.) Heavy lines with arrows indicate character of magnetic field.

coil parameters are shown to be related to the field energy and rock parameters by

$$\Delta\mathcal{R} = -\frac{1}{2}\frac{C}{I_0^2}\int E^2\,dv$$
$$\Delta\mathcal{L} = \frac{1}{2}\frac{\mu_0\chi\omega}{I_0^2}\int H^2\,dv \qquad (5\text{-}157)$$

Broding et al. evaluated the integrals in Eq. (5-157) by adopting a static approximation to the electromagnetic wave equation. Their solution showed that the radius of influence of the sonde (90 percent of signal) extended to a radius of one-half coil length for the susceptibility measurement and to a radius of 1.5 coil lengths for the conductivity measurement. In practice, χ is obtained from $\Delta\mathcal{L}$ by calibrating in formations of known χ, rather than by evaluating Eq. (5-157).

To avoid thermal drift, the sensing coil must be housed in an oven controlled by a thermostat to maintain a constant temperature greater than the maximum temperature in the well. Scott et al. (1981) describe the design improvements in a thermally stabilized sonde which achieves a system drift of less than 10×10^{-6} cgs susceptibility units per hour and an electronic noise level less than 1×10^{-6} cgs units.

Coils can also be arranged to produce a horizontal magnetic field in the formation. Zablocki (1974) points out that a horizontal field is preferable for anisotropic, horizontally bedded iron-rich formations. By packing the windings in a semitoroid (Fig. 5-50*b*), field intensities are made adequate for highly magnetic formations. Zablocki found that the nonlinear relation between magnetite content and susceptibility was compensated by a second nonlinear relation of reverse curvature between coil response and true susceptibility. The result was a highly linear correlation between sonde output and magnetite content.

5-8c Magnetic Field Measurements

Borehole magnetometers. Both proton precession magnetometers (Douglas and Millett, 1978) and fluxgate sensors (Hood and Dyck, 1975) have been incorporated into sondes. The principle of proton precession is described in more detail in the next subsection, where it is utilized as a rock property rather than a sensor. Precession magnetometers perturb the alignment of proton magnetic moment with the earth's field by applying a local field to a sensor containing water (a good source of protons). When the local field is removed, the protons precess about the earth's field at a frequency characteristic of the magnitude of the field. Hence, a magnetometer based on this principle provides an absolute field measurement. It is fairly insensitive to misalignment between sensor direction and field direction, although a single sensing coil does possess dead spots. Dead spots are eliminated by adding a second orthogonal coil or by winding a toroidal sensing coil (Hood and Ward, 1969).

A fluxgate sensor consists of a highly permeable core wrapped with a bias coil and a main coil (Hood and Ward, 1969). A steady current is applied to the bias coil to null most of the earth's field. An alternating current applied to the main coil produces a field large enough to alternately drive the core into positive and negative saturation and thereby produce a distorted field B in the core. The resulting differential voltage is proportional to the residual field not canceled by the bias coil. Fluxgate sensors are highly directional; the magnitude and direction of the residual field can be determined by deploying three orthogonal fluxgate sensors.

Interpretation. Interpretation of magnetic data requires the calculation of the fields produced by rock units possessing induced and remanent magnetization. Bhattacharya (1964) formulates the problem as follows: The observation point is at x, y, z, and a small volume of material of magnetic polarization M is located at x_0, y_0, z_0. The direction of M is given by direction cosines i, j, k, and that of the earth's field by l, m, n. From potential theory, the field due to the volume element is

$$d\mathbf{B} = \mathbf{M} \frac{\partial^2}{\partial s \, \partial \ell} \frac{1}{r} \, dx_0 \, dy_0 \, dz_0 \tag{5-158}$$

where

$$r^2 = (x - x_0)^2 + (y - y_0)^2 + (z - z_0)^2$$

and ∂s is an elemental length in the direction of the polarization vector, so that

$$\frac{\partial}{\partial s} = i \frac{\partial}{\partial x} + j \frac{\partial}{\partial y} + k \frac{\partial}{\partial z}$$

and $\partial \ell$ is an elemental length in the direction of the earth's field, so that

$$\frac{\partial}{\partial \ell} = l \frac{\partial}{\partial x} + m \frac{\partial}{\partial y} + k \frac{\partial}{\partial z}$$

The magnetic field produced by finite bodies is obtained by integrating Eq. (5-158) over the magnetic volume of interest. Grant and West (1965) give formulas for the vertical field component over a two-dimensional step (representing a fault), a ribbon (representing a dike or vein), and a tabular model. Bhattacharya (1964) gives an expression for the total field caused by prism-shaped bodies with vertical sides. By varying the direction and magnitude of the remanent polarization, a wide variety of magnetic fields can be produced. From such expressions the components of the total magnetic field in a nearby borehole can be computed.

It is desirable, however, to locate and characterize a magnetic body from a set of borehole data. We then face the problem of "inverting" the set of borehole data. Silva and Hohmann (1981) have addressed this problem by using the prism-based analytical model of Bhattacharya to calculate the field in a vertical borehole located three lengths from the center of a body measuring 2 by 3 by 6 lengths. Starting with an initial guess of prism parameter and location, they iteratively adjust the model parameters until the computed field agrees with the "measured" one. They examine the behavior of the iterative process with different initial guesses, add noise and errors to the "measurements," and reduce the data density. They conclude that a prism can be resolved to any desired accuracy if the data are noise-free. In the presence of noise or measurement errors, resolution degrades but the geometry and position of the prism can still be determined fairly well.

5-8d Nuclear Magnetism

Although neutron logs (Sec. 6-6) measure total hydrogen content, another method, nuclear magnetic resonance, can sometimes be used to measure the amount of hydrogen that is not bound chemically or to surfaces. Many atomic nuclei spin on their axes and consequently have magnetic moments. Because the atoms and molecules containing these nuclei are surrounded by other atoms and molecules, they are constantly colliding in thermal motion and therefore the directions of the spins and magnetic moments are random. The earth's magnetic field is not strong enough to align the spins of most materials. If, however, a strong enough field is applied, the nuclei tend to align in the direction of the field. A certain amount of time, called T_1, is required for this alignment. Of course, not all the nuclei are aligned with the field, but enough are to produce a net magnetization. The hydrogen nucleus (a proton) is magnetic; and if a field is applied, the hydrogen nuclei present in the material will tend to align with it.

If the field is applied approximately perpendicular to the earth's field for a time large compared to T_1, the hydrogen nuclei will tend to align perpendicular to the earth's field. If the applied field is then suddenly switched off, the thermal motion will take over, and the hydrogen then return to a disordered state with a characteristic *relaxation time*

T_2. In the process, the nuclei will precess about the direction of the earth's field, in the manner of a precessing gyroscope in a gravitational field, with a frequency (the Larmor frequency) that depends on the strength of the field. If a coil is nearby, the precessing nuclei will cause a change in magnetic flux in the coil, and they will therefore induce an alternating voltage at the same frequency. A decay curve from water-saturated sandstone is shown in Fig. 5-51.

Brown and Gamson (1960) developed a logging sonde based on an elongated vertical coil in a borehole. A large direct current was passed through the coil, and a horizontal field was induced. Then the current was switched off, and the same coil was used to sense the precession signal. The received signal was a damped sinusoid with a frequency of about 2 kHz and a time to decay by a factor of e, T_2, of about 50 ms.

In fact, T_2 is a function of the state of the material. T_2 for hydrogen in a solid is very short, and for hydrogenous fluids it can vary from tens to hundreds of milliseconds (Devereux, 1967) depending on the fluid's viscosity and its interaction with the surfaces nearby. T_2 was not used to obtain the amount of hydrogen; instead, a parameter called the free fluid index (FFI) was used. The FFI was assumed to be proportional, under ideal conditions, to the number of hydrogen atoms free to decay

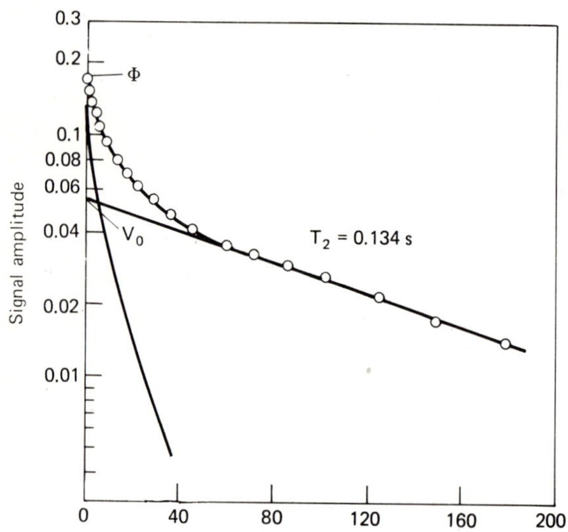

FIGURE 5-51 Magnetic thermal relaxation measurement for water protons contained in the pores of a consolidated sandstone. (From Seevers, 1966.)

with the T_2 characteristic of the fluid. It was defined (Hull and Coolidge, 1960) as

$$\text{FFI} = \frac{V_0}{KG} \tag{5-159}$$

where V_0 = the induced voltage at the beginning of the exponential decay

K = an environmental factor depending on temperature, angle of the borehole with the earth's field, and the strength of the earth's field at the point in question

G = a geometric and tool factor depending on coil geometry, hole diameter, coil current, and position of the coil in the hole

In practice, FFI is scaled to porosity, so that it represents the fraction of the total porosity ϕ_f which is filled with "free" water contributing to the decay. Then we can write

$$\text{FFI} = \phi_f \tag{5-160}$$

Since the largest signal would ordinarily come from the large amount of hydrogen in the borehole fluid, magnetic material was mixed with the fluid so its signal would decay very rapidly. Then the observed signal decay was extrapolated back to zero time to obtain V_0 (Fig. 5-51). Praxmayer (1967) calculated the range of investigation to be approximately equal to the diameter of the borehole.

Unfortunately, the FFI was quite difficult to identify with specific physical quantities. A number of applications were proposed, but the method did not catch on and was almost abandoned. Recently an improved version of the system has been developed (Herrick et al., 1979) and several new applications have been proposed (Cowgill et al., 1980; Neumann and Brown, 1981). Jackson et al. (1980) have developed a method that might make it possible to control the distance to the sensed region and improve the signal-to-noise ratio. With these new developments, nuclear magnetism may become a more commonly used technique.

REFERENCES

Alger, R. P.: "Interpretation of Electric Logs in Fresh Water Wells in Unconsolidated Formations," *Trans. 7th Ann. Logging Symp. SPWLA,* 1966. Reprinted in *Well Logging,* SPE, 1971.

Allaud, L. A., and M. H. Martin: *Schlumberger: The History of a Technique*, Wiley, New York, 1977.

Angoran, Y., and T. R. Madden: "Induced Polarization: A Preliminary Study of Its Chemical Basis," *Geophysics*, vol. 42, 1977, pp. 788–803.

Archie, G. E.: "The Electrical Resistivity Log as an Aid in Determining Some Reservoir Characteristics," *Trans. AIME*, vol. 146, 1942, pp. 54–62.

Arps, J. J.: "The Effect of Temperature on the Density and Electrical Resistivity of Sodium Chloride Solutions," *Petr. Trans. AIME*, vol. 198, 1953, pp. 327–330.

Bhattacharya, B. K.: "Magnetic Anomalies Due to Prism-Shaped Bodies with Arbitrary Polarization," *Geophysics*, vol. 29, 1964, pp. 517–531.

Brace, W. F., A. S. Orange, and T. R. Madden: "The Effect of Pressure on the Electrical Resistivity of Water Saturated Rocks," *J. Geophys. Res.*, vol. 70, no. 22, 1965, pp. 5669–5678.

——— and ———: "Further Studies of the Effects of Pressure on Electrical Resistivity of Rocks," *J. Geophys. Res.*, vol. 73, no. 16, 1968, pp. 5407–5420.

Bradshaw, A., and K. E. Schleicher: "The Effect of Pressure on the Electrical Conductance of Sea Water," *Deep Sea Res.*, vol. 12, 1965, pp. 151–162.

Broding, R. A., C. W. Zimmerman, E. V. Somers, E. S. Wilhelm, and A. A. Stripling: "Magnetic Well Logging," *Geophysics*, vol. 17, 1952, pp. 1–26.

Brown, R. J. S., and B. W. Gamson: "Nuclear Magnetism Logging," *Petr. Trans. AIME*, vol. 219, 1960, p. 201.

Calvert, T. J., R. N. Rau, and L. E. Wells: "Electromagnetic Propagation—A New Dimension in Logging," Soc. Petr. Eng. Paper 6542, 47th Ann. Calif. Reg. Meeting, 1977.

Carmichael, R. S.: "Magnetic Properties of Minerals and Rocks," in R. S. Carmichael (ed.), *Handbook of Physical Properties of Rocks*, Vol. II, CRC Press, Boca Raton, Fla., 1982.

Chew, W. C.: "Response of the Deep Propagation Tool in Invaded Boreholes," Soc. Petr. Eng. Paper 10989, 57th Ann. Fall Tech. Conf., 1982.

Clavier, C., A. Heim, and C. Scala: "Effect of Pyrite on Resistivity and Other Logging Measurements," *Trans. 17th Ann. Symp. SPWLA*, 1976, pp. HH1–HH34.

———, G. Coates, and J. Dumanoir: "The Theoretical and Experimental Bases for the 'Dual Water' Model for the Interpretation of Shaly Sands," Soc. Petr. Eng. Paper 6859, 52d Ann. Fall Tech. Conf., 1977.

Cowgill, D. F., J. K. Pitman, D. O. Seevers, and T. L. Dobecki: "Application of NMR to Characterization of Low Permeability Sandstone Reservoirs," Soc. Expl. Geophys. Ann. Meeting, 1980.

Dakhnov, V. N.: "Geophysical Well Logging," English trans., *Quart. Colo. School Mines*, vol. 57, no. 2, April 1962.

Devereux, O. F.: "Effect of Crude Oil on the Nuclear Magnetic Relaxation of Water Protons in Sandstone," *Nature,* vol. 215, 1967, p. 614.

Dickey, P. A.: "Increasing Concentration of Subsurface Brines with Depth," in "Geochemistry of Subsurface Brines," *Chem. Geol.,* vol. 4, no. 1–2, 1969, pp. 361–370.

Doherty, L. H.: "Electrical Conductivity of the Great Lakes," *J. Res. Natl. Bur. Std.,* vol. 67D, no. 6, 1963, pp. 767–771.

Doll, H. G.: "The SP Log: Theoretical Analysis and Principles of Interpretation," *Trans. AIME,* vol. 179, 1949(*b*), pp. 146–185. Reprinted in SPE Reprint Series No. 1, 1971.

———: "Introduction to Induction Logging and Application to Logging of Wells Drilled with Oil Base Mud," *Petr. Trans. AIME,* vol. 186, June 1949*a*, pp. 148–162.

———: "The Microlog—A New Electrical Logging Method for Detailed Determination of Permeable Beds," *Trans. AIME.,* vol. 187, 1950, pp. 155–164. Reprinted in SPE Reprint Series No. 1, 1971.

———: "The Laterolog: A New Resistivity Logging Method with Electrodes Using an Automatic Focusing System," *Trans. AIME,* vol. 192, 1951, pp. 305–316. Reprinted in SPE Reprint Series No. 1, 1971.

———: "The Microlaterolog," *Trans. AIME,* vol. 198, 1953, pp. 17–32. Reprinted in SPE Reprint Series No. 1, 1971.

Dorsey, N. E.: *"Properties of Ordinary Water Substance,"* Reinhold, New York, 1940.

Douglas, A. C., and M. R. Millett: "Total Intensity Magnetometer Logging as a Stratigraphic Tool in Tertiary Volcanic Rock," Lawrence Livermore Lab. Rept. UCRL–52617, 1978.

Dresser Atlas: *Log Review 1,* Dresser Atlas, Houston, 1974.

Drinkrow, R. L., and R. H. Duffin: "Scale Model Results for Inductive Logging in the Region of Single and Multiple Conductive Bodies," *Geophysics,* vol. 43, no. 4, 1978, pp. 804–818.

Duesterhoeft, W. C., Jr.: "Propagation Effects in Induction Logging," *Geophysics,* vol. 26, no. 2, 1961, pp. 192–204.

Evers, J. F., and B. G. Iyer: "Quantification of Surface Conductivity in Clean Sandstones," *Trans. 16th Ann. Logging Symp. SPWLA,* 1975, pp. L1–L11.

Freedman, R., and J. P. Vogiatzis: "Theory of Microwave Dielectric Constant Logging Using the Electromagnetic Wave Propagation Method," *Geophysics,* vol. 44, no. 5, 1979, pp. 969–986.

Freeze, R. A., and J. A. Cherry: *Groundwater,* Prentice-Hall, Englewood Cliffs, N.J., 1979.

Fuller, B. D., and S. H. Ward: "Linear System Description of the Electrical Parameters of Rocks," *IEEE Trans. Geosci. Elect.,* vol. GE-8, no. 1, 1970, pp. 7–18.

Gancy, A. B.: "Aqueous Solutions under Extreme Conditions. II. High Temperatures," in R. A. Horne (ed.), *Water and Aqueous Solutions,* Wiley-Interscience, New York, 1972, chap. 19.

Gianzero, S. C., and R. Rau: "The Effect of Sonde Position in the Hole on Responses of Resistivity Logging Tools," *Geophysics,* vol. 42, no. 3, 1977, pp. 642–654.

——— and B. Anderson: "A New Look at Skin Effect," *Log Analyst,* vol. 23, no. 1, January–February, 1982.

Gondouin, M., M. P. Tixier, and G. L. Simard: "An Experimental Study on the Influence of the Chemical Composition of Electrolytes on the SP Curve," *J. Petr. Tech.,* 1957. Reprinted in SPE Reprint Series No. 1, 1971.

——— and C. Scala: "Streaming Potential and the SP Log," *Petr. Trans. AIME,* vol. 213, August 1958, pp. 170–179.

Grant, F. S., and G. F. West: *Interpretation Theory in Applied Geophysics,* McGraw-Hill, New York, 1965.

Grant, I. S., and W. R. Phillips: *Electromagnetism,* Wiley, New York, 1975.

Guyod, H.: "Electrical Well Logging," *Oil Weekly,* August–December 1944. Series of 16 articles. Reprinted as Bulletin A132 by Welex, Fort Worth, Tex.

———: "Electric Analogue of Resistivity Logging," *Geophysics,* vol. 20, no. 3, 1955, pp. 615–629.

Hasted, J. B.: "Liquid Water: Dielectric Properties," in F. Franks (ed.), *Water—A Comprehensive Treatise,* Vol. 1, Plenum, New York, 1972, chap. 7.

Helgeson, H. C., and D. H. Kirkham: "Theoretical Prediction of the Thermodynamic Behavior of Aqueous Electrolytes at High Pressures and Temperatures," Part I, *Am. J. Sci.,* vol. 274, December 1974, pp. 1089–1198.

Hill, H. J., and A. E. Anderson: "Streaming Potential Phenomena in SP Log Interpretation," Soc. Petr. Eng. 33rd Ann. Fall Meeting Paper 1119-G, 1958.

Hood, P., and S. H. Ward: "Airborne Geophysical Methods," in H. E. Landsberg and J. Van Mieghem (eds.), *Advances in Geophysics,* Vol. 13, Academic, New York, 1969.

Hood, P. J., and A. V. Dyck: "Magnetic Drillhole Measurements in Mineral Exploration," in *Borehole Geophysics Applied to Metallic Mineral Prospecting: A Review,* Geol. Survey Canada, 1975 Paper 75-31, pp. 35–38.

Huchital, G. S., R. Hutin, Y. Thoraval, and B. Clark: "The Deep Propagation Tool (A New Electromagnetic Logging Tool), Soc. Petr. Eng. Paper 10988, 56th Ann. Fall Tech. Conf., 1981.

Hull, P., and J. E. Coolidge: "Field Examples of Nuclear Magnetism Logging," *Proc. Soc. Prof. Well Log Analysts First Logging Symp.,* 1960; *J. Petr. Tech.,* August 1960.

Jackson, J. A., L. J. Burnett, and J. F. Harmon: "Remote NMR Well Logging," *Soc. Expl. Geophys. Meeting,* 1980.

Jackson, P. D., D. Taylor-Smith, and P. N. Stanford: "Resistivity-Porosity-Particle Shape for Marine Sands," *Geophysics,* vol. 43, no. 6, 1978, pp. 1250–1268.

Jacobs, J. A.: "The Earth's Magnetic Field," in *Mining Geophysics,* Vol. II. *Theory,* Soc. Expl. Geophys., Tulsa, Okla., 1967.

Keller, G. V.: "'Electrical Properties of Rocks and Minerals," sec. 26 in S. P. Clark, Jr. (ed.), *Handbook of Physical Constants,* Geol. Soc. Am. Memoir 97, 1966.

————: "Electrical Properties of Rocks and Minerals," in R. S. Carmichael (ed.), *Handbook of Physical Properties of Rocks,* CRC Press, Boca Raton, Fla., 1982, chap. 3.

———— and F. C. Frischknecht: *Electrical Methods in Geophysical Prospecting,* Pergamon, Elmsford, N.Y., 1966.

Klein, J. D., and R. T. Shuey: "Nonlinear Impedance of Mineral-Electrolyte Interfaces," Part I, "Pyrite," *Geophysics,* vol. 43, no. 6, 1978, pp. 1222–1234.

Kunz, K. S., and J. H. Moran: "Some Effects of Formation Anisotropy on Resistivity Measurements in Boreholes," *Geophysics,* vol. 23, no. 4, 1958, pp. 770–794.

Lindsley, D. H., G. E. Andreasen, and J. R. Balsley: "Magnetic Properties of Rocks and Minerals" in S. P. Clark, Jr. (ed.), *Handbook of Physical Constants,* Geol. Soc. Am. Memoir 97, 1966.

Logan, J.: "Estimation of Electrical Conductivity from Chemical Analyses of Natural Waters," *J. Geophys. Res.,* vol. 66, no. 8, 1961, pp. 2479–2483.

Logn, O.: "Mapping Nearly Vertical Discontinuities by Earth Resistivities," *Geophysics,* vol. 19, no. 4, 1954, pp. 739–760.

Lynch, E. J.: *Formation Evaluation,* Harper & Row, New York, 1962.

MacInnes, D. A.: *The Principles of Electrochemistry,* Dover, New York, 1961.

McElhinny, M. W.: *Palaeomagnetism and Plate Tectonics,* Cambridge, New York, 1973.

Meador, R. A., and P. T. Cox: "Dielectric Constant Logging: A Salinity Independent Estimation of Formation Water Volume," SPE Paper 5504, 50th Ann. Fall Meeting, 1975.

Mitchell, J. K.: *Fundamentals of Soil Behavior,* Wiley, New York, 1976.

Moran, J. H.: "Comments on 'New Results in Resistivity Well Logging'," *Geophys. Prospecting,* vol. 24, no. 2, 1976, pp. 401–402.

———— and K. S. Kunz: "Basic Theory of Induction Logging and Application to Study of Two-Coil Sondes," *Geophysics,* vol. 26, no. 6, 1962, pp. 829–858.

———— and S. Gianzero: "Effects of Anisotropy on Resistivity-Logging Measurements," *Geophysics*, vol. 44, no. 7, 1979, pp. 1266–1286.

Nelson, P. H., and G. D. Van Voorhis: "Estimation of Sulfide Content from Induced Polarization Data," *Geophysics*, vol. 48, no. 1, 1983, pp. 62–75.

Neuman, C. H., and R. J. S. Brown: "Applications of Nuclear Magnetism Logging to Formation Evaluation," Soc. Petr. Eng. Paper SPE 10108, 1981.

Olhoeft, G. R.: "Electrical Properties of Rocks," in Y. S. Touloukian and C. Y. Ho (eds.), *Physical Properties of Rocks and Minerals*, McGraw-Hill, New York, 1981, chap. 9.

Parkhomenko, E. I.: *Electrical Properties of Rocks*, Plenum, New York, 1967.

Pirson, S. J.: *Geologic Well Log Analysis*, Gulf Publishing Co., Houston, Tex., 1977.

Poley, J. Ph., J. J. Nooteboom, and P. J. deWaal: "Use of V.H.F. Dielectric Measurements for Borehole Formation Analysis," *Log Analyst*, May–June 1978, pp. 8–30.

Praxmayer, T.: "Range in Nuclear Magnetic Logging," *Acta Geophys. Polonica*, vol. 15, 1967, p. 105.

Purcell, E. M.: *Electricity and Magnetism*, McGraw-Hill, New York, 1965.

Rau, R. N., and R. P. Wharton: "Measurement of Core Electrical Parameters at UHF and Microwave Frequencies," Soc. Petr. Eng. Paper 9380, 55th Ann. Fall Tech. Conf., 1980

Rink, M., and J. R. Schopper: "Interface Conductivity and Its Implications to Electric Logging," *Trans., 15th Ann. Logging Symp. SPWLA*, 1974, pp. J1–J15.

Roy, A.: "New Results in Resistivity Well Logging," *Geophys. Prospecting*, vol. 23, no. 3, 1975, pp. 426–448.

————: "Focused Resistivity Logs," in A. A. Fitch (ed.), *Development in Geophysical Exploration Methods*, Vol. III, Applied Science, London, 1982.

Sarma, V. V. J., and C. S. Rao: "Electrical Conductivity of Rain Water at Visakhapatnam, India," *J. Geophys. Res.*, vol. 77, no. 12, 1972, pp. 2197–2200.

Sato, M., and H. M. Mooney: "The Electrochemical Mechanism of Sulfide Self Potentials," *Geophysics*, vol. 25, 1960, pp. 226–249.

Schlumberger, C., and M. Schlumberger: "Electrical Logs and Correlations in Drill Holes," *Mining Metallurgy*, vol. 10, no. 275, 1929, pp. 515–518.

Schlumberger Ltd.: *Interpretation Handbook for Resistivity Logs*, Document 4, Schlumberger, New York, 1950.

————: *Resistivity Departure Curves (Beds of Infinite Thickness)*, Document 7, Schlumberger, New York, 1955.

————: *Log Interpretation:* Vol. 1, *Principles*, Schlumberger, New York, 1972.

Schmidt, G. W.: "Interstitial Water Composition and Geochemistry of Deep

Gulf Coast Shales and Sandstones," *Bull. AAPG*, vol. 57, no. 2, 1973, pp. 321–337. Reprinted in AAPG Reprint Series no. 11, Abnormal Subsurface Pressure.

Seevers, D. O.: "A Nuclear Magnetic Method for Determining the Permeability of Sandstones," *Proc. Soc. Prof. Well Log Analysts 7th Logging Symp.*, 1966.

Segesman, F.: "New SP Correction Charts," *Geophysics*, vol. 27, no. 6, 1962, pp. 815–828.

Sen, P. N., C. Scala, and M. H. Cohen: "A Self-Similar Model for Sedimentary Rocks with Application to the Dielectric Constant of Fused Glass Beads," *Geophysics*, vol. 46, no. 5, 1981, pp. 781–795.

Shuey, R. T.: *Semiconducting Ore Minerals*, Elsevier, New York, 1975.

Silva, J. B. C., and G. W. Hohmann: "Interpretation of Three-Component Borehole Magnetometer Data," *Geophysics*, vol. 46, no. 12, 1981, pp. 1721–1731.

Strangway, D. W.: "Magnetic Characteristics of Rocks," in *Mining Geophysics*, Vol. II, Soc. Expl. Geophys., Tulsa, Okla., 1967.

Stratton, J. A.: *Electromagnetic Theory*, McGraw-Hill, New York, 1941.

Swanson, B. F.: "Visualizing Pores and Nonwetting Phase in Porous Rock," *J. Petr. Tech.*, vol. 31, no. 1, 1979, pp. 10–18.

———: "Rationalizing the Influence of Crude Wetting on Reservoir Fluid Flow with Electrical Resistivity Behavior," *J. Petr. Tech.*, vol. 32, no. 8, 1980, pp. 1459–1464.

Talash, A. W., and P. B. Crawford: "An Improved Method for Calculating Water Resistivities from Chemical Analyses," *J. Petr. Tech.*, vol. 17, no. 12, 1965, pp. 1396–1398.

Thadani, S. G., and H. E. Hall, Jr.: "Propagated Geometric Factors in Induction Logging," *Proc. Soc. Prof. Well Log Analysts 22d Logging Symp.*, 1981, paper WW.

Waxman, M. H., and L. J. M. Smits: "Electrical Conductivities in Oil-Bearing Shaly Sands," *Soc. Petr. Eng. J., Trans. AIME*, vol. 243, June 1968, pp. 107–122. Reprinted in *Well Logging*, SPE, 1971.

——— and E. C. Thomas: "Electrical Conductivities in Shaly Sands. I. The Relation between Hydrocarbon Saturation and Resistivity Index. II. The Temperature Coefficient of Electrical Conductivity," *J. Petr. Tech.*, February 1974, pp. 213–225.

Winsauer, W. O., H. M. Shearin, Jr., P. H. Mason, and M. Williams: "Resistivity of Brine-Saturated Sands in Relation to Pore Geometry," *Bull. AAPG*, vol. 36, no. 2, 1952, pp. 253–277.

Wyllie, M. R. J.: "A Quantitative Analysis of the Electrochemical Component of the SP Curve," *J. Petr. Tech.*, vol. 1, 1949, pp. 17–27. Reprinted in SPE Reprint Series No. 1, 1971.

———: *The Fundamentals of Well Log Interpretation,* Academic, New York, 1963.

Yu, J. S., P. C. Reardon, and P. C. Lysne: "Dielectric Logging with Borehole and Filtrate Invasion: A Transfer Impedance with Fast-Convergence Integrations," Sandia Natl Labs., SAND82-1975, UC-32, Albuquerque, N.M., 1982.

Zablocki, C. J.: "Magnetite Assays from Magnetic Susceptibility Measurements in Taconite Production Blast Holes, Northern Minnesota," *Geophysics,* vol. 39, 1974, p. 174.

6

Nuclear Radiation Logging

6-1 INTRODUCTION

It has been found that interactions of two types of radiation: γ-rays and neutrons, can be used to infer properties of interest in logging. Figure 6-1 illustrates most of the standard logging methods and shows the type and energy of the particles involved.

Some minerals contain elements (potassium, thorium, uranium) that are natural sources of γ-radiation. Detection of these γ's indicates the presence of these sources (Fig. 6-1*a*, *b*). This is called *γ-ray logging.*

The number of γ's transmitted through a given thickness of the medium is a function of the density, γ energy, and atomic number Z of the medium. Consequently, the number of γ's detected at some distance from an artificial source can be used to infer these quantities (Fig. 6-1*c*, *d*). This is called *density logging,* or *selective γ-γ logging* or *γ-ray backscattering.*

The number of neutrons passing through a given thickness of the medium depends primarily on the hydrogen content of the medium, so the number of neutrons detected at some distance from a source (Fig.

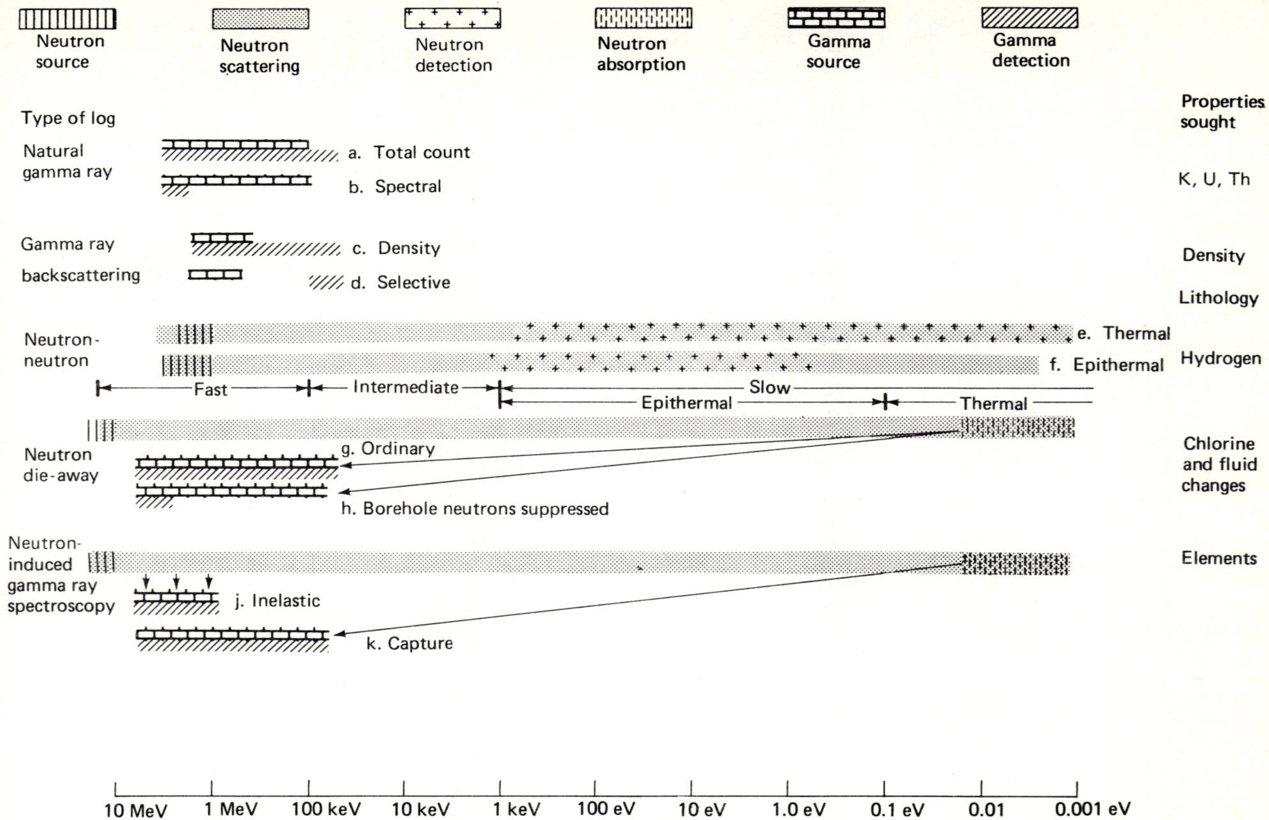

FIGURE 6-1 Schematic of the various types of nuclear radioactivity logging.

6-1*e, f*) can be used to infer the hydrogen content (or water content or porosity). This is called *neutron logging.*

When neutrons interact with matter, γ-rays are often emitted. The number and energy of the γ's are characteristic of the elements from which they are emitted, so they can be used to infer the presence of certain elements. There are two methods used to infer the quantity of certain elements; they are called *neutron die-away logging* (Fig. 6-1*g, h*) and *neutron-induced γ-ray logging* (Fig. 6-1*j, k*) or similar names.

In this chapter these various types of logging will be explained in detail.

6-1a Nuclides and Energy States

A *nuclide* is a specific nuclear species with a given number of protons Z and a given number of neutrons N. Its mass number $A = Z + N$. A nuclide is usually designated by its chemical symbol and mass number; for example, cobalt 60 is written as ^{60}Co. Each nuclide with the same Z but different A is called an *isotope* of the element. For example, ^{59}Co and ^{60}Co are two isotopes of cobalt. A nuclide (or an atom) can exist only in certain definite states each of which is characterized by a definite value of total energy. These states are often referred to as *energy levels.* Each nuclide has its own set of energy levels with energies characteristic of that nuclide. The level with lowest energy is called the *ground state;* the others are called *excited states.**

The energies of the states are often specified with respect to the energy of the ground state or the energies of the ground states of nuclides nearby in the periodic table. Energies are ordinarily measured in *electronvolts;* where one electronvolt is the energy needed to move one electronic charge through a potential difference of one volt. (It is equal to 1.6×10^{-19} J.) Convenient multiples are the kiloelectronvolt (keV) and megaelectronvolt (MeV). Kinetic energy and the total energy of particles and waves are expressed in these same units.

As we shall see, when a nuclide changes from one state to another with lower energy, it often emits a particle or γ-ray. After such a particle or ray has been emitted, it interacts with the matter surrounding its source, loses energy, and is eventually absorbed. Often the number of particles emitted or absorbed as a function of time or position, or the energy of some of these particles, can be used to infer some of the properties of the medium with which the particles are interacting.

*A good reference for these concepts is Wichmann, 1971.

6-1b Decay Processes

Nuclides can be either stable or unstable. If a nuclide is unstable, it will eventually decay by one or more of several possible processes (Meyerhof, 1967):

1 Alpha particle emission. An alpha particle (α) is a helium nucleus (^4He) made up of two protons and two neutrons. Alpha particles can generally be stopped by a sheet of paper.

2 Beta particle emission. A beta particle is either an electron (β^-) or a positron (β^+). A 1-MeV beta particle can be stopped by 2 mm of aluminum.

3 γ-Ray emission. A γ-ray (γ) is a quantum of light, often called a photon. γ-Ray emission occurs either when the nucleus absorbs one of the electrons orbiting the atom or when it changes energy level from an excited state to a lower excited state or the ground state. A 1-MeV γ-ray can penetrate several centimeters of lead. (An x-ray is also a quantum of light, but it is distinguished from a γ-ray by being emitted by the decay of an atom from one state to another rather than by the decay of a nucleus.)

4 Spontaneous fission. Some nuclides will fission spontaneously and, in doing so, emit one or more neutrons and split into two other nuclides called *fission products.* Neutrons can penetrate tens of centimeters of lead. The fission fragments, because of their high mass, can travel only a few micrometers in matter.

5 Neutron emission. If a nuclide is involved in a nuclear reaction, it is transformed into another nuclide. This *compound nucleus* is usually in a high-energy state. It will often decay either back to the original nuclide or to yet a third, and it can emit a neutron in the process.

Because α's and β's cannot penetrate much material, they cannot go very far in a formation or drilling fluid, and they cannot penetrate the housing of a logging sonde. Therefore, only γ's and neutrons are ordinarily used in logging.

When a nuclide decays by α- or β-emission, it is transformed into a different nuclide. If it decays by α-emission, Z and N of the new nuclide are each decreased by two units and A by four. If decay occurs by β^--emission (for example, conversion of a neutron to a proton), then Z is increased by one unit, N is decreased by one unit, and A is unchanged; if by β^+-emission, Z is decreased by one unit, N is increased by one unit, and A is again unchanged. This new nuclide is often left in an excited state. It then decays to the ground state and emits one or more γ-rays with energy equal to the difference in energy of the two states. Sometimes a nuclide in a high excited state will decay directly to the ground state and emit a γ-ray of fairly high energy; sometimes it will decay from state to state and emit a series of γ-rays. This same process often occurs after neutron emission.

Figure 6-2 shows the decay of ^{140}Ba. There can be either β-decay directly to the ground state of ^{140}La with 60 percent probability or β-decay to an excited state with 40 percent probability, followed by γ-decay to the ground state with emission of a 542-keV γ-ray or γ-decay to a series of states with emission of several γ-rays. The energies of the excited states and the probabilities of the different modes of decay are characteristic of the nuclide; consequently, the energies of the γ-rays are also characteristic of the nuclide that emits them. This is the basic principle of identification of elements by γ-ray energies.

Decay rates. The probability for an individual nucleus to decay within a short time interval dt is independent of any external influence such as temperature or pressure. All specimens of a given nuclide have the same probability of decay per unit time. The probability of one nucleus

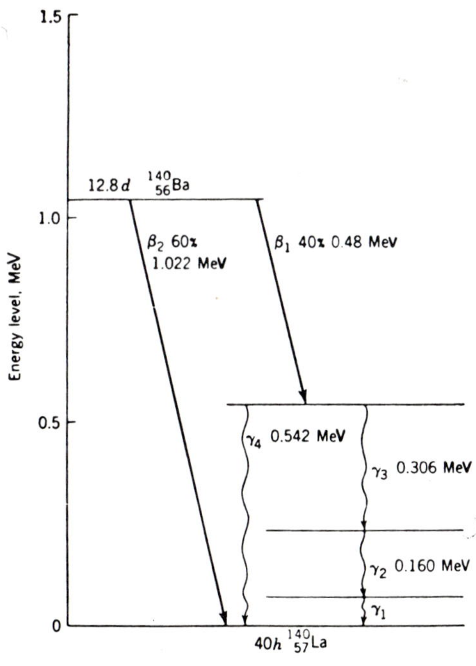

FIGURE 6-2 Decay scheme of ^{140}Ba. (Benedict et al., 1981.)

decaying by a given mode in time dt is $\lambda\, dt$, where λ is called the *decay constant* of the nuclide. Then the number of decays per unit time for a source initially having N_0 nuclei is

$$\frac{dN}{dt} = -\lambda N \qquad \text{and} \qquad N(t) = N_0 e^{-\lambda t} \tag{6-1}$$

When $N(t) = N_0/2$, we call that value of t the *half-life* of the radionuclide $\tau_{1/2}$, and

$$\tau_{1/2} = \frac{0.693}{\lambda}$$

The half-life, like the γ-ray energy, is characteristic of the nuclide. If it is short enough to measure, it can be used to identify the nuclide. The average time of survival of any individual nucleus, called the *mean life* τ, is

$$\tau = \frac{1}{\lambda}$$

We can calculate the *activity* of a sample, or the number of decays per unit time, as

$$\frac{dN}{dt} = \lambda N_0 e^{-\lambda t} \tag{6-2}$$

Hence, the activity, or strength, of a source declines exponentially with time. This is an important consideration in the selection of a radioisotope for use as a source in a logging sonde.

If the resulting nuclide also is radioactive, with a probability of decay λ_2 per unit time, the activity of the second nuclide is

$$\frac{dN_2}{dt} = -\lambda_2 N_2 + \lambda_1 N_1 \tag{6-3}$$

If the product nuclide 3 also is radioactive,

$$\frac{dN_3}{dt} = -\lambda_3 N_3 + \lambda_2 N_2 + \cdots \tag{6-4}$$

If we start with N_{10} nuclei of nuclide 1 and no nuclei of nuclide 2,

$$N_2 = \frac{N_{10}\lambda_1}{\lambda_2 - \lambda_1} [\exp(-\lambda_1 t) - \exp(-\lambda_2 t)] \tag{6-5}$$

We can proceed in the same way for a chain of decays. In such a chain, the first nuclide in any chain is the *parent* and the others are *daughters*. Decay proceeds until all nuclides have decayed to stable nuclides.

If τ of the parent is much longer than τ of any of the daughters and if the material is undisturbed for a long time,

$$\lambda_1 N_1 = \lambda_2 N_2 = \cdot \cdot \cdot \lambda_n N_n \tag{6-6}$$

the chain of n nuclides is said to be in *secular equilibrium* and the activities of all members of the chain are equal.

In geological processes it is not uncommon for some members of a decay chain to be transported into or out of a local volume by fluid movement, which causes the chain to be in disequilibrium. Under these circumstances, one can calculate the amount of any daughter only by solving Eqs. (6-2) to (6-4) subject to known amounts of some of the nuclides.

6-1c Statistics

Because nuclear decay is described by probability, there is a statistical distribution associated with decay which is very useful in analyzing experimental data. We assume that (1) the chance for any individual nucleus to disintegrate in any time interval dt is the same for all nuclei, (2) the probability of a disintegration is the same for any time interval of equal size, (3) the fact that one nucleus has disintegrated has no effect on the chance that other nucleus will disintegrate, and (4) the total number of nuclei and the total time are very large, so the probability that any given nucleus will decay in a specific dt is very small. Based on these assumptions, it can be shown (Evans, 1955) that the probability of x disintegrations in time t, $P_x(t)$, when the average number of disintegrations per unit time is a is

$$P_x(t) = \frac{(at)^x e^{-at}}{x!} = \frac{m^x e^{-m}}{x!} \tag{6-7}$$

where m is the mean number of disintegrations in time t.

This is called the *Poisson distribution,* and it has the characteristics that its variance is equal to its mean. Therefore, the standard deviation of a count of m is equal to $\sqrt{m}$. This means that if the true value of the mean count is m, the observed count will lie within $(m \pm \sqrt{m})$ 68 percent of the time and within $(m \pm 1.96\sqrt{m})$ 95 percent of the time.

Many of the processes we will discuss are characterized by the Poisson distribution. Therefore, we can usually calculate the uncertainty in a measurement to be the square root of the measured value. We can use this ability to decide, for example, how many total counts we need in order to reduce the uncertainty in the count a given fraction of the count. For example, the uncertainty of total count from a γ-ray log in a

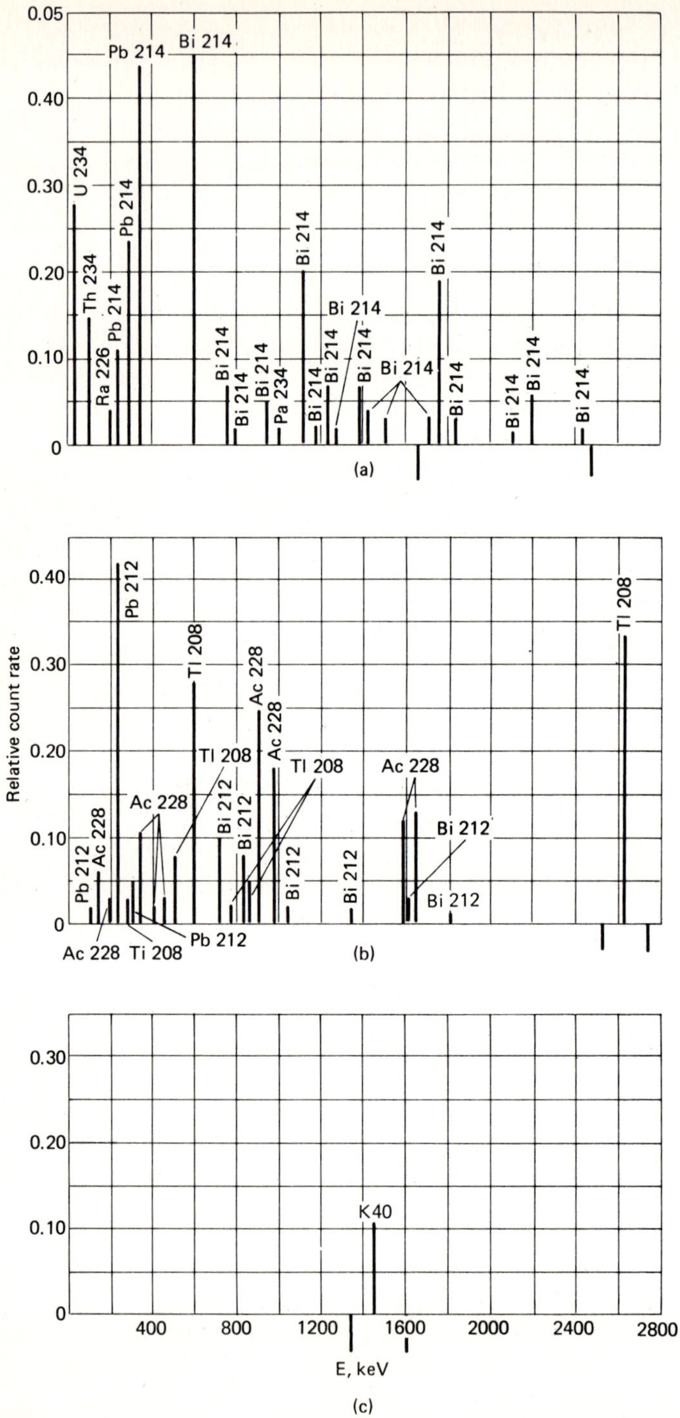

Relative count rate
E, keV
U 234
Th 234
Ra 226
Pb 214
Bi 214
Pa 234
Pb 212
Ac 228
Tl 208
Bi 212
Ti 208
K40
(a)
(b)
(c)

given depth interval is inversely proportional to the square root of logging speed; hence, halving the speed will reduce uncertainty by $\sqrt{2}$.

6-2 γ-RAY LOGGING

6-2a γ-Ray Sources

Natural γ-ray sources. There are three families of natural radionuclides that are common in the earth's crust: the uranium chains, the thorium chain, and potassium. In the case of uranium, two isotopes, ^{238}U and ^{235}U, are present in significant amounts, with ^{238}U making up 99.3 percent of the total.

Each isotope of uranium decays by α-emission to an isotope of thorium. Thorium also is radioactive, as are its daughters, and so a decay chain characteristic of the parent isotope is produced. The half-lives vary widely. The half-life of ^{238}U is orders of magnitude longer than that of any of its radioactive daughters, so in an undisturbed medium there is secular equilibrium. Many of the daughters are left in excited states and produce γ-rays; for example, ^{214}Bi produces a 1.765-MeV γ-ray. Figure 6-3*a* shows the prominent γ-rays produced by ^{238}U. The height of the line is proportional to the relative intensity of the γ-ray; its horizontal position is proportional to the energy.

^{232}Th has a decay chain similar to the decay chains of uranium. Again, many daughters are γ-emitters. Figure 6-3*b* shows the prominent γ's produced by natural thorium.

One of the less common isotopes of potassium, ^{40}K (its abundance is about 0.011 percent) decays to ^{40}Ca and emits a γ-ray of energy 1.46 MeV. It is shown in Fig. 6-3*c*. Because many rock-forming minerals contain potassium, it is the commonest natural γ-ray encountered in logging.

Concentrations of K, U, and Th vary widely in crustal rocks, as shown in Fig. 6-4. When only the total number of γ's is counted, the source isotope is not known; consequently, there is no way of knowing the amount of the isotope producing the observed number of counts. One method of expressing count is in units of equivalent concentration of uranium (eU), the number of γ's that would be emitted by 1 ppm natural uranium in secular equilibrium (International Atomic Energy

←

FIGURE 6-3 Prominent γ-ray energies from (*a*) the ^{238}U decay series, (*b*) the ^{232}Th decay series, and (*c*) ^{40}K. Height of the vertical line is proportional to the relative intensity of the γ-ray. (International Atomic Energy Agency, 1976.)

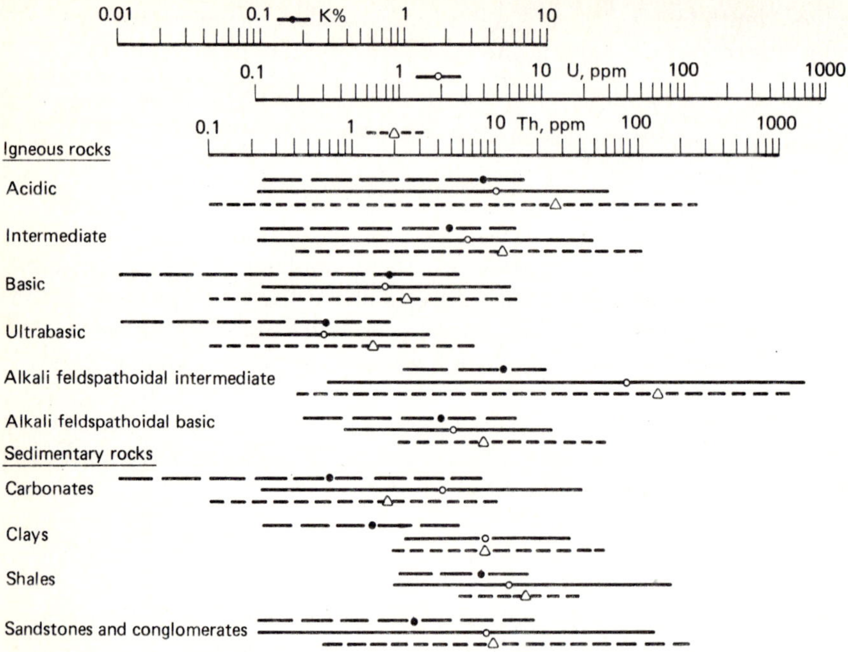

FIGURE 6-4 Range and mean values of K, U, and Th in igneous and sedimentary rocks. Scales have been shifted so that all three elements produce comparable count rates in an NaI detector. For example, 1 ppm U produces the same count rate as 0.95 percent K and 2.13 ppm Th. Left edges of some lines extend to concentrations less than given by scale. (After Wollenberg, 1984.)

Agency, 1976). The concentration scales in Fig. 6-4 have been converted to these units and have been translated so that the concentration along any vertical line gives comparable count rates in an NaI detector. The trends discussed below are based on the mean value denoted by the symbols in Fig. 6-4. The wide range of values measured for each element in each rock type shows that generalizations regarding K, U, and Th abundances will meet with many exceptions.

In the common igneous rocks, here characterized as acidic through ultrabasic, K and U contribute about equally to the total count rate and Th typically contributes about as much radiation as K and U combined. Concentrations of all three elements are about 10 times greater in acidic rocks than in the ultrabasics; the concentrations are in general proportional to SiO_2 content. Exceptions to this trend are found in the alkali feldspathoidal rocks; U and Th concentrations in excess of 100 ppm are not unusual. These alkaline rocks, however, comprise only a small fraction of crustal igneous rocks.

In sedimentary rocks, the relative mean abundances are less predictable than in igneous rocks. On the average, K is lower in effective concentration than U or Th, and Th contributes about the same level of activity as U. As a class, carbonates are the lowest in natural radioactivity of the sedimentary rocks. Most of their radioactivity can be expected to be caused by low concentrations of U. Generally, shales will have a higher level of natural radioactivity than other sediments; consequently, the γ-ray sonde is used to distinguish between shales and other sediments. As Fig. 6-4 shows, there is a great deal of overlap in concentration among the detrital sediments.

Artificial γ-ray sources. There are two commonly used artificial radionuclides. ^{60}Co β-decays to ^{60}Ni and emits two γ-rays of 1.17 and 1.33 MeV. ^{137}Cs β-decays to ^{137}Ba and emits one γ-ray of 663 keV. One chooses the source on the basis of one's energy requirements, and we shall see some of the reasons for a choice later.

6-2b Interaction of γ-Rays with Matter

There are three common ways in which γ-rays having energy of interest to us interact with matter and lose energy: photoelectric effect, Compton effect, and pair production. The energy at which each of these interactions is dominant depends not only on energy but also on the Z of the medium. Figure 6-5 shows the relationship. In high-Z materials

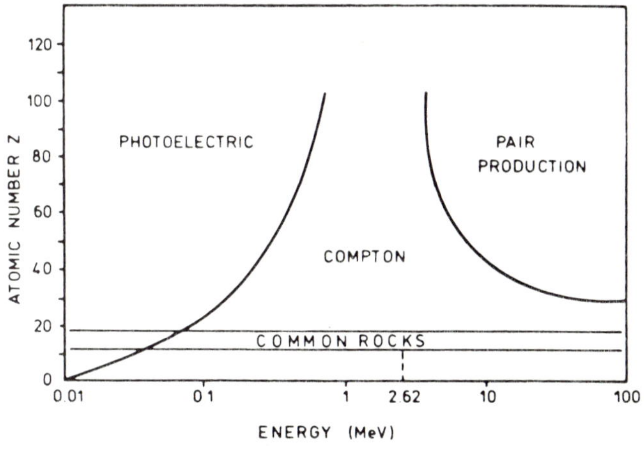

FIGURE 6-5 Relative importance of the three major types of γ-ray interaction. The lines show the values of Z and energy for which the two neighboring effects are just equal. (International Atomic Energy Agency, 1976.)

the photoelectric effect dominates at higher energies than in low-Z materials.

Photoelectric effect. At low energy, the photoelectric effect is the dominant interaction. Orbital electrons of an atom, like nuclides, have discrete energy levels. If the energy of a γ-ray exceeds the energy of such a level (called the *binding energy*), the γ can give up *all* of its energy to an orbital electron. The electron will leave its orbit (the atom will be *ionized*) and have a kinetic energy equal to the difference between its binding energy and the γ-ray energy. Figure 6-6*a* is a con-

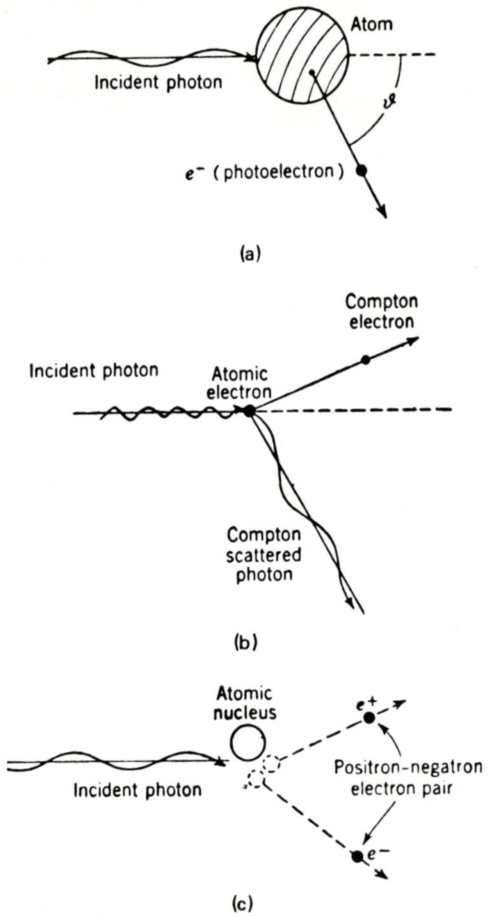

FIGURE 6-6 Conceptual sketches of (*a*) the photoelectric effect, (*b*) Compton scattering, and (*c*) pair production (Evans, 1955.)

ceptual sketch of the process. The interaction with an electron in a given orbit is most probable when the γ energy is exactly equal to the binding energy and decreases with increasing energy. Eventually the electron gives up its kinetic energy to other atoms in the material and drops back into its orbit. When it does so, it emits an x-ray of energy equal to the binding energy. These x-rays, which are characteristic of the medium that is stopping the γ-rays, interact with the medium again and again until all the energy is absorbed either at very low energy orbital levels or by other mechanisms that are not of interest to us.

Compton effect. At intermediate energies the Compton effect dominates. The γ-ray scatters (bounces) from an electron whose binding energy is negligible compared to the γ energy and gives up only *part* of its energy. Figure 6-6*b* is a conceptual sketch. The energy of the scattered γ and the kinetic energy of the electron (called the *Compton electron* or the *recoil electron*) can be calculated from conservation of energy and momentum. The energy of the scattered γ-ray is a function of the angle of scattering. The maximum energy that the electron can acquire is less than the total energy of the γ-ray. (It is, at most, $E_\gamma -$ 225 keV.) The recoiling electron then ionizes atoms in the medium, perhaps emitting x-rays, until it stops, and the scattered γ either is again Compton-scattered, sometimes repeatedly, or is absorbed in a photoelectric interaction.

Pair production. At high energies a third phenomenon takes over. If the γ-ray energy is greater than 1.02 MeV, it can be converted into an electron-positron pair when it is near a nucleus. This is called *pair production*, and it is sketched in Fig. 6-6*c*. Conservation of energy and momentum require that the electron and positron have equal kinetic energy; the kinetic energy of each is half the difference between the γ-ray energy and 1.02 MeV. (1.02 MeV is equal to exactly twice the rest mass, mc^2, of an electron.) The electron slows down as in the case of a Compton electron or photoelectron, but the positron very quickly meets an electron. The positron is antimatter, and when it interacts with an ordinary electron, they annihilate one another and produce two γ-rays. This occurs at essentially zero velocity, so the γ's travel in opposite directions and each has an energy of 0.51 MeV. These annihilation γ-rays then interact by Compton or photoelectric effect until they are absorbed.

Attenuation of γ-rays in bulk material. The probability of interaction of any particle (neutron, γ, etc.) with a single atom or nucleus is called the *cross section* (σ) for that interaction, and it is normally expressed in barns (10^{-24} cm^2). The probability of interaction in a given thickness of material is the *linear attenuation coefficient* $\mu = \sigma n$, where n is the number of atoms per cubic centimeter, so μ is expressed in cm^{-1}. The *mass attenuation coefficient* is μ/ρ, where ρ is the bulk density of the material. Mass attenuation coefficient (expressed in square centimeters per gram) is the most convenient way to express the attenuation properties of a medium because it takes into account all the elemental properties of the medium but permits the user to insert the bulk density. It is then possible to define the mass attenuation coefficient of a mixture of elements as

$$\frac{\mu}{\rho} = \frac{\mu_1 f_1}{\rho_1} + \frac{\mu_2 f_2}{\rho_2} + \cdots \frac{\mu_n f_n}{\rho_n} \tag{6-8}$$

where the f_i are the weight fractions of the elements.

As a γ-ray passes through matter, the individual attenuation processes are independent. The total linear attenuation coefficient μ is

$$\mu = \mu_E + \mu_C + \mu_P$$

where μ_C, μ_E, and μ_p are the linear attenuation coefficients for Compton scattering, photoelectric effect, and pair production, respectively. Mass attenuation coefficients add in the same manner. The fractional number of γ's removed from a beam having an initial count rate of J γ's per second in a distance dx is

$$dJ = -Jn\sigma\, dx = -J\mu\, dx$$

which is similar in form to the law of radioactive decay. And, in a similar manner

$$J = J_0 \exp(-n\sigma x) = J_0 \exp(-\mu x) \tag{6-9}$$

The mean path length, or *mean free path*, of the γ-ray $= 1/\mu$. Thickness of absorbers is sometimes expressed in units of mean free paths or in tenth-value layers. A tenth-value layer is the thickness required to decrease γ intensity by a factor of 10.

Figure 6-7 shows the mass attenuation coefficients for γ-rays in aluminum (which has similar Z and hence attenuation properties similar to those of most rocks) as functions of energy. The sum of the three coefficients is shown as a solid line.

Shielding. Shields are needed to prevent unwanted γ's from entering detectors or people. When γ-rays interact with a shield, the electrons that have been ionized by the γ-ray drop back into their original orbits and emit lower-energy γ-rays or x-rays. These, in turn, must be stopped.

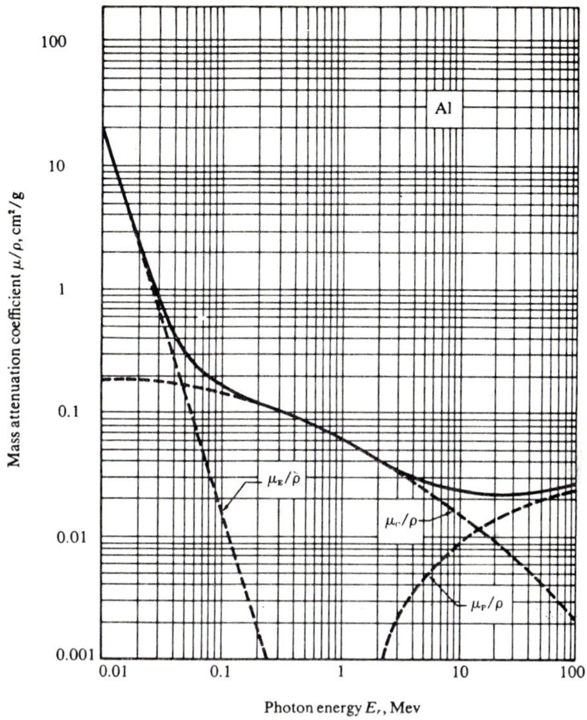

FIGURE 6-7 Mass attenuation coefficient for γ rays in aluminum as a function of γ-ray energy. The coefficients for photoelectric effect (μ_E/ρ), Compton effect (μ_C/ρ), and pair production (μ_p/ρ) are shown separately. To obtain the coefficients in cm^{-1}, multiply by the bulk density. For Al metal, that is 2.70 g/cm^3. (Meyerhof, 1967, after Evans, 1955.)

A procedure for doing so, called a *graded filter*, is to layer materials with very high probability of stopping the x-rays and emitting only x-rays much lower in energy. For example, a lead shield will give rise to many x-rays. A thin tantalum shell between the lead and some detector will absorb those x-rays and emit x-rays much lower in energy. Then an iron shell can be used to absorb those x-rays, and so on.

6-2c γ-Ray Detectors

There are three kinds of γ-ray detectors of interest in logging: gas-filled tubes, scintillators, and semiconductor detectors.

Gas-filled tubes. A gas-filled detector can be either a Geiger-Müller counter or a proportional counter, depending on applied voltage. In either case, the tube is a cylinder with a fine wire (the *anode*) at high positive voltage down the center. The other electrode (the *cathode*) is usually the wall of the tube.

A Geiger-Müller tube is usually filled with a noble gas, and the voltage of the anode is set so that a single event, caused by a γ-ray ionizing wall material or gas, causes an avalanche of ionization and a large pulse of current in the tube. If one plots voltage versus count rate for such a tube, there is a *plateau*, or a range of voltage (in the neighborhood of 1000 V for counters used in logging) at which this phenomenon occurs and at which the count rate is not a function of voltage. The anode voltage is adjusted to the midrange of the plateau. A small amount of another gas, called a *quenching gas*, is added to the noble gas to prevent repeated discharges being caused by electrons emitted from the outer cathode during the first discharge. The Geiger-Müller tube, since it produces a large discharge that is not a function of γ energy, simply counts arriving γ's.

A similar tube operated at a voltage below the plateau and with a different gas mixture can produce a discharge whose pulse height is proportional to the energy of the exciting γ. Tubes are rarely used in this proportional mode in γ-ray logging sondes.

Scintillators. A *scintillator* is a material, most commonly a sodium iodide crystal activated with a small amount of thallium, such that, when ionized by incoming particles, it emits visible light. The amount of light thus emitted is proportional to the energy of the ionizing particle. This light is detected by a photomultiplier tube that is in close optical contact with the crystal. The photomultiplier then produces a pulse whose amplitude is proportional to the energy of the detected particle. A sodium iodide crystal is dense and large enough (commonly 300 cm^3) to stop a fairly large fraction of the photons that enter it; that is, it is fairly *efficient.* Efficiency increases with crystal volume and decreases with photon energy.

Figure 6-8 shows a typical plot of the pulse height versus number of counts from a scintillator-photomultiplier combination. The pulse height is in arbitrary units (which will be explained later). The plot is called a *pulse height spectrum.* It was obtained from the γ-rays from decay of ^{24}Na interacting with the crystal. ^{24}Na emits γ's at 1.368 and 2.75 MeV with equal intensity. If detection were perfect, we would simply see two spikes at pulse heights corresponding to the peaks labeled with those energies. Those peaks are called the *photopeaks,* and they are caused not only by photoelectric effect (a main contributor) but also by Compton scattering followed by other interactions that absorb all of the energy of the scattered γ.

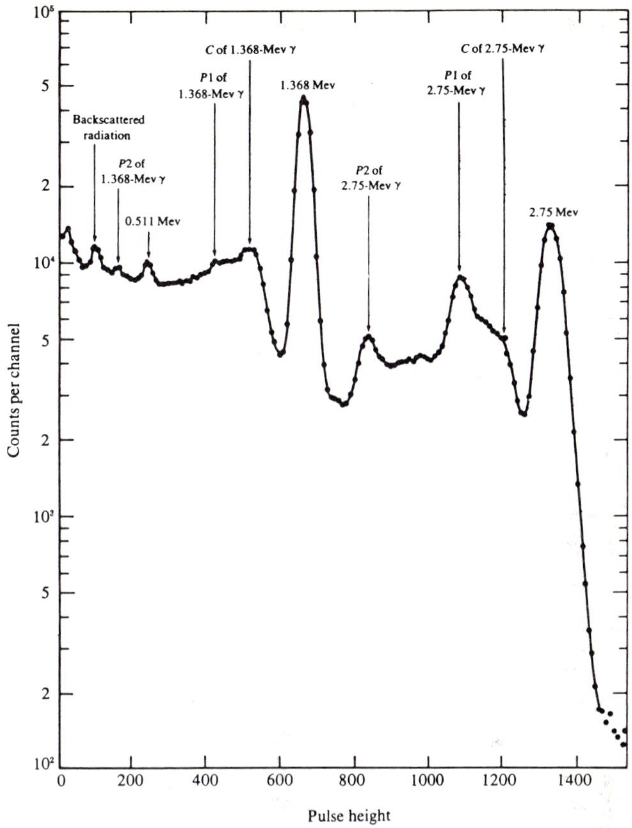

FIGURE 6-8 Pulse height spectrum of 1.368- and 2.75-MeV γ-rays from ^{24}Na in a sodium iodide detector. Detector was 7.6 cm in diameter by 7.6 cm long. (Meyerhof, 1967, after Heath, 1964.)

The peak labeled P1 is called the *first escape peak.* When pair production occurs, and the positron that is created annihilates to produce two γ's of 0.51 MeV each, either or both of these γ's can escape from the crystal. If one escapes (and this happens often) and all of the energy except that of the escaping γ is absorbed, a peak corresponding to an energy 0.51 MeV less than the photopeak results. If both escape, the *double escape peak* (P2) with energy 1.02 MeV less than the photopeak results. Both are shown for both primary γ's in the figure. Note that the escape peaks P1 and P2 for the higher-energy γ are larger with respect to the photopeak than those for the lower-energy γ. That is because pair production is more likely for the higher-energy γ.

The rest of the spectrum, apparently a continuum, is called the

Compton tail. Often a γ will lose some of its energy in the crystal, thereby causing light to be emitted, and then escape while still carrying some energy. Each γ loses an arbitrary amount of energy in the crystal, so it can yield a pulse with any energy less than that of the photopeak. These pulses cause the Compton tail. But small losses are more probable than large ones, so there are more small pulses than large ones. When a γ is Compton-scattered from an electron in the crystal, sometimes the scattered γ escapes without further energy loss. Therefore, only the electron energy is observed in the spectrum. The Compton edge, shown as C, is the maximum energy such an electron can have. The bigger the crystal, the more the photopeak dominates the other features of the spectrum because escape is less probable (and therefore the detector is more efficient).

Semiconductor detectors. Semiconductor detectors consist of single crystals of doped semiconductors; those used for γ's are most commonly germanium with lithium impurities [Ge(Li), spoken "jelly"]. If a γ-ray interacts with the crystal, it creates a number of electron-hole pairs that is proportional to its energy. A voltage is placed across the crystal, and a current proportional to the number of pairs, and therefore to the energy, is observed. Only a small amount of energy (about 3 eV) is needed to create one pair, so the resolution of the semiconductor detector is very good compared to that of the scintillator. Most of these crystals must be kept at very low (liquid nitrogen) temperature, and they are therefore difficult to use in boreholes. Recently pure undoped germanium detectors that can be used at higher temperatures have become available. Consequently, semiconductor detectors are beginning to be used in commercial logging sondes.

The peaks are extremely sharp, and they permit detailed investigation of complex spectra. Figure 6-9 shows the spectrum in a semiconductor detector from the same ^{24}Na γ's as in Fig. 6-8. Note that the peaks are much sharper. There are two reasons for this. First, the resolution of the semiconductor detector is much better than that of the scintillator. Second, most Compton-scattered γ's escape, so the Compton tail is much less important.

Although the density of these detectors is almost as high as that of a scintillator, the sensitive region of available detectors is very small, and therefore its stopping power, or efficiency, is much less than that of the scintillator. This is somewhat compensated for by its very high resolution, which increases the efficiency for identification of specific γ-rays.

The efficiency and resolution of scintillator and semiconductor detectors are a function of energy. For example, at 500 keV, a 7.5-cm-

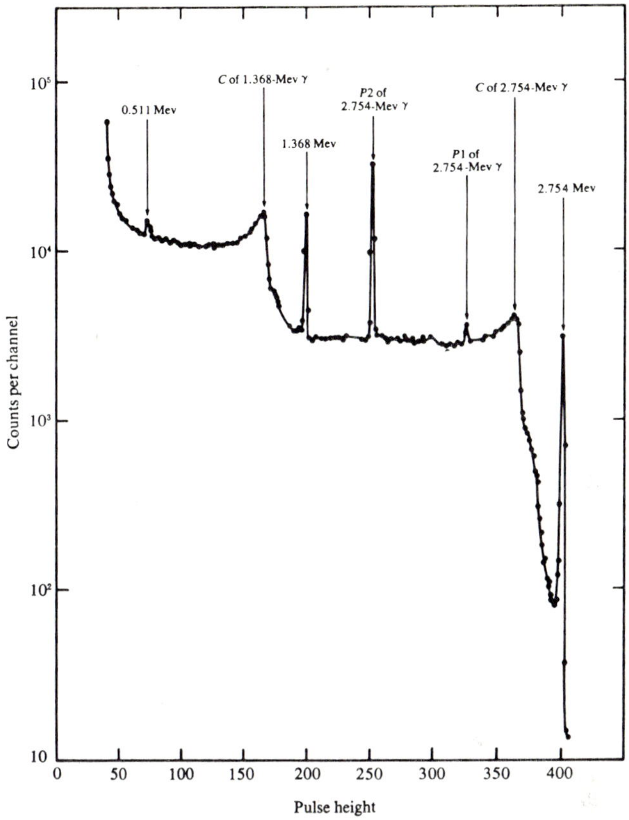

FIGURE 6-9 Pulse height spectrum of 1.368- and 2.75 4-MeV γ -rays from ^{24}Na in a lithium-drifted germanium detector. Detector was 1.9 cm in diameter and 0.5 cm thick. (Meyerhof, 1967, after Tavendale, 1963.)

diameter 7.5-cm-long NaI scintillator is 50 percent efficient, whereas a very large (54 cm^3) Ge(Li) is 20 percent efficient (Gorski, 1969). On the other hand, the resolution [expressed as *full width at half maximum* (FWHM)] of the NaI is 8 percent and that of the Ge(Li) is 0.3 percent. At 2 MeV, NaI is 20 percent efficient, whereas Ge(Li) is 4 percent efficient.

Dead time. During some period (typically a few hundred microseconds) following the discharge in a Geiger-Müller tube, the tube is unable to produce another pulse if another γ enters it. This period is called the *dead time* τ. (Dead time is much shorter for a photomultiplier.) If the second γ does not extend τ, the system is said to be *non-*

paralyzable. If τ is known, the true rate of arrival of γ's J_t can be calculated from the observed rate J_o as follows: The fraction of the time for which the counter is insensitive is $J_o\tau$. The number of counts lost per unit time is then $J_tJ_o\tau$. Therefore,

$$J_t - J_o = J_tJ_o\tau \qquad \text{and} \qquad J_t = \frac{J_o}{1 - J_o\tau} \tag{6-10}$$

On the other hand, if an arriving γ increases the dead time, so that τ starts over for each arrival, the system is *paralyzable.* A pulse is detected only when the time between pulses $\geq \tau$. If P_0 is the probability that with a true count rate J_t a pulse will *not* arrive during the time τ, then $J_o = J_tP_0$. The mean number of pulses that arrive in time τ is $J_t\tau$. P_0 has a Poisson distribution with $x = 0$, so from Eq. (6-7)

$$P_0 = \exp(-J_t\tau) \qquad \text{and thus} \qquad J_o = J_t \exp(-J_t\tau) \tag{6-11}$$

and the maximum possible observed rate is $J_{max} = 1/\tau e$.

One can measure the dead time of a paralyzable system by simply bringing a source closer and closer to the system until J_{max} is obtained and then calculating τ. One can measure the dead time of a nonparalyzable system through the use of two sources. If J_1 and J_2 are the count rates observed from each of the sources and J_3 is the rate with both sources, then (Scott, 1980)

$$\tau = 2\frac{J_1 + J_2 - J_3}{(J_1 + J_2)J_3} \tag{6-12}$$

Pileup. *Pileup* occurs when two γ's arrive at times such that their pulses in the detection system (usually a scintillator-photomultiplier) overlap. The resulting signal will appear to have been caused by a single γ whose energy is the sum of the energies of the two γ's. If the pulses are not digitized downhole, the long logging cable will stretch each individual pulse in time. Consequently, overlap is more probable and pileup problems are aggravated. There is no way to correct for pileup. Electronics to decrease pileup are available, but the only sure way to eliminate it is to reduce the count rate or shorten the pulses.

6-2d Data Recording

Two major types of electronics are used to analyze and record the output from detectors: ratemeters and pulse height analyzers.

Ratemeters. The ratemeter consists basically of a simple $\mathcal{RC}$ circuit. It gives no energy information. The output from either a Geiger-Müller tube or a scintillator is passed through a discriminator that eliminates pulses of energy too low to be of interest (one must be sure that the setting of the discriminator is always the same) and then through a circuit that feeds a charge q to the $\mathcal{RC}$ circuit at each pulse. Then the current entering the capacitor is $I_i = Jq$, where J is the pulse rate. The current leaving is $I_o = V/\mathcal{R}$. At equilibrium $I_i = I_o$, so $Jq = V/\mathcal{R}$, $V = J\mathcal{R}q$, and the output voltage is proportional to the pulse rate.

If we suddenly change the rate from J_1 to J_2, the time constant of the $\mathcal{RC}$ circuit becomes important and

$$V(t) = J_1\mathcal{R}q + (J_2 - J_1)\,\mathcal{R}q[1 - \exp(-t/\mathcal{RC})] \qquad (6\text{-}13)$$

and the output voltage lags the change in input pulse rate. Therefore, the ratemeter provides a sort of running average of the count rate by smoothing out statistical discontinuities. The uncertainty in the output of a ratemeter (based on Poisson statistics) is (Artsybashev and Ivanyukovich, 1975)

$$\Delta J = \left\{\frac{J}{2\mathcal{RC}}\left[1 - \exp\left(\frac{-2t}{\mathcal{RC}}\right)\right]\right\}^{1/2}$$

and for $t \gg \mathcal{RC}$

$$\Delta J = \left(\frac{J}{2\mathcal{RC}}\right)^{1/2} \qquad (6\text{-}14)$$

which shows that quadrupling the time constant will halve the uncertainty in the count. The logging speed (Sec. 6-1c) and time constant of the recording system should be so chosen that the uncertainty and averaging distance in the hole are acceptable to the user. The effect of this uncertainty on the accuracy of measured density will be discussed in Sec. 6-6f.

Pulse height analysis. There are two sorts of pulse height analysis. The simpler method uses a few discriminators, or *windows*, to sort the pulses according to their energy. Usually two to four windows are used, depending on the purpose. The pulses passing through each energy window are fed to a ratemeter (a different ratemeter for each window) and then displayed as desired.

The multichannel pulse height analyzer is based on digitization of the amplitudes of the pulses. The analyzer has a memory of, say, M channels (commonly 1024 nowadays). The expected maximum energy of the pulses is divided by M to provide M energy intervals, and each interval is assigned to one channel. The incoming pulse is amplified and shaped before it is digitized. There is no simple way to set up a correspondence between pulse height and energy because the entire system, from the scintillator through the cable, amplifiers, and digitizer, affects the pulse height. The user normally puts two or three radioactive sources of known energy near the detector and adjusts the gain of the various parts of the system until the peaks from those sources lie in the desired channels. So, for example, if the maximum energy were 10 MeV, one would choose 10.24 MeV and assign 10 keV per channel; channel 410 would then correspond to an energy band between 4.095 and 4.105 MeV. When a pulse of a given energy arrives, a bit is added to the total count in the channel corresponding to its assigned energy. At the end of a counting run, each channel contains a number of bits corresponding to the number of pulses with that energy. The number of counts in each channel is displayed on a screen with the channel number (or pulse height or energy) on the abscissa and counts per channel on the ordinate. Figure 6-8 is the display from such a pulse height analysis.

On-line computations. Once the analyzer is calibrated, many useful circuits can be added to it. Because the amplitude of the pulse from the scintillator changes with temperature and photomultiplier voltage, the peaks can drift after they have been set up. A digital stabilizer (or "stab amp") can be used to sense the channel in which a reference peak is located. If system instability causes the peak to change channels, the stabilizer assumes the drift to be linear with energy and changes the gain to bring the peak back into the right channel. This adjusts the gain for all channels.

Modern pulse height analyzers incorporate minicomputers, so that many convenient operations such as integration under part of a spectrum, subtraction of one spectrum from another, or overlays of displays can be performed with ease. Often the pulses are digitized downhole so that digital information can be transmitted over long logging cables and thereby eliminate pulse degradation. The output of the analyzer can be recorded on tape for further processing.

6-3 NATURAL γ-RAY LOGGING

A γ-ray probe (either a Geiger counter or, more commonly, a scintillator) is often used to estimate the distribution of natural γ-ray emitters in rock. Two kinds of sondes are used: a total γ-ray sonde that counts all γ's without discrimination and a spectral γ-ray sonde that counts the γ's whose energies lie in ranges corresponding to the windows for uranium, potassium, and thorium. In either case, one must consider the effect of attenuation in the medium and in the borehole fluid on the number of γ's detected.

6-3a γ-Ray Transport in Media with Cylindrical Symmetry

Count rate for unscattered γ-rays. It is convenient to consider a monoenergetic γ-ray source and avoid consideration of scattering phenomena (Czubek, 1962; Rhodes et al., 1961). At some observation point, the γ flux from a small source volume dV that emits M γ's per cubic centimeter per second is

$$dJ = MG\beta\, dV$$

where the factor G accounts for geometric spreading and β accounts for absorption. (The calculated J must, of course, be multiplied by the detector efficiency to obtain the observed J.)

Consider the geometry of Fig. 6-10, where a borehole containing material with a linear attenuation coefficient μ' is shown surrounded by an annulus with source density M and attenuation coefficient μ. The factor G becomes $1/4\pi r^2$ to account for spherical spreading, and β becomes a product of exponential terms of the sort shown in Eq. (6-9). The count rate at some point on the central vertical axis is

$$J = \int_V \frac{1}{4\pi r^2} \exp\left[-\mu(r - r') - \mu' r'\right] dV \tag{6-15}$$

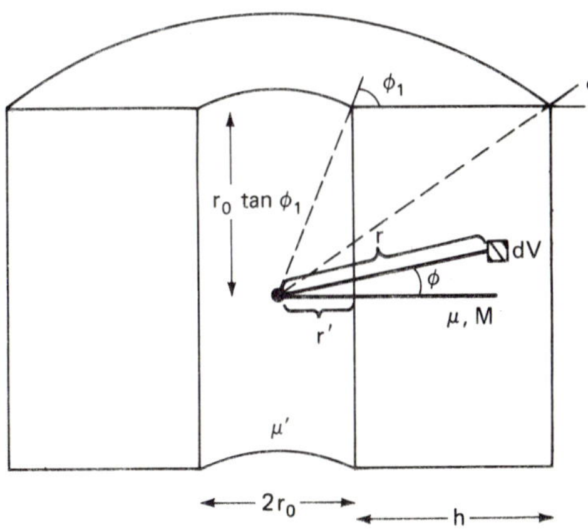

FIGURE 6-10 Annulus of radioactive material of activity M and attenuation coefficient μ surrounding a borehole of radius r'.

Rhodes et al. (1961) show that, for cylindrical symmetry, the volume integral can be expressed as three integrals over the angle φ:

$$\frac{\mu J}{M} = \int_0^{\varphi 1} \cos\varphi \exp(-x_1 \sec\varphi)\, d\varphi + \int_{\varphi 2}^{0} \cos\varphi \exp(-x_4 \sec\varphi)\, d\varphi$$

$$+ \int_{\varphi 1}^{\varphi 2} \cos\varphi \exp(-x_2 \csc\varphi + x_3 \sec\varphi)\, d\varphi \qquad (6\text{-}16)$$

where $x_1 = \mu' r_0$
$x_2 = \mu r_0 \tan\varphi_1$
$x_3 = (\mu - \mu')r_0$
$x_4 = \mu' r_0 + \mu h$

The x_i terms can be interpreted as the number of mean free paths existing along the ith linear segment. For example, x_2 gives the number of mean free paths along a vertical length within the source volume which extends from the midplane to the top.

Rhodes and Mott (1966) use this formulation to calculate response (correction) factors for horizontal beds of varying thickness and source energy. Such results are instructive for design purposes, but they cannot be used directly to correct data from a given sonde because of the monoenergetic limitation and because the effect of the sonde itself is ignored. One of the results, showing the volume of investigation of a point detector in a medium emitting 1.76-MeV γ-rays is shown in Fig. 6-11. About 75 percent of the γ's detected come from a volume of 14 cm radius and 25 cm vertical extent above and below the detector, as indicated by the dashed line in Fig. 6-11. The shape of the volume emitting that 75 percent fraction can be obtained by rotating the curve about the central axis and then mirroring it with respect to the midplane.

Spatial deconvolution. The development above shows that the γ-rays detected at any level in the borehole come not only from points in the medium directly opposite the detector but also from points above and below it. Consequently, if the amount of radioactive material in the medium is not homogeneous, the γ-ray count rate is not a direct measure of the source strength at the depth of the log. Several methods have been developed to infer the depth distribution of the source strength from the depth distribution of the count rate (the log). One of the first, and conceptually the easiest to understand, is due to Scott (1963), and it was developed (as most γ-ray data reduction methods have been) for analysis of uranium ore.

Consider a complex ore zone as sketched in Fig. 6-12. It could be

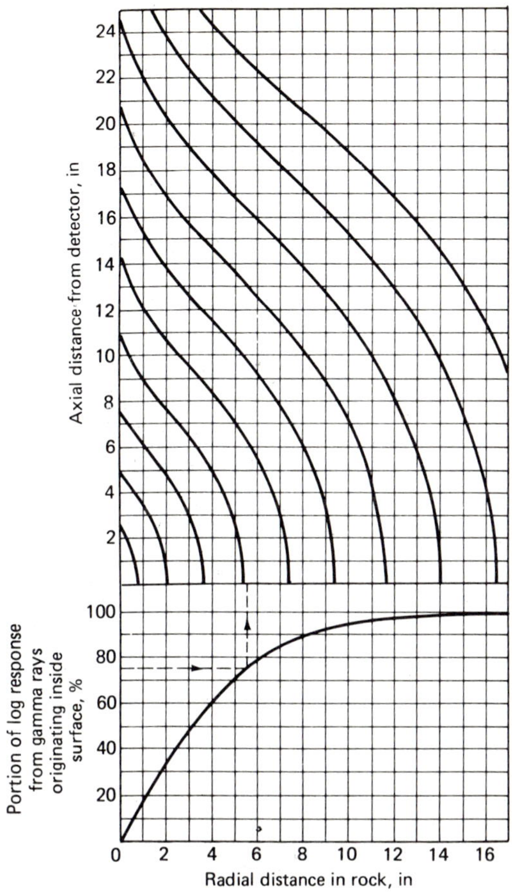

FIGURE 6-11 Volume of investigation of γ-ray sonde. Conditions: monoenergetic source of 1.76-MeV γ's, centralized detector, 15-cm-diameter uncased borehole containing 10 lb/gal (1.2 kg/L) mud. (Rhodes and Mott, 1966.)

considered to be composed of a series of layers each having a different thickness and ore grade (that is, source strength). One can construct a series of 15-cm-thick hypothetical layers with grades derived from the layers of the complex zone. These layers are shown in the lower part of Fig. 6-12. Imagine that each layer is independent and is placed between zones of zero source strength and that a separate γ-ray log is made of each one. The result is a series of symmetrical anomalies whose peak count rates are proportional to the strengths of the individual sources. Scott determined these anomalies; they are plotted as lines

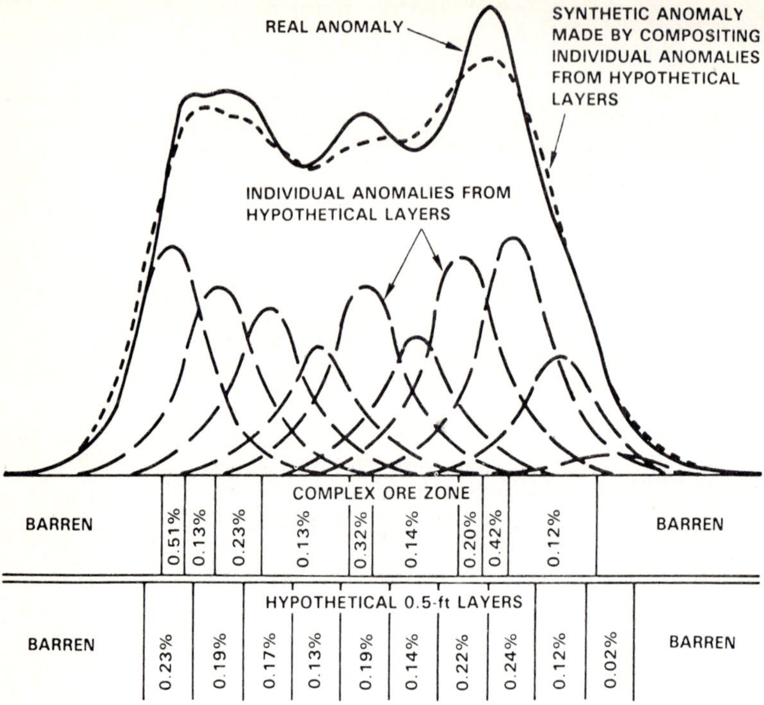

FIGURE 6-12 Complex uranium ore zone and equivalent hypothetical 15-cm-thick layers with associated γ-ray anomalies. (Scott, 1963.)

with long dashes in Fig. 6-12. These anomalies can be added to form a single anomaly, which is shown as the line with short dashes. Scott developed a computer program, GAMLOG, that adjusts the source strength of the individual anomalies until the summed anomaly matches, as closely as desired, the measured γ-ray log.

Czubek (1962) and Conaway and Killeen (1978) developed other methods, similar in principle but different in detail, to obtain the source distribution.

Grade-thickness product. Consideration of the γ flux emitted by a sequence of thin beds leads to a result of great practical importance for uranium exploration (Scott et al., 1961). Consider a small volume dV within a thin slab of thickness dT (Fig. 6-13) containing dW grams of radioactivity per gram of rock. Assuming the attenuation in the well bore is the same as that in the rock, the incremental count rate dJ is

$$dJ = \frac{[\exp(-\mu r')]B(\mu r')}{4\pi r'^2}\, c\, dW \qquad (6\text{-}17)$$

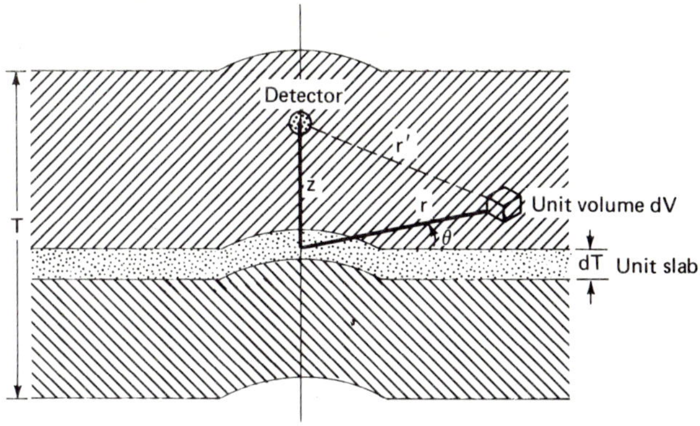

FIGURE 6-13 Geometry for derivation of grade-thickness relation. Thin bed of thickness dT lies within thick bed of thickness T.

where c is a constant and $B(\mu r')$ is a *buildup factor* that accounts for scattering. It is customary to express the ore content in terms of grade

$$G_\gamma = \frac{100\, dW}{\rho\, dV}$$

Using $dV = r\, dr\, d\theta\, dT$, the count rate from the entire slab becomes

$$J = \int_0^{2\pi} \int_{r0}^{\infty} \frac{c\rho G_\gamma}{100} \frac{[\exp(-\mu r')]B(\mu r')}{4\pi r'^2}\, r\, dr\, d\theta\, dT$$

Expressing result of the integral over r and θ as $f(\mu, r_0, z)$,

$$J = \frac{c\rho G_\gamma}{100} f(\mu, r_0, z)\, dT$$

Suppose we obtained a continuous log $J(z)$ of the γ's emanating from this thin bed. If we integrated the curve of the log with respect to z, the result (the area under the curve) would be

$$dA = \int_{-\infty}^{\infty} J\, dz = \frac{c\rho G_\gamma}{100} \left[\int_{-\infty}^{\infty} f(\mu, r_0, z)\, dz \right] dT \qquad (6\text{-}18)$$

There is, however, no need to perform the integral on f. If we express the result of the integral as $g(\mu, r_0)$, then

$$dA \propto 0.01 c\rho G_\gamma g(\mu, r_0)\, dT \qquad \text{or} \qquad G_\gamma\, dT = K\, dA$$

The final step is an integration over the total thickness of the bed T

$$A = \frac{1}{K} \int_0^T G_\gamma\, dT$$

to obtain, for the average grade in the bed,

$$G_\gamma T = KA \tag{6-19}$$

This shows that the product of the average grade and the thickness is proportional to the total area under the γ-ray log curve, assuming the bed is bounded above and below by formations devoid of radioactivity. The factor K is determined experimentally in calibration facilities (Mathews et al., 1978). Once K is known, Eq. (6-19) can be used to determine ore grade, since thickness T can be determined from the half-amplitude points of the log. Fink (1978) points out some of the ambiguities in the literature concerning the determination of K. Moore (1980) demonstrates that K is independent of the dip angle between the bed and the borehole.

6-3b Calibration and Correction

Field calibration. In all cases, probe stability must be assured. Ordinarily it is assured by using a secondary standard or *field calibrator*. This is usually a small long-lived radioactive source and a jig that holds the source in a fixed, reproducible position with respect to the sonde. The source produces a known count rate in the detector. This rate is tied to the primary calibration standard. If a spectral sonde is being used, then the source will be a mixture of isotopes that will produce a predicted number of counts in each window.

Base calibration. A standard must be established so that logs obtained with different sondes can be compared. There must be a γ-ray logging unit that is generally accepted and is traceable to a single source.

If an estimate of radioactive mineral content is desired, the amount of each radioactive element in the calibration standards must be known accurately. This becomes particularly important for the spectral or "KUT" sondes that present a concentration of potassium, uranium equivalent, and thorium equivalent in the formation.

These problems are resolved by construction of calibration facilities with test pits large enough to appear infinite for practical purposes. That is, the pit volumes must be a large fraction (usually greater than 95 percent) of the volume of investigation of the sondes.

Calibration of total-count γ logs. Total-count γ-ray logs have been used routinely in logging for decades. Some sondes count all incoming γ's, others count all above some selected energy threshold (Fig. 6-1*a*). In the former case, the lower energy limit is actually established by the characteristics of the sonde housing. In either case, a total count sonde will record most of the spectrum shown in Fig. 6-3. For many years, each logging company maintained its own γ-ray unit of measurement, which made it difficult to compare data obtained by different companies. This situation changed with the construction of the American Petroleum Institute (API) calibration facility in Houston, Texas.

The API facility is constructed of concrete with an admixture of radium to provide the ^{238}U decay series, monazite ore as a source of thorium, and mica as a source of potassium (Belknap et al., 1959). The concentrations, which were selected to be equivalent to twice those of an average shale in the midcontinent United States, were determined to be equivalent to 4.07% K, 24.2 ppm Th, and 13.1 ppm U. It was desired that the unit recorded by the log be such that 100 units would correspond to the deflection recorded opposite a shale formation; therefore, 200 API γ-ray units are defined as the difference in count rate recorded in concrete without the radioactive admixture and in the radioactive concrete.

The API standard is adequate for qualitative comparision among logs. It is, however, sometimes convenient to predict sonde response to known concentrations of K, U, and Th. For this purpose the count rate recorded by a total count γ sonde is considered to be linear and additive. The counts from each type of source can be converted to equivalent uranium concentration:

$$J = k(a\,\mathrm{K} + \mathrm{U} + b\,\mathrm{Th})$$

where k is a constant for a given sonde, a is the concentration of U (in ppm) that will give the same total count rate as 1% K, and b is the concentration of U that will give the same count rate as 1 ppm Th. For an NaI detector of dimensions 150 by 80 mm, with a discrimination threshold of 200 keV, for example, the value of a is 1.05 and b is 0.47 (IAEA, 1976). These values were used to translate the concentration scales of Fig. 6-4, so the comparisons shown in that figure are strictly valid only for a detector with the above dimensions.

Calibration of spectral γ-ray sondes. The windows used for spectral γ-ray sondes are ordinarily about 1.3 to 1.6 MeV for potassium, 1.6 to 2.4 MeV for uranium (the ^{214}Bi γ), and 2.4 to 2.8 MeV for thorium (the ^{208}Tl γ). Two of the three natural radioactive sources, uranium and thorium, produce counts not only in the window corresponding to that source but also in the windows corresponding to the other sources. These effects must be dealt with in calibration. Stromswold and Wilson (1981) give a good descripton of the calibration of a spectral γ sonde. Their method, often called *spectral stripping*, requires the use of observed count rates in three test pits (or models) of known concentration of potassium, uranium, and thorium. They employ three matrices, whose elements are defined as follows:

J_{ij} = count rate in the *i*th energy window in the *j*th calibration model after background subtraction and dead time correction.

C_{kj} = concentration of the *k*th radioactive element in the *j*th calibration model.

A_{ik} = a constant relating count rate in the *i*th energy window with concentration of the *k*th radioactive element.

Then the matrix can be written in terms of the elements as

$$J_{ij} = \sum_{k=1}^{3} (A_{ik}C_{kj}) \qquad (6\text{-}20)$$

The convention they use in defining the subscripts is

1 = potassium
2 = uranium
3 = thorium

Thus, for example, J_{13} is the count rate in the potassium energy window obtained in the thorium calibration model after background subtraction.

Their analysis technique calculates A^{-1}, the inverse of the matrix A, from the model data for a given sonde. Then concentrations can be obtained from field data by

$$C_i = \sum_{k=1}^{3} A_{ik}^{-1} J_k \qquad (6\text{-}21)$$

where C_i = concentration of the *i*th radioactive element
A_{ik}^{-1} = the *ik* element of the matrix A^{-1}
J_k = count rate in the *k*th energy window after background subtraction and dead time correction

This method is sometimes applied in a simplified form by use of *stripping ratios*. For example, if the number of counts in the thorium window is J_3 and (1) it is known from calibration that the ratio of counts in the uranium window caused by thorium to those in the thorium window is A_{23}, (2) if we assume that only thorium and uranium are present, and (3) that the uranium causes no counts in the thorium window, then

$$C_2 = A_{22}J_2 - A_{23}J_3$$

where C_2 is the uranium concentration, J_2 is the number of counts in the uranium window, and A_{22} is the number of counts in the uranium window caused by 1 ppm uranium. The stripping ratio A_{23} must be well known. Drift in the energy calibration (often caused by temperature) can perturb the results, so the calibration should be checked periodically.

Because of differences in concentration levels, calibration pits intended for logs to be run in uranium ore must have higher count rates than those intended for logs to be run in oil and gas lithologies. The user should be aware of the type of pit used to calibrate the sonde of interest.

Correction for water. Total-count γ-ray sondes are ordinarily calibrated eccentered (pushed to one side) in water-filled holes of specified diameter in the API test pits. When the sonde is run centralized or in a hole of different diameter or with a different borehole fluid, the attenuation by the borehole fluid is different from that in the calibration hole and a correction must be made. The correction is ordinarily developed by a logging company in its own test pit, and correction curves are published.

In the case of spectral γ-ray logs, Stromswold and Wilson (1981) have found that the correction should be performed after the calculations in Eq. (6-21) have been performed. They obtain separate correction factors for K, U, and Th for different hole sizes and then obtain corrected concentration by multiplying the uncorrected concentration from Eq. (6-21) by the correction factor. These factors can be quite large; for a borehole diameter 20 cm larger than the sonde diameter, the factor is close to 2.0.

Spectral fitting. When many γ-emitters, each with several peaks (examples will be given in Sec. 6-8b) are involved, and one wishes to identify the emitters and obtain a quantitative analysis, a more complex procedure is used (Hertzog, 1980). The spectrum is recorded by a pulse height analyzer with m channels ($i = 1, 2, \ldots, m$). The spectrum can be represented by an m-dimensional vector **B** and each channel by B_i.

It is assumed that the formation spectrum **B** can be described as a linear combination of *descriptor spectra* of n known materials $\mathbf{p}^j$ ($j = 1, 2, \ldots, n$). The vectors **P** include the response of the system to each of the spectra produced by these known materials, which are assumed to be the elements in the formation. Therefore, the **P** contain Compton tails, escape peaks, and effects of nonlinearity in the detector. The vectors **P** can be combined into an m by n matrix P such that

$$B_i = \sum_{j=1}^{n} P_i^j f_j \qquad i = 1, 2, \ldots, m \tag{6-22}$$

Each f_j represents the fractional contribution to the spectrum from element j.

The matrix is not square, so it cannot simply be inverted to solve for **f**. Furthermore, there are statistical errors in **B**. Hertzog chooses a "weighting matrix" W by least-squares fitting to a "typical spectrum" so that

$$\mathbf{f} = (P^T W P)^{-1} P^T W \cdot \mathbf{B}$$

In principle, this method compares measured spectra to a library of standard spectra and estimates how each spectrum in the library contributes to the measured spectrum. It is very much like spectrum stripping with m windows.

6-4 γ-RAY BACKSCATTERING AND DENSITY LOGGING

6-4a General Principles of Backscattering

If a γ-ray source and detector (usually rigidly attached in a logging sonde) are placed some distance apart in a borehole, some of the γ's from the source enter the formation and are Compton-scattered. Some of these scattered gammas return to the detector. They are called *backscattered* γ's (because the source and detector lie on the same side of the scatterer and so the γ's are scattered through more than 180°), and the process is called γ-ray backscattering. Czubek (1983) summarized much of the theory of this process.

As we saw in Sec. 6-2b, the attenuation property of bulk material is usually expressed as the mass attenuation coefficient μ/ρ. For Compton scattering, μ is a function of the γ-ray energy E and the number of electrons per atom of the scatterer Z. The number scattered per unit volume is also, of course, proportional to the number of atoms per unit volume; that is, to the bulk density ρ. The number of γ's removed from a beam with a 1-cm^2 cross-sectional area per unit volume is thus proportional to $(\mu/\rho)\rho$. The energy of the scattered γ is a function of the angle of scattering and the initial energy.

Not all of the γ's leaving the source in the direction of any volume element in the scatterer will get there. Some are Compton-scattered or are absorbed by photoelectric effect in some other volume element between the source and the volume of interest. Similarly, after a γ has been scattered toward the detector, it may be scattered again or absorbed by the photoelectric effect. (Generally, ^{60}Co or ^{137}Cs sources are used, so pair production can be neglected.)

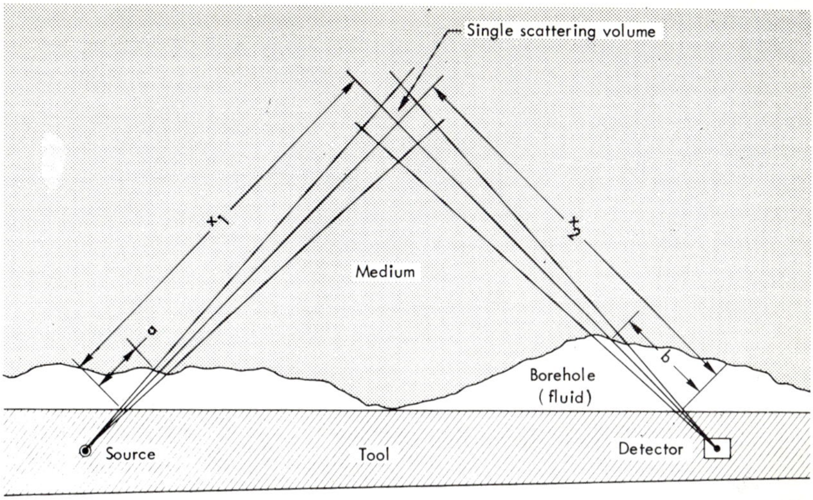

FIGURE 6-14 Idealization of a single scattering density sonde. (Hearst and Carlson, 1969.)

Figure 6-14 shows an idealized logging sonde in which a collimated beam of γ's enters a scattering volume and then the scattered γ's return to a collimated detector. Thus only those γ's scattered through a certain range of angles in a certain volume can be detected. Recall from Eq. (6-9) that the number of γ's remaining in a beam after it traverses a distance x is

$$J = J_0 e^{-\mu x} \quad . \tag{6-9}$$

where J is the count rate and the linear attenuation coefficient μ is a function of Z and energy. If we apply this expression to the situation in Fig. 6-14, with the gaps a and b = 0, and include geometrical spreading, then at a given angle θ

$$J(E) = \frac{KJ_0(E_1)}{x_1^2 x_2^2} \left(\frac{\mu}{\rho}\right)_{sv} \rho_{sv} \exp\left[-x_1\left(\frac{\mu}{\rho}\right)_1 \rho_1\right] \exp\left[-x_2\left(\frac{\mu}{\rho}\right)_2 \rho_2\right] \tag{6-23}$$

where $(\mu/\rho)_1$ refers to the mass attenuation coefficient of the material between the source (of energy E_1) and the scattering volume and is a function of source energy. $(\mu/\rho)_{sv}$ refers to a quantity that is similar to the mass attenuation coefficient of the material in the scattering volume. Under some circumstances, it might be different from that between the source and the scattering volume; but it refers only to the probability of γ's being scattered through the angle θ between the source beam and a line from the scattering volume to the detector. Since only *unscattered* gammas arrive at the scattering volume, it is also a function of the source energy. $(\mu/\rho)_2$ is the mass attenuation coefficient of the material between the scattering volume and the detector; it is evaluated at the energy of the scattered γ's (a function of θ). All of these coefficients are functions of the Z of the medium. K is a constant accounting for solid angles, scattering volume size, and so on.

A plot of measured values of $J(E)$ versus ρ is shown in Fig. 6-15, and typical calculated energy spectra of detected γ-rays for different values of Z are shown in Fig. 6-16. Both figures are for a scattering angle of about 270°. Note that the curve in Fig. 6-15 is double-valued! This can be justified intuitively if one realizes that if the density is zero, no γ's are backscattered, so J must be zero, and if the density is infinite, all γ's are removed from the beam, so J is again zero. Note that in Fig. 6-16 the spectra for different values of Z are different up to about 300 keV, above which they are about the same. The difference below 300 keV is caused by the difference in photoelectric cross section in materials of different Z; above 300 keV photoelectric cross section is small for all Z.

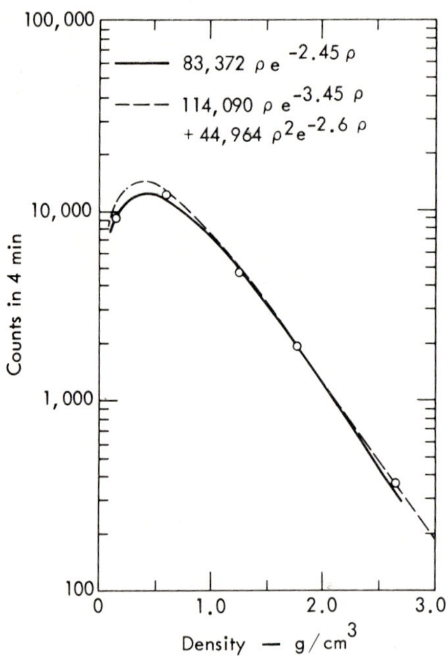

FIGURE 6-15 γ-Ray count rate J versus bulk density ρ for a single-scattering density sonde. Scattering angle 270°. (Hearst and Carlson, 1969.)

Above 300 keV, J is not a function of Z and depends primarily on ρ. Therefore, J can be used in a *density log* (Fig. 6-1*c*). Below 300 keV, J depends on Z as well as ρ, and by clever arrangement the dependence on ρ can be minimized. J below 300 keV can be used to determine Z, and this is called a *selective* γ-γ *log* (Fig. 6-1*d*).

6-4b Single-Scattering Density Logging

Single-detector density sonde. For a real system like that idealized in Fig. 6-14, one uses a band of energies and a range of angles, because very tight collimation would give a count rate so low that the uncertainty would be very large. The energies are chosen to be those that could arise from scattering through the angles used. (The selected energies are shown schematically in Fig. 6-1*c*.) The energy is selected either by analyzing the energy of the γ's and choosing a suitable window or by a lead shield and a graded filter (see Sec. 6-2c). The latter method takes advantage of the lead to eliminate the photoelectrons and also reduces pileup.

For a given sonde, x_1 and x_2 are constants. For most rocks, μ_c is a function only of energy (as long as $Z < 25$ or so), so for a given source energy and fixed scattering angle μ_1 and μ_2 also are constant. Therefore, assuming $\rho_1 = \rho_2 = \rho_{sv} = \rho$, Eq. (6-23) becomes

$$J = K_1 \rho \exp(-K_2\rho) \qquad (6\text{-}24)$$

where K_1 and K_2 are constants depending on the design of the sonde. This equation is illustrated by Fig. 6-15.

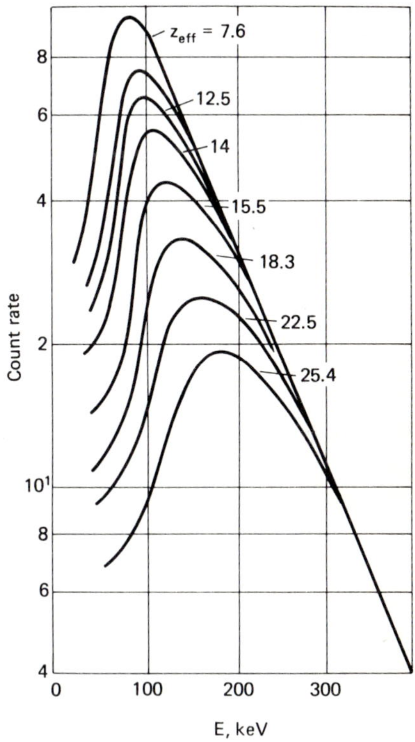

FIGURE 6-16 Spectra of γ-rays (count rate J versus energy) for different values of atomic number Z for a diffusion density sonde. (Artsybashev and Ivanyukovich, 1975.)

Because Eq. (6-24) is double-valued, there is some density at which the curve turns over. Below this density, count rate increases with increasing density; above it, count rate decreases with increasing density. Usually the side on which density decreases is used for density logging. It can be shown from Eq. (6-24) that the density at which turnover occurs is a function of the length of the sonde. Therefore, sondes can be designed to have this decrease at almost any desired density. Some density sondes are designed with variable source-detector spacings, so that the spacing can be selected in the field to suit the density expected (Artsybashev and Ivanyukovich, 1975).

In a real system one cannot collimate well enough and have a tight enough energy window to detect only singly scattered γ's and still have enough counts to be useful. At densities above about 2 g/cm^3 double scattering is quite common and the energies and angles are quite close to those for single scattering. Double scattering adds another term to Eq. (6-24), which becomes (Hearst and Carlson, 1969)

$$J = J_0[K_1\rho \exp(-K_2\rho) + K_3\rho^2 \exp(-K_4\rho)] \tag{6-25}$$

When calibrated with real data, this looks almost exactly like Fig. 6-15.

Note that, in Fig. 6-14, the sonde is placed against the wall of the borehole (eccentered). If we consider the gaps between the sonde and the wall of the hole, with x_1 and x_2 still constant,

$$J = K\rho_m \exp - \left[(x_1 - a)\left(\frac{\mu}{\rho}\right)_{m1} + (x_2 - b)\left(\frac{\mu}{\rho}\right)_{m2} \right]\rho_m \exp - \left[a\left(\frac{\mu}{\rho}\right)_{f1} + b\left(\frac{\mu}{\rho}\right)_{f2} \right]\rho_f \tag{6-26}$$

where $m1$ and $m2$ refer to the medium and $f1$ and $f2$ to the fluid, 1 refers to the initial energy, and 2 refers to the scattered energy. If the gaps are unknown, density cannot be determined.

Two-detector single-scattering sonde. One way to deal with the effect of the gap is to use a second detector. Figure 6-17 is an idealization of a two-detector sonde. The ratio of the counts in the two detectors is the ratio of two expressions like Eq. (6-26).

$$\frac{J_1}{J_2} = \frac{K\rho_m \exp - \left[(x_1 - a)\left(\frac{\mu}{\rho}\right)_{m1} + (x_2 - b)\left(\frac{\mu}{\rho}\right)_{m2} \right]\rho_m \exp - \left[a\left(\frac{\mu}{\rho}\right)_{f1} + b\left(\frac{\mu}{\rho}\right)_{f2} \right]\rho_f}{K\rho_m \exp - \left[(x_3 - a)\left(\frac{\mu}{\rho}\right)_{m1} + (x_4 - b)\left(\frac{\mu}{\rho}\right)_{m3} \right]\rho_m \exp - \left[a\left(\frac{\mu}{\rho}\right)_{f1} + c\left(\frac{\mu}{\rho}\right)_{f3} \right]\rho_f} \tag{6-27}$$

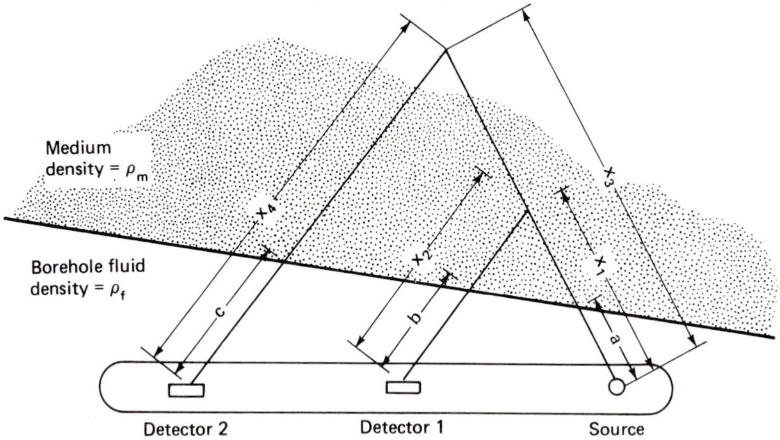

FIGURE 6-17 Idealization of a two-detector density sonde.

where the subscripts are the same as for Eq. (6-26) but with two possible values of μ for the scattered γ's.

If the sonde is parallel with the wall, so $a = b = c$, and the scattering angles are the same for both detectors, so $\mu_2 = \mu_3$, then

$$\frac{J_1}{J_2} = \frac{K_1}{K_2} \exp \rho \left[(x_3 - x_1)\left(\frac{\mu}{\rho}\right)_{m1} + (x_4 - x_2)\left(\frac{\mu}{\rho}\right)_{m2} \right]$$
$$= K_3 \exp(-K_4\rho) \qquad (6\text{-}28)$$

and ρ is easily obtained. Figure 6-18 shows Eq. (6-28). We see a "spine" for the situation of zero gap and a number of 45° "ribs" each of which corresponds to a line described by J_2/J_2 = const. Points on each rib correspond to different values of J_1 and J_2 arising from different values of gap and fluid density. To the right of the spine, $\rho_f < \rho_m$; to the left, $\rho_f > \rho_m$. When the sonde is not parallel and the gaps are unknown, no analytical solution has been published except in the case when the source is in contact with the wall (Hearst, 1976). We know of no two-detector single-scattering sondes in the United States, but some are in use in the Soviet Union (Varvarin et al., 1967).

6-4c Diffusion Density Logging

Because the count rate for a single-scattering density sonde is quite low, most commercial sondes are not well collimated, and the γ-rays are allowed to scatter many times. There is no tight energy window, but only an energy threshold, usually about 300 keV, so that all of the spectrum caused primarily by Compton scattering is accepted.

Such a situation can be described by the diffusion equation. If Γ is

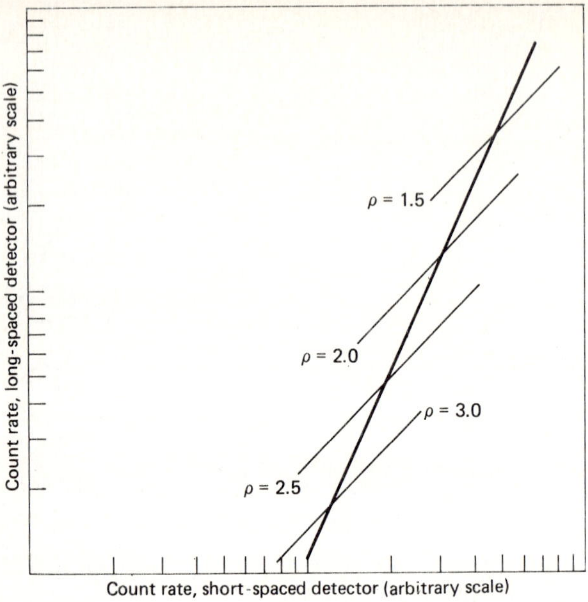

FIGURE 6-18 γ-Ray count rates J_1 and J_2 in each of two detectors of an idealized density sonde satisfying Eq. (6-28). Heavy spine corresponds to the zone-gap situation; ribs correspond to different gaps for constant density.

the number of γ's piercing the surface of a sphere of unit cross section per unit time, then j, the number of γ's flowing across unit area normal to the flow, per unit time, is

$$j = -D \text{ grad } \Gamma$$

where D is a constant called the diffusion coefficient. Then the net number of γ's leaving unit volume per unit time is

$$\text{div } j = -\text{ div } (D \text{ grad } \Gamma) = -D \nabla^2 \Gamma$$

The number of γ's absorbed (or scattered into an energy too low to be of interest) per unit volume per unit time is $\Gamma\mu$, where μ is the attenuation coefficient that applies both to absorption and to scattering to an energy lower than that of interest. (It is a sum of the absorption coefficient and the part of the scattering coefficient that results in low energy.) The number of γ's produced per unit volume per unit time is S. Since

$$\text{Production + scattering in } - \text{ absorption } - \text{ leakage} = 0$$

$$S - D \nabla^2 \Gamma - \mu\Gamma = 0 \tag{6-29}$$

This is the diffusion equation. Ordinarily, we use

$$\mu = \left(\frac{\mu}{\rho}\right)\rho$$

where μ/ρ is a constant for the medium of interest. D is proportional to $1/\rho$. Then the solution for the count rate J at a distance r from a point source in an infinite medium is usually given as (Voskoboinikov, 1957)

$$J = \frac{K_1\rho \exp(-K_2\rho r)}{r} \tag{6-30}$$

For constant r, this is the same as Eq. (6-24) for single scattering.

Single-detector diffusion sonde. Many solutions have been obtained for the geometry of a logging sonde in a borehole; an example is a cylindrical solution with Bessel functions (Voskoboinikov, 1957). One such solution, for a source and detector a distance r apart on the boundary between a half-space of density ρ and an infinitely absorbing half-space (Tittman and Wahl, 1964) seems to be in good agreement with data from a diffusion log.

$$J = \frac{K_1}{r^3}\left(1 + \frac{\rho r}{K_2}\right)\exp\left(\frac{-\rho r}{K_2}\right) \tag{6-31}$$

If one obtains the constants in Eq. (6-30) by fitting three or four calibration points to a curve of count rate versus density for a real single-detector density sonde, however, then all diffusion density sondes, both in the United States and in the Soviet Union, in practice, satisfy Eq. (6-30) (Ferronskiy et al., 1968). For example, we find that we can fit three calibration points with density ranging between 1.27 and 2.62 g/cm^3 to either Eq. (6-30) or Eq. (6-31). And in fact, an empirical calibration curve of the form of Eq. (6-30) is generally used.

Another method for determining density and other formation properties—using the ratio of counts in different parts of the γ-ray spectrum from a diffusion sonde—has been developed by Charbucinski (1983). It has evidently not yet been put into practice.

Two-detector diffusion sonde. Gaps affect the diffusion sonde as much or more than the single-scattering sonde. Again, two detectors are used to deal with this problem. (This is called a *compensated density log.*) No sophisticated theoretical analysis of a two-detector diffusion sonde has been published. Of course, if Eq. (6-30) is used to describe a single-detector sonde, a ratio of counts in two detectors will be described by a form similar to Eq. (6-28). Samworth (1980) used a semitheoretical approach to obtain an equation of the same form as (6-28). Sherman and Locke (1975) state that the true density ρ_t is given by

$$\rho_t = \frac{4\rho_l}{3} - \frac{\rho_s}{3}$$

where ρ_l and ρ_s are the apparent densities from the far (long-spaced) and near (short-spaced) detectors. But this also satisfies Eq. (6-28).

Scott (1977) also converted count rate in near and far detectors to density before applying a compensation equation. He pointed out that such a procedure makes the compensation equations independent of small changes in detector response and source strength because each detector can be routinely calibrated for density (see Sec. 6-3e).

Data from a commercial two-detector diffusion sonde, shown in Fig. 6-19, conform to a spine-and-ribs form like that in Fig. 6-18. In this figure the spine is again for zero gap. The ribs on the right side of the spine correspond to fluid density greater than matrix density, and those on the left correspond to fluid density less than matrix density.

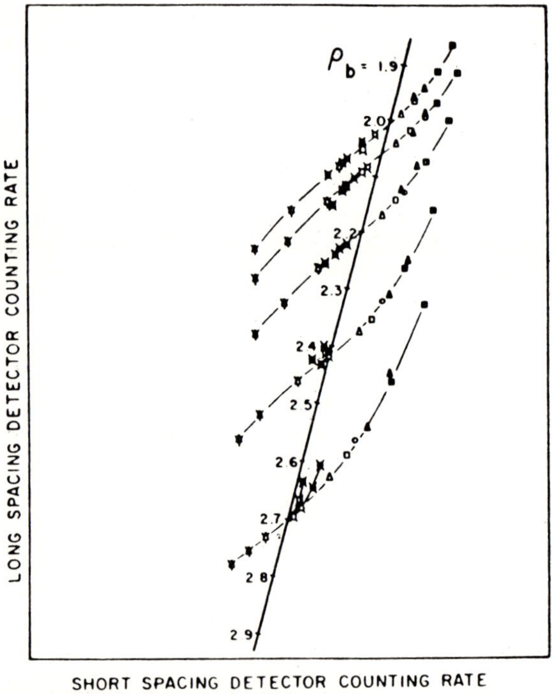

FIGURE 6-19 γ-Ray count rates J_1 and J_2 for a real two-detector diffusion density sonde. (Courtesy of Schlumberger.)

However, the ribs are not at an angle of 45° and are not straight, so J_1/J_2 is only approximately constant.

6-4d Density Equipment

Single-detector sondes. The only single-detector single-scattering sonde described in the U.S. literature (Hearst and Carlson, 1969) is called the rugosity insensitive density system (RIDS). It is so designed that source and detector ports are always in contact with the borehole wall, which eliminates the effect of gaps. Contact is assured by reading strain gages at the contact points; if there is no contact, the sonde is moved until contact is obtained. Both because of this need to adjust sonde position and because of the low count rate resulting from tight collimation, the RIDS is time-consuming to use. Consequently, its use is rare.

Commercial single-detector sondes are usually of the diffusion type. They are usually eccentered, and they work on the right-hand side of the curve in Fig. 6-15. Some collimation is used to select those γ's going into the formation. These sondes are quite sensitive to gap between the sonde and the borehole wall. The gap effect can be simply described as

$$J_o = B(G, \rho)J_c$$

where J_o is the observed count, J_c is the corrected count, and G is the gap. Figure 6-20 shows B as a function of G and ρ for a Soviet sonde in drilling mud of density 1.2 g/cm^3 and an American sonde in water, both with a source-detector spacing of approximately 40 cm. The effect would be much larger in a dry hole. It is interesting to note that the curves for the two sondes are quite similar. Most logging companies publish gap-correction curves, but not many measure gap while logging. One commercial sonde has a gap-measuring wheel attached to the sonde. In that case, the user can correct for gap. However, Artsybashev and Ivanyukovich (1975) point out that even if a correction is used, the gap reduces the sensitivity of the system. It is suggested that if a single-detector sonde is used in a rough hole, the user insist on measurement of gap.

Some sondes, especially those designed for small holes, where there is no room for collimation, are run-centered and then no collimation is needed. Both centered and eccentered logs must be corrected for hole size if the size is different from that for which the sonde was calibrated. Size is usually obtained from caliper logs. This correction is especially important for a centered sonde.

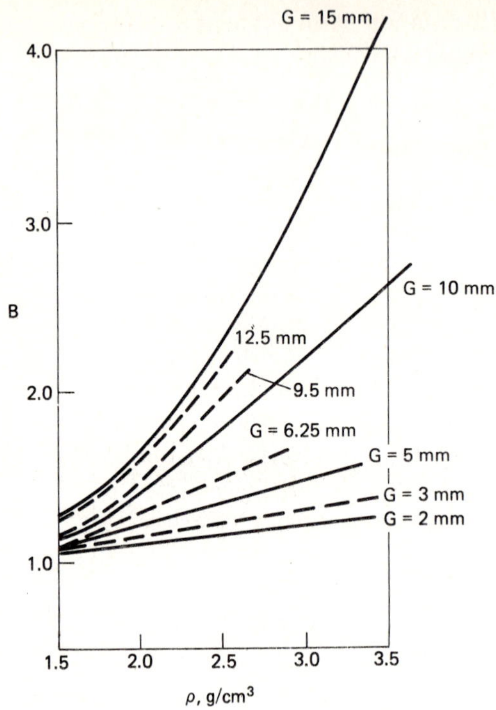

FIGURE 6-20 Count rate effect B as a function of gap G and density ρ for a Soviet (Artsybashev and Ivanyukovich, 1975) diffusion sonde (solid line) and a similar U.S. sonde (dashed line). Observed count = $B \times$ true count.

Two-detector sondes. All commercial two-detector sondes are of the diffusion type. Wahl et al. (1964) point out that the density can be obtained only if the sonde is parallel with the wall. The log is usually presented with both density and correction traces. If the correction is very large, the user can reject the data. However, if the sonde is not parallel with the wall, it is possible that the correction algorithm in the surface electronics will give a wrong answer and yet not give any indication of that in the correction trace. For example, for one commercial sonde, in a dry hole, a true bulk density of 2.59 g/cm^3 was recorded as 2.69 g/cm^3 with the source 2 cm from the wall and the detector in contact with the wall; whereas when the detector was 2 cm from the wall and the source was in contact, the recorded density was 2.39 g/cm^3. Since the system does not correct for the difference in spacing, the difference is not evident on the "correction" trace.

Most companies also publish curves for correcting compensated logs for hole size. Again, the size can be obtained from the caliper log. All two-detector systems fail in really rough holes.

6-4e Calibration

The best form of calibration for any density sonde is a set of holes *(test pits)*, with diameters approximating those of the holes to be logged, in materials of well-known density. Most logging companies have their own test pits containing different materials of known density for initial calibration of density sondes. For a single-detector sonde, a simple curve or straight line is fitted to the calibration points. For a compensated sonde, the spine-and-ribs patterns are determined empirically (Scott, 1977). Then surface electronics *(panels)* are programmed or hardwired to automatically apply the calibration to get density from count rate. Unfortunately, these test pits are not portable, so they cannot be used for field calibration.

It is very important that the sonde be checked in the field each time it is run. It is claimed (Artsybashev and Ivanyukovich, 1975) that a 0.1-mm change in the position of the source will change J by 1 percent. A small change in discriminator level setting can also make a substantial change in J. The best form of field calibration is the placement of the sonde on several large blocks of different known densities. At least three blocks are needed for complete calibration of a single detector because Eq. (6-30) is nonlinear in log J. If only a limited range of densities is wanted, two blocks will suffice. For a compensated sonde, only two blocks are required. It is desirable to test the gap correction by calibrating with several gaps between the sonde and the blocks. However, blocks that are large enough for this purpose are inconvenient to transport. Many ingenious *calibration fixtures* have been developed to avoid transporting blocks. Often they are designed simply to supply certain standard count rates to the detector or detectors. For example, Artsybashev and Popov (1972) describe a device in which the sonde is held a fixed distance above a scattering surface and the count or corrected density is recorded. Then lead plates are placed betweeen the sonde and the surface and the count is again recorded. These are then arbitrary count rates that can be used to scale the recorder. Other systems use sources of known activity at known distances.

These calibration fixtures often do not give the same spectrum as real formations or known-density blocks, and they often do not provide enough data points to show difficulties. The most reliable calibrator is two or three blocks of known density.

6-4f Corrections

Hole size, gap, and dead time. Calibration for hole size and gap effects is generally done once, at the service company's base, for each type of logging system. Then correction curves are published, and the user is expected to make the correction. Of course, the compensated sonde is supposed to correct automatically for gap. Many modern density sondes use scintillation detectors, so dead time is quite small and can either be neglected or accounted for in the surface electronics. The user, however, should learn if the density sonde being used contains a Geiger counter and, if so, be sure the dead time correction is made.

Electron density. Recall that μ is a function of Z, the atomic number, for Compton scattering. Therefore, the sondes really measure *electron density*

$$\rho_e = \rho_t \frac{2Z}{A} \tag{6-32}$$

where ρ_e is the electron density and ρ_t is the true bulk density. For most earth materials $2Z/A \simeq 1$, but for water $2Z/A = 1.11$. For a mixture

$$\rho_e = \rho_t \sum \frac{2Z_i}{A_i} f_i \tag{6-33}$$

where f_i is the fraction by weight of material *i*. The correction is significant for water and for ores of fairly high Z. ρ_e for a number of materials is shown in Table 6-1.

If the water content of the medium is known, one can correct the electron density for water. If one knows the fraction by weight f_w (for example, from a core) then

$$\rho_t = \frac{\rho_e}{1 + 0.11 f_w} \tag{6-34}$$

If one knows the fraction by volume of water I_H (for example, from a neutron log), then

$$\rho_t = \rho_e - 0.11 I_H$$

Unfortunately, at least one logging company applies this correction to its compensated density log automatically. It obtains I_H by assuming that any difference between measured density ρ_b and an input matrix density ρ_{ma} is caused by fresh-water-filled porosity and sets I_H equal to ϕ. If this is not the case, the user must uncorrect the compensated density. It may not be easy for the user to find out whether a particular service company performs this (possibly unwanted) correction, but the effort is worthwhile if high accuracy is required.

Statistics. The precision of J, and therefore of ρ, is a function of J. The count is Poisson-distributed. For a fixed time t, since the total count is Jt and the standard deviation of a number of counts is its square root, the fractional uncertainty in count $\Delta J/J$ is

$$\frac{\Delta J}{J} = \frac{1}{\sqrt{Jt}}$$

For a ratemeter, Eq. (6-14) gives $\Delta J/J = 1/\sqrt{2\tau}$. From Eq. (6-24),

$$\frac{d\rho}{\rho} = \frac{1 - K\rho}{\rho}\frac{dJ}{J} \tag{6-35}$$

A good value for K is about 2.5, so for a typical value of $\rho = 2$ g/cm^3, Eq. (6-35) gives $d\rho \simeq -2.5\ dJ/J$. This value can be used to choose T or the time constant τ to obtain an acceptable value of dJ/J and thus a desired accuracy in ρ. One of the most common sources of error in density logging is the use of too high a logging speed. Halving the speed will increase accuracy by a factor of $\sqrt{2}$. The user should always give thought to the best speed.

Background. In some situations, particularly for high-density materials, natural γ-ray background can be high enough to cause measured density to be spuriously low (Hearst and Carlson, 1969). The best procedure for correction is to run the log with and without the source and use the difference in count rates. A second possibility is to multiply the reading of a γ-ray log by an appropriate constant and subtract the result from the density sonde count rate. Scott (1977) discusses some of the problems of background compensation of a dual-detector sonde in detail. In any case, of course, the data must be processed off-line; the processing by the standard surface electronics will not be satisfactory.

Particle size. If the particle size of a nonhomogeneous medium is large enough, a correction is required. Umiastowski and Buniak (1977) show that the linear attenuation coefficient of a fluid-saturated granular medium is

$$\mu = \mu_f + \Lambda(1 - \phi)(\mu_{ma} - \mu_f)$$

where μ_f is the attenuation coefficient of the fluid and μ_{ma} is that of the matrix. Λ is a parameter depending on the porosity ϕ and a dimension-

TABLE 6-1 Densities, atomic numbers, molecular weights, and photoelectric cross sections of various materials. Adapted from Bertozzi et al. (1981).

Name	Formula	ρ_t	ρ_e $(2\rho_t Z/A)$	Z_{eff}	Molecular weight	P_e°	U†
Minerals							
Anhydrite	$CaSO_3$	2.960	2.957	15.6	136.14	5.05	14.95
Barite	$BaSO_4$	4.500	4.011	47	233.37	267	1070
Calcite	$CaCO_3$	2.71	2.708	15.7	100.09	5.08	13.77
Corundum	Al_2O_3	3.970	3.894	11.2	101.96	1.55	6.04
Dolomite	$CaCO_3 \cdot MgCO_3$	2.870	2.864	13.7	184.42	3.14	9.00
Feldspar or.	$KAlSi_3O_8$	2.55	2.52	13.39	278	2.86‡	
Feldspar pl.	$NaAlSi_3O_8$	2.62	2.59	11.55	262	1.68‡	
Feldspar pl.	$Ca_2Si_2O_8$	2.76	2.74	13.73	278	3.13‡	
Gypsum	$CaSO_4 \cdot 2H_2O$	2.320	2.372	16.4	172.18	3.42	8.11
Halite	NaCl	2.165	2.074	14.0	58.45	4.65	8.65
Hematite	Fe_2O_3	5.240	4.987	23.8	159.70	21.48	107
Illite		2.65	2.63	14.22		3.55‡	
Kaolinite		2.44	2.44	11.85		1.84‡	
Magnetite	Fe_3O_4	4.870	4.708	23.6	231.55	22.08	109
Montmorillonite		2.12	2.12	12.19		2.04‡	
Pyrite	FeS_2	5.000	4.834	22.0	119.98	16.97	82
Quartz	SiO_2	2.654	2.650	11.8	60.09	1.81	4.79

Liquids							
Water	H_2O	1.000	1.110	7.5	18.016	0.36	0.40
Salt water	120,000 ppm	1.086	1.185	9.4		0.81	0.96
Oil	(Typical)	0.850	0.948	5.5		0.12	
Rocks							
Sandstone (clean—saturated)		2.308	2.33	11.6		1.75	4.07
Average shale		2.65	2.645	14.0	3.42	6.52	
Anthracite coal		1.70	1.75	6.0		0.16	0.28
Bituminous coal		1.40	1.47	6.2		0.18	0.26
Granite				13.3–14§			
Pegmatite				13.0–13.9§			
Diabase				15.4–16.4§			
Chalcopyrite				24.6§			

*P_e is the average photoelectric cross section per electron.

†U is the average photoelectric cross section per unit volume.

‡Edmundson and Raymer, 1979.

§From Artsybashev and Ivanyukovich (1975); $\rho_t \simeq \rho_e$.

less grain size Y. For spherical grains of diameter d

$$Y = \frac{2d(\mu_{ma} - \mu_f)}{3}$$

For water-saturated sand with diameter less than 5 mm, $\Lambda \simeq 1.0$. For a diameter of 5 cm, Λ varies between 0.85 and 0.95, depending on porosity. For a diameter of 2 m, $\Lambda = 0.1$. This implies that a larger count will be obtained from a density log for a material with the same density and porosity but with large grain size. Umiastowski and Buniak (1977) performed an experiment with saturated sand of different grain sizes in a geometry with a source and detector on the surface of the sample. They found an increase in count rate of about 20 percent when the grain size was increased from 0.1 mm to 4 cm.

These results indicate that for a density log the observed density $\rho_o = \Lambda\rho_t$. Expressions for Λ as functions of grain dimensions are not given. If one were logging a medium with large grains, it would be well to calibrate in a medium of similar grain size and shape.

6-4g Depth of Investigation

The depth of investigation of a density sonde is a function of the sonde length and type (one- or two-detector), source energy, and formation density. It is generally determined by experiments in which the sonde is surrounded with concentric layers of material of constant density and the change in count rate is noted. The depth is commonly considered to be the thickness of material for which the count rate is 90 percent of that for infinite material (or the maximum count that can be obtained). Several values have been quoted in the literature. Longer sondes have deeper penetration but generally lower count rates and therefore less accuracy.

Gulin and Ponyatov (1969) find a depth of 9 to 12 cm depending on the type of sonde and density. Ferronskiy et al. (1968) get 5 cm for a ^{137}Cs source and a density of 2.6, 6 cm for a density of 1.8, and as much as 12 cm for a ^{60}Co source with the same densities. Sherman and Locke (1975) get about 12 cm for a compensated sonde and a density of about 2.1 g/cm^3.

It is possible to obtain the density of the material (such as cement, Chap. 16) behind casing if the system is calibrated with known densities behind the thickness of casing to be used. The sensitivity is greatly reduced because many of the γ's are absorbed in the casing. Generally, tolerances on casing thickness are quite loose, and variations in casing thickness are the chief source of error when logging through casing.

6-4h Selective γ-γ Logging

Density logging uses the energy region above 300 keV, where Compton scattering predominates. Selective γ-γ logging uses the lower-energy part of the spectrum, where photoelectric effect predominates (Fig. 6-16), to infer the medium's average atomic number and thereby learn something about its elemental composition and lithology (Fig. 6-1*d*). Even in this lower region, however, there is a substantial change in count rate as a function of density; and if one wishes to find changes in Z, one must take density into account. Two methods have been developed to do this.

The *inversion sonde* used in the Soviet Union is an ordinary density sonde so designed that the region of turnover (Fig. 6-15) occurs in the range of densities expected to be found in the formation. Artsybashev (1967) finds a source-detector spacing of the order of 4 cm to give a minimum sensitivity to density in the range 1.5 to 3.0 g/cm^3. An ordinary density sonde would have a spacing of 20 to 40 cm. The depth of penetration of the 4-cm-spaced sonde is very small, so the sonde is often placed several centimeters from the wall of the hole and only the surface is studied. Density can be obtained from another sonde and a correction applied.

Bertozzi et al. (1981) take the ratio of the counts below about 100 keV (soft) to those above about 300 keV (hard) to obtain an S/H ratio. Figure 6-21 is a plot of S/H versus Z_{eff} for their system. This method

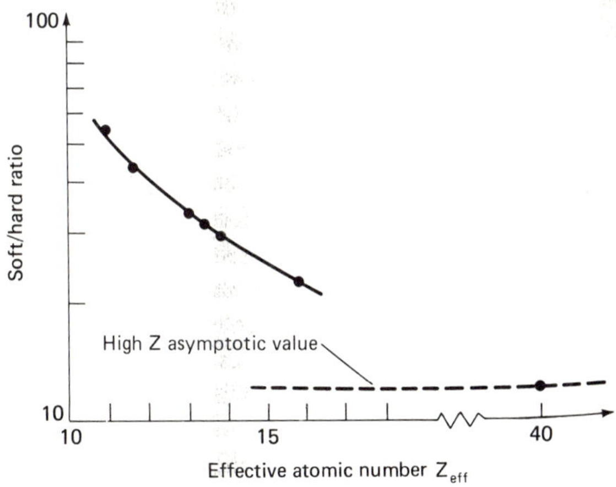

FIGURE 6-21 Ratio of γ count rate below about 50 keV to γ count rate above about 300 keV (soft/hard or S/H ratio) as a function of effective atomic number Z_{eff} (Bertozzi et al., 1981) for a special selective γ-γ sonde.

has the advantage that the hard counts can be simultaneously used to obtain density, and, in fact, Bertozzi et al. (1981) used a modification of a compensated sonde to obtain *S/H*.

In either system an effective Z is measured. Bertozzi et al. (1981) show that the photoelectric cross section is proportional to $E^{-3.15}Z^{4.6}$ with E in kiloelectronvolts. They obtain an effective Z by defining an index $P_e = (Z/10)^{3.6}$, and they combine the P_e's for a mixture to obtain a weighted average. Artsybashev and Ivanyukovich (1975) obtain similar Z_{eff} by taking

$$Z_{eff} = \left(\sum \frac{A_j}{Z_j f_j} \sum \frac{Z_i^4 f_i}{A_i} \right)^{1/3} \tag{6-36}$$

where f_i is the fraction by weight of the material. Table 6-1 includes values of Z_{eff}. Values of Z_{eff} by the two methods agree within a few percent. Note that Z_{eff} is *not* the average Z that is used to determine Z/A for correction for electron density.

The selective γ-γ method has not been used much in the United States up to now. In the Soviet Union it appears to be used primarily to obtain a quantitative analysis of ore grade when bulk density and Z of the mineral desired are fairly well known. Density changes slowly for low-grade ore, but apparently Z_{eff} changes fast enough for the analysis to work.

6-5 NEUTRONS

The neutron is a neutral particle of approximately the same mass as a proton (about 1800 times the rest mass of an electron). Unlike a γ-ray or a charged particle, which generally interacts with the orbital electrons of atoms in a medium, the neutron interacts almost exclusively with atomic nuclei. Consequently, neutron interactions in matter are much rarer than γ or charged particle interactions, and neutron ranges are much longer.

There is a fairly arbitrary classification of neutron energies in the literature, and it is convenient to use it (see Fig. 6-1*e*):

Fast neutrons	Above 500 keV
Intermediate	1 keV to 500 keV
Slow	Below 1 keV

Slow neutrons are further subdivided into

Epithermal neutrons	Approx. 0.1 eV to 1 keV
Thermal neutrons	Below approx. 0.1 eV

Thermal neutrons are so called because they are in thermal equilibrium with their surroundings.

6-5a Neutron Sources

Whereas γ-rays are often produced by the decay of the excited states of nuclei, neutrons are produced only by transmutation of one nuclide into another. There are three types of sources of interest in logging: spontaneous fission, mixtures of α-emitters with beryllium, and neutron generators.

Spontaneous fission. The spontaneous fission of ^{252}Cf is a prolific source of neutrons. The Cf atom, which is an artificial reactor product, splits spontaneously into two *fission products* and emits one or more neutrons. Most of the neutrons have energies between 250 keV and 2 MeV. This energy is too low for some processes of interest in logging, but it is quite satisfactory for others. The advantage of ^{252}Cf is its compactness and high neutron output. It is still an expensive and experimental source not in common use.

α-n Sources. The source in most general use is one in which an α-emitter such as Pu, Po, Am, or Ra is intimately mixed with beryllium. The α reacts with the Be to produce a neutron and also produce either three alpha particles or a ^{12}C nucleus. The notation for these reactions is:

$$^{9}\text{Be}\ (\alpha, n)\ 3\ ^{4}\text{He} \qquad \text{and} \qquad ^{9}\text{Be}\ (\alpha, n)\ ^{12}\text{C}$$

The neutrons produced have energies between 1 and 12 MeV, and most of them have energies below 4.5 MeV. These higher-energy neutrons have longer ranges than those from ^{252}Cf and therefore permit greater depths of investigation. This type of source is the most common in the industry. It is often described as Pu-Be, Am-Be, and so on, depending on the α-emitter.

Neutron generators. The reaction ^{3}H (d, n) ^{4}He, often called the deuterium-tritium (d, t) reaction, is exoergic. Deuterons (d, or ^{2}H), of 100 keV or more, impinging on tritium (^{3}H) produce neutrons with a minimum energy of 14.1 MeV. Because the threshold energy for the reaction is so low, it is possible to build portable neutron generators using it. The generator uses either a deuteron beam (from an ion source, accelerated by a potential difference of about 130 kV) impinging on a solid target with tritium adsorbed in it or a beam of mixed deuterium and tritium impinging on a scandium target (Shope et al., 1981). The neutron energy is a function of the direction of emission of the neutron and of the deuteron energy, but for our purposes it may be considered (and is usually spoken of) as 14 MeV.

The neutron generator has several advantages over the radioactive sources. First, it can be turned off, so it is not hazardous when it is not in use. Second, the energy is high, so that its depth of investigation is greater than that for radioactive sources and the 14-MeV neutrons can give rise to reactions (Sec. 6-5b) that are not copiously produced with radioactive sources. Third, it can be pulsed, and this permits a number of logging techniques that cannot be used with steady-state sources (Secs. 6-7 and 6-8). Fourth, the helium nucleus (α particle) produced recoils in a direction which is opposite to the neutron direction. One can select only those neutrons coincident with recoil α's going in a certain direction, and thus choose only those reactions caused by neutrons going in a desired direction (Peters, 1981). There are three disadvantages. First, the generator is not very reliable; even the best generators have a life of 100 to 200 h. Second, it is very expensive compared to an α-Be source. Third, generators are large compared to isotopic sources. Because of the higher neutron energy, the generator is not suitable for water-content logging.

One sometimes hears someone speak of thermal or epithermal neutron sources. This is a misconception. Thermal and epithermal neutrons, in logging situations, come only from the slowing down of fast neutrons from the sources discussed above. The only source of thermal neutrons other than slowing down (thermalization) of fast neutrons is a nuclear reactor or nuclear explosion, neither of which is suitable for use in a logging system. There are no direct sources of epithermal neutrons; they are always produced by slowing down.

6-5b Interaction of Neutrons with Matter

Because it is rare for neutrons to be absorbed until they have lost most of their energy, each neutron often has many interactions in the process of losing its energy. This process is called *slowing down.* The upper part of Fig. 6-22 is a conceptual sketch of the slowing down of a neutron. The type of interaction which is most probable is a function of the neutron energy. Figure 6-22 also illustrates the energy ranges in which the different interactions are likely and the energies of neutron sources. There are four major types of interaction: inelastic scattering, (n, x) reactions (where x is a charged particle), elastic scattering, and radiative capture. As in the case of γ-rays, the probability of scattering from or reaction with a single nucleus is called the cross section (σ) and is expressed in square centimeters. A common unit for σ is the barn; it is equal to 10^{-24} cm^2.

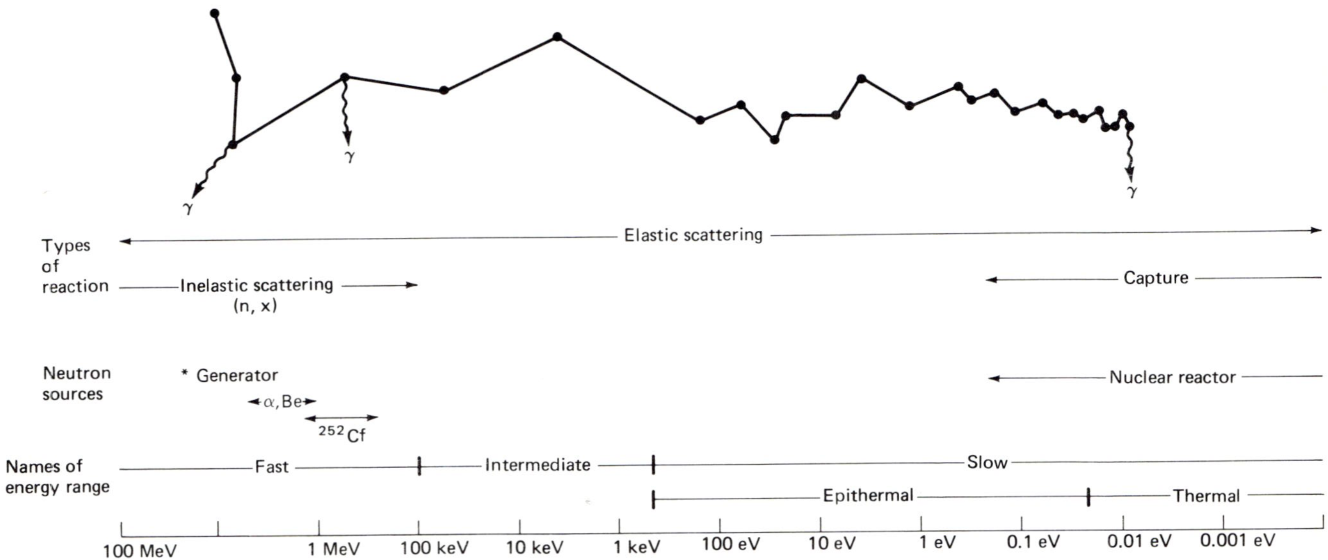

FIGURE 6-22 Neutron phenomena as a function of energy in million electron volts. At the bottom, energies produced by different sources are shown. In the center are shown the most common interactions with matter. At the top is a conceptual sketch of the slowing down of a neutron. The wiggly lines represent γ-rays produced by neutron reactions.

Inelastic scattering. A neutron sometimes interacts with a nucleus (with atomic number Z and mass number A) and forms a *compound nucleus* with mass number $A + 1$. This compound nucleus is in an excited state. It decays very rapidly to the ground state while emitting a γ-ray, and it then emits a neutron to again become an atom of the target nuclide. The reaction is written:

$$^{A}Z + n \rightarrow {}^{A+1}Z^{*} \rightarrow {}^{A}Z + n + \gamma$$

where * signifies an excited state. This type of reaction, called *inelastic scattering,* is common at neutron energies above a few million electron-volts. Figure 6-23 is a sketch of an inelastic scattering reaction. Inelastic scattering cannot take place unless the energy of the incident neutron is large enough to create an excited state of the compound nucleus. The cross sections for inelastic scattering for some common nuclides are given in Table 6-2.

The γ-rays that are produced are, of course, characteristic of the energy levels of the compound nucleus. Therefore, they can be used to identify the target nucleus. Cross sections and γ-ray energies for many earth materials are given in Caldwell et al. (1960).

(n, x) reactions. A nuclear reaction involving the release of a charged particle is identical in principle with inelastic scattering. The difference is that, when it decays, the compound nucleus emits a charged particle instead of a neutron, and so its Z changes. Therefore, one does not end up with the target nuclide. Sometimes characteristic γ-rays are emitted. Inelastic scattering and (n, x) reactions are sometimes lumped together as *nonelastic interactions.* Some of the cross sections in Table 6-2 are for calculated values and are for nonelastic interactions rather than

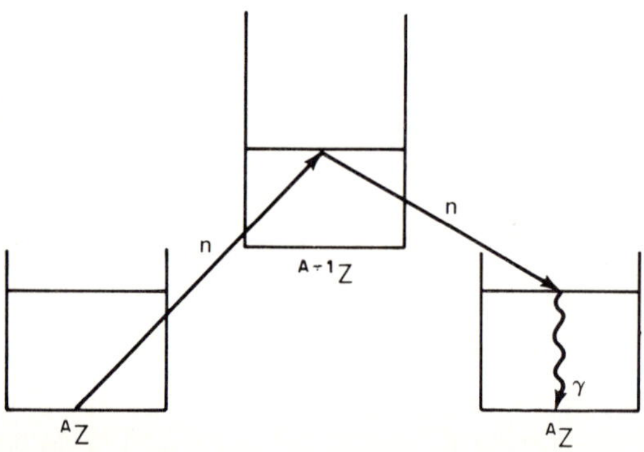

FIGURE 6-23 Energetics of inelastic scattering of a neutron.

TABLE 6-2 Neutron properties of some elements encountered in logging.

Material	$\sigma_{(n,x)}$, 14 MeV, barns	σ_{in}, 14 MeV, barns	σ_{ne} calc., barns	Inel γ energy, MeV	ξ	Coll to thermal	L_s, cm	σ_a, barns	Capture γ energies, MeV	Activ. $\tau_{1/2}$	Activ. energy
H			0.42		1.0	19		0.33	2.2		
B	0.6	0.7			0.17	89		759			
C	0.6	0.85		4.43	0.15	112	19	0.003	4.94		
N	0.8				0.13			0.07	5.26		
O	0.8		1.4	6.1, 7.0	0.12	154	18	0.0002			
Na					0.085	190		0.53	6.3 5.6 3.9	15 h	2.75
									3.8 3.6 3.1		1.37
Mg	1.0				0.081	235		0.06	3.9	9.5 m	1.02, 0.8
Al	1.0	1.06		2.2	0.072	290		0.23	7.7	2.3 m	1.78
Si	1.0			1.78	0.070	297		0.16	4.9 3.5	2.62 h	1.26
S	1.1			2.2	0.06			0.51	5.4 4.9 3.2		
Cl					0.056	343		33	8.6 7.8 7.0	37.5 m	2.16
									7.0 6.6 6.1		1.63
Ca	1.4			3.7	0.049	380		0.43	6.4 4.4 2.0		
Fe	1.4	1.4	2.6		0.035	539		2.6	7.6 7.3 6.0		
									5.9 4.2		
Cu	1.5							3.85	7.9 7.6 7.2		1.3, 1.0
											0.51
Pb	2.5	2.6	5.7	0.01				0.18			
H_2O							7.3		0.664		
SiO_2 (p = 2.65)							27				
$CaCO_3$							21–24				
Dolomite							19–20				

NOTES: $\sigma_{n,x}$ and σ_{in} are measured cross sections for (n, x) reactions and inelastic scattering (in barns.); σ_{ne} is the total cross section for both (n, x) and inelastic scattering calculated from Eq. (6-37). "Inel γ energy" is the characteristic energy (in MeV) of a common γ peak from inelastic scattering from the element. ξ is the average logarithmic energy decrement, and "Coll to thermal" is the number of collisions required in that element, on average, to thermalize a 14-MeV neutron. L_s is the slowing down length for Pu-Be neutrons to epithermal energy. σ_a is the cross section for radiative capture (in barns). Capture energies are the energies (in MeV) of common capture γ peaks. "$\tau_{1/2}$ activ." is the half-life for activation γ's, and "Activ. energy" is the energy of a common activation γ.

inelastic scattering because it is much easier to calculate the cross section for the sum of the reactions than for individual cross sections.

Elastic scattering. Perhaps the most common interaction is elastic, or "billiard ball," scattering. A compound nucleus is created in the ground state, and it immediately decays back to the original nuclide. All of the incident neutron's kinetic energy is converted to kinetic energy of the recoiling nucleus and of the emitted neutron. The reaction is written

$$^{A}Z + n \rightarrow {}^{A+1}Z \rightarrow {}^{A}Z + n$$

The cross section for elastic scattering for fast neutrons (Evans, 1955) is approximately equal to the cross section for nonelastic scattering:

$$\sigma_{el} \simeq \sigma_{ne} \simeq \pi \left(R + \frac{\lambda}{2\pi} \right)^2 \qquad (6\text{-}37)$$

where $R = R_0 A^{1/3}$ is approximately equal to the radius of the nucleus with mass number A. $R_0 = 1.4 \times 10^{-13}$ cm. $\lambda/2\pi$ is the wavelength of the neutron and is equal to $(4.55 \times 10^{-13}/E_n)$, where E_n is the neutron energy in million electronvolts.

When a neutron interacts elastically with a nucleus, the ratio of the final energy E_2 to the initial energy E_1 is

$$\frac{E_2}{E_1} = \frac{A^2 + 2A \cos \theta + 1}{(A + 1)^2} \qquad (6\text{-}38)$$

where A is the nuclear mass of the target and θ is the scattering angle in the center-of-mass system. A quantity of considerable utility in the theory of neutron transport is the *average logarithmic energy decrement*

$$\xi = \ln \overline{(E_1/E_2)} = \frac{\int_{E_1}^{E_{2\,\min}} [\ln (E_1/E_2)] P(E_2)\, dE_2}{\int_{E_1}^{E_{2\,\min}} P(E_2)\, dE_2} \qquad (6\text{-}39)$$

where $P(E_2)$ is the probability that a scattered neutron will have energy E_2. $P(E_2)$ is a function of θ, and therefore the integral is over the angle. The minimum possible value $E_{2\,\min}$ of E_2 is

$$E_{2\,\min} = \left(\frac{A - 1}{A + 1} \right)^2$$

and is the result of a head-on collision. For a target nucleus with $A = 1$ (hydrogen), $E_{2\,\min} = 0$! Consequently, hydrogen is much more effective in decreasing neutron energy than any other nuclide.

When the integral is performed, it is found that the average logarithmic energy decrement ξ is

$$\xi = \ln\,(\overline{E_1}/\overline{E_2}) = 1 + \frac{(A-1)^2}{2A}\ln\frac{A-1}{A+1} \tag{6-40}$$

For values of $A > 10$, this reduces to

$$\xi \simeq \frac{2}{A + 2/3}$$

Typical values of ξ are given in Table 6-2. Note that, in any mixture, hydrogen will greatly dominate the slowing-down process. Table 6-2 also shows the average number of collisions necessary to reduce the energy of a neutron from 14 MeV to thermal (0.01 eV) in each element. Note that hydrogen is about 5 times as effective at slowing down neutrons as boron and is even more effective than other available materials.

Radiative capture. When the neutrons become thermalized, the most probable interaction is radiative capture. A compound nucleus is formed in an excited state, and it decays to the ground state.

$${}^{A}Z + n \rightarrow {}^{A+1}Z^{*} \rightarrow {}^{A+1}Z + \gamma$$

The ground state is either stable or fairly long-lived. The *capture* γ has an energy characteristic of the final nuclide. If this nuclide is stable, there is no further decay. If it is not, it later (after a half-life of between a few microseconds and a number of years) decays, usually by β^--emission, to another nuclide. This nuclide is almost certainly in the excited state, and it emits one or more *activation* γ*'s*. These have energy characteristic of the new nuclide. For example

$${}^{23}\mathrm{Na} + n \rightarrow {}^{24}\mathrm{Na}^{*} \rightarrow {}^{24}\mathrm{Na} + \gamma \rightarrow {}^{24}\mathrm{Mg} + \beta^- + \gamma$$

Figure 6-24 is a sketch of the distinction between activation and capture γ's.

Cross sections and γ energies are tabulated in many handbooks, for example, Harvey and Hughes (1963), and some are shown in Table 6-2 together with half-lives and energies of some of the activation γ's. Note that the γ-ray energies are in the same range as those from inelastic scattering. Often the activation γ's are the same as those resulting from inelastic scattering from a nuclide with Z one unit greater or less than the one giving rise to the activation γ. This is due to the fact that

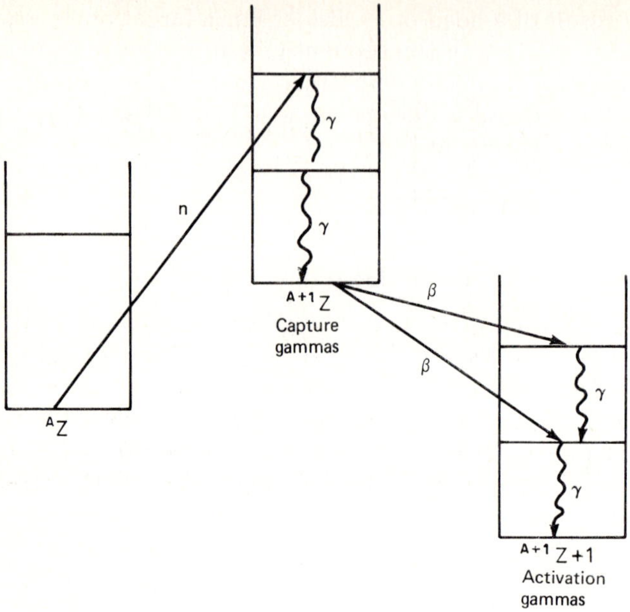

FIGURE 6-24 Comparison of capture and activation γ radiation.

the nuclide formed by the β-decay shown in Fig. 6-24 is often the same as the compound nucleus formed in the inelastic scattering process, so the same energy levels occur. The cross section varies a great deal from nuclide to nuclide. For example, the thermal capture cross section for boron is 4×10^5 times that for oxygen. Thus small amounts of one nuclide can have more effect on thermal neutron population than large amounts of another.

6-5c Interaction of Neutrons with Matter in Bulk

The interactions discussed above are characterized by their *microscopic cross section* σ—σ_s for scattering cross section, σ_a for absorption cross section, and so on. The unit of σ is usually barns, or 10^{-24} cm. When interaction with matter in bulk is considered, it is convenient to express this information in somewhat different terms. The *macroscopic cross section* Σ_s or Σ_a is often used, where

$$\Sigma = n\sigma$$

and n is the number of atoms per cubic centimeter. Therefore, the unit of Σ is cm^{-1}. Macroscopic cross sections combine by simple summing

$\Sigma = \sum \Sigma_i$. Σ is completely analogous to μ in γ-ray attenuation:

$$N = N_0 e^{-\Sigma x}$$

for a beam of neutrons passing through a thickness of material x, where N refers to the number of neutrons that has had no interaction. Then the *mean free path* λ is defined as

$$\lambda = \frac{1}{\Sigma}$$

We can have Σ_s, Σ_a, λ_s, λ_a, and so on. If the neutron density is N neutrons per cubic centimeter, the number of interactions per cubic centimeter is $Nv\Sigma$, where v is the velocity of the neutrons. (Unlike γ's whose velocity is always the velocity of light, neutrons have velocities that are related to their energies.) The quantity Nv is often called the *flux* denoted by Φ, and it has units of neutrons per square centimeter per second.

Much has been written about neutron slowing down. It is covered extensively in many texts on reactor theory (for example, Glasstone and Edlund, 1952). It will be discussed very briefly in Sec. 6-6. One quantity that is developed in that theory is of particular interest in neutron logging; it is the *slowing-down length* L_s defined as

$$L_s = \left(\frac{r^2}{6}\right)^{1/2} \tag{6-41}$$

where r is the rms crow-flight distance a neutron travels from the source to the point where it reaches the energy of interest. Physically, L_s is a measure of the ability of the bulk medium to reduce the neutron energy to a given final energy in a given distance. We will generally be discussing L_s for epithermal or thermal energy. It is a function of the initial and final neutron energies, the scattering cross sections of the materials in the medium as functions of energy, and the logarithmic energy decrement. Barsukov et al. (1965) give a formula which, while it cannot really be used to calculate L_s, is helpful conceptually:

$$L_s = \lambda_s \left(\frac{ja}{3}\right)^{1/2}$$

where λ_s is the scattering mean free path, a is a constant depending on the angular distribution of the scattered neutrons, and j is the number of collisions needed to slow down from initial energy E_1 to final energy E_2. As can be seen from Table 6-2, j is dominated by hydrogen content.

A number of authors (Kozhevnikov, 1963, for example) have calculated L_s for rocks and similar materials. A useful report is the work of Kreft (1974). He uses a method that takes into account elastic and inelastic scattering and absorption and a multigroup method that tracks the neutrons, on average, through their different energies as they slow down. Figure 6-25 shows Kreft's results for Ra-Be and ^{252}Cf neutrons in water-saturated sandstone, limestone, and dolomite. The final neutron energy is 0.215 eV. Note that the effect of hydrogen (as water) is much greater than that of either the rock matrix or the source energy.

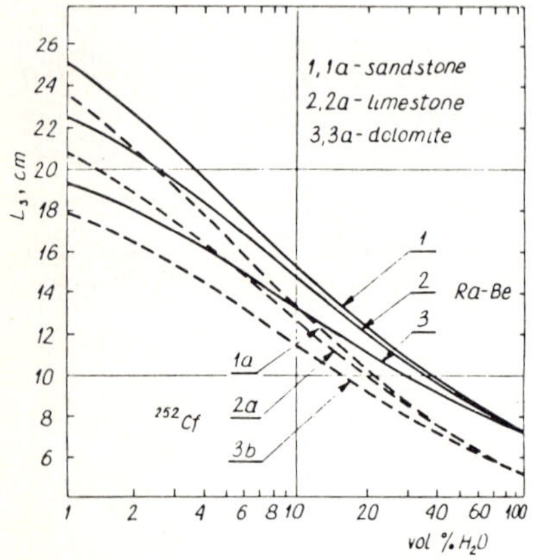

FIGURE 6-25 Slowing-down lengths L_s for limestone, sandstone, and dolomite as a function of the fraction by volume of water. (Kreft, 1974.)

A few other values of L_s are given in Table 6-2. Czubek (1981) points out that for a given material (that is, for a given matrix with a constant fraction by weight of water), the product ρL_s, where ρ is the bulk density, is a constant. That implies that if one knows L_s for a given material at one density, one can calculate it for any other density.

Shielding. Much more material is required to shield neutrons than γ-rays. The basic procedure for neutron shielding is to slow the neutrons down until they are thermalized and then capture them. Hydrogen, in the form of water or paraffin or plastic, is the most common neutron shield because of its high logarithmic energy decrement. Often, boron-loaded polyethylene is used so that the boron can capture the thermal neutrons. Under some circumstances, however, hydrogen is not the best shield. For example, if one wishes to shield a detector from a source, it is not necessary to absorb the neutrons; it is only necessary to scatter them once. The scattering mean path λ_s for 14-MeV neutrons in water is about 8 cm, in iron 2.75 cm, and in lead 3.25 cm. Therefore, if one wishes to minimize shield thickness, iron is better than water, whereas if one wishes to minimize weight, water is better. Note that boron itself is not a good shield for fast neutrons. It is useful only after the neutrons have been thermalized.

6-5d Neutron Detectors

Since neutrons do not cause ionization, the detectors used for γ-rays are often not satisfactory for neutrons. The most common neutron detectors for logging use neutron capture in the gas of a Geiger or proportional counter. The neutron is captured in a reaction, and the reaction products, which are charged, do cause ionization, which permits them to be detected.

BF_3 *counters.* A common detector uses the reaction $^{10}B\ (n, \alpha)\ ^{7}Li$, where both the α and the Li nucleus cause ionization in a tube filled with BF_3 gas. The tube is similar to a Geiger-Müller tube, and it has a similar plateau (Sec. 1.5). The gas pressure is about 86 kPa. Efficiency is fairly low.

^{3}He *counters.* A more modern detector is a similar tube filled with the artificial isotope ^3He. The reaction is $^3He\ (n, p)\ ^3H$, and both the proton and the triton cause ionization. The main advantage of this system is the ability to fill it to much higher pressure—1 MPa or more (Mills et al., 1962). The higher density makes its efficiency from 4 to 20 times greater than that of BF_3, depending on neutron energy.

Epithermal neutron detectors. To restrict detection to only neutrons of greater than thermal energy, a gas detector, which is sensitive to neutrons of any energy, is surrounded with a thin shield of Cd or In. The capture cross section for Cd increases with decreasing energy, and it has a very large increase for energies at about 0.18 eV. (The cross section is two orders of magnitude lower at 0.6 eV.) So a Cd shield will prevent thermal neutrons from reaching the detector with very little loss of neutrons above 1 eV. Sometimes one speaks of "epicadmium" neutrons.

γ-Ray detection. For some applications, the neutrons are not detected; only the γ-rays from scattering or capture are detected. Then the γ-ray detectors discussed in Sec. 6-2c are used.

6-6 NEUTRON LOGGING (FOR WATER CONTENT)

As neutrons from a point source interact with matter, they diffuse, slow down, and eventually are captured. Because the amount of hydrogen in the medium dominates the slowing-down process, it is possible to use the slowing down of neutrons to infer the amount of hydrogen present. When neutron logging was first developed, the neutron density was found by detecting the capture γ's. The number of thermal neutrons that produce the capture γ's, however, and the number of γ's produced, is strongly dependent on the presence of small amounts of strong absorbers such as boron and chlorine. In fact, Czubek (1979) describes the use of thermal neutron detection for quantitative analysis of ores of strong absorbers. Also, more efficient neutron detectors have become available. So in recent years, neutron density has been measured directly. Because they are less influenced by the presence of absorbers, epithermal neutrons are often detected.

But the capture γ's are still useful. γ-Ray spectroscopy is used to determine the presence of elements that emit capture and activation γ-rays after neutron irradiation.

6-6a Neutron Diffusion

There is no neutron equivalent of the single-scattering density system. All neutron logging systems involve diffusion and slowing down of neutrons. The theory of neutron slowing down and diffusion was first developed for nuclear reactors (Glasstone and Edlund, 1952). Two groups, one at Mobil Research Laboratories (Tittle et al., 1951; Tittle, 1961; Tittle and Allen, 1966; Allen et al., 1967) and one at the Moscow Oil Institute "Acad. I. M. Gubkin" (Barsukov et al., 1965) have published much of the theory of neutron logging. We shall describe it very briefly. Neutrons, like γ's, obey the diffusion equation (6-29) which describes conservation of neutrons. That is,

$$\text{Production} + \text{scattering in} - \text{absorption} - \text{leakage} = 0$$

It is generally expressed in terms of the neutron flux Φ (the product of the number of neutrons per cubic centimeter and their velocity—the units are neutrons per square centimeter per second)—and in the steady state is

$$D\,\nabla^2\Phi - \xi\Phi\Sigma_s + S = 0 \tag{6-42}$$

where $\xi\Sigma_s$ is the *slowing-down power* of the medium and is the probability that a neutron will have its energy reduced to a value lower than that of interest. (Capture is disregarded in this development; if it is present, Σ includes capture cross section.) For a point source in an infinite medium the usual solution quoted is

$$\Phi = \frac{Q}{4\pi\xi\Sigma_s L_s^2 r}\exp\left(\frac{-r}{L_s}\right) \quad \text{(Tittle, 1962)} \tag{6-43}$$

where the average logarithmic energy decrement ξ, the slowing-down length L_s, and the scattering cross section Σ_s are energy-dependent. In Eq. (6-43) they depend only on the final neutron energy. They are, of course, also functions of the material in the medium, especially the hydrogen. Q is the source strength in neutrons per second, and r is the distance from the source to the point in question.

This calculation assumes that all neutrons have the same velocity. Of course, that is not true; the neutrons are continually slowing down. The theory accounts for this situation by *multigroup diffusion theory* in which the neutrons are treated as being in several velocity (or energy) groups. Then the scattering cross section refers to the probability of a neutron being reduced in energy enough that it is then in a lower-energy group. Equation (6-42) becomes

$$D_1 \nabla^2 \Phi_1 - \xi_1 \Sigma_{s1} = 0 \qquad (6\text{-}44)$$
$$D_2 \nabla^2 \Phi_2 - \xi_2 \Sigma_{s2} + \xi_1 \Sigma_{s1} = 0$$

The absorption term for the first group is the source term for the second, and so on for n groups. Multigroup diffusion theory has been shown to agree well with experiment in both cylindrical and spherical geometry for media of different hydrogen content (Olgaard, 1965; Tittle, 1962). Figure 6-26 shows Tittle's (1961) result for neutrons in three media of different water content. But no simple analytical solution that could be inverted to find hydrogen content has been published.

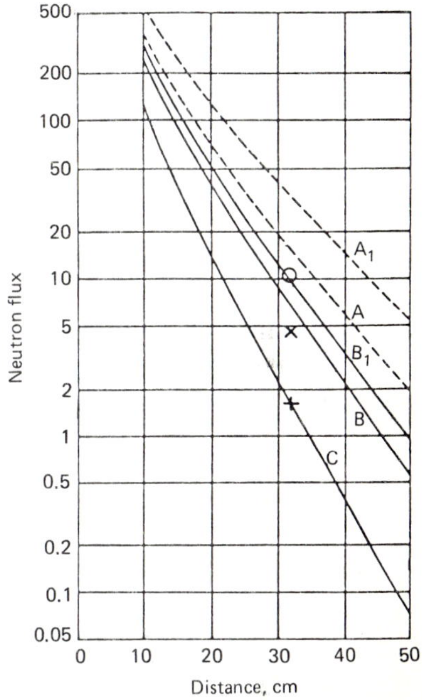

FIGURE 6-26 Theoretical epithermal neutron flux on the axis of a 6-in-diameter (15.24-cm-diameter) water-filled borehole in saturated limestone of porosity equal to 0.1 (curve A) and 0.3 (curve B) compared to that in an infinite medium of limestone of porosity 0.1 (curve A_1) and 0.3 (curve B_1) and in pure water (curve C). (Tittle, 1961.)

Monte Carlo calculations, which are computer simulations of slowing down of individual neutrons that take into account detailed cross sections and energy decrements as functions of scattering angle, have been used to obtain detailed predictions of flux versus position (Ellis et al., 1981; Sanders, 1983). They are more accurate than multigroup diffusion theory, but they take a good deal of computer time. However, they are the most practical method for complex geometry.

Another quantity of interest, the *slowing-down density* defined as the number of neutrons slowing down past a given energy per unit volume per unit time, is

$$q = \xi \Phi \Sigma_s$$

Two different expressions for q can be obtained from the literature. From Eq. (6-43),

$$q = \frac{Q}{4\pi L_s^2 r} \exp\left(\frac{-r}{L_s}\right) \tag{6-45a}$$

for a hydrogenous medium. Another version (Barsukov et al., 1965) is

$$q = \frac{Q}{8\pi L_s^3} \exp\left(\frac{-r}{L_s}\right) \tag{6-45b}$$

Two-detector diffusion theory. Allen et al. (1967) point out that the ratio of the flux at two different radii removes the dependence on ξ and Σ_s. For any form of Eq. (4-2) the ratio of the fluxes from a point source in an infinite medium at radii r_1 and r_2 is

$$\frac{\Phi_{r1}}{\Phi_{r2}} = K_1 \exp\left(\frac{r_1 - r_2}{L_s}\right) \tag{6-46}$$

The ratio of slowing-down densities q for any form of Eq. (6-45) also satisfies Eq. (6-46). With two-group diffusion theory and borehole geometry they obtain a very complex expression for the ratio of fluxes, and it agrees well with experiment. However, they show that for values of r greater than about 70 cm, Eq. (6-46) is a good approximation.

6-6b Neutron Logging for Water Content

At present, the two common methods for water content logging are the epithermal neutron log and the two-detector *compensated neutron log*. In the literature equations or plots are normally in terms of porosity ϕ, but in fact the sondes are sensitive primarily to the *hydrogen index* I_H. The hydrogen index is defined as the volume fraction of fresh water that would contain the same amount of hydrogen. In a rock of porosity ϕ and saturation S_w, containing fresh water and no other hydrogen, $I_H = \phi S_w$. If, for example, the rock were saturated with salt water, chlorine would displace some water and I_H would be less than ϕS_w. If shale, containing bound water, were present, I_H would be greater than ϕS_w. I_H of oil is not necessarily equal to I_H of the same volume of water. We shall use I_H in equations which, in the original, use ϕ.

Epithermal neutron sondes. Single-detector systems often employ epithermal neutron detection because thermal neutron flux is strongly affected by small amounts of chlorine, boron, and other strong absorbers. Most sondes are eccentered—forced against the borehole wall. In most cases, as in the case of density sondes, empirical calibration curves (Fig. 6-27; Tittman et al., 1966) are used to relate count rate J to hydrogen index I_H. The more complex theory is used to design, not to calibrate, the sondes. Notice that in Fig. 6-27 there are different calibration curves for sandstone, limestone, and dolomite because L_s as a function of water content is different for each rock medium.

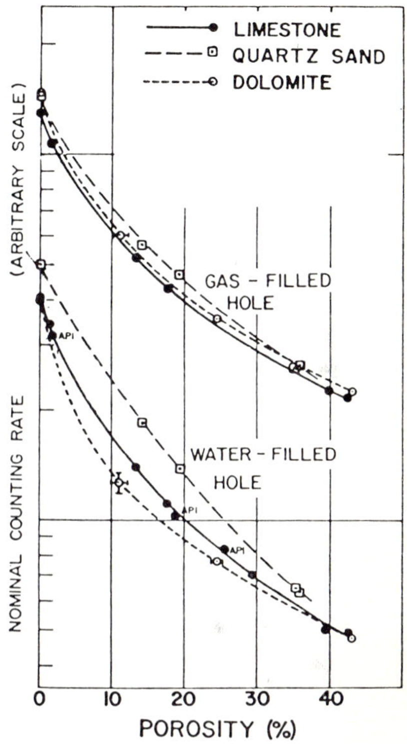

FIGURE 6-27 Empirical calibration curves for epithermal neutron sonde. (Tittman and Wahl, 1965. Copyright 1966, SPE–AIME.)

It is an experimental fact that for saturated media, for most eccentered epithermal neutron sondes, J is directly proportional to L_s (Edmundsen and Raymer, 1979; Czubek, 1981). Figure 6-28 shows a typical example. Note that all three materials plot on the same line. The proportionality of count rate to L_s alone is not consistent with theory [Eq. (6-45*a*)], which states that Φ is a function of L_s, ξ, and Σ, all of which vary with I_H. Czubek (1981) suggests that the proportionality to L_s could be explained by assuming J was a function of q rather than Φ, since q in Eq. (6-45) is a function only of L_s. This is a plausible argument, because q is a measure of the number of neutrons that slow down past epithermal energy per unit time, and it might therefore be associated with epithermal count rate.

But Eqs. (6-43) and (6-45) show that neither Φ nor q is directly proportional to L_s. There is always an L_s^2 or L_s^3 term in the denominator multiplied by exp $(-r/L_s)$. It can be shown that this proportionality to L_s is in part fortuitous. Most epithermal neutron sondes have a source-detector spacing between 30 and 40 cm. For r = 40 cm and values of L_s for earth materials (Fig. 6-25), exp $(-r/L_s)$ is, to very good approximation, directly proportional to L_s (Fig. 6-29). Although this is still unexplained by theory, it is a possible explanation of the experimental data.

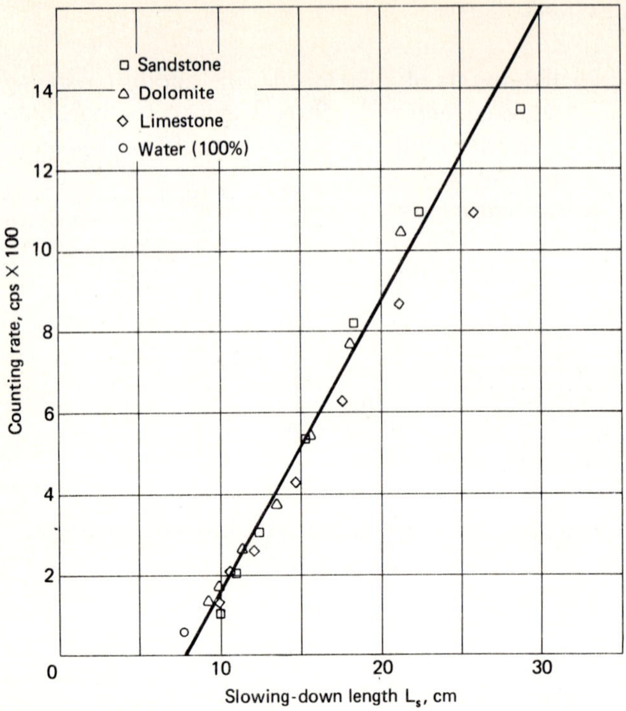

FIGURE 6-28 Count rate J versus slowing down length L_s for an epithermal neutron sonde. (Edmundsen and Raymer, 1979.)

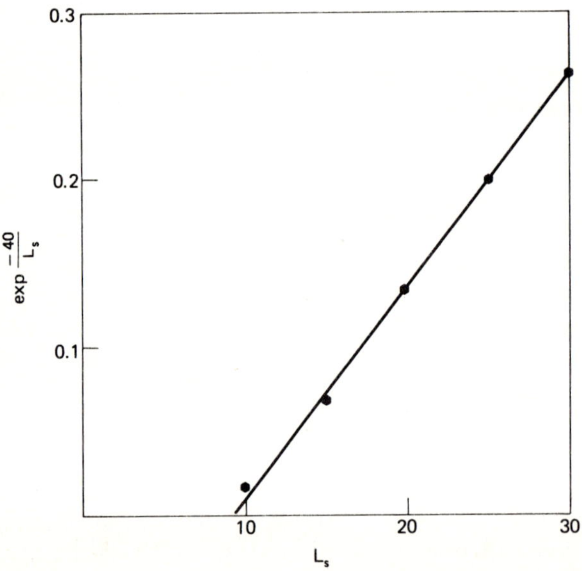

FIGURE 6-29 Exp $(-r/L_s)$ versus L_s for r = 40 cm.

Unfortunately, J is not directly proportional to L_s when there is a good deal of air-filled porosity. Figure 6-30 is a plot of J versus L_s for materials with a wide range of density and water saturation. L_s for the proportion of water to solid was determined from Fig. 6-25, and adjusted for the bulk density by Czubek's (1981) method (Sec. 6-5c). The points not lying near the line have a great deal of air-filled porosity (Hearst et al., 1981). Calculations (Ullo, 1981) confirm that while J is proportional to L_s for saturated material, it is not when gas is present.

Another formula (Wood et al., 1974) for J versus I_H is often plotted on exponential paper (Fig. 6-31). It states that

$$\exp(-K_1 I_H) = K_2(J_a - J_t) \tag{6-47}$$

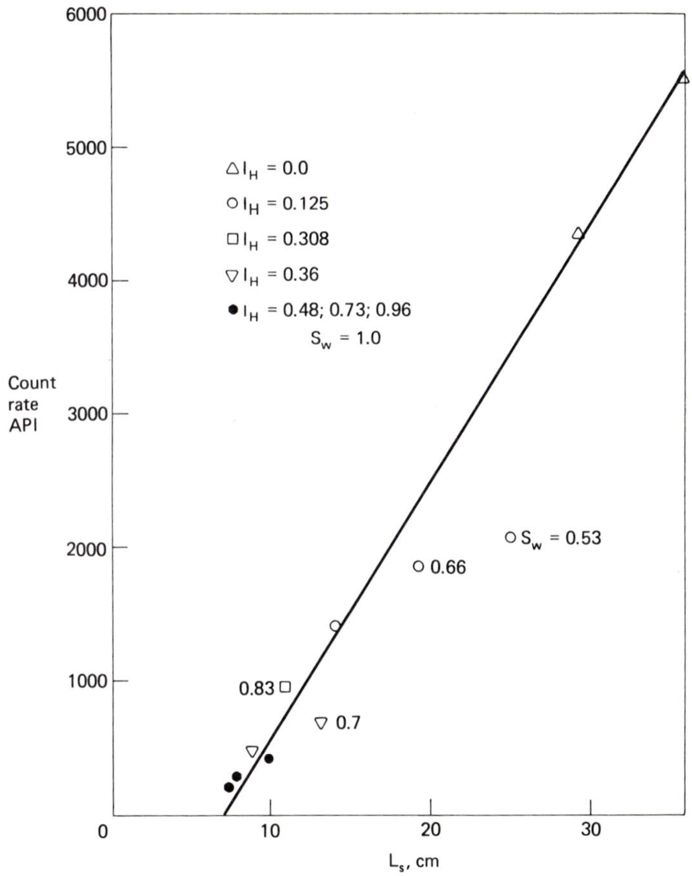

FIGURE 6-30 Count rate J versus slowing-down length L_s in mixtures of SiO_2 and Al_2O_3 for an epithermal neutron sonde. Values of S_w are shown opposite four points with I_H greater than 0. (Data from Hearst et al., 1981.)

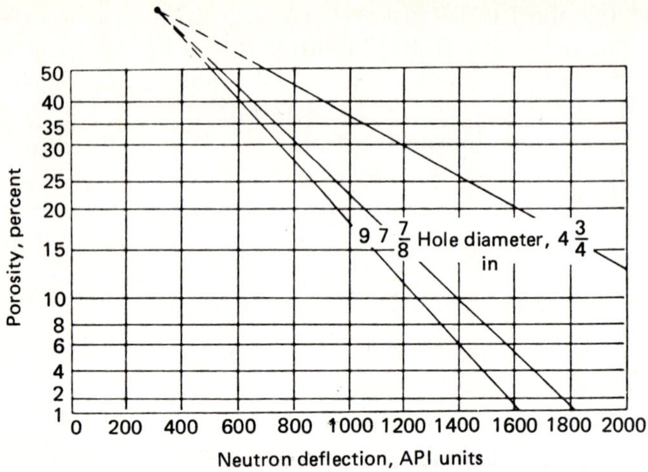

FIGURE 6-31 Exponential calibration of a neutron sonde. Count rate, or neutron deflection, versus exp (porosity). (Wood et al., 1974.)

where J_a is the measured count rate and J_t is "a constant representing that part of J_a attributable to material other than the formation." Wood et al. state that Eq. (6-47) can be derived from theory but do not give the theory. They also state than an empirical expression for the same data is

$$\log I_H = K_1 - K_2 J_a$$

which implies

$$I_H = K_1 \exp(-K_2 J_a) \tag{6-48}$$

No theoretical justification is claimed for this expression. It appears to have come from empirical results of neutron-γ logging (Scotty and Egan, 1952).

We may conclude from all of the above that no simple theory explains the real behavior of eccentered epithermal neutron sondes but empirical formulas can be found.

Two-detector neutron sondes. Most two-detector systems use thermal neutrons because at the distances recommended by Allen et al. (1967), 70 cm or greater, the epithermal flux is too small to obtain good statistics. The ratio of count rates in two detectors J_1/J_2 can be empirically related to I_H for different materials (Fig. 6-32). Edmundsen and Raymer (1979) show that J_1/J_2 is directly proportional to $\exp(-K/L_d)$, where L_d is the *migration length,* the thermal equivalent of L_s. Figure 6-33

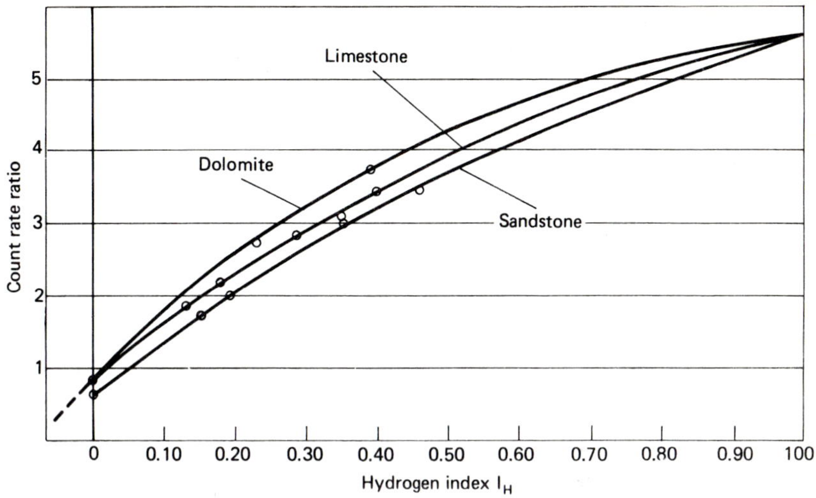

FIGURE 6-32 Ratio of count rates in two detectors versus hydrogen index I_H in several materials. (Edmundson and Raymer, 1979.)

shows the data from Fig. 6-32 plotted in the form of log J_1/J_2 versus $1/L_d$. We see that Eq. (6-46) is satisfied. Equation (6-46) is valid for either Φ_1/Φ_2 or q_1/q_2, so we do not know to which of these quantities (if either) J_1/J_2 is really proportional. As in the case of L_s for the epithermal neutron log, L_d unifies the curves for different materials.

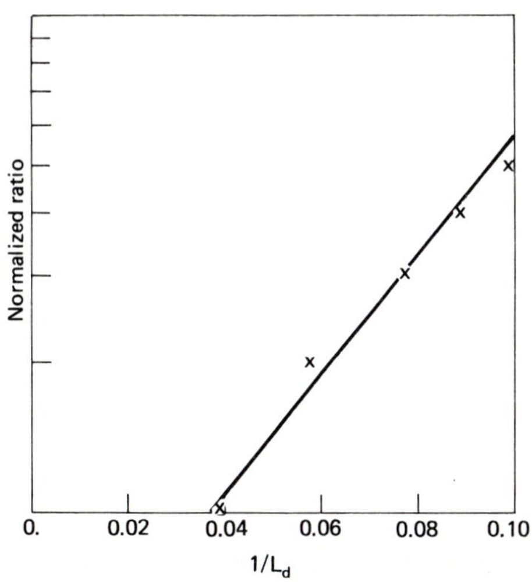

FIGURE 6-33 Ratio of count rates in two detectors versus $1/L_d$; same data as Fig. 6-32.

Allen et al. (1972) found that I_H obtained from real two-detector sondes by using thermal neutrons is often affected by small amounts of absorbers, and they cite anomalous results in some systems. Davis et al. (1981) deal with this problem by adding a pair of epithermal neutron detectors. But then shorter source-detector distances must be used and Eq. (6-46) is no longer exactly satisfied. They find that one must use spine-and-ribs plots similar to Fig. 6-19, with curved ribs, to obtain I_H.

Ellis et al. (1981) show that Monte Carlo calculations agree with two-detector as well as with single-detector data. They can account for many perturbations with these calculations.

It is important to note that most service companies state that two-detector systems cannot be used in dry holes, and we have seen one such system give nonsensical results in a dry hole. The count rate in dry holes evidently saturates the detector systems designed for wet holes. It would be possible, however, to build compensated neutron sondes for dry holes if it were so desired.

6-6c Calibration

One of the API test pits (Belknap et al., 1960) was designed for calibration of neutron sondes. It contains 6-ft-diameter (1.8-m-diameter) blocks of limestone with fresh-water-filled porosities of 0.019, 0.19, and 0.26. The API neutron unit is defined as 1/1000 of the count rate in the 0.19 block. Most companies also have their own test pits.

As in the case of density sondes, empirical smooth curves are passed through the calibration data (Fig. 6-27, for example). The surface electronics (panels) are programmed to calculate "porosity" from the count rate by using the empirical curves. Generally one dials in the medium, so that different calibrations for each medium can be used. Consequently, if the real medium is not the one assumed in the calibration curve, corrections are required.

Σ_s and L_s are functions of bulk density as well as water content. One reason different media have different calibration curves is that their matrix densities differ. Most companies do not include bulk density in their calibrations, but instead publish correction curves for different densities (Sec. 6-6d). Some users (Hearst, 1979; Hearst et al., 1981) calibrate in a facility that has several bulk densities at each of several values of I_H and then fit semiempirical calibration curves to both density and I_H.

Neutrons have longer range than γ's, so it is impractical to carry realistic calibrators to the field. Therefore, most loggers use a reference device called an *environmental calibrator* to check their equipment in the field. The device is a polyethylene fixture that is placed on the sonde and has one or more removable polyethylene blocks to change the neutron count rate and simulate several fixed values of I_H. It is so constructed that each position of the blocks corresponds to a certain number of API units. This method of checking sonde response has two advantages:

1. By converting count rate to API units, the method eliminates the effect of source strength, which permits different sources to be used and deals with the problem of source decay.
2. The method eliminates the effect of changes in discriminator settings and of drift in the electronics. This is acceptable within limits, but it is possible for the discriminator or the high voltage on the counter to be set to count γ's as well as neutrons and still give an apparently valid calibration. It is difficult for the user to determine whether the electronics have been set properly, and the calibration does not help a great deal.

6-6d Corrections

Gap. Like that in the γ-ray sonde, the count rate in the eccentered neutron sonde is affected by the gap between the sonde and the borehole wall. In an air-filled hole the effect, like that for the γ-γ sonde, is to increase the count rate

$$J = J_0 B(I_H, G)$$

Figure 6-34 shows B versus I_H and G. In water-filled holes, $B < 1$, and it *decreases* with increasing gap. This is quite reasonable, because a water-filled gap looks like more formation water. Most logging companies publish correction curves for gap, for both one- and two-detector sondes, but none routinely measure the gap. So the user must guess.

Hole size. Count rate is also affected by hole diameter, and correction curves for hole size are published. The effect can be quite large ($\Delta I_H \simeq 0.05$ for a diameter change of 10 cm; Tittman et al., 1966). Hole size can be obtained from a caliper log.

Material. The sonde has different calibration curves in different materials, and one must make a correction if the panel is not set for the right medium. These curves are published by all companies. This is no problem as long as one is logging limestone, sandstone, dolomite, or one or two other standard rock types. There are, however, no settings or correction curves for basalt, iron ore, or other nonsedimentary formations. If one has access to values of L_s for the media of interest, one can produce such corrections; otherwise, one must guess.

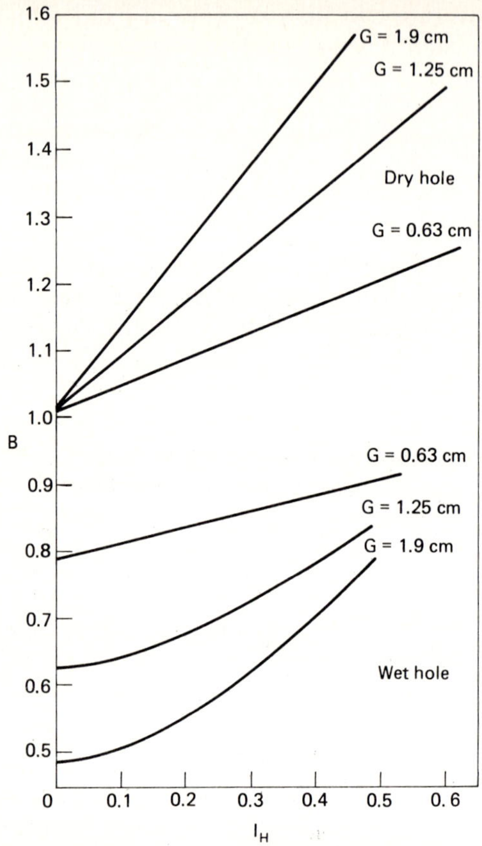

FIGURE 6-34 Count rate multiplier B versus hydrogen index I_H for different values of sonde-wall gap G in dry and water-filled boreholes.

Shale, chlorine, and so on. The neutron sonde measures I_H; most logs are plotted in terms of porosity ϕ. Chlorine, in salt water, with its high thermal cross section, has a strong effect on sondes that detect thermal neutrons (nowadays generally two-detector sondes), and correction curves are published for salt in water. (We discuss measurement of chlorine in Sec. 6-7.) Salt not only has a high capture cross section, but also displaces water in solution; a solution with 10 percent by volume NaCl has only 90 percent by volume water and only that amount of water will be shown on a neutron log. Consequently the sonde, while it measures the amount of water, does not measure the amount of saturated porosity when there is a substantial amount of solute. The correction curves include this effect. But the user may want to know the amount of water, not of porosity, and must then use care in applying the corrections.

In the case of shale, much of the water is present as bound water. The sonde measures this, but the water is not in porosity and so corrections are applied to measured porosity when shale is present. Again, the corrections must be used if one wants porosity but must be ignored if one wants total water.

Other hydrogenous materials, such as coal, will give spurious values of water but true values of I_H. If the distinction is kept in mind, the difficulties will be minimized. The issue of porosity will be discussed at length in Part 3.

Density. Most sondes are calibrated, not in terms of L_s, but rather in terms of I_H in a fully saturated medium. If the density differs from that of a saturated medium (for example, if gas is present), a correction (called by one company the "excavation effect"!) is applied. Ullo (1981) has calculated the effect on the two-detector system under many circumstances. Many companies now publish correction curves for density; often the curves are expressed in terms of S_w.

Statistics. The count rate in a neutron sonde, like that in a density sonde, is Poisson-distributed and the precision of $J = \sqrt{J}$. Because dI_H/dJ is much larger at low J than at high, the measured value of I_H is much more sensitive to fractional errors in J at low J than at high. Therefore, the choice of time constant and logging speed [Eq. (2-13)] is important. Some companies now use adaptive filtering (Schultz and Thadani, 1981) to average J over time. By averaging over longer times when J is low than when J is high, they compensate for the smaller precision at low values of J. Good filter design is required to minimize loss of vertical resolution. Averaging is more important in neutron logging than in density logging, although a similar, but smaller, change in sensitivity occurs for $d\rho/dJ$.

Mud cake, temperature, and casing. Some companies (Tittman et al., 1966; Wilson and Wichmann, 1974; Alger et al., 1972) publish correction curves for temperature, mud density, mud cake thickness, and other perturbations. These are not trivial but are specific to the equipment, and they are often automatically compensated for in the panel settings. The user must investigate the correction curves for the sonde he intends to use and judge the need for corrections.

Neutrons are not as strongly affected as γ's by steel casing, so the neutron sonde is often used for logging through casing. Logging companies provide correction curves for cased-hole logging. There is, however, usually cement behind the casing, and cement contains water of hydration. The nearer to the hole a certain volume of water is, the more effect the water will have on a sonde. Therefore, the cement, rather than the casing, will be the major cause of perturbation. Some correction curves allow for cement.

Depth of investigation. Depth of investigation of a neutron sonde is determined in much the same way as that of a density sonde. It is a function of both I_H and ρ, although I_H has a much greater effect than ρ. Ferronskiy et al. (1968) give, for a 50-cm sonde, a depth of investigation (the depth from which 90 percent of the data are obtained) of 97 cm in dry sand with $\rho = 1.65$ g/cm^3 and 58 cm for dry sandstone with $\rho = 2.65$ g/cm^3. For the $\rho = 1.65$ sand, with $I_H = 0.10$, the depth is 44 cm. Ferronskiy states that the depth of investigation is the product of L_s and a function of the sonde length. Sherman and Locke (1975) give 17 cm for 39 percent porosity sand, fully saturated. This appears consistent with Ferronskiy's result in similar sand with lower I_H.

6-7 NEUTRON DIE-AWAY LOGGING

There are two common ways of using a pulsed neutron generator (Sec. 6-5a). The first makes use of the change in neutron flux versus time after a pulse. This is called *neutron die-away*, or *neutron lifetime*, or *thermal decay time* logging. It was first developed to measure the chlorine content of formation fluid so that an ordinary neutron log could be corrected for the amount of chlorine (Sec. 6-6d). The second entails spectroscopy of the γ's emitted in the neutron reaction, and that will be discussed in Sec. 6-8.

6-7a Basics

Youmans et al. (1964) give a good description of die-away:

> When a short pulse of 14 MeV neutrons is emitted by a source in a borehole, the individual neutrons are slowed to thermal energy within a few μs. Thus a cloud of slow neutrons is formed around the source. . . . The life of each neutron is terminated by capture in the nucleus of some formation constituent which it encounters along its path. This event is marked by emission of capture gamma-rays. . . . At any instant, the number of capture gamma-rays in a given medium is proportional to the number of "live" neutrons.

The thermal neutrons are traversing the medium with average velocity v. The number present at any time is related to the macroscopic capture cross section Σ_a by (Sec. 6-5b)

$$N = N_0 \exp(-\Sigma_a x)$$

and, since $x = vt$ (where t is time)

$$N(t) = N_0 \exp(-\Sigma_a vt) \tag{6-49}$$

In reality, because the sonde is in a borehole, and the borehole fluid usually has a value of Σ_a different from that in the formation, the number of neutrons that reach a detector some distance from the source does not satisfy Eq. (6-49) until the neutrons in the borehole have ceased to affect the situation. The standard technique for measuring the rate of die-away is to take the ratio of γ-ray count rates at two different times some time after the neutron pulse in the borehole has decayed (Fig. 6-35).

$$\frac{J_1}{J_2} = \exp[\Sigma_a v(t_2 - t_1)] = \exp\left(\frac{\Delta t}{t_N}\right) \tag{6-50}$$

where $t_N = 1/\Sigma_a v$ is called the *lifetime* or *decay time*. In real systems, counts are summed over time gates of finite width, as shown in Fig. 6-35.

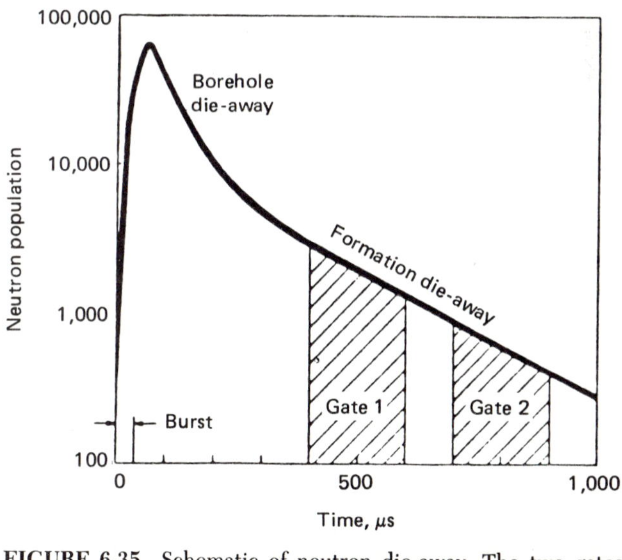

FIGURE 6-35 Schematic of neutron die-away. The two gates occur during the exponential part of the decay. (Youmans and Hopkinson, 1975. Copyright 1975, SPE–AIME.)

Properties of elements and rocks. Since macroscopic cross sections are additive in a formation containing water and hydrocarbons,

$$\Sigma_{\text{total}} = S_w\phi\Sigma_w + (1 - S_w)\phi\Sigma_h + (1 - \phi)\Sigma_{ma} \qquad (6\text{-}51)$$

where S_w is water saturation and Σ_w, Σ_h, and Σ_{ma} are the macroscopic capture cross sections for water, hydrocarbon (or other pore fluid), and matrix, respectively. Equation (6-51) and its permutations are the basic expressions for the use of neutron die-away.

If S_w is to be obtained from Eq. (6-51), values of Σ for fluid and rock are needed. Table 6-3 lists Σ_a for typical materials of interest. [To permit use of convenient numbers, Σ_a is generally listed in *capture units* (c.u.), which are 10^{-3} cm^{-1}.] The element of most interest in die-away logging is chlorine. It is abundant in salt water, and its t_N is much shorter than that for most other materials encounted in oilwells. Allen and Mills (1974) have published methods for laboratory measurement of these quantities. Rinehart and Weber (1975) designed a cell for field measurement of Σ_w. It simply uses a small tank, a steady-state neutron source, and a detector. The cell is calibrated with materials of known Σ.

6-7b Perturbations

Borehole effects. Locke and Smith (1975) show that a 27-cm-diameter borehole filled with fresh water (that is, no chlorine) removes all sensitivity of J_1/J_2 to Σ above a Σ of about 20 c.u. (Fig. 6-36). This is plausible because the lifetime of the neutrons in the hole is very long compared to that in the formation when the formation cross section is large, and so γ's entering the detector come from the borehole fluid and not the formation.

TABLE 6-3 Capture cross section for various materials. From Youmans et al. (1964).

Material	Σ, c.u.	Material	Σ, c.u.
Dry limestone ($CaCO_3$)	7	Shale	20–40
Limestone, 10% by vol. NaCl solution, 100,000 ppm	12.1	Dry anhydrite ($CaSO_4$)	12.1
		Dry dolomite $CaMg(CO_3)_2$	11.9
Limestone, 30% by vol. NaCl solution, 100,000 ppm	22.3	Methane (10.5 MPa, 38°C)	3.8
		Fresh water	22.2
Dry sandstone (SiO_2)	3.5	Crude oil	22.2
Sandstone, 10% by vol. NaCl solution, 100,000 ppm	8.9	NaCl soln., 30,000 ppm	31.7
		NaCl soln., 100,000 ppm	56
Sandstone, 30% by vol. NaCl solution, 100,000 ppm	19.8	NaCl	726
		Portland cement	13

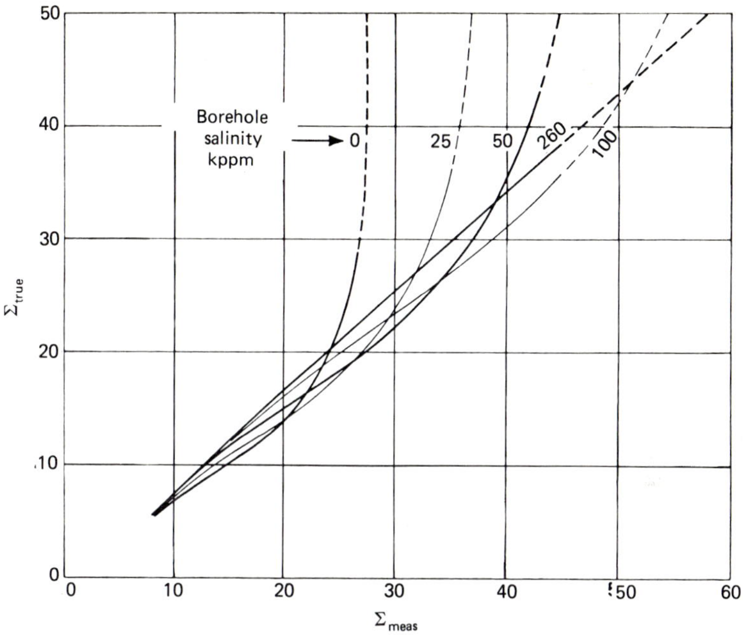

FIGURE 6-36 Two-group diffusion calculation of capture cross section that would be measured with a die-away sonde for different values of true capture cross section and different borehole fluids. (Locke and Smith, 1975.)

Youmans et al. (1970) state that the measured value of Σ is independent of borehole conditions *if* the lifetime in the hole is less than that in the formation. Polyachenko et al. (1969) say "A condition for pulsed neutron logging to be informative is $t_{hole} < t_{form}$." Consequently, either boreholes filled with fresh water (or oil!) which have long lifetimes must be avoided or special means must be used to deal with the problem.

Youmans and Hopkinson (1975) avoid the effects of long neutron life in a fresh-water-filled borehole by setting energy discrimination to select only γ's with energy greater than that of the 2.2-MeV capture γ in hydrogen. Then the neutrons in the hole will not give rise to any γ's that are counted, and the effect of the hole will be minimal. Schultz et al. (1983) use six time gates, and they fit separate exponentials to the borehole and formation components of the decay.

Diffusion. There is considerable controversy about the effects of neutron diffusion on measurement of lifetime. Equations (6-49) and (6-50) are correct for a beam of neutrons traveling through an absorber. But, as we have seen in Sec. 6-6, neutrons diffuse rather than travel in a beam. The quantity t in Eq. (6-50) is not really the time for the straight-line distance x; instead, it is the total time. Consequently, one must consider diffusion. The time-dependent version of the diffusion Eq. (6-42) is

$$\frac{1}{v}\frac{d\Phi}{dt} = -D\,\nabla^2\Phi + \Phi\Sigma_a \qquad (6\text{-}52a)$$

and since $\Phi = vN$,

$$\frac{dN(r, t)}{dt} = -D'\,\nabla^2 N + \Sigma_a vN \qquad (6\text{-}52b)$$

Then Eq. (6-49) is satisfied if $\nabla^2 N = 0$. If not, a diffusion time t_d is sometimes used (Wahl et al., 1970) to describe the effect of diffusion:

$$\frac{1}{t_d} = D'\frac{\nabla^2 N}{N}$$

where D' is proportional to L_s^2 and is therefore a function of medium properties, particularly I_H, the water content.

Polyachenko et al. (1969) calculate the effect of diffusion for a cylindrical geometry with a "black" borehole—one with lifetime in the hole equal to zero. They find

$$N(r, t) = e^{-\lambda t} \qquad \text{where} \qquad \lambda = \Sigma v + \Delta = \frac{1}{t_N} + \Delta \qquad (6\text{-}53)$$

and Δ is a function of the medium properties, especially L_s. They then find an optimum condition for which $\Delta \simeq 0$. With $I_H > 0.05$ and the hole radius greater than 10 cm, a source-detector distance of 70 cm will make Δ less than 5 percent of $1/t_N$. Locke and Smith (1975) also calculate diffusion effects (Fig. 6-36). For many source-detector distances used in die-away logging they find that a substantial diffusion correction is required. Logging companies that believe this publish curves to correct the casing, hole diameter, borehole fluid, and porosity.

Wichmann (1972*b*), however, studied the measured value of Σ under various conditions with different porosities (therefore, different values of L_s) and concluded that no diffusion effect exists. Cement, borehole fluid, borehole size, and porosity do not affect the measurement. Moreover, Randall et al. (1978) have made tests of measured Σ versus source-detector distance and specifically disagree with Polychenko et al. They find that the measured Σ is indeed a function of source-detector spacing but that "conclusive evidence for a diffusion correction to the normal [die-away] measurement of Σ is nonexistent."

Neumann (1978) disagrees with them and states that their data exhibit diffusion effects and are consistent with the data of Locke and Smith (1975).

These conflicting results cause a dilemma for users. Can Eq. (6-50) be used without correcting for diffusion? We propose a possible explanation. Youman's measurements were made in a situation in which Σ was quite low, $\approx$ 10 c.u. Figure 6-36 shows that, at low values of Σ and high borehole salinity (when borehole effects are small), the difference between Locke and Smith's (1975) calculated values of Σ_{meas} and Σ_{true} is a little over 2 c.u. According to Wichmann (1972*b*), the value of Σ of materials in the 10-c.u. range are not known well enough to be able to tell if he would have observed a change of 2 c.u. In Wichmann's experiment at high Σ the material was a boric acid solution, and diffusion would be small. Consequently, there is no publication of a clear test of the observability of diffusion effects.

Although this discussion, to some extent, explains the disagreement among researchers, it does not help the user. With normal values of the parameters in Eq. (6-51), a change in Σ of 2 c.u. discussed above can cause a change in calculated S_w of 0.15, a very significant amount. Caution is recommended. There are at least two methods for evading the problem. Some users deal with the problem by measuring *changes* in Σ when the fluid saturating the formation is changed but the borehole fluid is not (see Chap. 11). In that situation Fig. 6-36 shows that $\Sigma_{\text{true}} - \Sigma_{\text{meas}}$ is approximately constant for a $\Delta\Sigma$ of, say, 10 c.u. In that case the effects of diffusion are academic. Others only use die-away semi-quantitatively in fairly high porosity formations containing brine. They can distinguish between oil or gas and brine without need for accurate values of Σ.

6-7c Die-away Sondes

The differences among companies are in source-detector spacing, gate timing, energy discrimination, and details of construction.

Time gates. In Eq. (6-50) we saw that the lifetime could be determined from the ratio of the counts at two different times. One company (Wahl et al., 1970) uses two time gates, one immediately following the other. If the first gate is exactly t_N wide and the second is exactly $2t_N$ wide, it can be shown from Eq. (6-49) that $J_1/J_2 = 2$. A feedback loop sets the gate widths so that the ratio is indeed 2, and the width is then a measure of t_N. The first gate is delayed by $2t_N$ after the neutron pulse. The count rate in a third gate, of width $2t_N$, delayed by a further $2t_N$, is subtracted from J_1 and J_2 to account for background.

Another company (Youmans et al., 1964) has been using fixed-width gates at fixed times (Fig. 6-35). The company has developed a new method (Randall and Hopkinson, 1981) which determines Σ from the mean time to a pulse in a single, fixed-width, gate. This mean time is the sum of the arrival times of all the pulses arriving during the gate, divided by the number of pulses. This technique is said to improve statistics compared to the normal two-gate technique. Background can be subtracted if desired.

Two-detector sondes. Two-detector die-away sondes are now in common use. It appears that adding a second detector to a die-away sonde does not improve the measurement of Σ in the presence of borehole or diffusion effects. If Σ measured with one detector differs from Σ measured with the other, some authors (Dewan et al., 1973) conclude that only the diffusion effect is present and a correction curve should be used. At present, two detectors are generally used only to measure I_H, in the same manner as an ordinary compensated neutron log (Sec. 6-6). Preeg and Scott (1981), however, show by using Monte Carlo calculations that the value for the true Σ can in principle be obtained by using two detectors and many gates.

Depth of investigation. Allen et al. (1965), by using three-group diffusion theory, calculated the effects of invasion of the formation by borehole fluid. They found that if a salt-water-saturated formation is invaded by fresh water, an invasion depth of 10 cm has 50 percent of the effect of infinite invasion and a depth of 30 cm has 95 percent. If the formation is saturated with fresh water and invaded with salt water, the 50 percent depth is 20 cm and the 95 percent depth is 40 cm. Figure 6-37*a* and *b* illustrates this situation. Antkiw (1976) found experimentally that freshwater invasion of 20 cm has 90 percent of the effect of infinite invasion and saltwater invasion of 27 cm has 90 percent of the effect of infinite invasion. Hopkinson et al. (1974) observed that 90 percent of the data in a fresh-water-saturated formation comes from the inner 30 cm. These experiments are in good agreement with the calculations.

Evidently the effective depth of investigation of the die-away sonde is quite small. Invasion of the formation by borehole fluid is very significant. Clavier et al. (1971) state that invasion has such a large effect that S_w cannot be evaluated in uncased holes. Consequently, most applications of neutron die-away involve measuring the lifetime of the for-

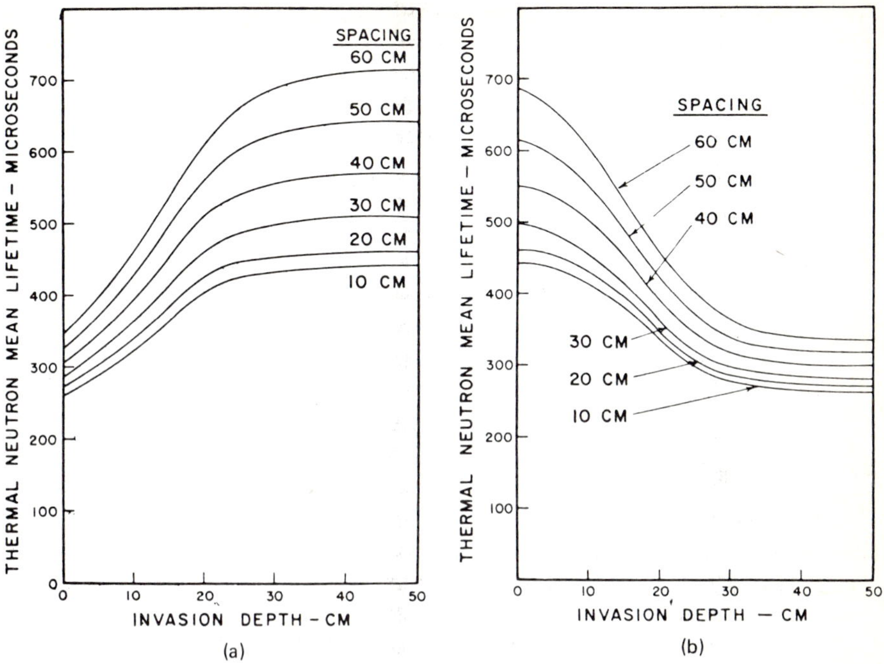

FIGURE 6-37 (*a*) Calculated effective thermal neutron mean lifetime as a function of freshwater invasion depth and source-detector spacing. The formation is a 20% porosity sandstone saturated with salt water. (Allen et al., 1965.) (*b*) Calculated effective thermal neutron mean lifetime as a function of saltwater invasion depth and source-detector spacing. The formation is a 20 percent porosity sandstone saturated with fresh water. (Allen et al., 1965.)

mation saturated with either borehole fluid or fluid in an annulus surrounding the casing.

6-8 SPECTROSCOPY OF NEUTRON-INDUCED γ-RAYS

Nuclei that have absorbed neutrons to produce a compound nucleus emit γ-rays at different times after the interaction (Sec. 6-5b). *Activation* γ's, for example, can be distinguished by their relatively long delay times, and it is possible to identify them by removing the neutron source and then observing the γ. Capture γ's are emitted within a few tens of microseconds after the interaction and are often observed in the presence of the neutron source, but then they cannot be distinguished from activation γ's (except by energy).

6-8a Capture and Activation Logging with a Steady Neutron Source

In practice, the capture γ's must be detected simultaneously with irradiation by an isotopic source. Many materials can be identified by the characteristic energy of the capture γ. Duffey et al. (1970) have tabulated many of the peaks that would result from irradiation of earth materials, and they have arranged the data by both energy and element. Some of the more prominent peaks are listed in Table 6-2. Caldwell et al. (1963) pointed out that some of the most prominent peaks in ordinary rocks will be 2.2 MeV from hydrogen, 4.9 and 3.5 MeV from silicon, and 6.4, 4.4, and 2.0 MeV from calcium. (It is, in fact, plausible to use the 2.2-MeV γ from hydrogen to obtain I_H, but it is rarely done in practice.)

Activation γ's are obtained by irradiating the medium with a neutron source and then removing the source, waiting a few seconds or minutes, and then detecting the γ's. This can be done either by inserting a source, removing it physically, and replacing it with a detector or by using a sonde with long source-detector spacing and a suitably slow logging speed. Since different activation products have different half-lives (some are shown in Table 6-2), the time of observation is one parameter that can be used to identify elements.

Moxham et al. (1972) used a Ge(Li) detector cooled by melting propane to look at both capture and activation γ's to analyze for many metallic elements. The count rate in the photopeak for an element's characteristic γ was proportional to the amount of the element present in a laboratory formation.

The choice of source is of some importance. Scott and Smith (1973) point out that if a Pu-Be or generator source (both of which produce neutrons at energy high enough to cause nuclear reactions) is used when logging for aluminum, the reaction ^{28}Si (n, p) ^{28}Al will produce the same activation γ's as the capture ^{27}Al (n, γ) ^{28}Al. Therefore, γ-rays from silicon will be interpreted as γ-rays from aluminum and the presence of extra aluminum will be inferred. If a ^{252}Cf source is used, the neutron energy is too low to cause the (n, p) reaction, and there is no problem.

The 14-MeV neutrons induce (n, p) reactions in many common elements, for example O, Si, and Al. The new nuclides β-decay back to the original nuclide while emitting a γ of the same energy that is emitted when a neutron is inelastically scattered from the original nuclide (Table 6-2). There is, however, a delay (the time required for β-decay) between the (n, p) reaction and the γ. Caldwell and Mills (1959) proposed using these (n, p) reactions to perform elemental analysis on rocks, and they tested the method in the laboratory on core. Wichmann (1972*a*) used a generator in a borehole. He counted all the γ's in a broad energy window and distinguished among elements by the number of γ's at a given time delay. The delay was simply the time between arrival of the generator and the arrival of the detector some distance below it as the long sonde was pulled up the hole.

Quantitative analysis is more difficult. The number of γ's of a given energy depends not only on the number of atoms of the isotope that produces them but also on the attenuation of both the γ's and the neutrons by other materials. It is much easier to obtain *ratios* of elements than absolute values of quantities, and that is the method generally used for quantitative analysis. If, for example, the ratio of iron to silicon in a formation is quite small but in parts of the formation is double that in other parts, it is most likely that the quantity of silicon will not have changed much but the quantity of iron will have doubled. In this way the grade of ore in different parts of a formation can be obtained fairly accurately by comparing the quantity of the desired element with the quantity of a much more common element. Czubek (1979) and Senftle (1980) give many examples of the use of this technique.

Uranium analysis. When an isotope (such as ^{235}U) fissions, most of the neutrons are emitted within about 10^{-24} s. A few, however, are emitted over the succeeding few hours. These are called *delayed neutrons.* It is possible to detect delayed neutrons from fission of ^{235}U by irradiating with a source and then removing the source and using a neutron detector (Igumnov, 1966). The half-life of the most abundant delayed neutrons is 22 s or less, so quick work is required.

6-8b Logging with a Pulsed Neutron Source

The major advantage of pulsing the source is the ability to recognize the γ's from inelastic scattering. Inelastic-scattering γ's are emitted within a few nanoseconds of the interaction. It is possible to distinguish among these interactions by using detection time gates and pulsing the source. Inelastic scattering occurs only during a neutron pulse; capture γ's are found within a few tens of microseconds; and activation γ's are found for some time thereafter. Figure 6-38 shows the gate timing used in a typical sonde. Note that the pulse rate is much greater than that for a die-away sonde (Fig. 6-35). The uses of these γ's can be divided into two categories: carbon/oxygen logging and logging for other elements. Practical application of the latter category will be discussed in Chap. 15.

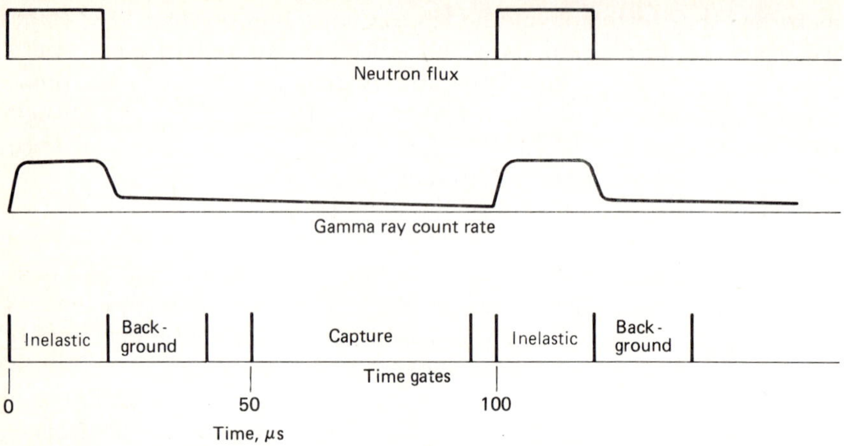

FIGURE 6-38 Typical timing of gates for a sonde using inelastic scattering.

Carbon/oxygen logging. Because hydrocarbons contain carbon, it is useful to develop methods of detecting carbon with a logging sonde. When a neutron is scattered inelastically from carbon, a 4.44-MeV γ is emitted; when a neutron is scattered from oxygen, a 6.1-MeV γ is emitted (see Table 6-2). Therefore, the ratio of 4.4-MeV γ's to those of 6.1 MeV should be the same as the ratio of carbon to oxygen in the formation. Moreover, calcium emits a 3.7 MeV γ, and silicon one of 1.78 MeV. Consequently, a calcium/silicon ratio can also be obtained. Comparison of the two ratios permits the user to distinguish carbon in calcium carbonates from that in hydrocarbons.

As we learned in Sec. 6-2c, however, not all of the γ's emitted at a given energy are counted at that energy. We find escape peaks (Fig. 6-8), and many of the γ's are reduced in energy by Compton scattering before being detected. Consequently, it is common that γ's emitted at, for example, 6.1 MeV give rise to pulses in the detector that are interpreted as 4.4 MeV. Furthermore, if only a narrow energy band is detected, the number of γ's detected is small and statistical error is large.

To minimize statistical error, carbon/oxygen logging systems often use wide windows, rather than detailed pulse height analysis, to obtain the ratios. Normally, the generator is pulsed at 20,000 pps with a neutron pulse width of the order of 5 μs (Schultz and Smith, 1974). Typical windows are 1.5 to 1.9 MeV for Si, 2.5 to 3.3 MeV for Ca, 3.2 to 4.6 MeV for C, and 4.9 to 6.6 MeV for O. The spectrum of inelastic scattering γ's is taken during the neutron pulse or a few microseconds thereafter. The spectrum taken in a time window of the same length, but with the generator turned off, is considered to be that of capture γ's and is subtracted from the first spectrum. Figure 6-39 illustrates typical observed spectra and the windows.

Because many oxygen γ's are scattered into the carbon window, the measured C/O ratio is greater than 1 even if no carbon is present. Figure 6-40*a* shows measured laboratory C/O ratios versus quantity of carbonate and hydrocarbons (Hopkinson et al., 1982). In SiO_2 the C/O

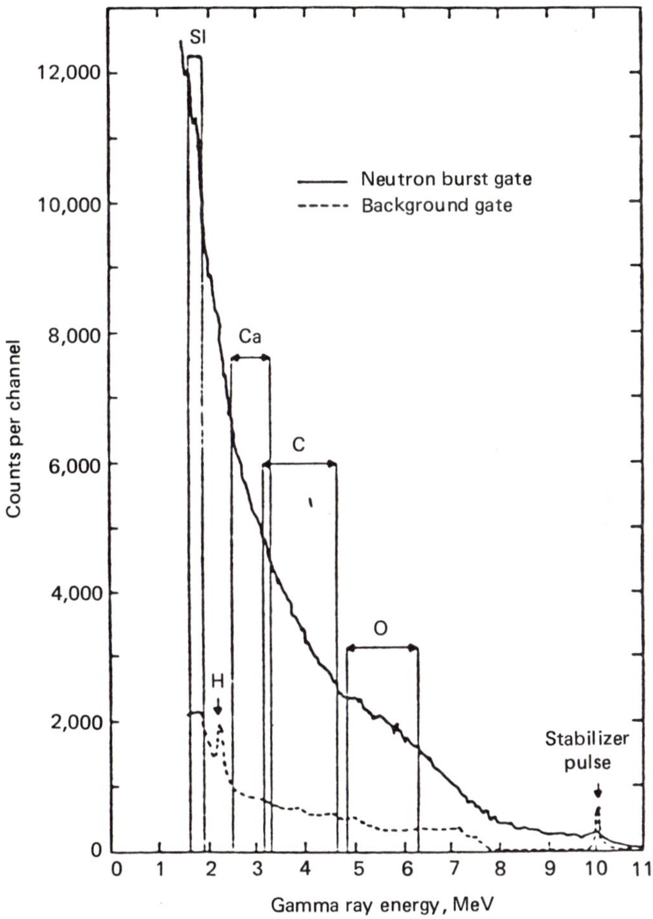

FIGURE 6-39 Spectra of γ-rays from a carbon/oxygen sonde in a laboratory formation consisting of 35% porosity sandstone saturated with oil. The spectra taken during the neutron pulse (inelastic) and with the generator off (capture) are shown. The windows for selection of γ-ray energies also are shown. (Schultz and Smith, 1974. Copyright 1974, SPE–AIME.)

atomic ratio is, of course, zero; in $CaCO_3$ the atomic ratio is 0.33. Clearly, the measured γ-ray ratio has far less range than the atomic ratio. The situation is much the same for the Ca/Si atomic ratio, shown in Fig. 6-40*b*, which is often used as a correction to the C/O data (Lawrence, 1981).

Error bars in Fig. 6-40 show that the statistical uncertainty in the ratio, even for a modern continuously moving sonde, is quite small. The range of the ratio, however, also is small. (The total change in ratio from SiO_2 to $CaCO_3$ is 13 percent of the mean in Fig. 6-40.) Hearst (1979) has been able to increase this range by at least 50 percent by using a 15-cm-diameter detector instead of a detector of the usual 5- to 7-cm diameters. Fewer γ's escape from the crystal, so the number of γ's

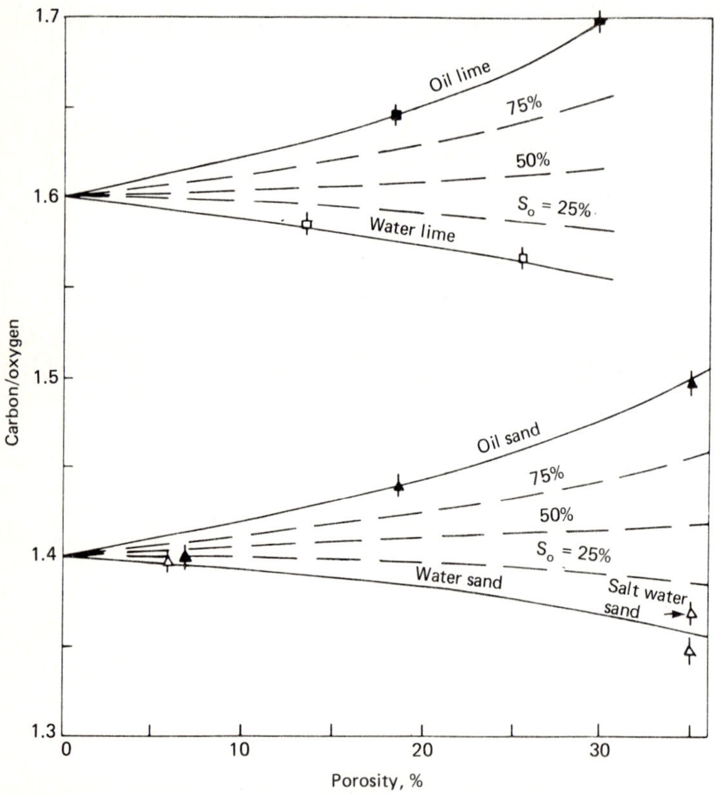

FIGURE 6-40*a* Measured carbon/oxygen spectral ratio versus saturated porosity and oil saturation for sandstone and limestone. Error bars are shown. (Hopkinson et al., 1982. Copyright 1982, SPE–AIME.)

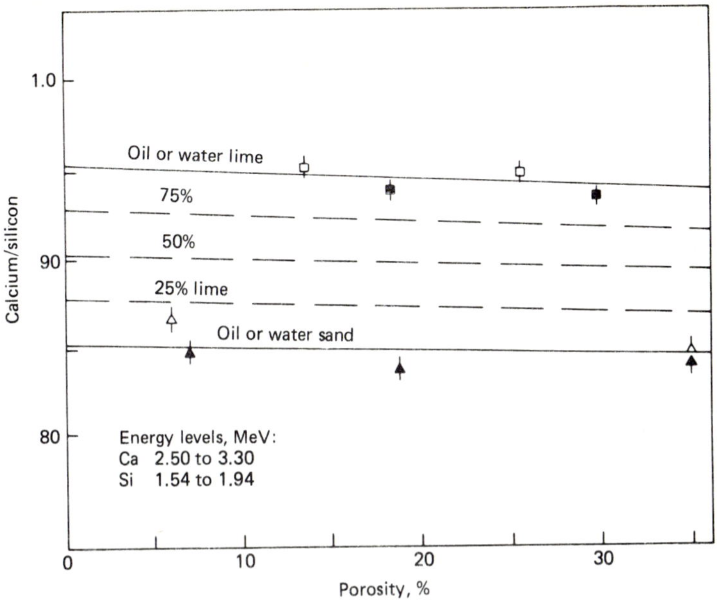

FIGURE 6-40*b* Measured calcium/silicon spectral ratio for mixtures of limestone and sandstone. (Hopkinson et al., 1982. Copyright 1982, SPE–AIME.)

counted at the wrong energy is reduced and we get a much cleaner spectrum. Unfortunately, this detector is too large for ordinary wells.

Spectral fitting. Hertzog (1980) uses γ's from both inelastic scattering and capture, with spectral fitting (Sec. 6-3b), to perform quantitative analyses for many elements. In particular, he obtains ratios of Cl to H for salinity and H to Si and Ca to get water content. In the oilfield, both Fe and Al have been used to obtain a measure of shaliness. Spectral fitting also provides a more accurate C/O ratio than window analysis because it can account explicitly for the scattered γ's and the escape peaks directly.

Uranium. A pulsed neutron source permits a much better way of detecting delayed neutrons from fission of uranium than the radioactive source described in the preceding section. Givens et al. (1976) pulse the generator and then wait until all the neutrons from the source have died away. They then use a neutron detector to observe the delayed neutrons from fission. Of course, the number of neutrons detected depends not only on the amount of uranium (the source term) but also on the amount of water and absorbers, just as in any other neutron logging situation. They have found an ingenious solution to this problem. The reaction

$$^{17}O\ (n, p)\ ^{17}N \rightarrow {}^{17}O + \beta \rightarrow {}^{16}O + n$$

has a low cross section compared to fission of uranium. But it does produce delayed neutrons. Therefore, it can be used, in the laboratory, to calculate the correction to the number of delayed neutrons for variables such as water content. The variations can be measured with other logging devices, and the corrections can be applied. The method is described in more detail in Chap. 15.

REFERENCES

Alger, R. P., S. Locke, W. A. Nagel, and H. Sherman: "The Dual-Spacing Neutron Log-CNL, *J. Petr. Tech.*, vol. 24, 1972, p. 2073.

Allen, L. S., W. R. Mills, and R. L. Caldwell: "The Effects of Fluid Invasion in Pulsed Neutron Logging," *Geophysics*, vol. 30, 1965, p. 389.

———, C. W. Tittle, W. R. Mills, and R. L. Caldwell: "Dual-Spaced Neutron Logging for Porosity," *Geophysics*, vol. 32, 1967, p. 60.

———, W. R. Mills, K. P. Desai, and R. L. Caldwell: "Some Features of Dual-Spaced Neutron Porosity Logging," *Trans. Soc. Prof. Well Log Analysts 13th Logging Symp.*, 1972, paper P.

——— and W. R. Mills: "Measurement of the Thermal Neutron Absorption Cross Section of Rock Samples by a Pulsed Source Method," *Trans. Soc. Prof. Well Log Analysts 15th Logging Symp.*, 1974.

Antkiw, S.: "Depth of Investigation of the Dual-Spacing Thermal-Neutron Decay Time Logging Tool," *Trans. Soc. Prof. Well Log Analysts 17th Logging Symp.*, 1976.

Artsybashev, V. A.: "Effect of Sonde Length on Rock-Density Measurements by the Gamma-Gamma Method, *Izv. Vyssh. Ucheb. Zaved. Geol. i Razvedka*, no. 2, 1967, p. 119.

——— and E. P. Popov: "Optimal Design of Devices for Testing and Adjustment of Gamma-Gamma Logging Apparatus," in E. M. Fillipov (ed.), *Nuclear Geophysical Methods*, Nauka, Novosibirsk, 1972, UCRL-Trans-11311.

——— and G. A. Ivanyukovich: "Density Gamma-Gamma Logging in Ore Deposits (Methodological Recommendations)," Atomizdat, Moscow, 1975, UCRL-Trans-11260.

Barsukov, O. A., N. M. Blinova, S. F. Vybornykh, Yu. A. Gulin, V. N. Dakhnov, V. V. Larionov, and A. I. Kholin: *Radioactive Investigations of Oil and Gas Wells*, trans. by J. O. H. Mulhaus, Macmillan, New York, 1965.

Belknap, W. B., J. T. Dewan, C. V. Kirkpatrick, W. E. Mott, A. J. Pearson, and W. R. Rabson: "API Calibration Facility for Nuclear Logs," *Am. Petr. Inst. Drilling Production*, 1960, p. 289.

Benedict, M., T. Pigford, and H. Levi: *Nuclear Chemical Engineering,* McGraw-Hill, New York, 1981.

Bertozzi, W., D. V. Ellis, and J. S. Wahl: "The Physical Foundation of Formation Lithology Logging with Gamma Rays," *Geophysics,* vol. 46, 1981, p. 1439.

Caldwell, R. C., and W. R. Mills, Jr.: "Activation Analysis in Petroleum Exploration Research," *Nuc. Inst. Meth.,* vol. 5, 1959, p. 312.

————, W. R. Mills, Jr., and J. B. Hickman, Jr.: "Gamma Radiation from Inelastic Scattering of 14-MeV Neutrons by the Common Earth Elements," *Nuc. Sci. Eng.,* vol. 8, 1960, p. 173.

————, W. F. Baldwin, J. D. Bargainer, J. E. Berry, G. N. Salaita, and R. W. Sloan: "Gamma-Ray Spectroscopy in Well Logging," *Geophysics,* vol. 28, 1963, p. 617.

Charbucinski, J.: "A Universal Gamma-Gamma Method for Simultaneous Determination of Rock and Ore Properties," *Int. J. Appl. Rad. Isotopes,* vol. 34, 1983, p. 353.

Clavier, C., W. R. Hoyle, and D. Meunier: "Quantitative Interpretation of Thermal Neutron Decay Time Logs," Parts I and II, *J. Petr. Tech.,* vol. 23, 1971, p. 743.

Conaway, J. G., and P. G. Killeen: "Computer Processing of Gamma-Ray Logs Iteration and Inverse Filtering," in *Current Research,* Part B, Geol. Surv. Can. Paper 78-1b, 1978, pp. 83–88.

Czubek, J. A.: "Quantitative Interpretation of Gamma-Ray Logs," *Proc. Nuc. Geophys. Conf.,* Crakow, Poland, 1962. Reprinted in *Gamma Ray, Neutron, and Density Logging,* SPWLA, 1978.

————: "Modern Trends in Mining Geophysics and Nuclear Borehole Logging Methods for Mineral Exploration," in P. J. Hood (ed.), *Geophysics and Geochemistry in the Search for Metallic Ores,* Geol. Survey of Canada, Econ. Geol. Rept. 31, 1979, p. 231.

————: Personal communication, 1981.

————: "Advances in Gamma-Gamma Logging," *Int. J. Appl. Rad. Isotopes,* vol. 34, 1983, p. 153.

Davis, R. R., J. E. Hall, Y. L. Boutemy, and C. Flaum: "A Dual Porosity CNL Logging System," Soc. Petr. Eng. Paper SPE 10296, 1981.

Dewan, J. T., C. W. Johnstone, L. Jacobson, W. B. Wall, and R. P. Alger: "Thermal Neutron Decay Time Logging Using Dual Detection," *Trans. Soc. Prof. Well Log Analysts 14th Logging Symp.,* 1973.

Duffey, D., A. El-Kady, and F. E. Senftle: "Analytical Sensitivities and Energies of Thermal-Neutron-Capture Gamma-Rays," *Nuc. Inst. Meth.,* vol. 80, 1970, p. 149.

Edmundson, H., and L. L. Raymer: "Radioactive Logging Parameters for Common Minerals," *Trans. Soc. Prof. Well Log Analysts 20th Logging Symp.,* 1979, paper O.

Ellis, D. V., J. Ullo, and H. Sherman: "Comparison of Calculated and Measured Response of a Dual-Detector Epithermal Neutron Porosity Device," Soc. Petr. Eng. Rept. SPE 10294, 1981.

Evans, R. D.: *The Atomic Nucleus*, McGraw-Hill, New York, 1955.

Ferronskiy, V. I., A. I. Danilin, V. T. Dubinchuk, V. S. Goncharov, V. A. Polyakov, Yu. B. Seletskiy, and N. Ya. Flekser: "Radioisotope Investigative Methods in Engineering Geology and Hydrogeology," ATOMIZDAT, Moscow, 1968. Translation published by U.S. Geol. Survey, Denver, 1969.

Fink, J. B.: "On K-Factors and Gamma Log Calculations," *Geophysics*, vol. 43, 1978, p. 1546.

Givens, W. W., W. R. Mills, C. L. Dennis, and R. L. Caldwell: "Uranium Logging Using a Pulsed 14-MeV Neutron Source and Detection of Delayed Fission Neutrons," *Geophysics*, vol. 41, 1976, p. 468.

Glasstone, S., and M. C. Edlund: *"The Elements of Nuclear Reactor Theory,"* Van Nostrand, Princeton, N.J., 1952.

Gorski, L.: "Solid-State Detectors and Electronics in Gamma and x-Ray Spectrometry: Review of Present Knowledge and Techniques," *Proc. Nuc. Tech. Min. Exploration Exploitation*, International Atomic Energy Agency, Kracow, 1969.

Gulin, Yu. A., and G. I. Ponyatov: "Results of Mathematical Modeling of Gamma-Gamma Logging under Oil-Well Conditions," *Yadernaya Geofiz.*, 1969, p. 3. UCRL-Trans-10591.

Harvey, J. A., and D. J. Hughes: "Neutrons," in D. E. Gray (ed.), *American Institute of Physics Handbook*, McGraw-Hill, New York, 1963.

Hearst, J. R: "Effects of Mudcake and Sonde Angle on a Simple Two-Detector Density Sonde," *Log Analyst*, vol. 17, May–June 1976, p. 11.

————: "Laboratory Studies of an Improved Carbon/Oxygen Log," *Log Analyst*, vol. 19, November–December 1978, p. 26.

————: "Calibration of a Neutron Log in Partially Saturated Media," *Trans. Soc. Prof. Well Log Analysts 20th Logging Symposium*, 1979, paper B.

———— and R. C. Carlson: "The RIDS, a Density Logger for Rough Holes," *Geophysics*, vol. 34, 1969, pp. 222–234.

————, P. W. Kasameyer, and L. A. Dreiling: "Calibration of a Neutron Log in Partially Saturated Media. Part II. Error Analysis, "*Trans. Soc. Prof. Well Log Analysts 22d Logging Symp.*, 1981, paper QQ.

Heath, R. L.: "Scintillation Spectrometry Gamma-Ray Spectrum Catalog," U.S. Atomic Energy Comm. Rept. IDO-16880-2, 1964.

Hertzog, R. C.: "Laboratory and Field Evaluation of an Inelastic Neutron Scattering and Capture Gamma Ray Spectrometry Tool," *Soc. Petr. Eng. J.*, vol. 20, 1980, p. 327.

Hopkinson, E. C., A. H. Youmans, and R. B. Johnson, Jr.: "Depth of Investigation of the Neutron Lifetime Log," *Trans. Soc. Prof. Well Log Analysts 15th Logging Symp.*, 1974, paper AA.

————, W. H. Fertl, and D. W. Oliver: "The Continuous Carbon/Oxygen Log—Basic Concepts and Recent Field Experiences," *J. Petr. Tech.*, vol. 34, May 1982, pp. 2441–2448.

Igumnov, S. A.: "The Use of Delayed Neutrons to Determine Uranium Content in Bore Holes without Coring," *Izv. Vyssh. Ucheb. Zaved. Gorn. Zhur.*, vol. 9, no. 2, 1966, p. 3.

International Atomic Energy Agency (IAEA): "Radiometric Reporting Methods and Calibration in Uranium Exploration," Tech. Rept. Ser. 174 Vienna, Austria, 1976.

Killeen, P. G.: "Gamma Ray Spectrometric Methods in Uranium Exploration—Application and Interpretation," in Peter J. Hood (ed.), *Geophysics and Geochemistry in the Search for Metallic Ores,* Geol. Survey Canada, Econ. Geol. Rept. 31, 1979, pp. 163–229.

Kozhevnikov, D. A.: "Calculation of the Neutron Characteristics of Rocks," *Trudy Moskov. Inst. Neft. i Gazovoi Prom.*, no. 14, 1963, p. 54.

Kreft, A.: "Calculation of the Neutron Slowing Down Length in Rocks and Soils," *Nukleonika,* vol. 19, 1974, p. 145.

Lawrence, T. D.: "Continuous Carbon/Oxygen Log Interpretation Techniques," *J. Petr. Tech.*, vol. 33, 1981, p. 1394.

Locke, S., and R. Smith: "Computed Departure Curves for the Thermal Neutron Decay Time Log," *Trans. Soc. Prof. Well Log Analysts 16th Logging Symp.*, 1975.

Meyerhof, W. E.: *Elements of Nuclear Physics,* McGraw-Hill, New York, 1967.

Mills, W. R. Jr., R. L. Caldwell, and I. L. Morgan: "Low Voltage He^3-Filled Proportional Counter for Efficient Detection of Thermal and Epithermal Neutrons," *Rev. Sci. Instr.*, vol. 33, 1962, p. 866.

Moore, D. C.: "Interpretation of Total Gamma Logs in Thin and Dipping Beds," *Geophysics,* vol. 45, 1980, p. 1847.

Moxham, R. M., F. E. Senftle, and G. R. Boynton: "Borehole Activation Analysis by Delayed and Capture Gamma Rays Using a ^{252}Cf Neutron Source," *Economic Geology,* vol. 67, 1972, p. 579.

Neuman, C. H.: "Discussion of 'A Study of the Effects of Diffusion on Pulsed Neutron Capture Logs'," *J. Petr. Tech.*, vol. 30, 1978, p. 1794.

Olgaard, P. L.: "On the Theory of the Neutronic Method for Measuring the Water Content of Soil," Danish Atom. Energy Comm. Riso Rept. 97, 1965.

Peters, C. W.: "Neutron Activation Analysis of Turbine Deposits," Consolidated Controls Corp. Rept. CS-1958, 1981.

Polyachenko, A. L., T. E. Gavrina, V. G. Tseitlin, and T. A. Shaposhnikova: "Analysis of Amplitude, Time, and Space Characteristics of the Gamma-Quanta Distribution in Pulsed Neutron-Gamma Logging," *Yadernaya Geofiz.*, vol. 7, 1969, p. 40. UCRL-Trans-10597.

Preeg, W. E., and H. D. Scott: "Computing Thermal Neutron Decay Time (TDT) Environmental Effects Using Monte Carlo Techniques," Soc. Petr. Eng. Paper SPE 10293, 1981.

Randall, R., and E. C. Hopkinson: "Improved Methods of Signal Processing for Pulsed Neutron Capture Logging," *Trans. Soc. Prof. Well Log Analysts 22d Logging Symp.*, 1981, paper JJ.

————, ————, and A. H. Youmans: "A Study of the Effects of Diffusion on Pulsed Neutron Capture Logs," *J. Petr. Tech.*, vol. 30, 1978, p. 1788.

Rhodes, D. F., R. A. Stallwood, and W. E. Mott: "Intensity of Unscattered Gamma Rays inside Cylindrical Self-Absorbing Sources," *Nuc. Sci. Eng.*, vol. 9, 1961, p. 41.

———— and W. E. Mott: "Quantitative Interpretation of Gamma-Ray Spectral Logs," *Geophysics*, vol. 31, 1966, p. 410.

Rinehart, C. E., and H. J. Weber: "Measuring Thermal Neutron Absorption Cross Sections of Formation Water," *Trans. Soc. Prof. Well Log Analysts 16th Logging Symp.*, 1975.

Samworth, J. R.: "Slimline Dual Detector Density Logging—A Semitheoretical but Practical Approach to Correction and Compensation," *J. Petr. Tech.*, vol. 32, 1980, p. 1316.

Sanders, L. G.: "The Application of Monte Carlo Computations to Formation Analysis by Neutron Interactions," *Int. J. Appl. Rad. Isotopes*, vol. 34, 1983, p. 173.

Schultz, W. E., and H. D. Smith: "Laboratory and Field Evaluation of a Carbon/Oxygen (C/O) Well Logging System," *J. Petr. Tech.*, vol. 26, October 1974, pp. 1103–1110.

———— and S. G. Thadani: "Applications of Digital Filtering Techniques to Nuclear Well Logs," *Trans. Soc. Prof. Well Log Analysts 22d Logging Symp.*, 1981, paper UU.

————, H. D. Smith, J. L. Verbout, J. R. Bridges, and G. H. Garcia: "Experimental Basis for a New Borehole Corrected Pulse Neutron Capture System (TMD)," *Proc. Soc. Prof. Well Log Analysts 24th Logging Symp.*, 1983, paper CC.

Scott, H. D., and M. P. Smith: "The Aluminum Activation Log," *Trans. Soc. Prof. Well Log Analysts 14th Logging Symp.*, 1973, paper F.

Scott, J. H.: Computer Analysis of Gamma-Ray Logs," *Geophysics*, vol. 28, 1963, p. 457.

————: "Borehole Compensation Algorithms for a Small-Diameter, Dual-Detector Density Well-Logging Sonde," *Trans. SPWLA 18th Ann. Logging Symp.*, 1977, paper S.

————: "Pitfalls in Determining the Dead Time of Nuclear Well-Logging Probes," *Proc. Soc. Prof. Well Log Analysts 21st Logging Symp.*, 1980, paper H.

————, P. H. Dodd, R. F. Droullard, and P. J. Mudra: "Quantitative Evaluation of Gamma-Ray Logs," *Geophysics*, vol. 26, 1961, p. 182.

Scotty, C. B., and E. F. Egan: "Neutron Derived Porosity—Influence of Bore Hole Diameter," *Petr. Trans. AIME*, vol. 195, 1952, p. 203.

Senftle, F. E.: "Application of Gamma-Ray Spectral Analysis to Subsurface Mineral Exploration," in G. K. Muecke (ed.), *Short Course in Neutron Activation Analysis in the Geosciences,* Mineral. Assoc. Canada, Halifax, 1980.

Sherman, H., and S. Locke: "Depth of Investigation of Neutron and Density Sondes for 35-Percent-Porosity Sand," *Trans. 5th Formation Eval. Symp. Can. Well Logging Soc.*, 1975.

Shope, L. A., R. S. Berg, M. L. O'Neal, and B. E. Barnaby: "Operation and Life of the Zetatron: A Small Neutron Generator for Borehole Logging," *IEEE Trans. Nuc. Sci.*, vol. NS-28, 1981, p. 1696.

Stromswold, D. C., and R. D. Wilson: "Calibration and Data Correction Techniques for Spectral Gamma-Ray Logging," *Proc. Soc. Prof. Well Log Analysts 22d Annual Logging Symp.*, 1981.

Tavendale, A. J.: "Characteristics of Lithium-Ion Drift Germanium Gamma Spectrometers," *Proc. European Nuc. Energy Agency "Nuclear Electronics" Conf.*, 1963.

Tittle, C. W.: "Theory of Neutron Logging I," *Geophysics,* vol. 26, 1961, p. 27.

————: "How to Compute Neutron Transmission and Diffusion," *Nuclear-Chicago Tech. Bull. 12,* Des Plaines, Ill., 1962.

———— and L. S. Allen: "Theory of Neutron Logging II," *Geophysics,* vol. 31, 1966, p. 214.

————, H. Faul, and C. Goodman: "Neutron Logging of Drill Holes: The Neutron-Neutron Method," *Geophysics,* vol. 16, 1951, p. 626.

Tittman, J., and J. S. Wahl: "The Physical Foundation of Formation Density Logging (Gamma-gamma)," *Geophysics,* vol. 30, 1965, p. 284.

————, H. Sherman, W. A. Nagel, and R. P. Alger: "The Sidewall Epithermal Neutron Porosity Log," *J. Petr. Tech,* vol. 18, 1966, pp. 1351–1362.

Ullo, J. J.: "Response of the Dual Spacing Neutron Log (CNL) to Gas," Soc. Petr. Eng. Paper SPE 10295, 1981.

Umiastowski, K., and M. Buniak: "Influence of the Rock Heterogeneity on the Results of Gamma-Gamma Logging," *Proc. Symp. Nuc. Tech. Min. Res.*, IAEA, 1977.

Varvarin, G. B., G. A. Kuznetsov, and E. M. Fillipov: "Double-Beam Down-Hole Gamma-Gamma Density Meter," *Geol. i Geofiz., Akad. Nauk, SSSR Siberskoe Otd.*, 1967, p. 89.

Voskoboinikov, G. M.: "Theoretical Basis of Selective Gamma-Gamma Core Sampling," *Bull. Acad. Sci. USSR Geophys. Ser.*, no. 3, 1957, p. 94.

Wahl, J. S., J. Tittman, C. W. Johnstone, and R. P. Alger: "The Dual Spacing Formation Density Log," *J. Petr. Tech.*, vol. 16, 1964, p. 1411.

————, W. B. Nelligan, A. H. Frentrop, C. W. Johnstone, and R. J. Schwartz: "The Thermal Neutron Decay Time Log," *Soc. Petr. Eng. J.*, vol. 19, 1970, p. 365.

Wichmann, A.: *Quantum Physics,* McGraw-Hill, New York, 1971.

Wichmann, P. A.: "Neutron Activation for Elemental Determination in Boreholes," *Log Analyst*, vol. 13, January–February 1972, p. 10.

———: "Notes on the Accuracy of the Neutron Lifetime Measurement," *Trans. Soc. Prof. Well Log Analysts 13th Logging Symp.*, 1972.

Wilson, B. F., and Wichmann, P. A.: "The Compensated Neutron Log and the Effects of Environment," Soc. Petr. Eng. Paper SPE 5118, 1974.

Wilson, R. D., and D. C. Stromswold: "Spectral Gamma-Ray Logging Studies," Bendix Field Eng. Corp., Grand Junction, Colo., 1981.

Wollenberg, H. A.: "Naturally Occurring Radioelements and Terrestrial Gamma-Ray Exposure Rates: An Assessment Based on Recent Geochemical Data," Lawrence Berkeley Lab. Rept., 1984.

Wood, R. D., P. A. Wichmann, and H. B. Watt: "Gamma-Ray Neutron Log," *Log Review I*, Dresser Atlas, Houston, Tex., 1974.

Youmans, A. H., E. C. Hopkinson, R. A. Bergen, and H. L. Oshry: "Neutron Lifetime: A New Nuclear Log," *J. Petr. Tech.*, vol. 16, 1964, p. 319.

———, D. W. Bishop, and P. A. Wichmann: "Application of the Neutron Lifetime Log in New Wells," *Trans. Soc. Prof. Well Log Analysts 11th Logging Symp.*, 1970.

——— and E. C. Hopkinson: "The Influence of Instrument Design on Neutron Lifetime Measurements," *Soc. Petr. Eng. J.*, vol. 15, 1975, pp. 161–168.

7

Acoustic Logging

The velocity and attenuation of sound waves in rocks depend on many properties of the rock, for example, density, porosity, saturation, and the amount of fracturing. Consequently, the use of sound waves as probes of rock has proved to be quite valuable. Unfortunately, the very properties of the sound waves that make them such a useful probe also give rise to extremely complex waveforms when a transmitter and receiver are placed in a borehole that penetrates material with interesting properties. The logging industry is in the midst of an effort to learn how to use all of this information. One can say that acoustic logging is almost unique among logging methods because far more information is available than can be interpreted.

7-1 WAVES IN ELASTIC MEDIA

7-1a Stress and Strain in Elastic Media

Elastic media. An elastic medium can be described as one that recovers completely after being deformed. Although the earth is not elastic, it is almost elastic for small displacements, and acoustic logging sondes produce very small displacements. Most of the phenomena in acoustic logging can be described mathematically by using the theory of elastic waves in elastic media. These are actually strain waves, and we refer to them as such.

Stress. Stress is defined as force per unit area. If the force is perpendicular to the element of area on which it is exerted, it is said to be a *normal stress;* if it is tangential to the element of area, it is called a *shear stress.* If we consider a small element of volume, with dimensions δx, δy, δz, inside a stressed body, the stresses acting on the six faces of the element can be resolved into components of normal and shear stress as shown in Fig. 7-1. Subscripts denote the x, y, and z axes. σ_{xy}, for example, denotes a shear stress applied parallel to the x axis and acting on a surface perpendicular to the y axis, and σ_{xx} denotes a normal stress in the x direction. We shall consider tensile stress to be positive.

Strain. Telford et al. (1981) give a clear description of strain. It is condensed here. When an elastic body is subjected to stresses, changes in shape and dimensions, called *strains,* occur. Consider a rectangle *PQRS*, with edges δx and δy, in the xy plane (Fig. 7-2). When stresses are

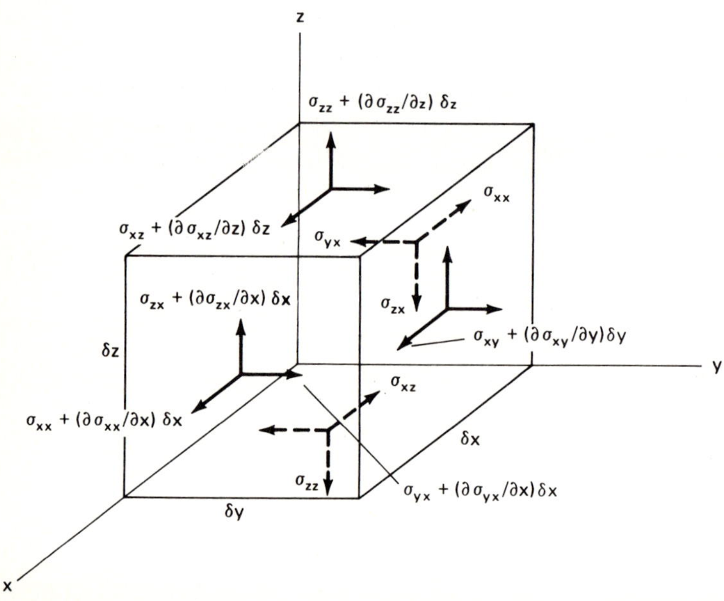

FIGURE 7-1 Sketch of stresses acting on a small rectangular parallelepiped. (After Officer, 1958.)

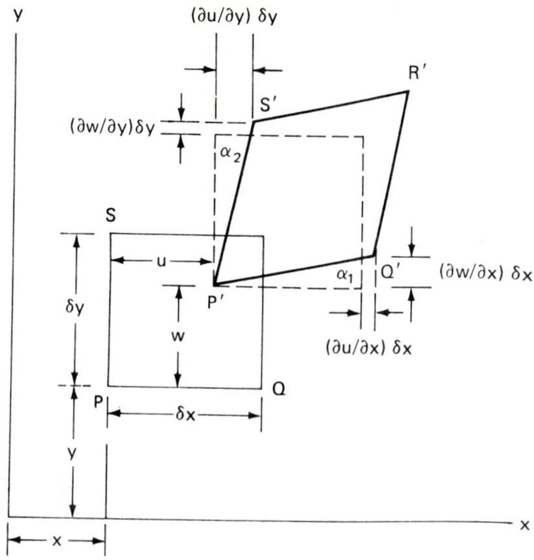

FIGURE 7-2 Illustration of the concept of two-dimensional strain. (After Wang, 1953.)

applied, point P moves to P'. The x and y components of PP', called *displacements*, are u and w. If the other vertices have the same displacements as P, the rectangle is displaced as a whole by u and w; there is no change in size or shape and there is no strain. If, however, u and w are different for the different vertices, there are changes in size and shape, and so there is strain.

If $u = u(x, y)$ and $w = w(x, y)$, the coordinates of the vertices are:

$$P \quad (x, y) \qquad P' \quad (x + u, y + w)$$

$$Q \quad (x + \delta x, y) \qquad Q' \quad \left(x + \delta x + u + \frac{\partial u}{\partial x}\delta x,\ y + w + \frac{\partial w}{\partial x}\delta x\right)$$

$$S \quad (x, y + \delta y) \qquad S' \quad \left(x + u + \frac{\partial u}{\partial y}\delta y,\ y + \delta y + w + \frac{\partial w}{\partial y}\delta y\right)$$

$$R \quad (x + \delta x, y + \delta y) \qquad R' \quad \left(x + \delta x + u + \frac{\partial u}{\partial x}\delta x + \frac{\partial u}{\partial y}\delta y,\ y + \delta y + w + \frac{\partial w}{\partial x}\delta x + \frac{\partial w}{\partial y}\delta y\right)$$

The changes in u and w are much smaller than δx and δy, so we shall assume that quantities like $\partial u/\partial x$ are small enough that their powers and products can be neglected. With this assumption, we see that:

1 PQ increases in length by the amount $(\partial u/\partial x)\ \delta x$ and PS by the amount $(\partial w/\partial y)\ \delta y$. Hence, $\partial u/\partial x$ and $\partial w/\partial y$ are fractional increases in the length in the direction of the axes. These changes are called normal strains. The normal strain in the x direction is referred to as ε_{xx} and that in the y direction as ε_{yy}.

2 The right angle at P decreases by the amount $\alpha_1 + \alpha_2 = \partial w/\partial x + \partial u/\partial y$. The amount a right angle changes when stress is applied is a measure of the change in shape. It is called the *shear strain* and is denoted in this case by ε_{xy}.

3 The rectangle as a whole has been rotated counterclockwise through the angle $\alpha - \alpha_2 = \partial w/\partial x - \partial u/\partial y$. This rotation about the z axis is referred to as the rotation and can be denoted θ_z.

Extending to three dimensions, recognizing that $\varepsilon_{xy} = \varepsilon_{yx}$, and denoting the displacement in the z direction by ζ, the elementary strains are as follows.
Normal strains:

$$\begin{aligned} \varepsilon_{xx} &= \frac{\partial u}{\partial x} \\ \varepsilon_{yy} &= \frac{\partial w}{\partial y} \\ \varepsilon_{zz} &= \frac{\partial \zeta}{\partial z} \end{aligned} \tag{7-1}$$

Shear strains:

$$\begin{aligned} \varepsilon_{xy} = \varepsilon_{yx} &= \frac{\partial w}{\partial x} + \frac{\partial u}{\partial y} \\ \varepsilon_{yz} = \varepsilon_{zy} &= \frac{\partial \zeta}{\partial y} + \frac{\partial w}{\partial z} \\ \varepsilon_{zx} = \varepsilon_{xz} &= \frac{\partial u}{\partial z} + \frac{\partial \zeta}{\partial x} \end{aligned} \tag{7-2}$$

The simple rotation about the axes is

$$\begin{aligned} \theta_x &= \frac{\partial \zeta}{\partial y} - \frac{\partial w}{\partial z} \\ \theta_y &= \frac{\partial u}{\partial z} - \frac{\partial \zeta}{\partial x} \\ \theta_z &= \frac{\partial w}{\partial x} - \frac{\partial u}{\partial y} \end{aligned} \tag{7-3}$$

The changes in dimensions result in volume changes. The change in volume per unit volume is called the *dilatation* and is represented by Δ. The change in volume of a rectangular parallelepiped with edges δx, δy, and δz is approximately $(\varepsilon_{xx} + \varepsilon_{yy} + \varepsilon_{zz})\, \delta x\, \delta y\, \delta z$, so

$$\Delta \simeq \varepsilon_{xx} + \varepsilon_{yy} + \varepsilon_{zz} = \frac{\partial u}{\partial x} + \frac{\partial w}{\partial y} + \frac{\partial \zeta}{\partial z} \tag{7-4}$$

Hooke's law. It is often true that in an elastic medium with small strains the strain is directly proportional to the stress that produced it. This relation is known as Hooke's law. The total strain is the sum of the strains produced by the individual stresses; therefore, each strain is a linear function of all the stresses (and vice versa). When the medium is isotropic Hooke's law can be expressed as:

$$\sigma_{ii} = \lambda\Delta + 2\mu\varepsilon_{ii} \qquad i = x, y, z \tag{7-5a}$$

$$\sigma_{ij} = \mu\varepsilon_{ij} \qquad i, j = x, y, z;\ i \neq j \tag{7-5b}$$

Elastic moduli. The quantities λ and μ are known as Lamé's constants. Since μ is a measure of resistance to shear, it is known as the *shear modulus* or *rigidity modulus*. Various combinations of these constants are also often used. Two are defined in a situation in which all stresses are zero except σ_{xx} (a condition of *uniaxial stress*). Then ε_{xx} is positive and $\varepsilon_{yy} = \varepsilon_{zz}$ are negative. We then define
Young's modulus

$$E = \frac{\sigma_{xx}}{\varepsilon_{xx}} \tag{7-6}$$

Poisson's ratio

$$\nu = -\frac{\varepsilon_{yy}}{\varepsilon_{xx}} = -\frac{\varepsilon_{zz}}{\varepsilon_{xx}} \tag{7-7}$$

The *bulk modulus* K is useful for a medium subjected only to a hydrostatic pressure P. Then

$$\sigma_{xx} = \sigma_{yy} = \sigma_{zz} = -P \qquad \sigma_{xy} = \sigma_{yz} = \sigma_{xz} = 0$$

and we define

$$K = -\frac{P}{\Delta} \tag{7-8}$$

Using Hooke's law, the standard relations among the elastic moduli are obtained:

$$E = \frac{\mu(3\lambda + 2\mu)}{\lambda + \mu} \tag{7-9a}$$

$$\nu = \frac{\lambda}{2(\lambda + \mu)} \tag{7-9b}$$

$$K = \frac{3\lambda + 2\mu}{3} \tag{7-9c}$$

Fluids cannot sustain shear stresses. Therefore, μ for a fluid is always zero and $K = \lambda$.

NOTE: The symbols λ, E, K, and μ are used here to denote elastic constants because the notation is standard. Elsewhere in this section the symbols λ and E are used for other purposes. The reader must beware of this use of the same symbol for different quantities in the same section. We shall specify the meaning when there is an ambiguity.

7-1b Waves in Continuous Media

The wave equation. If the stress field in a body is uniform, there are no unbalanced external forces acting on the volume element of Fig. 7-1 except for body forces such as gravity. But if there is a stress gradient, then there are unbalanced forces. Still following Telford et al. (1981), we see from Fig. 7-1 that the stresses on one face of the volume element are σ_{xx}, σ_{yx}, and σ_{zx} and that those on the opposite face are

$$\sigma_{xx} + \frac{\partial \sigma_{xx}}{\partial x}\,\delta x \qquad \sigma_{yx} + \frac{\partial \sigma_{yx}}{\partial x}\,\delta x \qquad \text{and} \qquad \sigma_{zx} + \frac{\partial \sigma_{zx}}{\partial x}\,\delta x$$

Consequently, the net unbalanced forces on that pair of faces are

$$\frac{\partial \sigma_{xx}}{\partial x}\,\delta x \qquad \frac{\partial \sigma_{yx}}{\partial x}\,\delta x \qquad \text{and} \qquad \frac{\partial \sigma_{zx}}{\partial x}\,\delta x$$

Similar expressions hold for the other faces. The net unbalanced force per unit volume in the x direction is then

$$\frac{\partial \sigma_{xx}}{\partial x} + \frac{\partial \sigma_{xy}}{\partial y} + \frac{\partial \sigma_{xz}}{\partial z}$$

and similarly for the other three directions. By Newton's second law the product of mass and acceleration is equal to the net unbalanced force, so

$$\rho\,\frac{\partial^2 u}{\partial t^2} = \frac{\partial \sigma_{xx}}{\partial x} + \frac{\partial \sigma_{xy}}{\partial y} + \frac{\partial \sigma_{xz}}{\partial z} \tag{7-10a}$$

By using Hooke's law to convert stress to strain, and expressing strain in terms of displacement, we get

$$\rho\,\frac{\partial^2 u}{\partial t^2} = (\lambda + \mu)\,\frac{\partial \Delta}{\partial x} + \mu\,\nabla^2 u \tag{7-10b}$$

and for the other axes,

$$\rho \frac{\partial^2 w}{\partial t^2} = (\lambda + \mu) \frac{\partial \Delta}{\partial y} + \mu \nabla^2 w \tag{7-10c}$$

$$\rho \frac{\partial^2 \zeta}{\partial t^2} = (\lambda + \mu) \frac{\partial \Delta}{\partial z} + \mu \nabla^2 \zeta \tag{7-10d}$$

If we differentiate Eq. (7-10*b*) with respect to x, (7-10*c*) with respect to y, and (7-10*d*) with respect to z, and add the results, we obtain

$$\frac{\partial^2 \Delta}{\partial t^2} = \frac{\lambda + 2\mu}{\rho} \nabla^2 \Delta \tag{7-11a}$$

for the dilatation. Then if we differentiate Eq. (7-10*c*) with respect to z and (7-10*d*) with respect to y, subtract, and perform similar operations for the other pairs, we obtain

$$\frac{\partial^2 \theta_i}{\partial t^2} = \frac{\mu}{\rho} \nabla^2 \theta_i \tag{7-11b}$$

for each component of the rotation. Equations (7-11*a*) and (7-11*b*) are specific examples of the *wave equation*, whose general form is

$$\frac{\partial^2 \psi}{\partial t^2} = v^2 \nabla^2 \psi \tag{7-12}$$

where v is the velocity of propagation of the wave. Equation (7-11*a*) respresents a wave in which the rotation is zero; it is called an *irrotational, compressional, dilatational,* or *P wave.* The *P* wave is observed as alternate zones of compression and rarefaction, propagating in the direction of motion of the wave. Equation (7-11*b*) represents a wave in which the dilatation is zero. It is called an *equivoluminal, distortional, shear,* or *S wave.*

When ψ is a function of only one variable, say, x,

$$\frac{\partial^2 \psi}{\partial t^2} = v^2 \frac{\partial^2 \psi}{\partial x^2} \tag{7-13}$$

and we have a plane wave. The general solution of Eq. (7-13), called the *wave function,* is any function of $x - vt$ provided ψ and its first two derivatives have no discontinuities. There are, therefore, infinitely many particular solutions such as $\exp[k(x - vt)]$ and $(x - vt)^n$. A solution that holds for a pulse of any shape is the Fourier series

$$F(x - vt) = \sum_{n=1}^{\infty} [A_n \cos nk(x - vt) + B_n \sin nk(x - vt)] \tag{7-14}$$

where A_n and B_n are the Fourier coefficients for the nth harmonic term or *mode,* the *wavenumber* $k = 2\pi/\lambda$, and λ now represents the wavelength (in this case, the fundamental or longest wavelength, generally the length of the initial pulse). The *period* of the fundamental wavelength is

$$T = \frac{\lambda}{v} \tag{7-15a}$$

its *frequency* is

$$f = \frac{1}{T} = \frac{v}{\lambda} \tag{7-15b}$$

and consequently

$$v = f\lambda \tag{7-15c}$$

The *angular frequency* $\omega = 2\pi f$ is even more commonly used.

NOTE: We are using λ to mean a property of the wave now, not an elastic constant.

The solution to the wave equation can also be written in terms of a spherical radius r as, for example,

$$r\psi = F(r - vt) \tag{7-16}$$

Any sum of solutions to the wave equation is itself a solution, so wave functions can be *superposed.* When wave functions are superposed, the amplitudes add algebraically; consequently, the sum at any point in space or time can be greater or less than the value of an individual amplitude. Thus we can have *constructive* or *destructive* interference. For example, if two identical waves are superposed, the sum is a wave of the same form with double the amplitude. If, however, two waves that are exactly out of phase (that is, are such that the amplitudes are of opposite sign so the peak of one coincides with the trough of the other) are superposed, the result is zero everywhere even though each wave has finite amplitude.

There is an immense literature on solutions of the wave equation with different initial conditions and boundary conditions; in fact, most theory of acoustic waves is just that (Kolsky, 1963; White, 1965; and Officer, 1958 are good examples). Generally, we shall state the boundary conditions for situations of interest to us and then describe the solution and its significance for well logging applications.

Plane waves. The propagation of dilatational and distortional waves can be illustrated by consideration of the propagation of plane waves. Assume a wave propagated in the x direction. Then the displacements u, w, and ζ will be functions only of x and t. Then Eqs. (7-11) reduce to to the following:

$$\rho \frac{\partial^2 u}{\partial t^2} = (\lambda + \mu) \frac{\partial^2 u}{\partial x^2} + \mu \frac{\partial^2 u}{\partial x^2} = (\lambda + 2\mu) \frac{\partial^2 u}{\partial x^2} \qquad \text{(7-17)}$$

$$\rho \frac{\partial^2 w}{\partial t^2} = \mu \frac{\partial^2 w}{\partial x^2} \qquad \text{(7-18}a\text{)}$$

$$\rho \frac{\partial^2 \zeta}{\partial t^2} = \mu \frac{\partial^2 \zeta}{\partial x^2} \qquad \text{(7-18}b\text{)}$$

Equation (7-17) represents a wave whose velocity is $(\lambda + 2\mu)/\rho$ [or $(K + 4\mu/3)/\rho$], the *P* wave velocity, with its displacement u in the direction of motion (x direction). This is referred to as a dilatational or compressional vibration. Equations (7-18) represent waves whose velocity is μ/ρ, the *S* wave velocity, with displacement (w or ζ) at right angles to the direction of propagation. These are referred to as *transverse* or *shear waves*. We shall see in Sec. 7-2 that velocity for fractured or porous rocks depends strongly on the state of stress, the fluid saturation, and sometimes temperature. Typical velocities of *P* and *S* waves in rocks are given in Table 7-1. Since λ is always positive, the *P* wave

TABLE 7-1 Compressional wave v_p and shear wave v_s velocities for several low-porosity rocks at a confining pressure pressure of 100 MPa. From Christensen (1982) and Clark (1966).

Material	v_p, km/s	v_s, km/s
Sandstone (Catskill, N.Y.)	5.3	
Shale	5.3	
Serpentinite	5.7	2.8
Solenhofen limestone	5.7	3.1
Graywacke (New Zealand)	5.8	
Westerly granite	5.8	3.4
Granite (Stone Mtn., Ga.)	5.9	3.5
Slate	5.9	3.3
Quartz monzonite	6.0	3.6
Granite (Barre, Vt.)	6.1	3.5
Graywacke (Czechoslovakia)	6.2	3.6
Marble	6.3	
Quartz diorite	6.5	3.7
Albitite	6.6	3.6
Dolomite	6.7	3.4
Norite	7.1	3.8
Dunite (Webster, N.C.)	7.5	4.2
Dunite	8.2	4.7
Sillmanite	9.6	5.1
Steel	5.9	
Cement	2.7–4.9	

velocity v_p is always greater than the S wave velocity v_s. Since μ in a fluid is always zero, there is no S wave in a fluid.

If a shear wave is traveling horizontally, its transverse motion can be resolved into horizontal and vertical components; in fact, some shear waves can be polarized so that only one of these components exists. A component whose motion is in the vertical plane is called an *SV* wave, and one whose motion is in the horizontal plane is called an *SH* wave.

Energy in the wave. The energy of a wave in a given medium is proportional to the square of the peak displacement, or *amplitude*, of the wave. Telford et al. (1981) show that the amount of energy in a wave passing through unit area in unit time is

$$E = \tfrac{1}{2}\rho v \omega^2 A^2 \tag{7-19}$$

As a spherical wave spreads out from its source, the energy must be distributed over an area equal to that of a sphere whose radius is equal to the distance from the source. Therefore, the energy of a spherical wave is inversely proportional to the square of the distance from the source, and the amplitude, the square root of the energy, is inversely proportional to the distance from the source.

Besides geometric spreading losses, there are energy losses from absorption, scattering from fractures, vibration of pore fluids, and many other causes. These losses appear to be exponential with distance (Dutta, 1981; Toksöz et al., 1979; Winkler and Nur, 1982). Consequently, for a spherical wave, geometric spreading and attenuation losses combine to give

$$E = E_0 \left(\frac{r_0}{r}\right)^2 e^{-\alpha r} \tag{7-20}$$

where the *absorption coefficient* α will be studied in Sec. 7-3b.

Group velocity. The velocity v in Eq. (7-14) is known as the *phase velocity* because it is the distance traveled in unit time by a point of constant phase, such as a peak or trough. If v is not constant but is instead a function of frequency [and therefore of n in Eq. (7-14)], then the energy in the wave does not necessarily travel with a single velocity v. This phenomenon is known as *dispersion*. Then a pulse travels at a velocity U, the *group velocity*, shown in Fig. 7-3. It can be shown (Officer, 1958) that if v is a function of λ,

$$U = v - \lambda \frac{dv}{d\lambda} = v + \omega \frac{dv}{d\omega} \tag{7-21}$$

When v decreases with frequency, v is greater than U as shown in Fig. 7-3, where individual cycles overtake and pass through the wave packet that is traveling with group velocity U. On the other hand, if v increases with frequency, then U is greater than v.

Surface waves. If a wave is propagating parallel to and near the surface of an elastic medium, it is called a *surface wave.* The stresses and strains must be continuous across the surface. In particular, at a free surface the stress must be zero. The solution of the wave equation with this boundary condition gives a Rayleigh wave with velocity v_r slightly less than v_s (an analytical expression is given by White, 1965) and displacements for propagation in the x direction (Wasley, 1973) of

$$u = A\frac{\omega}{v_r}\left\{\exp\left[-z\omega\left(\frac{1}{v_r^2}-\frac{1}{v_p^2}\right)^{1/2}\right] - 2\left[\left(\frac{1}{v_r^2}-\frac{1}{v_p^2}\right)\left(\frac{1}{v_r^2}-\frac{1}{v_s^2}\right)\right]^{1/2}\right.$$
$$\left.\times\left(\frac{2}{v_r^2}-\frac{1}{v_s^2}\right)^{-1}\exp\left[-z\omega\left(\frac{1}{v_r^2}-\frac{1}{v_s^2}\right)^{1/2}\right]\right\}\sin\omega\left(t-\frac{x}{v_r}\right) \qquad (7\text{-}22a)$$

$$\zeta = -A\omega\left(\frac{1}{v_r^2}-\frac{1}{v_p^2}\right)^{1/2}\left\{\exp\left[-z\omega\left(\frac{1}{v_r^2}-\frac{1}{v_p^2}\right)^{1/2}\right.\right.$$
$$\left.\left.-\frac{2}{v_r^2}\times\left(\frac{2}{v_r^2}-\frac{1}{v_s^2}\right)^{-1}\exp\left[-z\omega\left(\frac{1}{v_r^2}-\frac{1}{v_s^2}\right)^{1/2}\right]\right]\right\}\cos\omega\left(t-\frac{x}{v_r}\right) \qquad (7\text{-}22b)$$

Now the parametric equations for an ellipse can be given as

$$\xi_1 = c_1 \sin\eta \qquad \xi_2 = c_2 \cos\eta \qquad (7\text{-}23)$$

Therefore, on comparing these expressions with Eq. (7-22), we see that η represents the arguments $\omega(t - x/\mathrm{v}_r)$, the ξ_1 and ξ_2 can be identified with the displacements u and ζ, and the c's correspond to the coefficients of the trigonometric functions. It is a theorem of geometry that the c's are the lengths of the semiaxes of the ellipse. Hence, Eq. (7-22) shows that the path of a particle is an ellipse. It can be shown that $c_1 < c_2$, so the major axis of the ellipse is perpendicular to the surface.

If we set z equal to zero and recognize that $\mathrm{v}_p > \mathrm{v}_s - \mathrm{v}_r$, we see that u is negative. Therefore, the motion is *retrograde* as shown in Fig. 7-4. Equations (7-22) also show that the amplitude decreases exponentially with the depth z.

There are other kinds of surface waves, and the reader can find details in Grant and West (1965).

Waves in a plate. In a plate of finite thickness that is infinite in the other two dimensions, the velocity of a wave depends on wavelength.

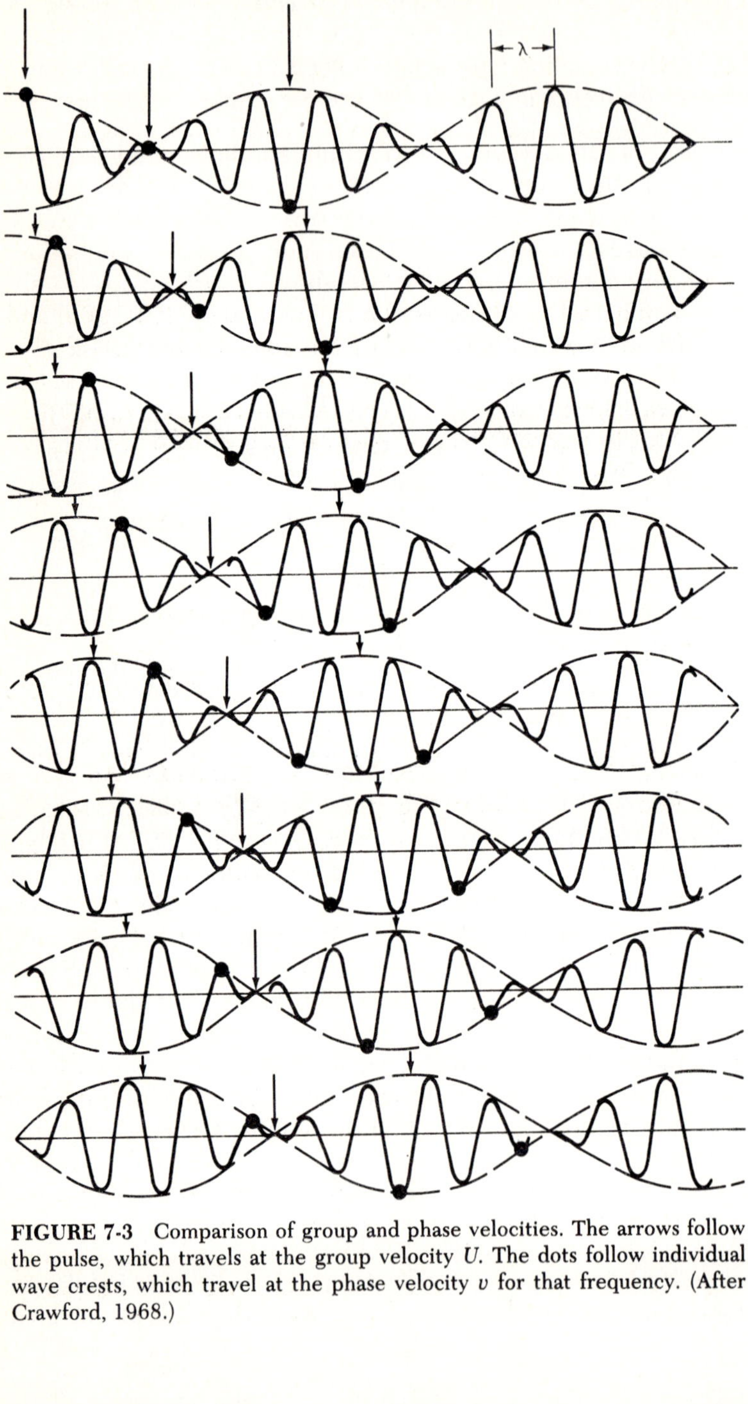

FIGURE 7-3 Comparison of group and phase velocities. The arrows follow the pulse, which travels at the group velocity U. The dots follow individual wave crests, which travel at the phase velocity v for that frequency. (After Crawford, 1968.)

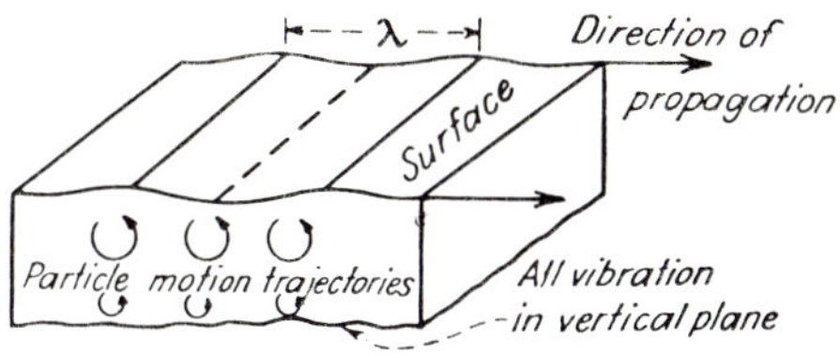

FIGURE 7-4 Characteristics of a Rayleigh wave. (Dobrin, 1976.)

If the wavelength is very short compared to the thickness, then over much of the thickness the wave does not sense boundaries and propagates as in an infinite solid. If the thickness is a few tens of wavelengths, propagation is like that of a Rayleigh wave. If the wavelength is long compared to the thickness, then propagation is that of a *plate wave* or *longitudinal wave* with velocity (Kolsky, 1963)

$$v_\ell = \left[\frac{4\mu(\lambda + \mu)}{\rho(\lambda + 2\mu)}\right]^{1/2} \tag{7-24}$$

If the wavelength is comparable to the thickness, the velocity is a complex function of position in the plate.

7-1c Reflection and Refraction

The boundary conditions at an interface require that stress and strain be continuous. If either a *P* wave or an *S* wave is incident at an angle other than normal incidence at a boundary, it can be shown that both a *P* wave and an *S* wave will be reflected from the boundary and, if the medium on the far side of the boundary is a solid, under many circumstances both a *P* wave and an *S* wave will be transmitted across the boundary. Thus both *P* and *S* waves undergo *mode conversion* at the boundary.

Huygens's principle and head waves. The reflection and refraction of many waves can be visualized by use of Huygens's principle. (Dobrin, 1976, gives a good description of the principle.) The principle is that every point on a wavefront is a source of a new wavefront that travels out from that point as a spherical shell. If the wavefront has a large enough radius, the wave can be treated as a plane wave.

Figure 7-5 shows a plane wave with velocity v_1 incident at angle θ_1 on a plane interface. A wave with velocity v_2 is reflected at angle θ_2, and a wave with velocity v_3 is refracted at angle θ_3 into the second medium.

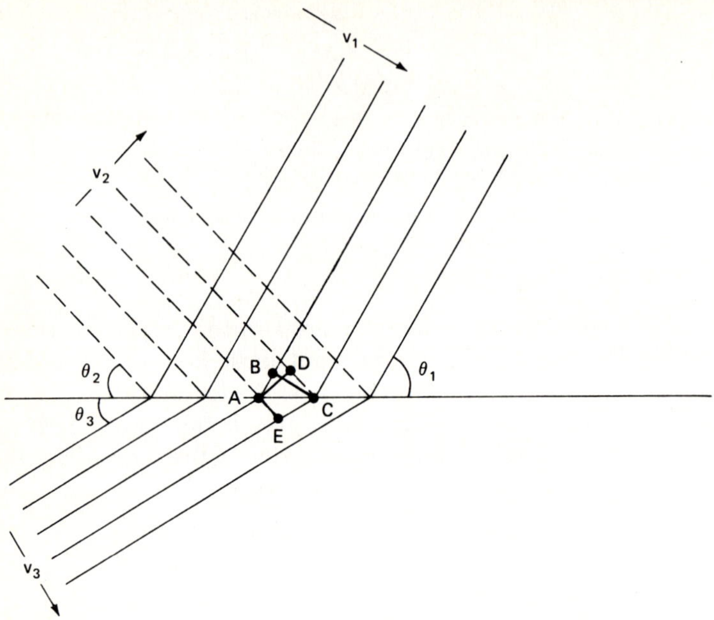

FIGURE 7-5 Reflection and refraction of a compressional wave incident on a plane surface.

(The velocities are, of course, characteristic of the medium.) The distance $BC = v_1 \Delta t$, $AD = v_2 \Delta t$, and $AE = v_3 \Delta t$, where Δt is some travel time of each wave. By simple trigonometry,

$$\sin \theta_1 = \frac{BC}{AC} \qquad \sin \theta_2 = \frac{AD}{AC} \qquad \text{and} \qquad \sin \theta_3 = \frac{AE}{AC}$$

Consequently,

$$\frac{\sin \theta_1}{v_1} = \frac{\sin \theta_2}{v_2} = \frac{\sin \theta_3}{v_3} \tag{7-25a}$$

This is Snell's law, familiar from optics. For a reflected wave of the same type (P or S) as the incident wave, $v_1 = v_2$ and so $\theta_1 = \theta_2$. For an incident P or S wave there can be both P and S waves reflected and refracted, so finally,

$$\frac{\sin \theta_{p1}}{v_{p1}} = \frac{\sin \theta_{p2}}{v_{p2}} = \frac{\sin \theta_{s1}}{v_{s1}} = \frac{\sin \theta_{s2}}{v_{s2}} \tag{7-25b}$$

where v_{p1}, v_{p2}, v_{s1}, and v_{s2} are the P and S velocities in the first and second media, respectively.

If $v_{p2} > v_{p1}$, $\sin \theta_1$ can become equal to or greater than v_{p1}/v_{p2}. When $\sin \theta_1 = v_{p1}/v_{p2}$, $\sin \theta_3$ becomes equal to 1 and θ_3 becomes 90°. This means the refracted wave does not penetrate the medium but travels along the interface. This angle, $\sin^{-1} (v_{p1}/v_{p2})$, is known as the *critical angle* for refraction. For any angle greater than the critical angle, there is no refraction into the second medium, and the wave is *totally reflected.* At the critical angle, a wave called the *head wave* can travel along the interface at velocity v_{p2} and propagate energy back into the first medium. Although its mathematical description is quite complex (Grant and West, 1965), in practice the head wave acts like a wave traveling along the interface with each point being the source of Huygens's wavelets that can propagate into medium 1. Then the least-time path can be from a source in medium 1, along the interface, and to a detector in medium 1 (Fig. 7-6).

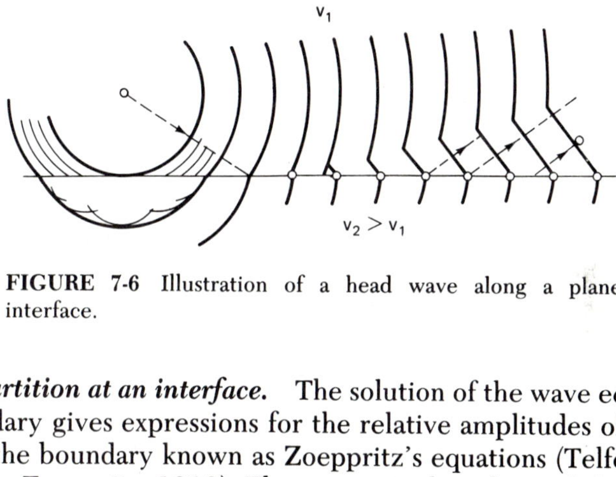

FIGURE 7-6 Illustration of a head wave along a plane interface.

Energy partition at an interface. The solution of the wave equation at the boundary gives expressions for the relative amplitudes of the four waves at the boundary known as Zoeppritz's equations (Telford et al., 1981, from Zoeppritz, 1919). They are quite lengthy, and they are not quoted here. The products of density and velocity, called *acoustic impedances,* are denoted by Z. Figure 7-7 shows the energy partition for an incident compressional wave at a boundary having specific velocities and densities as a function of angle of incidence.

Zoeppritz's equations reduce to a simple form for normal incidence, since $\cos \theta_1 = \cos \theta_2 = 1$, and there is no tangential stress. Then the *reflection coefficient* and *transmission coefficient* for amplitude are

$$R = \frac{A_1}{A_0} = \frac{Z_2 - Z_1}{Z_2 + Z_1} \tag{7-26a}$$

$$T = \frac{A_2}{A_0} = \frac{2Z_1}{Z_2 + Z_1} \tag{7-26b}$$

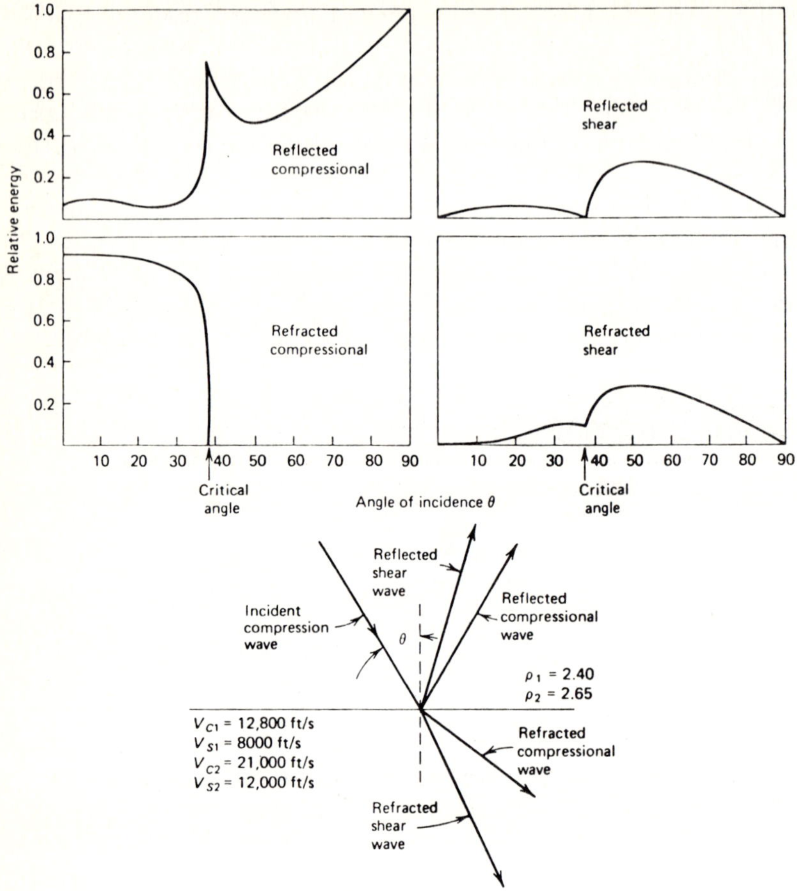

FIGURE 7-7 Partition of energy in a compressional wave incident at a boundary between materials having specified velocities and densities as a function of the angle of incidence. (Dobrin, 1976.)

The fractions of energy reflected and transmitted also are called the reflection and transmission coefficients; the reader should be sure which is meant in any discussion. The latter are [recalling Eq. (7-19)]:

$$R^2 = \frac{v_{p1}\rho_1\omega^2 A_1^2}{v_{p1}\rho_1\omega^2 A_0^2} = \frac{(Z_2 - Z_1)^2}{(Z_2 + Z_1)^2} \tag{7-27a}$$

$$T^2 = \frac{v_{p2}\rho_2\omega^2 A_2^2}{v_{p1}\rho_1\omega^2 A_0^2} = \frac{4Z_1 Z_2}{(Z_2 + Z_1)^2} \tag{7-27b}$$

Equation (7-27) is a good approximation for angles of incidence up to about 20° (Telford et al., 1981).

Note that, although the energy fractions R^2 and T^2 do not depend on which side of an interface the wave strikes, the reflected amplitude does depend on the side. If $Z_1 > Z_2$, A_1 is negative. Then the reflected wave is 180° out of phase with the incident wave, and an incident compression is reflected as a rarefaction. This is particularly important at a free surface, where $Z_2 = 0$ and $A_1 = -A_0$.

This development assumes reflectors that are infinite in the plane perpendicular to the direction of travel of the wave. If the reflector is comparable in size to the wavelength of the wave, less energy will be reflected. If the reflector is small compared to the wavelength, the wave will essentially ignore it, and will "go around." There will be no reflection.

7-1d Waves in a Borehole

The boundary conditions for waves in a borehole are similar to those at an interface; stress and strain must be continuous at the borehole wall. In practice this means shear stress must vanish, because a liquid cannot sustain a shear stress. White (1965) made a comprehensive study of the propagation of waves in a borehole.

The waves that have so far been of most interest in logging are head waves. The source of the signal is usually centered in the borehole, so a wave travels through the fluid to the borehole wall. Velocities v_p and v_s in the wall are generally greater than v_f, the velocity in the fluid, so both a *P* and *S* head wave propagate along the interface and back to a detector elsewhere in the hole. These are the first two waves observed at a detector. Sometimes there is also a "hybrid" or "leaky" mode observed between the *P* and *S* arrivals.

Biot (1952) showed that two other types of wave propagate in a borehole. The first of these, which he termed *conical* waves or *reflected type waves*, are now usually called *normal modes* or *pseudo-Rayleigh waves*. They are analogous to standing waves in an organ pipe, and there are infinitely many of them. Their group velocity is between v_s and v_f, and it varies with frequency (that is, the wave is *dispersive*). The second is a *Stoneley wave* or *tube wave*, and it is analogous to a Rayleigh wave at a plane interface. Its group velocity is slightly less than v_f, and it is only slightly dispersive. Both of these waves have much higher amplitude than the head waves.

Other authors (White and Zechman, 1968; Peterson, 1974; Paillet and White, 1982) have studied waves in the borehole, but Paillet (1981) has given the description that seems the most useful for giving physical insight into the situation. He replaces the cylindrical geometry of a logging sonde centered in a hole by two plane parallel walls with a fluid between them, and he then computes the amplitudes of different waves as functions of distance between them and properties of the fluid and solid. This approximation appears to be valid as long as the wavelength is large compared to the diameter of the borehole.

Paillet discusses four modes of wave motion that are commonly observed in an acoustic log. The first two are compressional and shear head waves, traveling along the interface between the formation and the borehole fluid. Another is the *tube wave,* which is a form of surface wave propagating along the wall of the hole. It has a velocity approximately equal to that in the fluid. Finally, there are *normal modes*, which are trapped or guided waves propagating in the borehole. They exist only for certain frequencies, and their velocity is a function of the frequency and the particular mode. Tube waves and normal modes are often called *fluid waves.* Figure 7-8 shows how these waves combine to give the final signal.

Paillet points out that, because the thickness of the annulus between the sonde and the borehole wall is comparable to the wavelength of the frequencies used in an acoustic log (Sec. 7-3), the energy spectrum of the head waves is sharply peaked at certain frequencies and is different for *P* and *S* waves. This implies that it may be possible to distinguish *P* waves from *S* waves by frequency, and this method has been tried with some success (Ingram et al., 1981).

Paillet computes the amplitude of tube wave and normal modes for different annular thickness in the same medium. The waveform is quite different for different thicknesses. When the annulus is very thin, the tube wave dominates, and very clear fluid arrivals are observed. For a larger hole, the first and second normal modes dominate, and they arrive earlier than the tube wave. In some cases, although the normal mode dominates, its speed is approximately the same as that of the tube wave, and the two cannot be distinguished by arrival time.

The type of vibration, however, is different for tube waves and normal modes (Paillet and White, 1982). This difference becomes important when one tries to use the fluid wave to detect fractures (Sec. 14-1).

7-2 PROPAGATION OF BODY WAVES IN POROUS ROCK

7-2a Velocity

We have been discussing the propagation of stress waves in elastic media. Rocks, however, are not really elastic. If the rock were made up only of a matrix of a few minerals, with no pores or fractures, its behav-

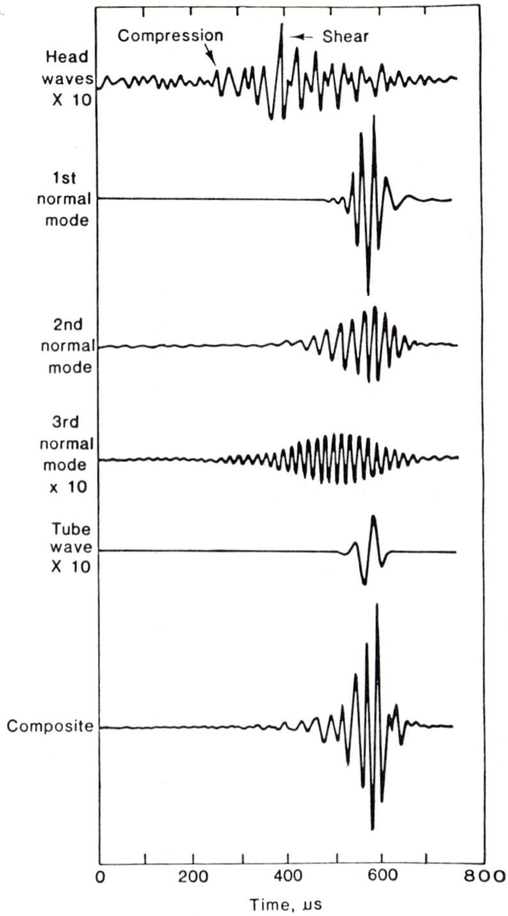

FIGURE 7-8 Signals for head wave, three normal modes, the tube wave, and the resulting combined signal. (Paillet, 1981.)

ior might be truly elastic. But the rock contains pores, fractures, joints, boulders, and other discontinuities. While the velocity of propagation of P and S waves of course depends on the velocity in the rock matrix, it is also affected by pressure, porosity, pore fluid, pore shape, fraction of the pore containing a given fluid (that is, saturation), fractures, and temperature. We shall discuss these in turn.

Matrix velocity. Table 7-1 shows the variation of P and S wave velocity at a pressure of 100 MPa in common low-porosity rocks (Christensen, 1982; Clark, 1966). The range is generally between 5.5 and 7.5

km/s for P waves and 3.0 to 4.5 km/s for S waves. This range is fairly small and would probably of itself not be of much interest for obtaining information about rock properties by measurement of velocity. Furthermore, King (1966) and others have demonstrated that, at low stress levels, there is often significant anisotropy in rocks, which could give rise to velocity variations comparable to the range of low-porosity high-pressure data.

The matrix velocity, then, is only the starting point to which velocities in common porous rocks can be compared.

Effects of pressure. Acoustic velocity in porous rock is a strong function of *differential* or *effective stress.* This quantity, ΔP, is the difference between applied external pressure and the pressure in the pore fluid. In rock mechanics ΔP is often the variable that best describes the stress state in the rock. A typical example of the effect of applied and differential pressure is shown in Fig. 7-9. The velocity increases with increasing differential pressure, rapidly at low pressure, and then more slowly. At some differential pressure a more or less constant terminal velocity is attained. A fair fraction of logging situations are at depths corresponding to pressures that give terminal velocity. At small constant dif-

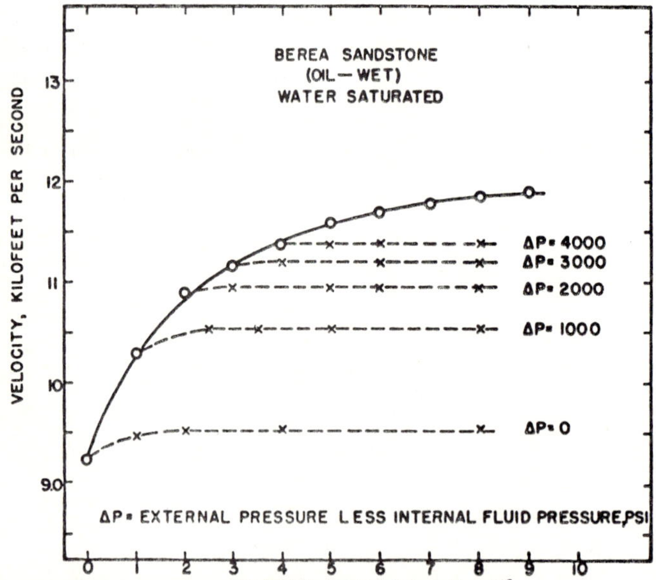

FIGURE 7-9 Velocities in a saturated sandstone as a function of pressure. (Wyllie et al., 1958.)

ferential pressure, the velocity increases with external pressure at first and then becomes approximately constant. At large constant differential pressure, the velocity is constant for all external pressures.

This behavior is generally explained as an effect of the closure of microcracks in the rock with increasing differential pressure. The stress tending to close the cracks is obviously the differential pressure, not the external pressure, so the explanation is plausible.

Effects of porosity. The most obvious effect of porosity on propagation of stress waves is reduction of velocity. A famous equation, suggested by Wyllie et al. (1956, 1958) relates velocity in bulk material v_b to velocity in the matrix v_{ma} and fluid v_f in rock of porosity ϕ by

$$\frac{1}{v_b} = \frac{\phi}{v_f} + \frac{1 - \phi}{v_{ma}} \tag{7-28}$$

Equation (7-28) is often known as the *time-average formula* because it is the formula for the velocity obtained from the average time for a wave to propagate through a slab of total thickness 1, of which the fraction ϕ is fluid and the fraction $(1 - \phi)$ is the matrix. (In other words, the propagation is through alternating layers of solid and fluid.) The equation is also often known as Wyllie's equation.

Equation (7-28) is the most commonly used equation in acoustic logging. In most of the standard techniques for acoustic logging applications Eq. (7-28) is used to obtain porosity (one of the most common objectives of running the log) and then various corrections are applied to the result.

In fact, Wyllie et al. (1958) point out that the application of Eq. (7-28) is very restricted. It was derived for the *terminal velocities* of hand specimens of consolidated sedimentary rock. Therefore, it applies only to pressures high enough for terminal velocity to have been attained. It is applicable only to *intergranular* porosities up to about 0.35. The rock should be fully saturated with the fluid, and v_f must be at least the velocity for water; in other words, it does not apply to gas-filled pores. The rock must be neither fractured nor vuggy. Figure 7-10 shows the typical scatter about Eq. (7-28) under those conditions.

Geertsma (1961) calculated velocity in a homogeneous medium with well-separated compressible, spherical pores. He showed that, for small porosity,

$$v_p = A + B\phi \tag{7-29}$$

which, he states, agrees with data over a wider range of properties and effective stress than does Eq. (7-28). He points out that Eq. (7-28) is equivalent to a special case of Eq. (7-29) with the appropriate choice

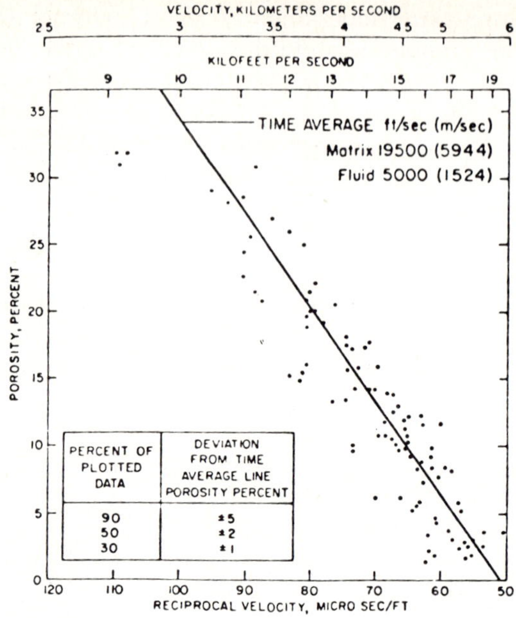

FIGURE 7-10 Velocity-porosity data determined in the laboratory for water-saturated sandstones compared with the time-average relation [Eq. (7-28)] for a quartz-water system. (Gregory, 1977.)

of *A* and *B*, and he attributes the success of Eq. (7-28) to this equivalence.

Raymer et al. (1980) have fitted data from many samples of sandstone, limestone, and dolomite and developed a more complex equation for transforming travel time, Δt, to porosity. The method eliminates matrix velocity as a parameter. Independence of travel time and matrix velocity seems somewhat nonphysical.

Wyllie et al. (1958) go on to discuss the effect of vugs and fractures on velocity. They find that generally vugs (small, widely scattered holes) and small fractures do not affect the terminal velocity. They suggest that in rocks with vugs and fractures the transmitted energy tends to follow the path of maximum velocity and bypass the vugs and fractures. Of course, large fractures perpendicular to the path of the signal cannot be bypassed but will attenuate the signal and give low apparent velocities. Equation (7-28) does not apply to vugular and fracture porosity. Because fractures and vugs often form after deposition, rather than during deposition, they are referred to as secondary rather than primary features. Hence, it is often stated that Eq. (7-28) does not apply if secondary porosity is present.

Effects of saturation. The variation of velocity with liquid saturation of pore space appears to be completely unsystematic at atmospheric pressure. At higher pressures (about 50 MPa) the phenomena become more systematic.

Biot (1956) and Geertsma (1961) developed a theory to account for the propagation of acoustic waves in fully and partially saturated media. The theory has been verified in detail for fully saturated media (Berryman, 1980). In general, the theory predicts that

$$v_{pW} > v_{pD} \tag{7-30a}$$

$$v_{sW} < v_{sD} \tag{7-30b}$$

where the second subscript W refers to fully saturated pores and D to fully dry pores. Equation (7-30) is generally observed experimentally to be true for v_p, but it is often violated for v_s, particularly at low effective stress (Gregory, 1976; Domenico, 1976; King, 1966, Toksöz et al., 1976).

Often v_p is less at partial saturation than for either dry or fully saturated material. These changes are neatly explained by Winkler and Nur (1982).

> As water is added to the rock, the overall density increases, thus reducing the velocity of both P and S waves [Eq. (7-18)]. As total saturation is reached, the pores in the rock become less compressible. This increases the effective bulk modulus of the rock [$\lambda + \frac{2}{3}\mu$] but has little effect on the effective shear modulus [μ]. Therefore, P wave velocity increases at total saturation while S wave velocity reaches a minimum.

This phenomenon permits a low value of $\mathbf{v}_p/\mathbf{v}_s$ to be used as a qualitative indicator of the presence of gas.

Gregory (1976) used a large number of measurements in sedimentary rock to find general relations among velocity, saturation, and porosity for both partially and fully saturated rock. He found that at low porosity (<0.1), both v_p and v_s increase with saturation, and both are higher at full saturation than in dry rock. At medium porosity (0.1 to 0.25), v_p increases slowly with saturation and v_s decreases. At high porosity (>0.25) v_p first decreases at very low saturation, and then increases to slightly over its original level at full saturation and v_s just decreases. Consequently, Gregory finds that Eq. (7-30b) is satisfied at porosity greater than 0.1 but at high porosity Eq. (7-30a) works only at the end points of saturation.

Toksöz et al. (1976) have advanced a plausible explanation for the observed change in velocities with saturation. They give a mixing law for effective values of K (or λ) and μ as functions of K and μ of the matrix and the fluid, with multiplicative correction terms that depend on α, the *aspect ratio* of the pores. The pores are considered to be in the shape of oblate spheroids, and the aspect ratio is the ratio of the length of the short axis to that of the long axis of the pore. They then calculate v_p and v_s for saturated and dry materials with pores of different aspect ratios. They point out that for a given concentration, the thinner pores (small α) affect both compressional and shear velocities much more than the spherical pores. They explain the result as follows:

> For spherical pores, the velocities of both compressional and shear waves are higher when the medium is saturated with gas than when saturated with water. This is because the density decreases more rapidly than both K and μ upon gas saturation, resulting in higher velocities [Eq. (7-18)]. For $\alpha = 0.1$, shear velocity is slightly higher and compressional velocity is lower for gas saturation than it is for water. At low aspect ratios ($\alpha = 0.05$ to 0.01) both compressional and shear velocities are lower for the gas-saturated case although the net change is greater for compressional waves than for shear.

This last is because the effective bulk modulus changes much more rapidly than the shear modulus as the saturation changes from water to gas. Consequently, at low aspect ratio, where porosities are small and the density does not change much, the change in K (or λ) dominates. These results agree with those of Gregory (1976) if we assume low α for low porosity and intermediate α for intermediate and high porosity. That is plausible because the pore space of low-porosity rock tends to be dominated by cracks, whereas that of higher-porosity rock tends to be comprised of rounder pores.

Toksöz et al. (1976) also find that they get different values of effective moduli for different manners of fluid mixing. If the gas is immiscible in the liquid, its effect is very large. If the gas dissolves, it has much less effect.

They summarize their measurements as follows:

> 1 The elastic moduli and seismic velocities of rock decrease with increasing porosity for a given matrix. For a given concentration of pores, the flatter (thinner) pores have a relatively greater effect on velocities than the rounder or spherical pores. . . .
>
> 2 The bulk and shear moduli are lower for the dry or gas-saturated case than for the water- or oil-saturated case for all pore shapes. The velocities, however, behave in a more complicated way because of density dependence. . . . For typical sedimentary rock models [with mixtures of pore sizes], generally the compressional velocities for the dry . . . cases are lower than for the water-saturated, and shear velocities are either about the same or higher (especially for highly porous rocks) for gas-saturated rocks.

3 The increase of the compressional and shear-wave velocities in saturated and dry rocks as a function of effective pressure is primarily determined by the closing of the thin pores as a function of pressure.

This work appears to give a plausible explanation for the observed (King, 1966) phenomena.

Eastwood and Castagna (1983) have shown that the model described by Toksöz et al. (1976) agrees with experiment. They have used the mixing law to calculate v_p/v_s for binary and ternary mixtures of common minerals and then cross-plotted it against v_s. They compared measured to calculated positions on the cross-plot to determine the lithology.

Effects of temperature. Table 7-2 shows the results of several studies (Christensen, 1982; Timur, 1977) on the effects of temperature on P and S wave velocities in many rocks that contain water. Much of the effect is due to temperature-induced changes in fluid velocity v_f. Generally, velocity decreases with increasing temperature, and the decrease can range from a fraction of a percent per 100° to as much as 10 percent. Therefore, the effect of temperature, while not trivial, is usually small compared to the effects of porosity and saturation.

TABLE 7-2 Dependence of compressional wave v_p and shear wave v_s velocities on temperature, in degrees Celsius, for several rock types at a confining pressure of 50 Mpa. From Christensen (1982).

	v_p, km/s, when temperature is as follows:				v_s, km/s, when temperature is as follows:			
Material	25	100	200	300	25	100	200	300
Dry sandstone $\phi = 0.05$	4.58	4.38	4.14		2.7		2.6	
Wet sandstone $\phi = 0.05$	4.51	4.23	3.87		2.9			
Basalt	5.57	5.59	5.50	5.40	3.23	3.20	3.18	3.16
Solenhofen limestone	6.03	5.93	5.78	5.69	2.99	2.96	2.93	
Argillaceous limestone	6.13	6.10	6.19		3.21	3.21	3.25	
Texas pink granite	6.29	6.23	5.86	5.57	3.35	3.28	3.13	
San Marcos gabbro	6.79	6.78	6.75	6.53	3.48	3.50	3.48	3.35
Dunite	7.54	7.38	6.78		3.88	3.83	3.50	

7-2b Attenuation

The mechanism causing attenuation of stress waves in rock is still not well understood, but it is the subject of much current research. Some definitions are needed:

The change in energy per cycle is

$$\frac{\Delta E}{E} = \frac{A_1^2 - A_2^2}{A_1^2}$$

where A_1 and A_2 are the amplitudes at successive cycles. The logarithmic decrement is defined as

$$\delta = \ln \frac{A_1}{A_2} \simeq \frac{A_1^2 - A_2^2}{2A_1^2} = \frac{\Delta E}{2E} \tag{7-31}$$

If we recall Eq. (7-20) where, for a plane wave,

$$E = E_0 e^{-\alpha r} \tag{7-20}$$

where α is an *attenuation coefficient,* we find that

$$\delta = \alpha(r_2 - r_1) = \alpha\lambda = \frac{\alpha v}{f}$$

where r_1 and r_2 are the radii to the peaks of two successive cycles. It is conventional to define a quantity Q, the *quality factor:*

$$\frac{1}{Q} = \frac{\delta}{\pi} = \frac{\alpha\lambda}{\pi} = \frac{\alpha v}{\pi f} \tag{7-32}$$

Therefore, a high Q implies small attenuation.

Attenuation is affected by the same rock properties as velocity. It is found, however, that α is often equal to a constant times the frequency, and therefore Q is independent of frequency. Consequently, Q is often more convenient to use than α. [It has recently been learned (Spencer, 1981) that Q is in fact a function of frequency below 500 Hz in water-saturated rocks at zero confining pressure.] Q depends on fracture frequency, saturation, porosity, and stress state. Table 7-3 gives values of attenuation of P and S waves for common rocks. Note the wide variation in Q found by different workers for the same rock type. In general, low Q is associated with material that appears to be weak.

Effects of pressure. It is generally agreed that attenuation decreases with increasing applied pressure (Dutta, 1981) or effective pressure (Toksöz et al., 1979). Attenuation decreases rapidly as pressure is applied and then levels off at high pressure.

TABLE 7-3 Quality factor for compressional waves, Q_p, and shear waves, Q_s, for several rocks.* (Table continues; footnotes on next page.)

Material	Q_p	Q_s	Lab (L) or in situ (I)	Reference†
Aluminum	35,000		L	Winkler and Nur, 1982
Vycor	435		L	Winkler and Nur, 1982
Lucite	23		L	Winkler and Nur, 1982
Granite	47–53	43–47	L	Levykin, 1965
Granite	185	349	I	Vasil'ev and Gurevich, 1962
Granite	21–84	28–142	I	Vasil'ev and Gurevich, 1962
Granite	125		L	Birch and Bancroft, 1938‡
Sierra granite	185–204		L	Winkler and Nur, 1982
Gneiss	42–57	35–51	L	Levykin, 1965
Gabbro	98	108	L	Levykin, 1965
Serpentinite	58	95	L	Levykin, 1965
Limestone	112–157	184–448	L	Vasil'ev and Gurevich, 1962
Solenhofen limestone	112	188	L	Peselnick and Zeitz, 1959‡
I-1 limestone	165		L	Peselnick and Zeitz, 1959‡
Hunton limestone	65		L	Born, 1941‡
Spergen LS (ϕ = 0.15–sat.)	15	12	L	Toksöz et al., 1979
Chalk	157	224	I	Vasil'ev and Gurevich, 1962
Pierre shale	32	10	I	MacDonal et al., 1958‡
Clay shale	31	101	I	Vasil'ev and Gurevich, 1962
Ice	21–105	15–157	I	Vasil'ev and Gurevich, 1962
Saturated sandstone	39	14	I	Vasil'ev and Gurevich, 1962
Amherst sandstone	52		L	Born, 1941‡
Berea sandstone (sat.)	10		L	Toksöz et al., 1979‡
Berea sandstone (dry)	103–140		L	Winkler and Nur, 1982
Navajo SS (ϕ = 0.125–sat.)	7	6	L	Toksöz et al., 1979
Navajo SS (dry)	21		L	Bruckshaw and Mahanta, 1954‡

TABLE 7-3 *(continued)*

Material	Q_p	Q_s	Lab (L) or in situ (I)	Reference†
Boise SS (ϕ = 0.25–sat.)	7	6	L	Toksöz et al., 1979
Massilon sandstone	107–138		L	Winkler and Nur, 1982
Saturated sand and clay	100–150	39	I	Vasil'ev and Gurevich, 1962
Saturated clay	52	8–12	I	Vasil'ev and Gurevich, 1962

*$1/Q = \alpha v/\pi f$, where α is the attenuation coefficient (Sec. 7-2b) and f is the frequency. Confining pressure is 0.1 to 1.0 MPa.

†See references listed at end of chapter.

‡Cited in Johnston et al., 1979.

Effects of porosity. Attenuation generally increases with increasing porosity (Dzeban, 1970; Toksöz et al., 1979), as can be seen in Table 7-3. The effect of porosity decreases with increasing pressure.

Effects of saturation. The effect of saturation is well summarized by Winkler and Nur (1982):

> Both shear and . . . compressional attenuation are minimum in dry rock. Attenuation for both modes increases substantially when the rock is partially saturated, with bulk compression being affected about twice as much as shear attenuation. As total saturation is achieved, bulk compressional attenuation is significantly reduced, while shear attenuation reaches a maximum.

This work was done at frequencies between 500 and 9000 Hz. Figure 7-11 illustrates the behavior. Johnston et al. (1979) show that above 100 kHz attenuation increases rapidly when water is first added to a dry rock and then levels off.

There are major disagreements among researchers as to the mechanism of attenuation at amplitudes and frequencies of interest in logging. Two main mechanisms are proposed:

1 Viscous fluid flow (Biot, 1956), also known, depending on scale, as *sloshing* and *squirting*

2 Solid friction in the rock, or *frame decrement,* known also as *jostling*

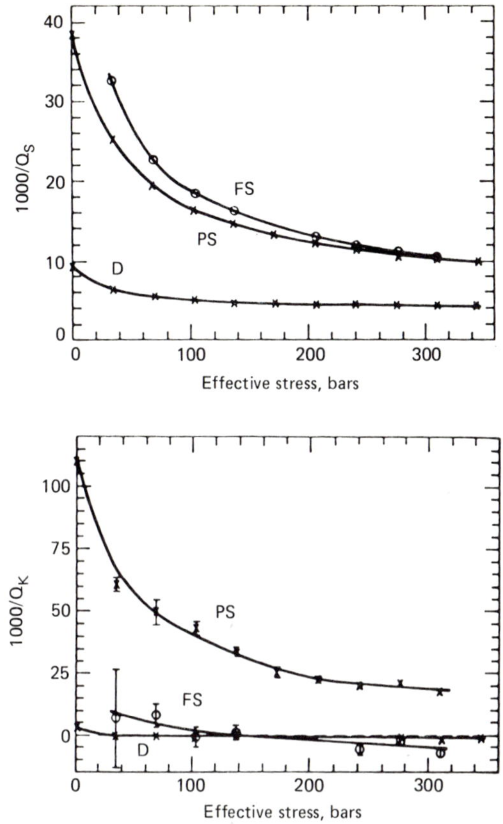

FIGURE 7-11 (*a*) Shear and (*b*) compressional attenuation versus effective stress in dry (D), partially (≃ 90%) saturated (PS), and fully saturated (FS) Massilon sandstone. Porosity = 0.23. (Winkler and Nur, 1982.)

These two mechanisms are characterized by somewhat different behavior of attenuation as a function of saturation, amplitude, and frequency.

Winkler and Nur (1982) have given a good explanation for the observed behavior. Figure 7-12 illustrates their concept.

> For fully saturated cracks, bulk compression causes every crack to be compressed by the same amount, so there is no substantial fluid flow. Pure shear causes compresssion of some cracks and dilation of others, resulting in fluid flow between cracks. Therefore in fully saturated rock we expect significant *S* wave attenuation and much less bulk compressional attenuation. For partially saturated cracks, bulk compression causes every crack to compress and contribute to fluid flow and/

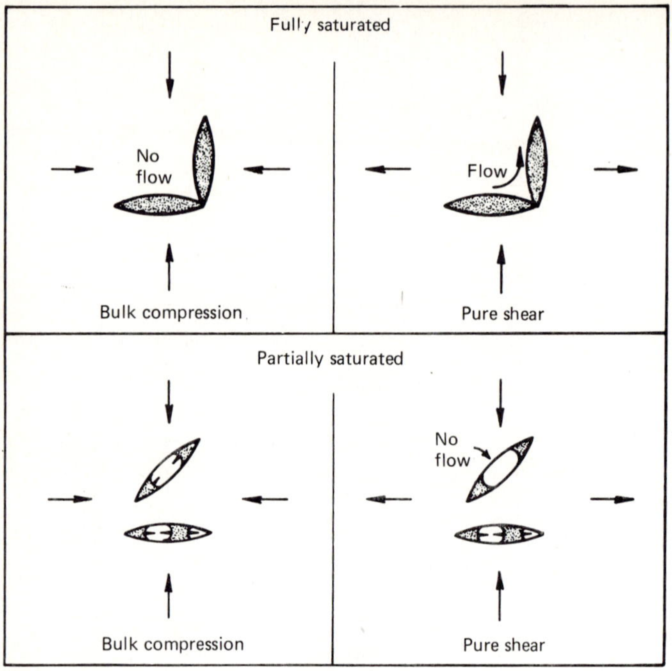

FIGURE 7-12 Conceptual model (Winkler and Nur, 1982) of pore fluid attenuation mechanisms involving local fluid flow. In fully saturated rock, pure shear induces more intercrack flow than does bulk compression. In partially saturated rock, bulk compression induces more intracrack flow than pure shear.

> or thermoelastic energy losses. In pure shear some cracks will be oriented such that there is no normal stress on the crack faces. These cracks will not be compressed and will not contribute to the energy loss. Averaged over all crack orientations, we expect about 2.5 times as much compressional attenuation as shear attenuation.

They then conclude that fluid flow mechanisms dominate attenuation at frequencies used in logging. Dutta (1981) agrees. Johnston et al. (1979), however, find that at high frequencies "the primary mechanism for attenuation is friction on grain boundaries and thin cracks."

Schopper (1983) suggests that the great increase in attenuation observed when a small amount of water is added to dry rock (Johnston et al., 1979) is caused by a boundary layer effect. The matrix crystal structure continues into the liquid, developing an ice-like structure of high viscosity, and causing attenuation.

Happily, the phenomena, rather than the mechanism, are most important for our purposes.

Effect of temperature. It is generally agreed (Johnston et al., 1979) that attenuation is independent of temperature, except that temperature changes can lead to changes in the state of stress (Paulson, 1983).

7-3 ACOUSTIC LOGGING EQUIPMENT

7-3a Transducers

Piezoelectric transducers. In the most common transducers in acoustic logging, the piezoelectric effect is used to convert mechanical wave energy into electrical signals. Many kinds of crystal have a lattice structure such that, when a stress σ is applied to a slab of the crystal in the proper direction with respect to the crystal axes, the centers of positive and negative charge in the crystal are separated and a surface charge (called a polarization charge) is induced. The expression for the charge is

$$P = h\sigma$$

where h is the piezoelectric strain coefficient. If an electric field E is applied in the same direction, a strain ε is induced, where

$$\varepsilon = Eh'$$

and h' is called the piezoelectric stress coefficient. Since, as we have seen in Sec. 7-1, there are three compressional and three shear components of stress and each stress component can give rise to three components of polarization or field, there are in the most general case 18 piezoelectric stress and 18 strain coefficients. Transducers, however, are usually so designed that either a single coefficient or a linear combination of coefficients relates field to stress.

Some piezoelectric crystals, for example, quartz, give a maximum effect when stress is applied along one axis, that on which the coefficient is a maximum, because the other coefficients are of opposite sign. Others, such as tourmaline, give the largest signal when stress is hydrostatic, because all coefficients are of the same sign. The choice of crystal, then, depends on the application. In an accelerometer, in which a crystal is compressed against a test mass, a crystal like quartz is best. In a hydrophone, in which pressure changes are to be observed, a crystal like tourmaline is preferred.

A class of dielectrics called *ferroelectrics* exhibit a property which is called *electrostriction* but for all practical purposes is the same as piezoelectricity. They have the advantage, however, that the property is observed only when electric dipoles, or *domains*, within the substance are aligned or *poled* by heating the material to a temperature at which electrostrictive properties vanish (the Curie temperature) and then cooling it in the presence of an electric field in the direction in which one desires the electrostrictive property to be present. Therefore, one may construct a transducer of almost any desired configuration. For example, the transmitter for a logging sonde is often made by poling a ferroelectric tube in the radial direction. Then it will emit pulses radially when electrical pulses are applied. The most commonly used ferroelectric is lead zirconate titanate (PZT).

The transmitter is generally excited by an electrical pulse, either by discharging a capacitor or transformer through the transducer or by charging the transducer (which is itself a large capacitor) slowly and then short-circuiting it and discharging it suddenly. In either case, no matter how sharp the pulse provided, the transmitter "rings." That is, it is producing a quasi-sinusoidal packet that decays with time. There are two ways to explain the phenomenon. Since the transmitter is a mechanical system, it has resonant frequencies. When it is pulsed, it tends to vibrate at those frequencies, and it continues to do so until energy is lost by damping. This gives rise to the damped harmonic motion that is the observed output of a transmitter. Another description states that a single, very sharp, pulse is supplied and the transmitter in fact moves only once. A signal is transmitted into the medium surrounding the transmitter, and another signal, whose strength depends on the reflection and transmission coefficients [Eqs. (7-26) and (7-27)], is reflected back into the transmitter (Crawford et al., 1972). This signal in turn is reflected from the other side of the transmitter and then back through the first surface into the medium. These multiple reflections continue until the energy is too small to be observed.

When particularly high energies are required and electrical insulation is a problem, thin PZT slabs can be stacked with sides of the same polarity adjacent (Suhler et al., 1979). Then the same voltage can be placed in the opposite sense across each alternate thin slab, and much higher electric fields (volts per centimeter) can be induced for the same total voltage.

Suhler et al. (1979) also developed a transducer arrangement that could detect both signals with motion radial to the sonde axis and signals with motion parallel to the axis. This was done by stacking PZT slabs polarized parallel to the axis of the transducer (thickness polarized) and slabs polarized transverse to the axis. Separate output leads were used to differentiate between the signals.

Magnetostrictive transducers. In another type of transducer the *magnetostrictive* effect is used. If a magnetic flux B cuts a magnetostrictive material, the resulting strain is

$$\varepsilon = KB^2$$

Magnetostriction is strong in all ferromagnetic materials (iron, nickel, cobalt) and in materials called ferrites: nonmetallic polycrystalline ferromagnetic materials. As in the case of electrostrictive materials, the domains (magnetic in this case) must be aligned. Magnetostrictive materials are not as convenient to use as electrostrictive in most cases. But they can handle more energy, and they are therefore sometimes used for large transmitters.

Other types of transducers. The first transmitter used for a velocity log was a spark gap (Vogel, 1952), but spark gaps are no longer in common use. Kitsunezaki (1980) has designed special transducers for shear wave logging. A very common receiver for seismic work is a *geophone,* which consists of a spring-mounted slug of magnetic material inside a coil or else a movable coil inside a permanent magnet. (The best response of a geophone is at frequencies much lower than the frequency for piezoelectric and magnetostrictive transducers.) In either case, when the geophone is moved, a voltage proportional to the velocity is induced across the output of the coil. Kitsunezaki uses a similar device as a transmitter. He immerses a horizontal moving-coil device in an elastic bag inside a liquid-filled borehole. When a large pulse is put into the coil, the mechanism is displaced abruptly, like a piston, and a horizontal wave moves into the water. This wave strikes the wall of the hole and causes a transverse, or S, wave to travel parallel to the wall of the hole. His detector is like an ordinary geophone suspended horizontally in the borehole fluid.

7-3b Sonde Design

Single-transmitter sondes. Commonly, a radial piezoelectric, electrostrictive, or magnetostrictive transducer is placed along the axis of an acoustic sonde. It is usually coupled to the borehole fluid either by being in direct contact with the fluid, perhaps covered with an elastic material, or immersed in a liquid-filled elastic bag that in turn is in contact with the fluid. Receivers are mounted in much the same way. The borehole fluid must be a liquid for most sondes to function. Care must be taken to acoustically isolate the transmitter from the receiver so signals are not transmitted through the sonde. That is not a trivial problem, but it has generally been solved.

The frequency used in logging sondes is a compromise between the desire for the highest possible frequency to get the best possible time resolution and the lowest possible frequency to minimize attenuation. Commonly, frequencies between 5 and 40 kHz are used.

The first *continuous velocity* (CVL) sondes (Summers and Broding, 1952) had a single large electrostrictive transmitter and one or two smaller receivers. Many similar sondes have now been developed. As the sonde moves up the hole, the transmitter is pulsed at regular intervals, and it generates an acoustic wave which travels through the borehole fluid as a *P* wave. When it reaches the wall, it also gives rise to an *S* wave and fluid waves (Sec. 7-1d). The *P* and *S* waves travel in the rock surrounding the hole as head waves (Fig. 7-13); the fluid waves propagate down the wall. In the more common two-receiver case some sort of electronic timing device is triggered at a preselected amplitude of the signal from each receiver. The distance between receivers, divided by the time difference between triggers, is the velocity of the first arrival.

$$\frac{\ell}{t_2 - t_1} = v$$

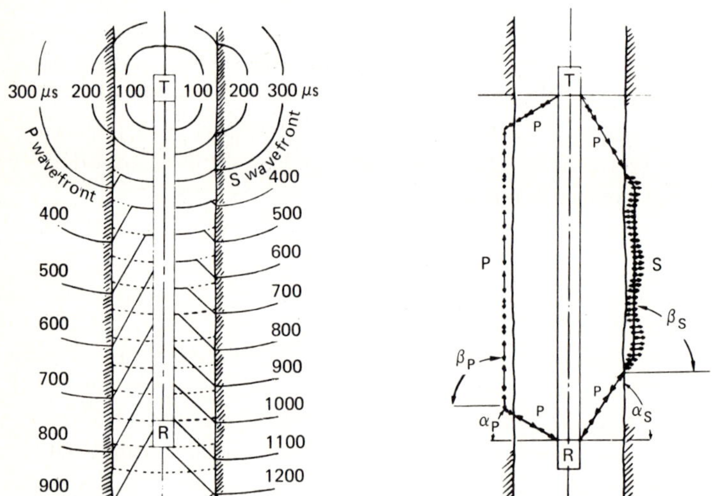

FIGURE 7-13 Illustration of compressional and shear head waves traveling in a borehole wall. (*a*) Wavefronts; (*b*) ray path. (Geyer and Myung, 1971.)

where ℓ is the distance between receivers. This technique removes the effect of the travel time in the borehole fluid, provided the sonde is parallel to the borehole wall.

The method can give incorrect results if the trigger level is not set at the same point on the same cycle of each of the two signals. Sometimes the first cycle does not trigger the system at all. This problem is called *cycle skipping.* If it occurs on only a fraction of the arrivals, or if one knows approximately what velocities to expect, then the user is not deceived by bad data. If, however, the trigger occurs at, say, the half height on an arrival in the first receiver, and the peak of the arrival in the second receiver, inaccuracy that is not visible will arise. If high accuracy is required, the user must be sure that the triggers are set optimally. Often, in modern systems, the trigger is at the first zero crossing after a preselected amplitude. This provides better timing accuracy.

Borehole compensated systems. This method described above fails if the sonde is not parallel to the borehole wall or if there are increases in hole diameter (such as washouts). To deal with those problems, the *compensated acoustic log* was developed (Kokesh et al., 1965). It has two transmitters and at least two receivers (Fig. 7-14). If there is a change in hole diameter Δd opposite a receiver, and we designate the time from transmitter 1 to receiver 2 as t_{12}, and so forth:

$$t_{12} - t_{11} = \frac{\ell}{v} + \frac{\Delta d}{v_f} \qquad t_{21} - t_{22} = \frac{\ell}{v} - \frac{\Delta d}{v_f}$$

where v_f is the velocity in the fluid, then averaging gives

$$\frac{t_{21} - t_{22} + t_{12} - t_{11}}{2} = \frac{\ell}{v} \qquad (7\text{-}33a)$$

as desired. If instead the sonde is tilted in the hole, as in Fig. 7-14, it can be shown that

$$\frac{t_{21} - t_{22} + t_{12} - t_{11}}{2} = \frac{\ell \cos \varphi}{v} \qquad (7\text{-}33b)$$

where φ is the angle of tilt in the hole. Since sondes are of the order of 5 m long and holes are of the order of 15 cm or less in diameter, $\cos \varphi$ is very close to 1 and the angle can normally be ignored.

It is also possible to use a sonde with a single transmitter as a borehole compensated system. If the transmitter (above the receiver) is fired once when it is in the position that the lower receiver will be in at a later time t, fired again when it is in the position that the upper receiver will be in at time t, and fired once more at time t, the first two signals are mirror images of the signals from the lower transmitter in the compensated case. Then the signals, suitably delayed, can be combined to be equivalent to those of a compensated system.

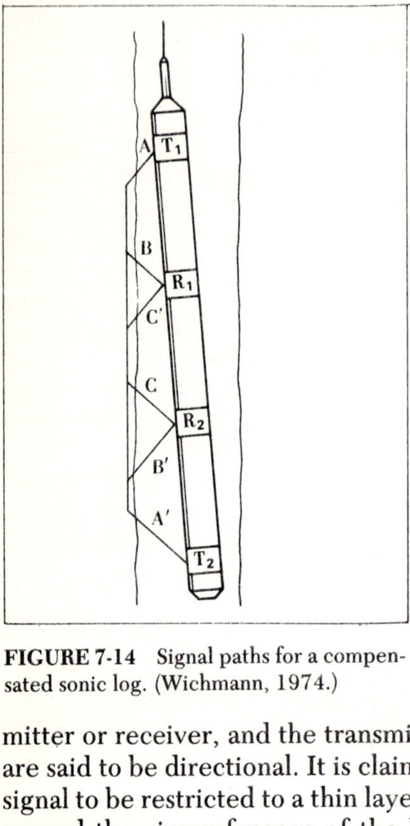

FIGURE 7-14 Signal paths for a compensated sonic log. (Wichmann, 1974.)

There are many variations on the continuous velocity log and many lengths, but all rely on approximately the same principle. Very often the entire signal or some characteristic such as amplitude is displayed, in addition to the arrival time. These displays will be discussed in the next section.

The circumferential sonde. A new and interesting sonde, called the circumferential acoustical device (CAD) has recently been developed (Vogel and Herolz, 1981). It has four arms positioned 90° apart around the circumference of the hole at the same depth. Each arm carries a transmitter or receiver, and the transmitters, which are long PZT cylinders, are said to be directional. It is claimed that this arrangement causes the signal to be restricted to a thin layer of rock normal to the borehole axis around the circumference of the hole (Fig. 7-15). It is reported that only shear waves, traveling around the circumference of the hole, and water waves are observed.

Dry hole sondes. The sondes described above require liquid-filled boreholes. In contrast, the Dry Hole Acoustic Log can be used in dry holes, but it cannot be used while moving (Hearst and Carlson, 1970). It has a transmitter and one or two receivers that are built into horns that are pressed firmly against the wall of the hole. The transmitter is pulsed, and the signal goes through the horns directly into the wall of the hole, with no mud path. Because of the solid-solid contact, the signal-to-noise ratio is substantially smaller than that for conventional systems. The signals are weak enough, if the rock is not competent, that signals must be digitally averaged to minimize the effect of noise. In

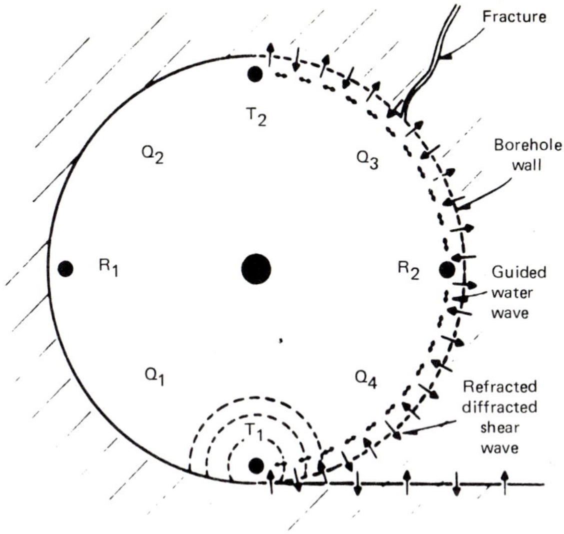

FIGURE 7-15 Transducers and wave paths for the circumferential acoustical device. (Vogel and Herolz, 1981. Copyright 1981, SPE–AIME.)

addition, because it is not surrounded with liquid, the acoustic isolator between the transmitter and receiver is less effective than the isolators for conventional sondes. This sometimes causes the appearance of "sonde signals" in the data. They can be reduced by filtering, because their frequency is usually lower than the frequency of data, but they are always a worry.

King et al. (1978) developed a sonde with retractable pistons, and Suhler et al. (1979) built a wireline sonde by using the transducers described in the preceding section to obtain both *P* and *S* wave velocity in competent formations. They can be used in smaller holes than the DHAL; but because the pistons must be retracted and extended for each reading, they are much slower to operate.

Fielding precautions. It is vital that acoustic logs, unless designed otherwise, be well centralized. Otherwise, signals reflected from opposite sides of the hole can interfere destructively and give poor data. A caliper log is desirable to warn the user of parts of the hole where poor caliper, and therefore questionable centering, is present.

More information can be obtained by using several transmitter-receiver spacings. Until recently, several runs with different sondes were required to do this, but now systems with many receivers at different spacings are sometimes available.

If the borehole fluid contains bubbles, it is possible that the signal in the fluid will be severely attenuated. This problem can be eliminated by waiting a while until the bubbles clear.

Occasionally the acoustic velocity in the borehole fluid is greater than that in the formation. We have found that a good log can still be obtained by surrounding the sonde's isolator with a section of rubber tire of a size such that it fills the hole. Then the waves in the fluid and hole are suppressed and the waves in the formation become visible.

An altered zone surrounding the borehole can give rise to spurious arrivals that are faster or slower than the true velocity depending on the characteristics of the altered zone. The situation can be recognized by varying the source-detector spacing.

Finally, if a variable-intensity log (Sec. 7-3c) is being recorded, the user must make sure that the film exposure and processing are done well. It is wise to have a short portion of the log processed before the entire log is run. One should insist on a good neutral gray area before the first arrival, which should appear as a clear or dark area depending on its sign.

Calibration. No method has been devised to calibrate velocity sondes in materials in which velocity is truly well known. But it is very easy to calibrate Δt of the electronics, and this is the standard calibration procedure. The only way to check the entire system is to operate it in casing, whose velocity is presumably known (but for which amplitudes can be affected by cement properties; see Chap. 16), or in formations in which the sound speed has been previously measured.

7-3c Presentation of Signals

The original presentation of the CVL was simply an ordinary trace of v or, quite often, Δt, versus depth. This presentation is still common in the logging industry. It has been recognized, however, that acoustic logs contain more information than just the first arrival, and several methods of presenting the information have been developed.

The full-wave display. A typical signal, with P, S, and borehole waves is shown in Fig. 7-16. A sonde is pulsed perhaps every few centimeters as it is brought up the hole. Selected pulses are displayed on the screen of an oscilloscope, and a film that records the displays is stepped incrementally past the face of the oscilloscope at a rate proportional to the logging speed (Fig. 7-16). Thus pulses are displayed at positions on the film corresponding to the depth of the sonde. If every pulse were displayed, there would be nothing but confusion, so the pulse spacing must be chosen to make the signals visible.

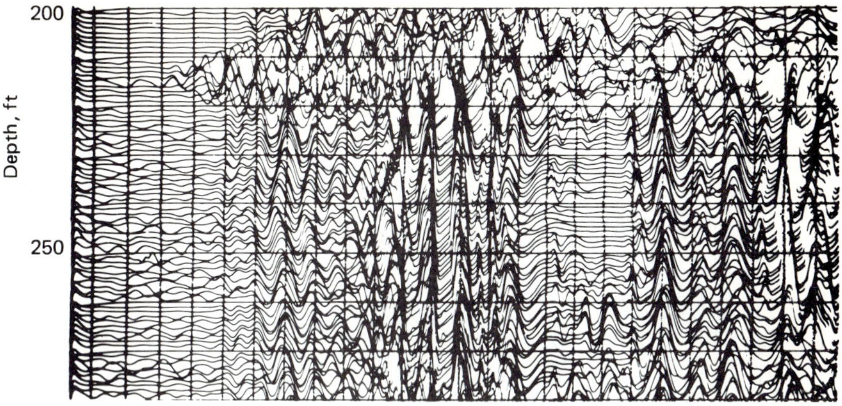

FIGURE 7-16 A "wiggle-trace" full-wave acoustic display. (Geyer and Myung, 1971.)

Variable-intensity display. A very useful presentation is the variable-intensity display (Walker, 1964; Geyer and Myung, 1971). The signal is swept across the display oscilloscope without vertical deflection. The intensity of the trace is modulated with the signal so that a zero signal gives a gray area on the film, a very strong negative signal gives clear film, and a very strong positive signal gives a very dark area. Thus one can see both amplitude and time for the entire record (Fig. 7-17). Arrivals are seen as light and dark bands on the film. Good control of signal amplitudes, beam intensity, and film processing are required for a really good display, but when it is good it is very, very good.

It is more difficult to determine the precise time of the first arrival from one of these displays than from a conventional CVL, because there is no obvious first break. Commercial systems ordinarily had only one receiver. It was common practice to run a conventional system for first arrivals and the more sophisticated, single-receiver system for later data. Recently (see below) multiple-receiver systems have become available and the problem is gone.

Amplitude logs. Because fractures attenuate acoustic signals, especially shear signals, it has long been thought desirable to try to use the amplitude of the signal to locate fractures (Summers and Broding, 1952). Various types of amplitude logs have been developed (for example, Anderson and Walker, 1961; Morris et al., 1964). The display is generally a plot of amplitude (in fairly arbitrary units) versus depth, as in a conventional log. The publications rarely describe precisely how the amplitude to be displayed is chosen. One would like to use shear wave amplitude. The method of choice appears to be to delay a gate to a time at which one expects the shear arrival and then obtain either a

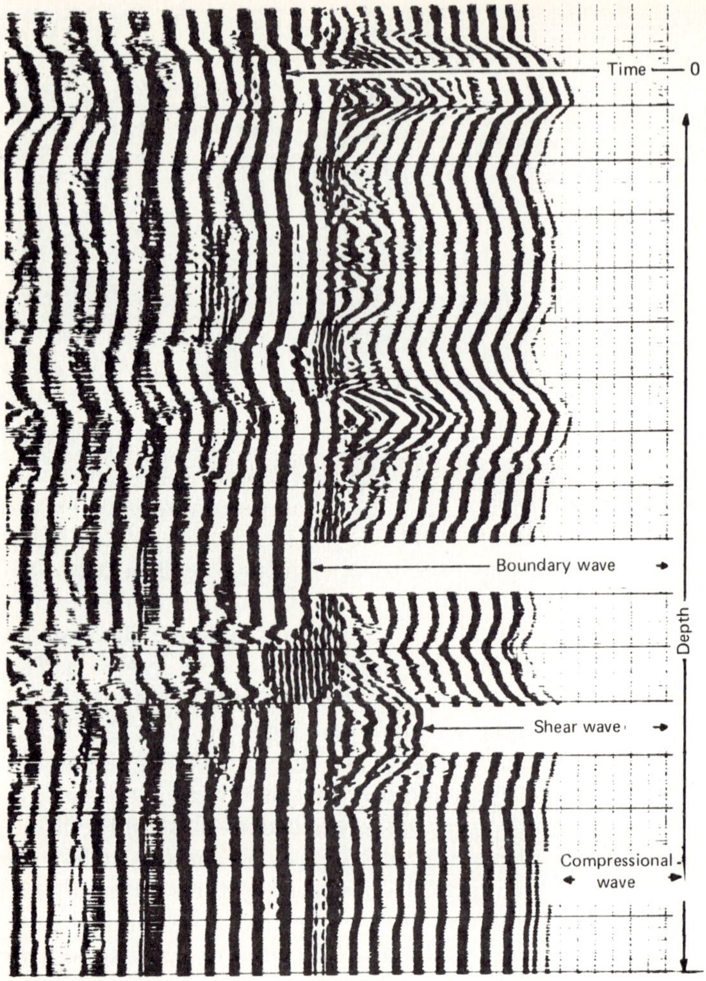

FIGURE 7-17 A variable-intensity acoustic log display. (Myung and Sturdevant, 1970.)

peak or mean amplitude electronically and record it. The major difficulty is in the choice of the appropriate gate time.

Full-wave digital recording. Recently it has become possible to record the entire wave digitally for later processing. The process has become known as *full-wave acoustic logging.* Multiple-receiver sondes can be used, and then the user, at leisure, can produce any of the displays described above. One could, for example, use a wiggle trace display to select the appropriate delay for an amplitude log. It is even possible to produce a variable-intensity display on color film. The amplitude is color-coded, and then low- and high-amplitude regions are strikingly visible.

7-4 DATA ANALYSIS

It is routine to obtain *P* wave velocity from continuous velocity logs. Many applications of acoustic logs require specific interpretation of log data in terms of porosity, cement bond quality, or fracturing. Some of these will be treated in Part 3. There are two problems in data analysis, however, which are sufficiently general to be treated in this section: explanation of features on variable-intensity logs and attempts to automatically obtain *S* wave velocity. We shall discuss them in turn.

7-4a Events on Variable-Intensity Logs

Wave arrivals. Figure 7-17 shows the three main features of a variable-intensity log. The first signal (marked compressional wave) is low intensity, and therefore low amplitude, as is expected for the first cycle of a *P* wave. The shear wave can be identified by a change in the slope of the band. As the sonde goes from a formation with characteristic v_p and v_s to another, both Δt_p and Δt_s change. But because v_s is ordinarily some slowly varying fraction of v_p, the change in Δt_s is greater than that in Δt_p. Consequently, the slope of the part of the band connecting the values of Δt_s in the two formations is steeper than that connecting the values of Δt_p. This greater slope serves to identify the *S* wave arrival. The same band can then be followed in regions where Δt is not changing, and the shear wave can still be identified. Note, however, that the best accuracy that can be obtained is about half a period of the signal—one can tell which band (corresponding, recall, to a peak in the signal) is *P* and which *S*, but not where between the bands the *S* wave arrived. Note that the shear bands can completely obscure the later *P* bands at some depths.

Later arrivals, whose slopes change very little with depth, can be identified as borehole waves. We know that the velocity of the normal modes is fairly close to v_s, so their arrivals may be obscured by the *S* waves. But the velocity of the tube wave is very close to that in the fluid, and it is not affected much by the formation. These are the straight, late-time bands. They can easily be identified by comparing their arrival time with that expected from a direct wave in the fluid; there is little difference.

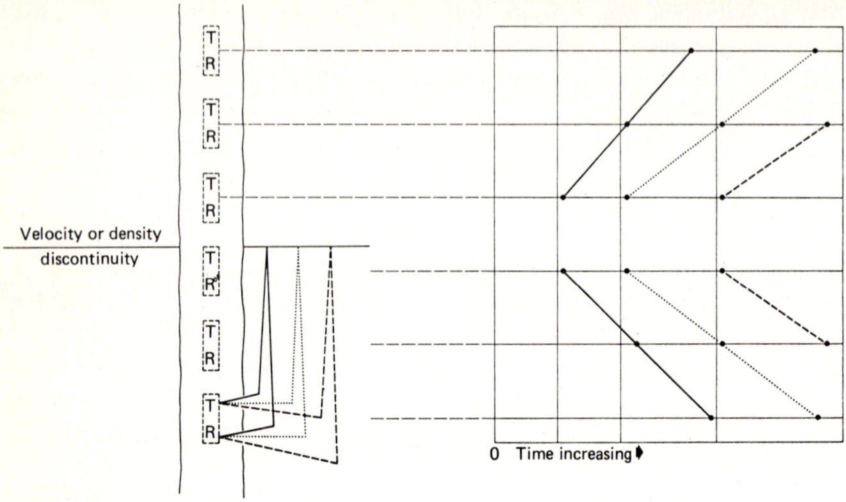

FIGURE 7-18 Conceptual sketch of reflections of waves from an acoustic log at an interface. (Bird, 1965.)

Horizontal discontinuities. Horizontal discontinuities—either interfaces or fractures—are particularly noticeable on variable-intensity logs. They can often be seen at very late times in the log. There are two types of events on the log caused by these discontinuities: reflection events and transmission events.

Figure 7-18 shows a sonde at several positions above and below a discontinuity. The discontinuity can be a velocity or density mismatch [see discussion of (Eq. 7-26)], caused by a contact between formations, or a fracture. In any case, it can reflect acoustic waves. The solid line represents a *P* wave, the dotted line an *S* wave, and the dashed line a boundary wave. Each is reflected from the interface, and each arrives at the receiver at a characteristic time

$$t = \frac{2z + \ell}{v}$$

where ℓ is the transmitter-receiver spacing and z is the distance from the interface to the nearest transducer. These times are shown on the right side of Fig. 7-18 for six positions of the sonde. These lines give a characteristic V pattern with one side of the V on each side of the interface. One can, of course, determine the velocity from the slope of the lines connecting these arrivals

$$v = \frac{2\Delta z}{\Delta t}$$

assuming the interface is horizontal.

When the transducers straddle the interface, a different phenomenon occurs. A signal travels from the transmitter to the interface as, say, a *P* wave, is converted to an *S* wave at the interface, and then continues to the receiver as an *S* wave. *S* waves are converted to *P* waves in the same manner. Consequently, the wave arrives at time Δt_p when the transmitter is just below the boundary and then arrives later and later as the sonde gets farther across the boundary. Finally it becomes Δt_s just as the receiver reaches the boundary. The same thing happens in reverse for the *S* wave. The arrival times for the two waves would be

$$\Delta t_p = \frac{z}{v_p} + \frac{\ell - z}{v_s} \qquad \text{and} \qquad \Delta t_s = \frac{z}{v_s} + \frac{\ell - z}{v_p}$$

where z is the distance from the transmitter to the interface.

These two sets of arrivals give rise to a characteristic X pattern straddling the interface. The slopes of the lines are

$$\frac{\Delta z}{\Delta t} = \pm \frac{v_p v_s}{v_p - v_s}$$

When both situations are present (which is quite common) the V and X patterns sum to a W pattern, which is indicative of a discontinuity. Figure 7-19 shows some clear examples of the W pattern.

Of course, similar patterns can be seen from discontinuities that are not exactly horizontal. Walker (1964) points out that the velocity can still be approximated by

$$v = \frac{\Delta z}{\Delta t} \cos \theta$$

where θ is the angle of the interface with the horizontal.

It is sometimes possible to observe discontinuities that do not intersect the borehole. Figure 7-20 shows a reflection from what is believed to be a near-vertical fracture 30 to 45 m from the borehole in a salt dome.

The Borehole Televiewer. An ingenious use of the variable-intensity recording is in the *Borehole Televiewer* (Zemanek et al., 1969). A closely spaced transmitter-receiver pair is rotated as the sonde is pulled

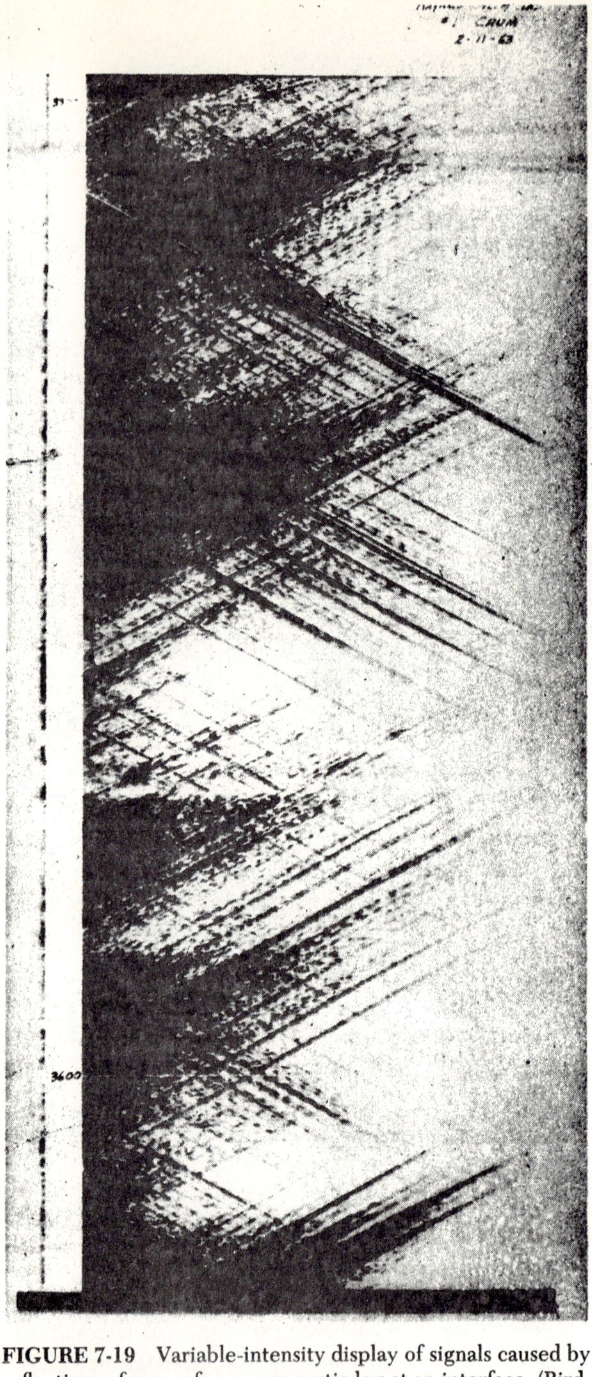

FIGURE 7-19 Variable-intensity display of signals caused by reflections of waves from an acoustic log at an interface. (Bird, 1965.)

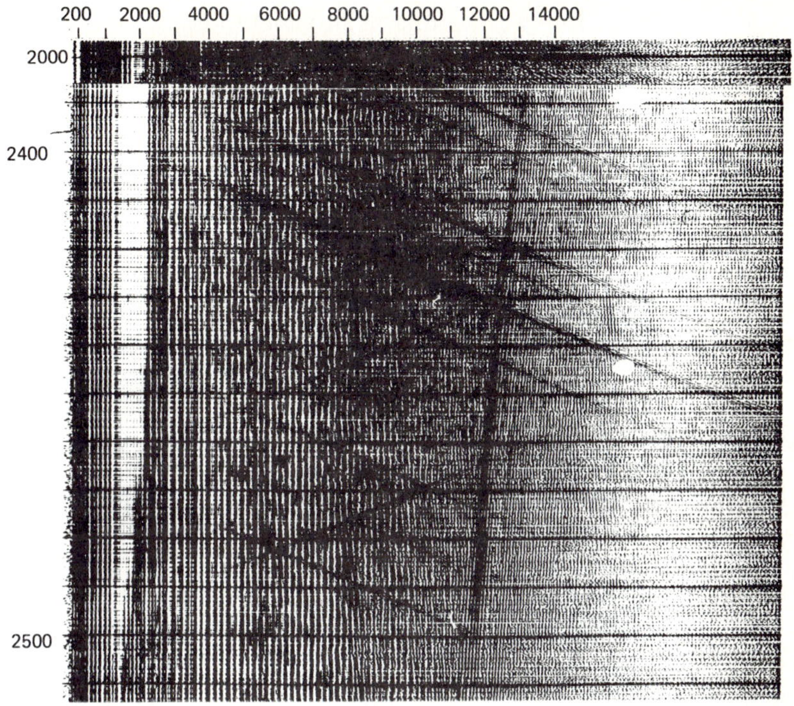

FIGURE 7-20 Variable-intensity display of signals reflected from an apparent near-vertical fracture 35 to 40 m from a borehole in salt (LLNL).

up the borehole, and the reflected signal is displayed as a function not of time but of azimuth. Ordinarily the display is of the amplitude of the first arrival. A hard borehole wall gives a strong reflection; a cave or fracture gives a weak one. Thus fractures can be displayed pictorially. The use of the televiewer is discussed in Secs. 14-2 and 19-4.

7-4b Shear-Wave Logging

We have seen that it is often possible to recognize shear arrivals on the variable-intensity display. Also, however, it is often *not* possible to recognize the shear arrival, because it is obscured by other modes. Moreover, it would be most desirable to be able to obtain shear velocity automatically. Much active research is being devoted to this problem; Willis (1983) gives a detailed description of many of the methods in use. We shall describe some of the techniques that are being tried.

Analog methods. One of the first analog methods, before full-waveform digitization was available, was tried by Elliott and Grine (1968). They delayed the signal from the first of two receivers by the time required for the *P* wave to travel from the first to the second receiver, adjusted the gain of this first signal to make the amplitudes of the *P* waves at both receivers identical, and then subtracted the signal from the second receiver from that from the first receiver. This, if the waveforms were the same, would cancel the *P* wave signal. The resulting difference would be a waveform very much like the *S* wave. There is the problem, however, that the time delay between *P* arrivals is not the same as that between *S* arrivals, and therefore the result is not really the *S* waveform. They evidently dealt with this problem by using short distances between the receivers, so that the time discrepancy would be much less than one period of the wave.

Their sonde in fact had three receivers, so that this procedure could be applied to two pairs. Then the result was two *S* waveforms. They applied a conventional CVL picking technique to these two waveforms to get the time difference between those *S* wave arrivals and then the *S* velocity.

Very recently Zemanek et al. (1984) built a continuous logging sonde containing a version of Kitsunezaki's (1980) shear-wave source and a similar detector (Sec. 7-3a). The system seems to work significantly better than conventional full-wave logs in several different formations.

Cross-correlation. When full-wave digitization became available, cross-correlation methods, already common in the seismic processing industry, could be applied to acoustic logs. The cross-correlation function is a measure of the similarity between two signals. It can be considered the scalar product of two vectors each of which is a string of amplitude-time points corresponding to one of the signals; that is, it is the projection of one data set on the other. It is obtained by multiplying corresponding values of the two sets and summing the products:

$$\Phi_{ab}(\tau) = \sum_k a_{k+\tau} b_k \tag{7-34}$$

where τ is a time delay, or *lag*, between the two sets. Figure 7-21 illustrates the cross-correlation between two identical sets of data which start four units of time apart for different values of τ. On the left the sets are not shifted and Φ is small. In the middle the sets are shifted by four units of time, so they overlie precisely, and Φ is large. On the right the shift is greater than the time between signal starts and Φ again is small. If the functions are not identical, but are similar, the peak is not

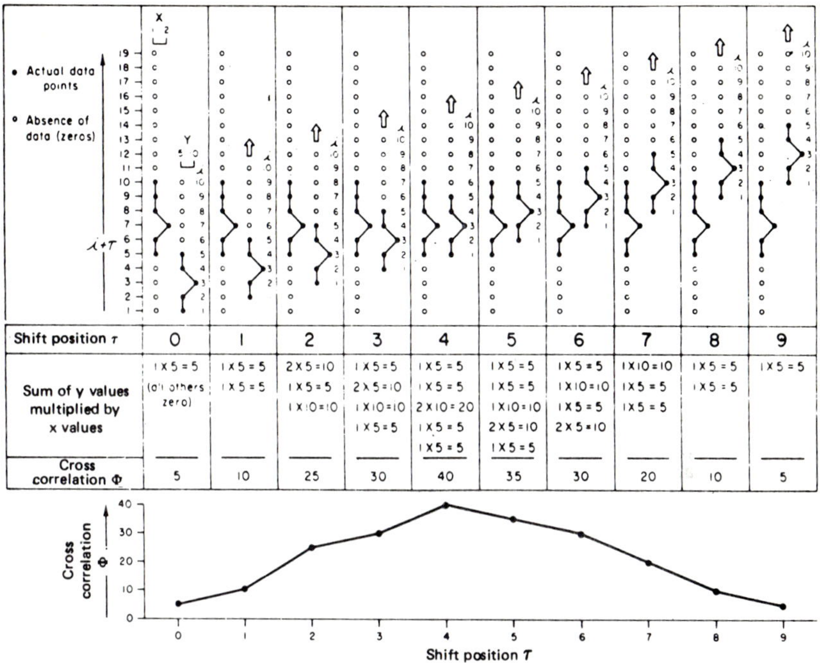

FIGURE 7-21 Cross-correlation of two model borehole logs. Optimum correlation is at maximum value of cross-correlation function ϕ at shift position $\tau = 4$. (Rudman and Lankston, 1973.)

as pronounced, but it is usually largest when the signals have been shifted to the proper time delay.

Scott and Sena (1974) used the above method to pick a shear arrival. They chose a time window with a length equivalent to several cycles of a waveform. They then chose lag times corresponding to a wide range of plausible velocities of interest. At each position of the window on the first waveform they cross-correlated the two signals at an appropriate lag time. They say, "As the equation (7-34) is evaluated for larger and larger values of the lag time τ, the time window can be visualized as stepping down the two waveforms so that later and later arrivals are bracketed between the summation limits." Figure 7-22 shows the windows at the optimum positions, and Fig. 7-23 shows the cross-correlation plot and arrival times. It can be seen that while the S arrivals are not at all clear on the waveforms, the peak is fairly sharp on the correlation.

Aron et al. (1978) extended this method to simultaneous correlation of signals from four receivers. Four receivers give sharper correlation peaks and, therefore, more precise velocities. They stepped the

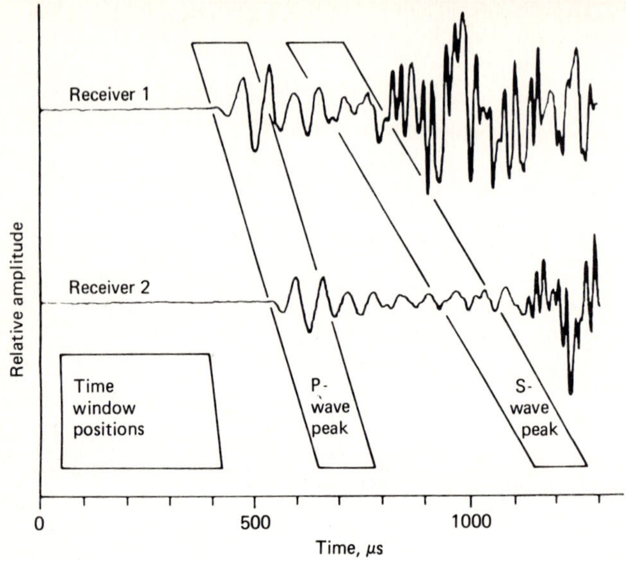

FIGURE 7-22 Waveforms from two acoustic probe receivers and the time windows used to compute cross-correlation. (Scott and Sena, 1974.)

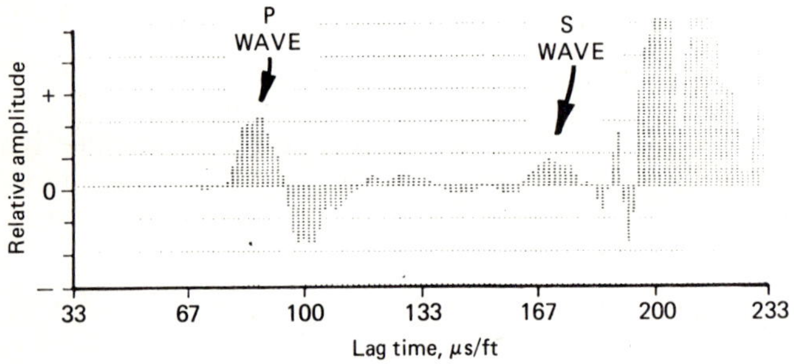

FIGURE 7-23 Plot of cross-correlation of the two acoustic waveforms in Fig. 7-22 showing *P* and *S* wave peaks. (Scott and Sena, 1974.)

window on each waveform in intervals of half the period of the waves. Then, at each window setting, they used lag times corresponding to all plausible velocities. This procedure gave an *S* wave velocity at each window setting. Finally, they optimized the choices of *S* arrival time by demanding that arrival times at one depth be consistent with those at nearby depths (a "continuity condition over depth") and by requiring velocities to be plausible.

Siegfried and Castagna (1982) use *semblance* instead of cross-correlation to obtain both *P* and *S* wave velocities. The semblance is

$$S = \frac{\Sigma_k (a_{k+\tau} + b_k)^2}{2(\Sigma_k a_{k+\tau}^2 + \Sigma_k b_k^2)} \tag{7-35}$$

and it can be shown to contain information about relative amplitude of the two waves. It can be defined as "the energy of the average of the two waveforms divided by the average of their energies."

Siegfried and Castagna then create *P* and *S* travel time curves by picking peaks in the semblance curves, which are similar in appearance to cross-correlations. The windows are chosen interactively, rather than automatically, if desired. This permits the user to pick the best window for the specific situation. They use several types of frequency, spatial, and velocity filtering techniques, similar to those used in seismic data processing, before producing the semblance curves.

Willis (1983) described a similar method for obtaining the *P* wave velocity. He then cross-correlated the *P* waveform with the rest of the record and identified the *S* arrival as that with the first correlation peak whose energy is greater than or equal to that of the *P* waveform. This second waveform is then cross-correlated with the record from the second receiver (with appropriate lags) to obtain the best fit to the *S* velocity. Willis stated that the method is better than that of Scott and Sena (1974), and the choice of lag is automated.

If $v_s < v_f$, so that the shear wave cannot be observed, it may be possible to process the Stoneley wave velocity to obtain v_s. The method is still in the experimental stage (Stevens and Day, 1983; Cheng et al., 1983), and it has evidently not yet been applied to a complete log.

Elastic moduli. The elastic moduli can obviously be expressed in terms of the velocities and density of the medium. For example,

$$\nu = \frac{(v_p/v_s)^2 - 0.5}{(v_p/v_s)^2 - 1} \tag{7-36a}$$

$$K = \rho\left(v_p^2 - \frac{4v_s^2}{3}\right) \tag{7-36b}$$

$$\mu = \rho v_s^2 \tag{7-36c}$$

Many service companies, in fact, calculate the moduli from velocities and present "elastic modulus logs" or even "strength logs." These moduli, of course, require knowledge of v_s, which, as we have seen, is difficult to obtain. It has often been shown (for example, Simmons and Brace, 1965) that, even with good velocity data, the "dynamic" moduli obtained at very small stresses from the velocities are not necessarily the same, or even very close to, the static moduli obtained from stressing rock samples. The user should be cautious in the use of elastic properties obtained from logs.

REFERENCES

Anderson, W. L., and T. Walker: "Application of Open Hole Acoustic Amplitude Measurements," Soc. Petr. Eng. Paper SPE 122, 1961.

Aron, J., J. Murray, and B. Seeman: "Formation Compressional and Shear Interval-Transit-Time Logging by Means of Long Spacings and Digital Techniques," Soc. Petr. Eng. Paper SPE 7446, 1978.

Bentz, A.: *Lehrbuch der angewandten Geologie Bd. 1,* Enke Verlag, Stuttgart 1961.

Berryman, J. G.: "Confirmation of Biot's Theory," *Appl. Phys. Letters,* vol. 37, 1980, p. 382.

Biot, M. A.: "Theory of Propagation of Elastic Waves in a Fluid-Saturated Porous Solid. I. Low-Frequency Range," *J. Acous. Soc. Am.*, vol. 28, 1956, p. 168.

———: "Propagation of Elastic Waves in a Cylindrical Bore Containing a Fluid," *J. Appl. Phys.*, vol. 23, 1952, p. 997.

Birch, F., and D. Bancroft: "Elasticity and Internal Friction in a Long Column of Granite," *Bull. Seis. Soc. Am.*, vol. 28, 1938, p. 243.

Bird, J. M.: "An Explanation of the 3-D Acoustical Logging System for the Exploration Geophysicist," preprint, Mexican Geophys. Soc. Conv., 1965, 29 pp.

Born, W. T.: "Attenuation Constants of Earth Materials," *Geophysics,* vol. 6, 1941, p. 132.

Cady, W. G.: *Piezoelectricity,* Dover, New York, 1964.

Cheng, C. H., M. N. Toksöz, and K. M. Tubman: "Determination of Shear Wave Velocities in 'Soft' Formations," Soc. Exp. Geophys. 53d Ann. Meeting, 1983.

Christensen, N. I.: "Seismic Velocities," in R. S. Carmichael (ed.), *Handbook of Physical Properties of Rocks,* CRC Press, Boca Raton, Fla., 1982.

Clark, S. P. (ed.): *Handbook of Physical Constants,* Geol. Soc. Am., New York, 1966.

Crawford, F. S.: *Waves,* McGraw-Hill, New York, 1953.

Crawford, G. E., W. A. Hoyer, and M. M. Spann: "Frequency Response and Resonance in Acoustic Logging," *Proc. Soc. Prof. Well Log Analysts 13th Logging Symp.*, 1972, paper O.

Dobrin, M. B.: *Introduction to Geophysical Prospecting,* McGraw-Hill, New York, 1976.

Domenico, S. N.: "Effect of Brine-Gas Mixture on Velocity in an Unconsolidated Sand Reservoir," *Geophysics*, vol. 41, 1976, p. 882.

Dutta, N. C.: "Seismic Loss Mechanisms," *Geophysics*, vol. 46, 1981, p. 806.

Dzeban, L. P.: "Elastic Wave Propagation in Fractured and Vuggy Media," *Izv. Earth Phys.*, no. 10, 1970. Translation by D. G. Fry.

Eastwood, R. L., and J. P. Castagna: "Basis for Interpretation of v_p/v_s Ratios in Complex Lithologies," *Proc. Soc. Prof. Well Log Analysts 24th Logging Symp.*, 1983, paper G.

Elliott, J. W., and D. R. Grine: Acoustical Well Logging Methods and Apparatus, U.S. Patent 3,390,377, 1968.

Geertsma, J.: "Velocity-Log Interpretation: "The Effect of Rock Bulk Compressibility," *Soc. Petr. Eng. J.*, vol. 1, 1961, p. 235.

Geyer, R. L., and J. I. Myung: "The 3-D Velocity Log, a Tool for In-Situ Determination of the Elastic Moduli of Rocks," *Proc. 12th Symp. Rock Mech., AIME*, New York, 1971, p. 71.

Grant, F. S., and G. F. West: *"Interpretation Theory in Applied Geophysics,"* McGraw-Hill, New York, 1965.

Gregory, A. R.: "Fluid Saturation Effects on Dynamic Elastic Properties of Sedimentary Rocks," *Geophysics*, vol. 41, 1976, p. 895.

————: Aspects of Rock Physics from Laboratory and Log Data That Are Important to Seismic Interpretation, Am. Assoc. Petr. Geol. Memoir 26, 1977.

Hearst, J. R, and R. C. Carlson: "Well-Logging Research for Geonuclear Technology," *Nuc. Appl. Tech.*, vol. 8, 1970, p. 276.

Ingram, J. D., C. F. Morris, E. E. MacKnight, and T. W. Parks: "Shear Velocity Logs Using Direct Phase Determination," *Proc. 51st Ann. Meeting Soc. Expl. Geophys.*, 1981.

Johnston, D. H., M. N. Toksöz, and A. Timur: "Attenuation of Seismic Waves in Dry and Saturated Rocks. II. Mechanisms," *Geophysics*, vol. 44, 1979, p. 691.

King, M. S.: "Wave Velocities in Rocks as a Function of Changes in Overburden Pressure and Pore Fluid Saturants," *Geophysics*, vol. 31, 1966, p. 50.

————, M. R. Stauffer, and B. J. Pandit: "Quality of Rock Masses by Acoustic Borehole Logging," *Proc. 3d Int. Symp. Rock Mech.*, 1978, p. 156.

Kitsunezaki, C.: "A New Method for Shear-Wave Logging," *Geophysics*, vol. 45, 1980, p. 1489.

Kokesh, F. P., R. J. Schwartz, W. B. Wall, and R. L. Morris: "A New Approach to Sonic Logging and Other Acoustic Measurements," *J. Petr. Tech.*, vol. 17, 1965, SPE-991, p. 282.

Kolsky, A.: *Stress Waves in Solids*, Dover, New York, 1963.

Levykin, A. I.: "Longitudinal and Transverse Wave Absorption and Velocity in

Rock Specimens at Multilateral Pressures up to 4000 kg/cm^2," *Izv. Earth Phys.*, ser. no. 2., 1965, p. 21.

McDonal, F. J., F. A. Agonil, R. L. Mills, R. L. Sengbush, R. G. Van Nostrand, and J. E. White: "Attenuation of Shear and Compressional Waves in Pierre Shale," *Geophysics*, vol. 23, 1958, p. 421.

Morris, R. L., D. R. Grine, and T. E. Arkfeld: "Using Compressional and Shear Amplitudes for the Location of Fractures," *J. Petr. Tech.*, vol. 16, 1964, p. 623.

Myung, J., and W. M. Sturdevant: *Introduction to the Three Dimensional Velocity Log*, Birdwell, Tulsa, Okla., 1970.

Nur, A.: personal communication, 1983.

Officer, C. B.: *Introduction to the Theory of Sound Transmission*, McGraw-Hill, New York, 1958.

Paillet, F. L.: "Predicting the Frequency Content of Acoustic Waveforms Obtained in Boreholes," *Proc. Soc. Prof. Well Log Analysts 22d Logging Symp.*, 1981.

——— and J. E. White: "Acoustic Modes of Propagation in Boreholes and Their Relationship to Rock Properties," *Geophysics*, vol. 47, 1982, p. 1215.

Paulson: Ph.D. thesis, Berkeley, 1983.

Peselnick, L., and I. Zeitz: "Internal Friction of Fine-Grained Limestones at Ultrasonic Frequencies," *Geophysics*, vol. 24, 1959, p. 285.

Peterson, E. W.: "Acoustic Wave Propagation Along a Fluid-Filled Cylinder," *J. Appl. Phys.*, vol. 45, 1974, p. 3340.

Raymer, L. L., E. R. Hunt, and J. S. Gardner: "An Improved Sonic Transit Time to Porosity Transform," *Proc. Soc. Prof. Well Log Analysts 21st Logging Symp.*, 1980.

Rudman, A. J., and R. W. Lankston: "Stratigraphic Correlation of Well Logs by Computer Techniques," *Am. Assoc. Petr. Geol. Bull.*, vol. 57, 1973, p. 577.

Schopper, J. R.: "Inelastic Absorption in Porous Rocks Related to Common Petrophysical Properties," *Proc. Soc. Prof. Well Log Analysts 24th Logging Symp.*, 1983.

Scott, J. H., and J. Sena: "Acoustic Logging for Mining Applications," *Proc. Soc. Prof. Well Log Analysts 15th Logging Symp.*, 1974, paper P.

Siegfried, R. W., and J. P. Catagna: "Full Waveform Sonic Logging Techniques," *Proc. Soc. Prof. Well Log Analysts 23d Logging Symp.*, 1982.

Simmons, G., and W. F. Brace: "Comparison of Static and Dynamic Measurements of Compressibility of Rocks," *J. Geophys. Res.*, vol. 70, 1965, p. 5649.

Spencer, J. W.: "Stress Relaxations at Low Frquencies in Fluid-Saturated Rocks: Attenuation and Modulus Dispersion," *J. Geophys. Res.*, vol. 86, 1981, p. 1803.

Stevens, J. L., and S. M. Day: "Shear Velocity Logging in Slow Formations Using the Stoneley Wave," *Soc. Exp. Geophys. 53d Ann. Meeting*, 1983.

Suhler, S. A., W. R. Peters, E. C. Schroeder, and T. W. Owen: "Modular Borehole Seismic Reflection Probe," Southwest Res. Inst. Rept. 14-4924, San Antonio, Tex., 1979.

Summers, G. C., and R. A. Broding: "Continuous Velocity Logging," *Geophysics*, vol. 27, 1952, p. 598.

Telford, W. M., L. P. Geldart, R. E. Sheriff, and D. A. Keys: *Applied Geophysics*, Cambridge, Cambridge, 1981.

Timur, A.: "Temperature Dependence of Compressional and Shear Wave Velocities in Rocks," *Geophysics*, vol. 42, 1977, p. 950.

Toksöz, M. N., C. H. Cheng, and A. Timur: "Velocities of Seismic Waves in Porous Rocks," *Geophysics*, vol. 41, 1976, p. 621.

———, D. H. Johnston, and A. Timur: "Attenuation of Seismic Waves in Dry and Saturated Rocks. I. Laboratory Measurements," *Geophysics*, vol. 44, 1979, p. 681.

Vasil'ev, Yu. I., and G. I. Gurevich: "On the Ratio between Attenuation Decrements and Propagation Velocities of Longitudinal and Transverse Waves," *Izv. Geophys.*, Ser. 1962/12, p. 1695.

Vogel, C. B.: "A Seismic Velocity Logging Method," *Geophysics*, vol. 27, 1952, p. 586.

——— and R. A. Herolz: "The CAD, a Circumferential Acoustical Device for Well Logging," *J. Petr. Tech.*, vol. 33, 1981, pp. 1985–1987.

Walker, T.: *The Interpretation of the Fracture Finder Micro-seismogram Log*, Welex, Houston, Tex., 1964.

Wang, C.: *Applied Elasticity*, McGraw-Hill, New York, 1953.

Wasley, R. J.: *Stress Wave Propagation in Solids*, Dekker, New York, 1973.

White, J. E.: *Seismic Waves, Radiation, Transmission, and Attenuation*, McGraw-Hill, New York, 1965.

——— and R. E. Zechman: "Computed Response of an Acoustic Logging Tool," *Geophysics*, vol. 33, 1968, p. 302.

Wichmann, P. A.: "Acoustilog," in *Log Review 1*, Dresser Atlas, Houston, Tex., 1974.

Willis, M. E.: "Seismic Velocity and Attenuation from Full Waveform Acoustic Logs," thesis, MIT, 1983.

Winkler, K. W., and A. Nur: "Seismic Attenuation: Effects of Pore Fluids and Frictional Sliding," *Geophysics*, vol. 47, 1982, p. 1.

Wyllie, M. R. J., A. R. Gregory, and L. W. Gardner: "Elastic Wave Velocities in Heterogeneous and Porous Media," *Geophysics*, vol. 21, 1956, p. 41.

———, ———, and G. H. F. Gardner: "An Experimental Investigation of Factors Affecting Elastic Wave Velocities in Porous Media," *Geophysics*, vol. 23, 1958, p. 459.

Zemanek, J., R. L. Caldwell, E. E. Glen, S. V. Halcomb, S. J. Norton, and A. J.

D. Strauss: "The Borehole Televiewer—A New Logging Concept for Fracture Location and Other Types of Borehole Inspection," *J. Petr. Tech.*, vol. 21, 1969, p. 762.

———, F. A. Angonil, D. M. Williams, and R. L. Caldwell: "Continuous Acoustic Shear Wave Logging," *Proc. Soc. Prof. Well Log Analysts 25th Logging Symp.*, 1984.

Zoeppritz, K., "Über Reflexion und durchgang seismicher Wellen durch Unstetigerslflaschen," Berlin, 1919.

8

Borehole Gravimetry

One of the most urgent requirements in logging is extending the radius of investigation. Whereas resistivity and induction logs can be designed to have a range of many meters, all nuclear and most acoustic logs have an effective range of only 10 to 30 cm. In recent years, the ability to make accurate measurements of gravity in boreholes (gravity differences of 10^{-6} cm/s^2) gives loggers another long-range tool. Gravity measurements are used to obtain a value of formation density. Comparison of densities from gravity logs with those from density logs is particularly useful in inferring structure (Sec. 19-1). Borehole gravimetry is a method new to the industry, and is expensive, but its use is becoming more common (Hearst and Carlson, 1982).

In this chapter we shall describe the principles and pitfalls of borehole gravimetry. In Sec. 19-1 we shall describe some applications.

8-1 THE GRAVITATIONAL FIELD AND POTENTIAL

Newton's universal law of gravitation states that a particle with mass m that is a distance $\mathbf{r}$ from another particle with mass M experiences a force **F**, where

$$\mathbf{F} = -\frac{GmM\hat{\mathbf{r}}}{r^2} \tag{8-1}$$

and $\hat{\mathbf{r}}$ is a unit vector in the direction from M toward m, and G is the universal gravitational constant. The accepted value of G is $6.673 \times 10^{-11}\ \mathrm{m^3/(kg)(s^2)}$. For an extended body, Eq. (8-1) becomes

$$\mathbf{F} = -Gm \int_V \frac{\rho(r)\hat{\mathbf{r}}}{r^2}\, dV \tag{8-2}$$

where dV is an element of volume of material with density ρ at position r. It is convenient to think of a gravitational field caused by mass M, acting on mass m, and pervading all space. Then the field vector

$$\mathbf{g} = \frac{\mathbf{F}}{m} = -\frac{GM\hat{\mathbf{r}}}{r^2} \tag{8-3}$$

is the force per unit mass exerted on a particle in the field of mass M. For an extended body

$$\mathbf{g} = -G \int_V \frac{\rho(r)\hat{\mathbf{r}}}{r^2}\, dV \tag{8-4}$$

where **g** has the dimensions of acceleration and is generally called the *acceleration due to gravity*, the *gravitational acceleration*, or just the *gravity*. At the surface of the earth $\mathbf{g} \simeq 9.8\ \mathrm{m/s^2} = 980\ \mathrm{cm/s^2}$. The most common unit of gravity is the milligal, defined as $0.001\ \mathrm{cm/s^2}$. At the surface, then, $\mathbf{g} \simeq 980$ gal.

It is also convenient to define a gravitational *potential*. A common definition is that the potential U is equal to the work done moving a particle of unit mass from infinity to a distance **r** from mass M:

$$U = \int_\infty^r \mathbf{F} \cdot \mathbf{dr} = -\int_\infty^r \frac{GM}{r^2}\, dr = -\frac{GM}{r} \tag{8-5}$$

An alternative way to determine U is to recognize that a vector field with $1/r^2$ dependence can be shown to be the gradient of a potential. Then

$$\mathbf{g} = -\nabla U$$

Of these quantities, U, **g**, and **F**, only the force **F** is an observable quantity. The one most commonly discussed, however, is the gravity or gravitational field **g**.

Vertical is defined as the direction of the gravitational field at a point. Gravitational fields caused by anomalous bodies may act in various directions with respect to the vertical. But gravimeters respond only to the vertical component of the field of the anomalous body. We always assume that the effect of a local anomaly $\Delta\mathbf{g}$ is much less than $\mathbf{g}$, so the direction of g is not affected by the local masses. Consequently, it is no longer necessary to treat g as a vector.

Because the earth is not truly spherical, and because it rotates, the gravitational field is a function of latitude and elevation of the point at which it is measured. In borehole gravimetry we are not concerned with variation of g over large areas, so we shall not discuss these variations; they are discussed in detail in many texts (for example, Grant and West, 1965.) We shall discuss only the variations that are of interest in borehole gravimetry.

8-1a The Field of a Spherical Shell

Consider a spherical shell with outer radius a and inner radius b (Fig. 8-1). Let us calculate the potential U at a point P a distance z from the center. (Any point P is, of course, on a radius.) This development follows that of Marion (1965).

$$U = -G \int_V \frac{\rho(r')}{r} dV = -2\pi\rho G \int_0^{2\pi} d\varphi \int_b^a r'^2 \, dr' \int_0^\pi \frac{\sin\theta \, d\theta}{r} \tag{8-6}$$

By the law of cosines, $r^2 = r'^2 + z^2 - 2r'z \cos\theta$. Differentiating gives us

$$2r\,dr = 2r'z \sin\theta \, d\theta \qquad \text{and} \qquad \frac{\sin\theta \, d\theta}{r} = \frac{dr}{r'z}$$

The integral over φ gives a factor of 2π; then

$$U = -\frac{2\pi\rho G}{z} \int_b^a r' \, dr' \int_{r1}^{r2} dr \tag{8-7}$$

where r_1 and r_2 are the minimum and maximum values of r. If P is outside the shell, that is, $z > a$,

$$U = -\frac{2\pi\rho G}{z} \int_b^a r' \, dr' \int_{z-r'}^{z+r'} dr = -\frac{4}{3}\frac{\pi\rho G}{z}(a^3 - b^3) \tag{8-8}$$

The mass of the shell is $\frac{4}{3}\pi\rho(a^3 - b^3)$, so

$$U\,(z > a) = -\frac{GM}{z} \tag{8-9}$$

It is important to note that U is a function *not* of r, but only of z. Therefore, the potential, and consequently the field, outside a spherical shell

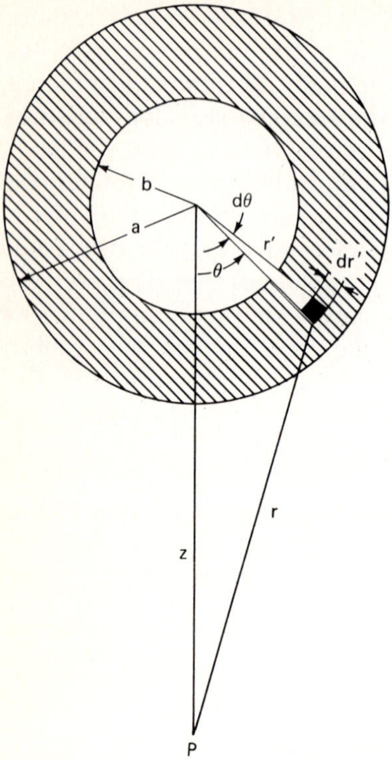

FIGURE 8-1 Volume element of a spherical shell.

of mass M is identical with that of a mass M concentrated at the center of the shell.

If the point is inside the shell ($z < b$), it is, of course, still on a radius and is between the center and the inner shell. Then

$$U = -\frac{2\pi\rho G}{z}\int_b^a r'\,dr' \int_{r'-z}^{r'+z} dr = -2\pi\rho G(a^2 - b^2) \quad (8\text{-}10)$$

and U is *constant* inside the shell. Within the shell ($b < z < a$), we can treat the situation as the sum of two shells (potentials add!) and

$$U = -\frac{2\pi\rho G}{z}\left(\int_b^z r'\,dr' \int_{z-r'}^{z+r'} dr + \int_z^a r'\,dr' \int_{r'-z}^{r'+z} dr\right) \quad (8\text{-}11a)$$

$$U = -4\pi\rho G\left(\frac{a^2}{2} - \frac{b^3}{3z} - \frac{z^2}{6}\right) \quad (8\text{-}11b)$$

If we evaluate g for each case, since $g = -\nabla U = -dU/dz$, for $z > a$ [Eq. (8-9)]

$$g(z > a) = -\frac{GM}{z^2} \quad (8\text{-}12)$$

which is not a function of a or b. For $z < b$ [Eq. (8-10)]

$$g(z < b) = 0 \tag{8-13}$$

there is no field inside the shell. And for $(b < z < a)$ [from Eq. (8-11b)]

$$g(b < z < a) = \frac{4\pi\rho G}{3}\left(\frac{b^3}{z^2} - z\right) \tag{8-14}$$

Change in gravity across a shell. Measuring gravity at two depths in a hole is equivalent to measuring it above and below a spherical shell. If we consider a solid sphere of density ρ, then $b = 0$ and, within the sphere [from Eq. (8-14)],

$$\frac{dg}{dz} = -\frac{4}{3}\pi\rho G \tag{8-15}$$

If g is measured above and below a shell that comprises part of the solid sphere, then the change in gravity due to the shell *alone* is

$$dg = -\frac{GM}{z^2} - 0 = -\frac{4}{3}\pi\rho G\frac{a^3 - b^3}{z^2} \tag{8-16}$$

If we let $b = a - dz$, then

$$dg = -\frac{4}{3}\pi\rho G\frac{a^3 - a^3 + 3a^2\,dz - 3a\,dz^2 + dz^3}{z^2} \tag{8-17a}$$

$$\simeq -\frac{4}{3}\pi\rho G\frac{3a^2\,dz}{z^2} \tag{8-17b}$$

and if $a \simeq z \gg dz$ (for a shell near the surface of the earth), then

$$\frac{dg}{dz} \simeq -4\pi\rho G \tag{8-17c}$$

Then Eq. (8-15) can be expressed as the sum of two terms: the part due to the shell (8-17c) and the part due to the rest of the sphere.

$$\frac{dg}{dz} = -\frac{4}{3}\pi\rho G = -4\pi\rho G + \frac{8}{3}\pi\rho G \tag{8-18a}$$

This second term, representing the part of the change of gravity with respect to z caused by the rest of the sphere (the earth), is the change that would be observed if the shell were absent, that is, above the surface. This term is often called the *free air gradient,* but we prefer to call it the *vertical air gradient.* The common symbol for the free air gradient is F, but because that is also used for force, we shall use Γ to represent the vertical air gradient. Then

$$\frac{dg}{dz} = -4\pi\rho G + \Gamma \tag{8-18b}$$

In the real earth, of course, ρ is not constant and the second term does not represent the true vertical air gradient. Ordinarily, Γ can be calculated as a function of latitude and elevation, and standard formulas are available (Grant and West, 1965). Such formulas take into account both the asphericity and the rotation of the earth, as well as the earth's true mean density. In reality, however, the vertical gradient at the surface is also affected by subsurface structure (Hammer, 1970). Because calculated values of Γ do not take that structure into account, it is sometimes desirable to measure Γ by measuring g at two or more points above the surface.

If Γ is known,

$$\rho = \frac{1}{4\pi G}\left(\Gamma - \frac{dg}{dz}\right) \tag{8-19}$$

Equation (8-19) is the classical expression for density obtained from measurement of gravity at two stations in a borehole (Hammer, 1950).

8-1b The Field of a Cylindrical Shell

Consider a volume element of a cylindrical shell (Fig. 8-2). A mass element can be expressed as $dM = \rho r\, dr\, d\varphi\, dz$. Then the total gravitational field for a point on the axis (Heiskanen and Vening Meinesz, 1958) is

$$dg_t = \frac{G\, dM}{r^2 + z^2}$$

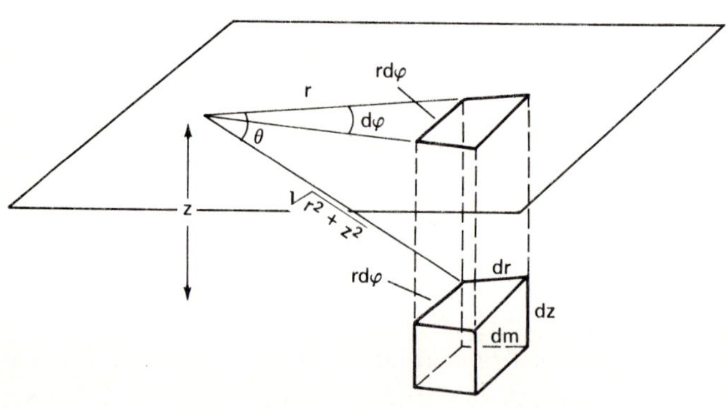

FIGURE 8-2 Volume element of a cylindrical shell.

and the vertical component $dg = dg_t \cos\theta$ is

$$dg = \frac{Gz\,dM}{(r^2 + z^2)^{3/2}} = \frac{G\rho\,d\varphi\,r\,dr\,z\,dz}{(r^2 + z^2)^{3/2}}$$

If we integrate between the limits $0 < \varphi < 2\pi$, $r_1 < r < r_2$, and $z_2 < z < z_1$, then we have the attraction of a cylindrical ring on its axis

$$g = G\rho \int_0^{2\pi} \varphi \int_{z_1}^{z_2} z\,dz \int_{r_1}^{r_2} \frac{r\,dr}{(r^2 + z^2)^{3/2}} \tag{8-20}$$

$$g = 2\pi\rho G\left(\sqrt{r_1^2 + z_2^2} - \sqrt{r_1^2 + z_1^2} - \sqrt{r_2^2 + z_2^2} + \sqrt{r_2^2 + z_1^2}\right) \tag{8-21}$$

Equation (8-21) is used for computation of many shapes and especially for terrain corrections (Sec. 8-9).

8-1c The Field of an Infinite Slab

If we substitute $r_1 = 0$ and $r_2 = \infty$ in Eq. (8-20), we have the expression for the vertical component of the field of an infinite, homogeneous, isotropic slab.

$$g = 2\pi\rho G(z_2 - z_1) = 2\pi\rho Gt \tag{8-22}$$

where t is the thickness of the slab. Note that g is *not* a function of z. Therefore, the field is constant above or below the slab (Fig. 8-3). Furthermore, since the integration over θ results only in a constant, any angular part of a slab with infinite radius produces a field at the center that is not a function of height. (Visualize a pie-shaped slab.) In particular, the field of a half slab is not a function of height. Often a fault can be represented as half of an infinite slab. Then the field of this fault,

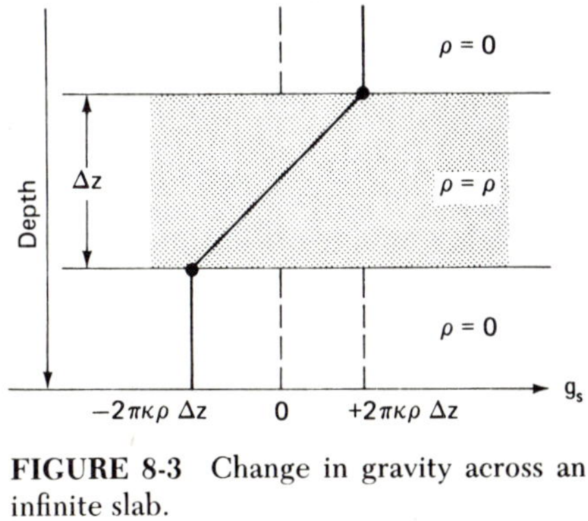

FIGURE 8-3 Change in gravity across an infinite slab.

directly above the fault scarp, is not a function of distance above the scarp!

If we use Eq. (8-22) to calculate the change in gravity across a slab,

$$dg = 4\pi\rho Gt = -4\pi\rho G\, dz$$

which is identical with Eq. (8-17*c*). In fact, this is the classical derivation for Eq. (8-19). It is reassuring that a method considering a spherical earth gives the same result.

8-1d Field of Arbitrary Shapes

Several formulas are available for calculating the fields of bodies of arbitrary shape. They are suited primarily for computer calculations. They are quite complicated, and the details are beyond the scope of this book. Talwani et al. (1959) derive formulas for the field at any point on the surface caused by a two-dimensional polygon which has infinite length into the paper. Hearst (1981) uses this formula to compute the field of any two-dimensional geologic cross section (describing all space) at any point in space. Hearst also computes the effect of any cylindrically symmetrical cross section at any point on its axis.

Plouff (1976) computes the effect of any three-dimensional prism with polygonal top and bottom and vertical sides. He uses this technique to calculate changes in gravity caused by one or more such prisms, but he does not attempt to fill all space with the three-dimensional prisms.

8-2 NONUNIQUENESS OF THE GRAVITATIONAL FIELD

We have seen some cases in which the source of the gravitational field cannot be deduced from the field. For example, Eq. (8-12) shows that the field outside a spherical shell is independent of the shell's size and density; only the total mass is important. An interesting example of such ambiguity is the field before and after detonation of an underground nuclear explosion. The explosion creates a cavity that is often 100 m or more in diameter. The cavity is formed by compaction and displacement of the material that originally occupied the volume now occupied by the cavity. After the explosion, then, there is a cavity surrounded by a compacted spherical shell. Outside this shell, the field is identical with that before the explosion, and the large cavity cannot be detected by gravity.

We have also seen that the field inside a shell is the same whether or not the shell is in fact present and that the field outside an infinite slab does not depend on the location of the measuring point. This ambiguity is in fact quite general. Skeels (1947) has shown the ambiguity for the general case of the gravitational field at the surface as diagrammed in Fig. 8-4.

It can be shown (Grant and West, 1965) that the potential U at a point x_1, y_1, z_1 on the plane $z = z_1$ above a plane $z = 0$, with an arbitrary mass distribution M below plane $z = 0$ but no mass above it (Fig. 8-4), is given by the integral over the plane

$$U = \frac{1}{2\pi} \int_S \frac{1}{r} \frac{\partial U}{\partial z} \, dS$$

where $\partial U/\partial z = g(0)$ is the gravity on the $z = 0$ plane, r is the distance from the point (x_1, y_1, z_1) to the point $(x, y, 0)$ on the plane, and the integration is carried out over the entire surface of the plane. Therefore, the gravity on the $z = z_1$ plane due to the mass distribution M is

$$g(x_1, y_1, z_1) = \frac{\partial U}{\partial z_1} = \frac{1}{2\pi} \int_S \frac{\partial}{\partial z_1} \frac{g(0)}{r} \, dS \tag{8-23}$$

since $g(0)$ is not a function of z. Now let us remove the mass distribution M and instead place a thin layer of variable density ρ and thickness Δt on the $z = 0$ plane. The density ρ is a function of x and y, so an element

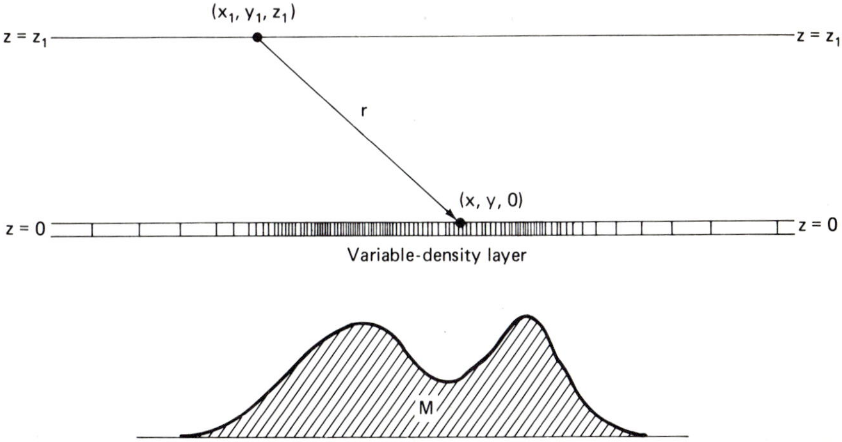

FIGURE 8-4 The mass distribution M and the indicated surface distribution at the plane $z = 0$ produce identical gravity effects at a higher plane $z = z_1$. (Skeels, 1947.)

of mass of area dS may be written as $dm = \rho\,\Delta t\,dS$. The attraction of this element at point (x_1, y_1, z_1) will be

$$dg = G \frac{\partial}{\partial z_1} \frac{\rho\,\Delta t}{r}\,dS$$

and the attraction of the whole infinite sheet is

$$g(x_1, y_1, z_1) = G \int_S \frac{\partial}{\partial z_1} \frac{\rho\,\Delta t}{r}\,dS \tag{8-24}$$

Equations (8-23) and (8-24) will be identical if we put

$$\rho(x, y, 0) = \frac{g(0)}{2\pi G\,\Delta t}$$

This means that a layer of variable density on the plane $z = 0$ produces the same gravitational effect on the higher plane $z = z_1$ as the mass distribution M. Moreover, we can choose any plane between M and z_1 as the $z = 0$ plane, so there are many possible thin layers each of which has the same effect. Nettleton (1976) gives a good example of non-uniqueness. Figure 8-5 shows three configurations with the same mass that give rise to the same gravity distribution.

In conclusion, we must recognize that there is no unique interpretation of a given set of gravity data. There are always many possibilities.

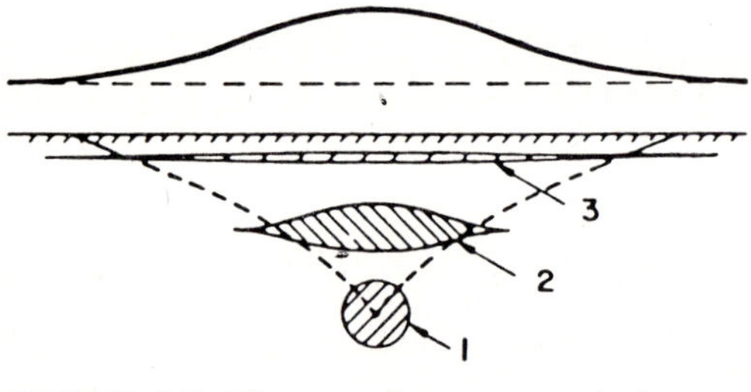

FIGURE 8-5 Three configurations with the same mass that give rise to the same gravity distribution. (Nettleton, 1976.)

8-3 MEASUREMENT OF GRAVITY IN BOREHOLES

In principle, the direct measurement of gravity is not difficult. The difficulty of constructing a useful gravimeter is due entirely to the requirement for very high precision. Although the gravity field is quite strong and pervasive, it is quite homogeneous and the perturbations, usually the objective of the measurement, are quite small compared to the total. Thus, to be useful, gravity measurements must be precise to at least one part in 10^6, and often to one part in 10^8.

The borehole gravimeter commonly in use is a modification of the LaCoste and Romberg standard geodetic gravimeter. It incorporates what is called a *zero-length spring* (LaCoste, 1934). It measures relative, not absolute, values of gravity. The usual expression for the force on a spring is $F = K(L - L_0)$, where K is the spring constant and L and L_0 are the lengths of the spring with loads F and 0, respectively. A zero-length spring, on the other hand, is prestressed, so that a force is required to separate the coils, and it is so designed that $F = KL$. (LaCoste and Romberg, 1942, describe the device in detail.) The spring is used to support a mass at the end of a beam (Fig. 8-6). The gravitational torque on the beam, caused by the weight of the mass, is

$$T_w = -mgd \cos \alpha$$

where m is the mass, d is the length of the beam, and α is the angular displacement of the beam. The torque exerted by the spring is

$$T_s = Ksr$$

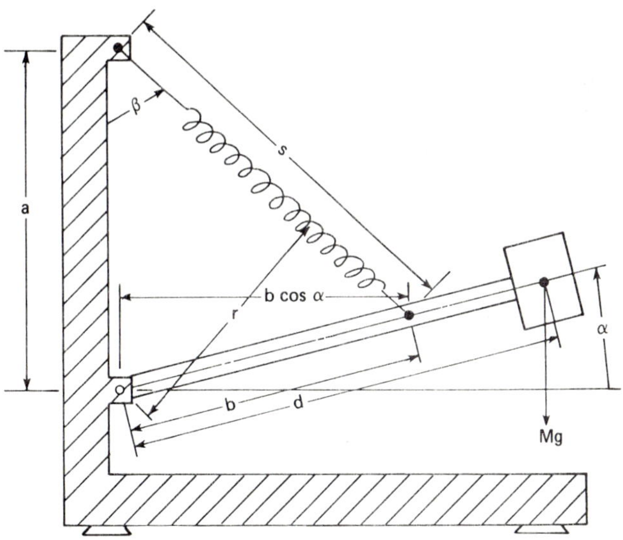

FIGURE 8-6 Schematic of the zero-length spring (LaCoste and Romberg) gravimeter.

where $s \sin \beta = b \cos \alpha$ and $r = a \sin \beta$. Thus

$$T_s = Kba \cos \alpha$$

and the net torque is

$$T = T_s + T_w = (Kba - mgd) \cos \alpha$$

In operation, at each measuring station a is adjusted through a complex system of levers and springs (LaCoste and Romberg, 1942, 1945) driven by a screw to set $Kba = mgd$. Then the net torque is zero, and the torque is very insensitive to small angular displacements of the beam. The number of turns of the screw is a measure of a, and thus of the relative value of g (and of the absolute value of g if the system has been calibrated against an absolute standard).

The gravimeter must be kept at constant temperature, and it is gimbal-mounted for remote leveling (LaCoste, 1977). The nulling screw of the meter is rotated by the operator by using a motor-driven gear train, and the train also turns a step switch that is followed by a synchro motor at the surface (McCulloh et al., 1967). The synchro motor in turn drives a counter that is so calibrated that an approximate value of g can be obtained from the counter units. The position of the beam near the null position is determined by observing the reading of a galvanometer which is driven by a pair of photosensors illuminated by light reflected from a mirror attached to the gravity meter beam. The pen of a strip chart recorder also is driven by the output of the photosensors. When the meter is in exact null position, the trace is, except for minor fluctuations, parallel with the direction of chart motion; when it is not, the trace is at an angle to the chart direction. In practice, two readings are taken with the trace on either side of the null and the angle between the two traces on the strip chart is recorded. The exact value of gravity can then be calculated from the counter reading and the slopes of the traces on either side of the null. This is a faster technique than obtaining a perfect null.

In a more recent, slim-hole, gravimeter built on the same principle, the nulling data are transmitted electronically. The slim-hole meter has an outside diameter of 10.5 cm. It is thermostatted at 124°C. (The sensing element must be kept within about 0.1°C to avoid changes in calibration, so care must be taken to keep it connected to power at all times.) The meter can be operated in a borehole inclined as much as 14° to the vertical.

Schmoker (1978) studied the precision of the original meter and determined that the gravity difference between two stations in a borehole can be measured to 10 μgal, or one part in 10^8. LaFehr et al.

(1979) found that the slim-hole meter is precise to 3 μgal in the laboratory and 7 μgal in a borehole.

8-4 FIELD OPERATIONS

Lowering the gravimeter. Rasmussen (1973) gives a good overview of the standard field operation of a borehole gravimeter. It can be operated on a standard seven-conductor logging cable. It is important to give serious consideration to the method of lowering the meter into the hole. The meter should always approach the depth station from the same direction to avoid errors caused by backlash in springs. Error in measurement of depth is often the largest error in a borehole gravity survey. Standard logging trucks can measure depth to 1 percent or better, but often that is not good enough. Caton (1983) pointed out that an error of 5 cm in the distance between stations 7 m apart will give a density error of 0.01 g/cm^3. If gravity stations are to be spaced less than 15 to 30 m apart, it is desirable to flag the cable and measure the distance between flags directly (with tension on the cable). It is possible to measure relative depth between stations to less than 1 cm with equipment developed by EDCON (1977). Caton (1983) has used automated drawworks to improve depth control.

Even slight vibration of the meter will make the readings useless. In some holes it is necessary to clamp or spring-load the meter to the borehole wall, or to clamp the logging cable to the top of the hole casing, to prevent the transmission of wind-induced noise to the meter, especially if the readings are at depths less than a few hundred meters.

Because the gravimeter is expensive, the hole should be in excellent condition with no possibility of caving. Many meter owners require a cased or lined hole.

Choice of stations. The choice of which gravity stations to use in a borehole is an important operational decision. Hearst (1977) discusses this in some detail and determines that, the finer the spacing between stations, the better an object will be resolved regardless of the distance from the hole. One must certainly select a station spacing equal to or less than the vertical resolution desired for density data. Beyond that criterion, additional data points at least allow one to average out local perturbations in the gravity field that tend to degrade the precision of the measurement. If it is known that the structure in the vicinity of the borehole consists of flat beds, it is advantageous to locate stations at bed boundaries. The usual interpretation method produces the average density between gravity stations, and the choice of stations according to the lithology produces a truer curve of density versus depth than if arbitrary stations were chosen.

One normally reduces the difference in gravity readings to density with some form of Eq. (8-19). If we know the precision of the gravity readings, we can calculate the smallest measurable density contrast for any station spacing:

$$\delta\rho = \frac{1}{4\pi G}\frac{\delta g}{\Delta z} \tag{8-25}$$

where δg is the smallest gravity difference we can measure and Δz is the station spacing. This is illustrated in Fig. 8-7. Equation (8-25) shows that the resolution in average density is inversely proportional to station spacing Δz. This does *not* mean that we get better resolution with wider spacing, because $\delta\rho$ refers to *average* density between stations. But it does mean that one can learn about a large volume of material with more accuracy than one can learn about a small volume.

Vertical air gradient. The first term in Eq. (8-19) involves Γ, the vertical air gradient. A typical value of Γ (in mid-latitudes) is about 0.31

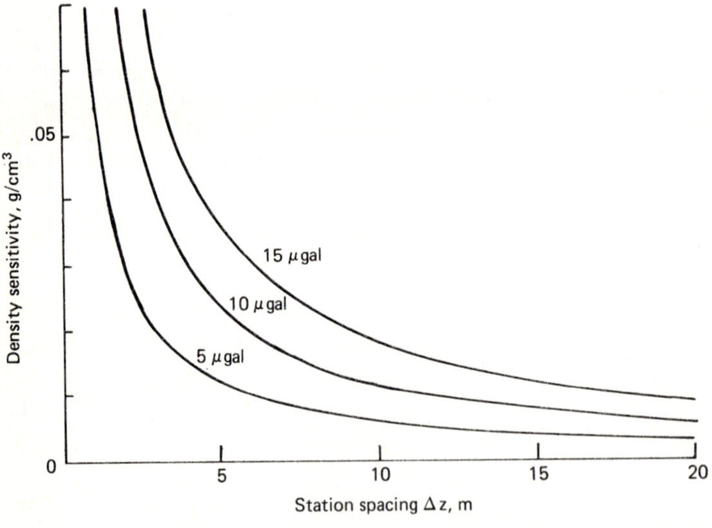

FIGURE 8-7 Minimum observable average density difference $\delta\rho$ for a minimum resolvable gravity difference δg as a function of station spacing. (Hearst and Carlson, 1982.)

mgal/m. Most workers calculate this term from the latitude and elevation of the top of the well (Heiskanen and Vening Meinesz, 1958) or by assuming that a layer of known density material exists at some depth interval and applying Eq. (8-19) to find Γ. We have found it is better to measure Γ. This can be done either by calculation *(upward continuation)* from surface gravity data (Grant and West, 1965) or by direct measurement on a tower. In the latter case, it is important that all measurements be made well above the surface (Fajklewicz, 1976) because effects of nearby terrain will otherwise invalidate the data.

Two other useful discussions of fielding methods have been published by Jones (1972) and Brown et al. (1975). EDCON's (1977) borehole gravity manual is a fine presentation of current field techniques.

8-5 CORRECTIONS

There are many corrections that must be applied to the gravity data, and each correction is obtained from field data. Terrain correction requires a knowledge of the elevation and density of nearby terrain to an accuracy depending on the distance to the terrain feature. Often, a survey of nearby terrain is needed for good results. Nearby mud pits and subsurface workings can also cause errors. It is important to record all information about any nearby changes in the surface and also record the density of nearby objects. The tide correction requires a knowledge of the time of day to about 5 min and position to 24 km. Measured depths must be corrected for hole deviation—the measured depth difference in a hole is equal to the true vertical difference only for a vertical hole. A directional survey is often useful.

8-5a Terrain Corrections

The terrain correction is most complex, most time-consuming, and most important for depths less than a few hundred meters. The method (Beyer and Corbató, 1972) is based on Eq. (8-21). A series of concentric circles of standardized radii (for example, Hammer, 1939) are plotted on a topographic map of the terrain surrounding the hole. Therefore, the map is divided into ring-shaped zones. Each ring is subdivided azimuthally into n compartments, where n is again standardized. Then the average elevation of each compartment is estimated, usually by visual estimation by an experienced operator. The terrain effect g_{te} caused by the material above or below the measuring datum in each compartment is

$$g_{te} = \frac{2\pi\rho G}{n}\left(\sqrt{r_1^2 + Z_2^2} - \sqrt{r_1^2 + Z_1^2} - \sqrt{r_2^2 + Z_2^2} + \sqrt{r_2^2 + Z_1^2}\right) \quad (8\text{-}26)$$

which is analogous to (8-21). In Eq. (8-26) $Z_1^2 = (B - C)^2$ and $Z_2^2 = (A - C)^2$, where A is the terrain compartment elevation, B is the datum elevation (the zero-depth elevation), and C is the gravity station elevation. With these conventions, g_{te} will be correct whether a terrain feature represents a mass excess or a mass deficiency with respect to the datum (that is, whether the terrain feature is a hill or a valley) and whether the station is above or below the datum or ring elevations. The value of g_{te} is subtracted from all gravity values.

Hearst et al. (1980) have studied the precision in the terrain elevations needed to obtain a given accuracy in g_{te} for a given terrain element. If the total elevation error of all terrain is allowed to give rise to a total error in estimated density no greater than 0.01 g/cm^3, the error in density caused by an individual compartment may be no greater than $0.01/\sqrt{\Sigma n}$, where Σn is the total number of compartments. The closer the compartment to the hole and the closer the elevation of the center of mass of the compartment to the measuring station, the smaller the elevation error that will give this acceptable density error. For example, one of the six compartments of Hammer zone D, with inner radius at 53 m and outer radius 170 m, may have an elevation error of 2.5 m for a station depth of 150 m or an elevation error of 20 m for a station depth of 450 m. One of the 12 compartments of Hammer zone H, with an inner radius of 1529 m and an outer radius of 2615 m may have an elevation error of 138 m, and depth is not important. Inclined zones have a different terrain effect than flat zones at the same mean elevation. Details can be found in Hearst et al. (1980).

Schmoker (1980) has studied the terrain effects of very nearby cultural features (water tanks, excavations, drilling pads) and finds them to be substantially larger than 0.01 g/cm^3 when the gravity measurements are made at depths close to the datum. This reinforces the statement that these features should be recorded and the appropriate correction calculated. Beyer (1979) has made a study of the terrain correction for both borehole and tower gravity measurements and shows many useful examples.

Plouff (1966) developed a convenient method for obtaining the terrain correction from elevations obtained by using the method described above for nearby terrain and from data digitized on a latitude-longitude grid at greater distances from the hole. Tapes of elevations digitized on 1/250,000 topographic maps may be obtained from the National Cartographic Information Center, but the accuracy is not good enough for shallow holes and nearby terrain.

8-5b Tide Correction

All gravity measurements must be corrected for earth tide effects. These effects are functions of time, and they are large enough to cause serious errors if not corrected for. Computer programs for calculation of earth tides are available, or they can be written from the original formulations by Longman (1959). The input required consists of Greenwich mean time, date, and the geographic coordinates of the measuring points.

8-5c Drift

Correction for drift of the gravity meter is obtained by repeated readings at the same gravity station. The gravimeter readings at the same station are plotted as a function of time, and the change appropriate to the time of any station reading is applied to that reading as a drift correction.

Some gravimeters are said to be very temperature-sensitive (Caton, 1981), although the manufacturer will evidently supply a correction circuit upon request. Caton has developed a method for correcting for this sensitivity by least-squares fitting a drift curve to data from three or more surveys at the same depths. Jageler (1981), however, states that the drift as a function of temperature can be determined for each individual meter and the problem of temperature compensation avoided by simply waiting with the meter downhole until the meter has reached a temperature at which the drift is known to be acceptable.

8-5d Other Corrections

If the vertical air gradient required for Eq. (8-19) is measured, the measurement must be corrected for terrain, tide, and drift. Also, if subsurface structure is found to have a substantial effect on the data, the measured vertical air gradient data should be appropriately corrected.

Corrections for diameter of the hole are ordinarily unnecessary, except for large hole diameters with readings within 1 m or so of the top or bottom of the hole, the water level, or major diameter changes. In that case the borehole can be treated as a terrain zone.

A nonvertical hole, if its angle is not greater than that permitted for the gravimeter, does not cause errors in the gravity data. But the angle does affect dz used in Eq. (8-19). If the deviation from the vertical θ of the hole is known, then the corrected value $dz' = dz \cos \theta$. If θ is not known, it must be determined from a directional survey. At 14°, a 2° error in angle can give a 1 percent error in the cosine, and therefore a 1 percent error in the calculated density. At smaller angles, the effect is considerably less. If the beds dip, the effect is quite different, but we consider that to be part of the data analysis rather than a correction.

8-6 RANGE OF INVESTIGATION

8-6a Minimum Resolvable Mass

It was shown in Sec. 8-1c that the value of g is unaffected by vertical distance from a measuring point to an infinite slab. Equation (8-25) and Fig. 8-7 show density resolution for an infinite slab. For a mass point, however, the vertical component of gravity, from Eq. (8-12) and geometry, is

$$g = \frac{GMz}{(x^2 + z^2)^{3/2}} \tag{8-27}$$

where x is the horizontal distance and z is the vertical distance from the measuring point to the point of mass M. Figure 8-8 shows the gravitational effect of a point mass at different horizontal radii as a function of

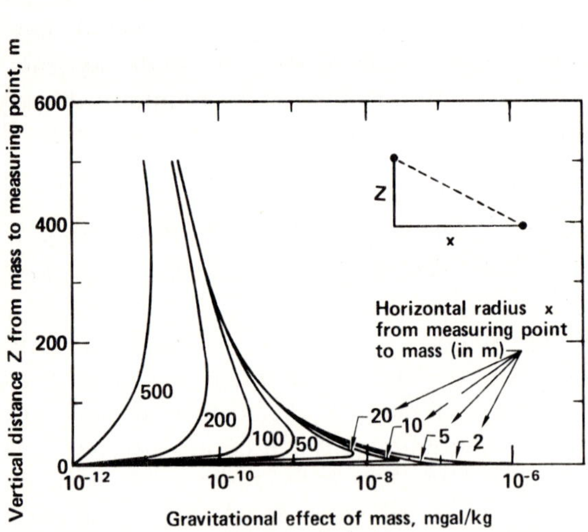

FIGURE 8-8 Vertical component of gravity caused by a point mass at horizontal radius x from a borehole as a function of vertical distance z from the measuring point to the mass. (Hearst, 1977*a*.)

vertical distance z. The maximum effect is found at $z = x/\sqrt{2}$ and therefore the maximum sensitivity is at an angle from the vertical of $\tan^{-1}\sqrt{2}$, or about 55°.

Since any gravimeter has a minimum resolvable δg, the minimum visible mass for any x must be that which gives a vertical component of gravity equal to δg at $z = x/\sqrt{2}$. This minimum mass is

$$M_m = 3.86 \times 10^5 x^2 \, \delta g$$

where M_m is in kilograms, x is in meters, and δg is in milligals. A reasonable value for δg is 0.01 mgal (Schmoker, 1978a); and so for $x = 1$ m, M_m is about 4000 kg, or the mass of an elephant, and for $x = 100$ m, about 4×10^7 kg, or that of a battleship.

Note that Fig. 8-8 shows the effect of a point at *constant horizontal distance* as a function of vertical distance. Clearly, the smaller the horizontal distance, the greater the effect on the vertical component. (The greatest effect is, of course, that of a point directly below or above the meter.)

Thus far we have been considering the effect of a mass anomaly on the vertical component of gravity. When the apparent density, or difference in vertical component between two measuring stations, is used, dg/dz in Eq. (8-19) can have either a positive or negative sign depending on the location of the mass point with respect to the measuring stations. Consequently, a mass anomaly whose density is *greater* than that of the surrounding medium (a positive mass anomaly) can cause a *decrease* in the apparent density in some parts of the borehole. The sign of an apparent density anomaly can sometimes be used to infer the position of an anomaly of known true density.

8-6b Change in Density of a Layer

Consider a horizontal circular disk with a borehole on its axis. The disk has a density ρ, and the infinite horizontal layer in which the disk is the central part has a density $\rho + \Delta\rho$. We can use Eqs. (8-21) and (8-22) to calculate the gravity difference across the layer containing the disk and Eq. (8-19) to calculate the apparent density ρ_a of an infinite horizontal slab, without the disk, having the same gravity difference across it. Then the gravity anomaly caused by the presence of the boundary between the disk and the slab is $\Delta\rho_a$, equal to $\rho - \rho_a$.

LaFehr (1983) gives the very useful Fig. 8-9 which, for three values of $\Delta\rho_a$, shows the radius and thickness of a disk of density contrast $\Delta\rho$ that would give the anomaly. Note that, the thicker the bed, the farther away one can detect a given density change $\Delta\rho$. For example, if one can detect an anomaly $\Delta\rho_a$ of 0.02 g/cm^3 and the bed is 40 m thick,

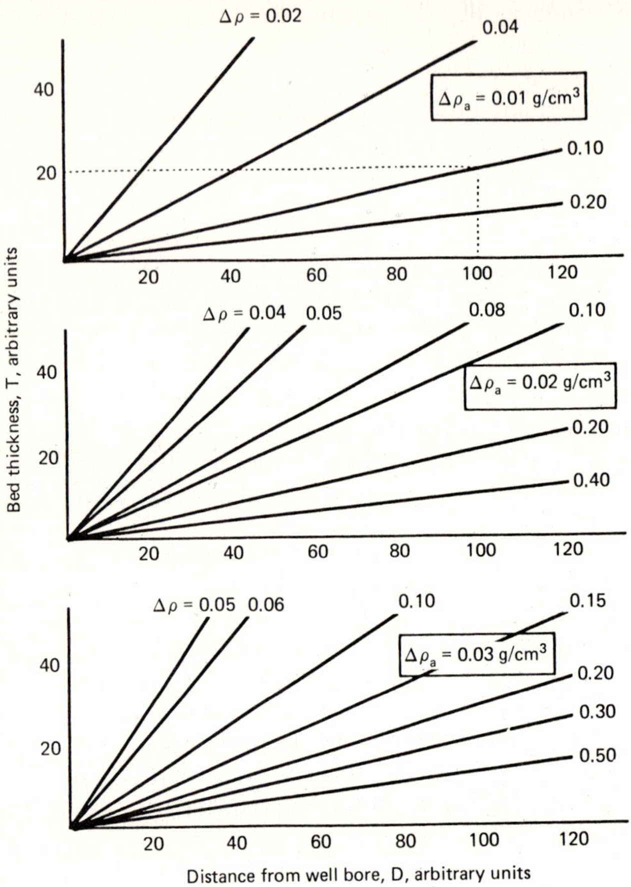

FIGURE 8-9 Relation between distance from well bore and bed thickness for different changes in density $\Delta\rho$ occurring at the boundary of a circular disk within an infinite layer for three different apparent density anomalies $\Delta\rho_a$. (LaFehr, 1983.)

one can detect a change of 0.1 g/cm^3 at a radius of 90 m, but if the bed is only 20 m thick, the change can be detected only out to 45 m.

LaFehr gives a similar figure for semi-infinite rectangular slabs with a density change some distance from a borehole.

8-7 COMPARISON OF A DENSITY LOG AND A GRAVITY LOG

A γ-γ density log (Sec. 6-4) is often used in conjunction with borehole gravity data. When the γ-γ log densities have been determined, they are averaged over the gravity station intervals, and then the difference between log densities and gravimetric densities [obtained from Eq. (8-19)] is computed. For nominal values of the constants, Eq. (8-19) becomes

$$\rho = 11.9\left(0.31 - \frac{\Delta g}{\Delta z}\right) \tag{8-19}$$

for gravity in milligals, density in grams per cubic centimeter, and distance in meters. The difference in density is compared to that which would be caused by some hypothetical model—either a simple geometrical shape such as a sphere or a fault or a complex subsurface structure—and the model is varied until the agreement between the observed and calculated values is good enough to satisfy the interpreter.

As shown in Sec. 8-2, such interpretations are not unique. It is always possible (at least in principle) to find infinitely many models that can cause the same density or gravity difference within the error of the measurement. Consequently, the interpreter must have some idea of what he is looking for, and some knowledge of the range of plausible models, before sensible models can be made.

It is always possible that the difference between the density log and the borehole gravity data is caused by poor depth correlation. Depth discrepancies between logs run on different passes in a hole are not uncommon. If the density log and gravimeter are run by different contractors, or even with different cables, the problem is compounded. As in any case where depth accuracy is important, the use of permanent reference points in the hole, for example, casing collars, hole bottom, or a distinctive lithologic bed, is recommended.

8-7a Horizontal Beds

The range of investigation of a density log is only about 12 cm (Sec. 6-4g). This short range explains the virtue of a gravity log even in the case of seemingly infinite homogeneous bedding. Sometimes the density observed by the γ-γ log is not representative of the true density of the material near the hole. If the user thinks the beds are infinite and homogeneous, or knows the structure and can correct the gravimetric density for that structure, then the gravity data can be used to infer the true density. If the true density is lower than the γ-γ density, it is likely that some of the original pore fluid was gas, which has been displaced by borehole fluid. This method is routinely used to locate gas zones (Jones, 1972; Jageler, 1976) and even, in formations of high porosity, oil.

On the other hand, sometimes the true density is greater than the γ-γ density. This could be caused by high-density ore bodies located at depths between gravity-measuring stations.

8-7b Simple Structures

A more sophisticated method is to assume $\Delta\rho$ is caused by a simple structure. The most comprehensive work to date involving postulated simple structures is that of Snyder (1976). Rather than the density, Snyder used the *borehole Bouguer anomaly,* the difference between measured gravity and the gravity that would be observed if the world were indeed comprised of infinite homogeneous horizontal slabs with density equal to the γ-γ density. Hammer (1974) showed, for example, that the anomaly in the vertical component of gravity caused by an infinite horizontal line source is

$$\Delta g = \frac{2G\sigma z}{x^2 + z^2} \tag{8-28}$$

where σ is the mass per unit length, x is the horizontal distance from the hole to the line, and z is the vertical distance from the gravity station to the line. Figure 8-10 shows the anomaly for a line source with a linear density of π megagrams per meter of length (this corresponds to an infinite cylinder of density 1 g/cm^3, 2 m in diameter). The anomaly reaches a maximum at a distance $z = x$ above or below the line. The maximum value of Δg is

$$\Delta g_m = \frac{G\sigma}{x}$$

so the diameter of a cylinder of density contrast $\Delta\rho$ at a distance x can be calculated from

$$d = \left(\frac{4x\,\Delta g_m}{\pi\rho_c G}\right)^{1/2} \tag{8-29}$$

Snyder showed that one can distinguish between, for example, a cylinder and a truncated bed by the shape of the anomaly and can then estimate the parameters of the perturbing shape from the size of the anomaly. He gives formulas for calculating the size of a perturbing sphere or cylinder from only ρ_c and Δg—without including the value of x. These formulas are not correct; Eq. (8-29) and a similar one for the sphere should be used.

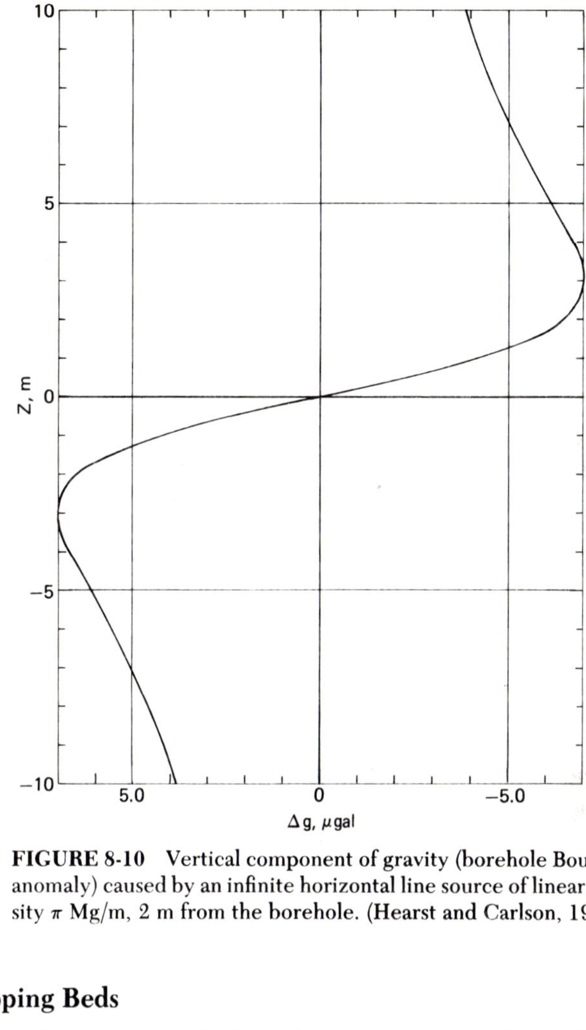

FIGURE 8-10 Vertical component of gravity (borehole Bouguer anomaly) caused by an infinite horizontal line source of linear density π Mg/m, 2 m from the borehole. (Hearst and Carlson, 1982.)

8-7c Dipping Beds

Beds are rarely infinite and often dip. Hearst (1977*b*) called the contrast between the gravimetric density ρ_g observed in a dipping bed and the average density ρ_{sur} in the formation surrounding the bed the apparent density contrast, ρ_{ca}:

$$\rho_{ca} = \rho_g - \rho_{sur}$$

and he called the contrast between the actual bed density ρ, measured by a γ-γ density log and that of the surroundings, the true density contrast ρ_{ct}:

$$\rho_{ct} = \rho - \rho_{sur}$$

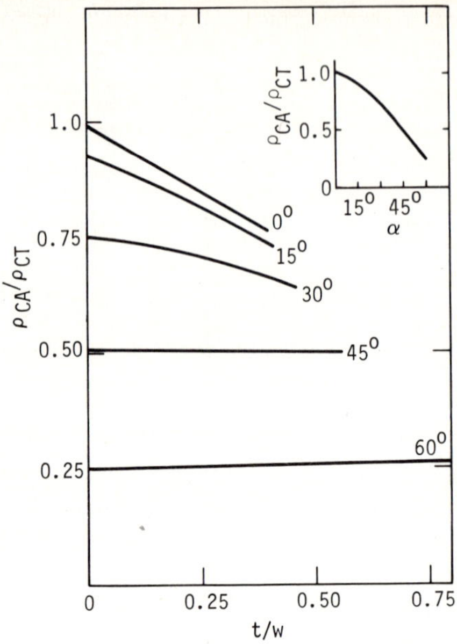

FIGURE 8-11 Ratio of apparent density contrast ρ_{ca} to true density contrast ρ_{ct} as a function of bed thickness t divided by horizontal width of bed w for various values of dip (Hearst, 1977*b*). Inset: ratio as a function of dip for an infinite bed.

He then showed that

$$\rho_{ca} = \rho_{ct} \cos^2 \theta \tag{8-30}$$

where θ is the dip angle.

For a horizontal bed of finite extent

$$\rho_{ca} = \rho_{ct} \left[1 - \frac{2 \tan^{-1} (t/w)}{\pi} \right] \tag{8-31}$$

where t/w is the ratio of bed thickness to its horizontal extent. The effect is more complicated for dipping beds of finite lateral extent; it is shown in Fig. 8-11.

Brown and Lautzenhiser (1982) have used a modification of Eq. (8-30) to estimate dip when the density of the surroundings ρ_{sur} is unknown. From Eq. (8-30)

$$\rho_g = \rho_{sur} \sin^2 \theta + \rho \cos^2 \theta \tag{8-32}$$

Then if, for some bed, $\rho_g = \rho$, it follows from Eq. (8-32) that $\rho_{sur} = \rho$. Then the dip of the beds can be obtained from Eq. (8-32). It is also possible to obtain ρ_{sur} and the dip from the slope of a plot of $\rho_g - \rho$ versus ρ. Since

$$\rho_g - \rho = (\rho_{sur} - \rho)\sin^2\theta \tag{8-33a}$$

$$\frac{d(\rho_g - \rho)}{d\rho} = -\sin^2\theta \tag{8-33b}$$

Consequently, if one fits a line to that slope, one can obtain θ, and thence ρ_{sur}, from Eq. (8-32).

REFERENCES

Beyer, L. A.: "Terrain Corrections for Borehole and Tower Gravity Measurements," *Geophysics*, vol. 44, 1979, p. 1584.

——— and C. E. Corbató: "A FORTRAN IV Computer Program for Calculating Borehole Gravity Terrain Corrections," U.S. Geol. Survey Open-File Rept., 1972. (Available from NTIS as PB 208-679.)

Brown, A. R., N. F. Rasmussen, C. O. Garner, and W. G. Clement: "Borehole Gravity Logging Fundamentals," preprint Soc. Expl. Geophys. 45th Ann. Meeting, Denver, Colo., 1975.

——— and T. V. Lautzenhiser: "The Effect of Dipping Beds on a Borehole Gravimeter Survey," *Geophysics*, vol. 47, 1982, p. 25.

Caton, P. W.: "Improved Methods for Reducing Borehole-Gravity Data—Applications and Analyses of Reduced Gravity Plots," *Trans. Soc. Prof. Well Log Analysts 22d Logging Symp.*, 1981, paper RR.

———: "Borehole Gravity Logging Developments: A Status Report," *Soc. Expl. Geophys. 53d Ann. Meeting*, 1983, p. 23.

EDCON: "Borehole Gravity Meter Manual," EDCON, Denver, Colo., 1977.

Fajklewicz, Z. J.: "Gravity Vertical Gradient Measurements for the Detection of Small Geologic and Anthropogenic Forms," *Geophysics*, vol. 41, 1976, p. 1016.

Grant, F. S., and G. F. West: *Interpretation Theory in Applied Geophysics*, McGraw-Hill, New York, 1965.

Hammer, S.: "Terrain Corrections for Gravimeter Stations," *Geophysics*, vol. 4, 1939, p. 184.

———: "Density Determinations by Underground Gravity Measurements," *Geophysics*, vol. 15, 1950, p. 637.

———: "The Anomalous Vertical Gradient of Gravity," *Geophysics*, vol. 35, 1970, p. 153.

———: "Approximation in Gravity Interpretation Calculations," *Geophysics*, vol. 39, 1974, p. 205.

Hearst, J. R: "On the Range of Investigation of a Borehole Gravimeter," *Trans. Soc. Prof. Well Log Analysts 18th Logging Symp.*, 1977*a*, paper E.

———: "Estimation of Dip and Lateral Extent of Beds with Borehole Gravimetry," *Geophysics*, vol. 42, 1977*b*, p. 990.

———: "BIFUR III, A Program for Calculating Borehole and Surface Gravity Caused by Two-Dimensional Structure in Cartesian or Cylindrical Coordinates," Lawrence Livermore Natl. Lab. Rept. UCID-19030, 1981.

——— and H. L. McKague: "Structure Elucidation with Borehole Gravimetry," *Geophysics*, vol. 41, 1976, p. 491.

———, J. W. Schmoker, and R. C. Carlson, "Effects of Terrain on Borehole Gravity Data," *Geophysics*, vol. 45, 1980, p. 234.

Heiskanen, W. A., and F. A. Vening Meinesz: *The Earth and Its Gravity Field*, McGraw-Hill, New York, 1958.

Jageler, A. H.: "Improved Hydrocarbon Reservoir Evaluation through Use of Borehole-Gravimeter Data," *J. Petr. Tech.*, vol. 28, 1976, p. 709.

———: personal communication (a comment at Soc. Prof. Well Log Analysts 22d Logging Symp.,) 1981.

Jones, B. R.: "The Use of Downhole Gravity Data in Formation Evaluation," *Proc. Soc. Prof. Well Log Analysts 13th Logging Symp.*, 1972, paper M.

LaCoste, L. J. B.: "A New Long-Period Vertical Seismograph, *Physics*, vol. 5, 1934, p. 178.

———: Method and Apparatus for Levelling an Instrument in a Well-Bore, U.S. Patent No. 4,040,189, 1977.

——— and A. Romberg: Force Measuring Device, U.S. Patent 2,293,437, 1942.

——— and ———: Force Measuring Instrument, U.S. Patent 2,377,889, 1945.

LaFehr, T. R.: "Rock Density from Borehole Gravity Surveys," *Geophysics*, vol. 48, 1983, p. 341.

———, R. H. Merkel, and A. T. Herring: Evaluation and Applications of the New LaCoste and Romberg Borehole Gravity Meter" (abstract only), *Geophysics*, vol. 44, 1979, p. 369.

Longman, I. M.: "Formulas for Computing Tidal Acceleration," *J. Geophys. Res.*, vol. 64, 1959, p. 2351.

Marion, J. B.: *Classical Dynamics of Particles and Systems*," Academic, New York, 1965.

McCulloh, T. H., L. J. B. LaCoste, J. E. Schoellhamer, and E. H. Pampayan: "The U.S. Geological Survey—LaCoste and Romberg Precise Borehole Gravimeter System—Instrumentation and Support Equipment," U.S. Geol. Survey Prof. Paper 575-D, 1967, p. D92.

Nettleton, L. L.: *Gravity and Magnetics in Oil Prospecting*," McGraw-Hill, New York, 1976.

Plouff, D.: "Digital Terrain Corrections Based on Geographic Coordinates," preprint Soc. Expl. Geophys. 36th Ann. Meeting, 1966.

———: "Gravity and Magnetic Fields of Polygonal Prisms and Application to Magnetic Terrain Corrections," *Geophysics,* vol. 41, 1976, p. 727.

Rasmussen, N. F.: "Borehole Gravity Survey Planning and Operations," *Trans. Soc. Prof. Well Log Analysts 14th Logging Symp.*, 1973, paper Q.

Schmoker, J. W.: "Accuracy of Borehole Gravity Data," *Geophysics,* vol. 43, 1978, p. 538.

———: "Terrain Effects of Cultural Features upon Shallow Borehole Gravity Data," *Geophysics,* vol. 45, 1980, p. 1869.

Skeels, D. C.: "Ambiguity in Gravity Interpretation," *Geophysics,* vol. 12, 1947, p. 43.

Snyder, D. D.: "The Borehole Bouguer Gravity Anomaly—Application to Interpreting Borehole Gravity Surveys," *Trans. Soc. Prof. Well Log Analysts 17th Logging Symp.*, 1976, paper AA.

Talwani, M., J. L. Worzel, and M. Landisman: "Rapid Gravity Computations for Two-Dimensional Bodies with Application to the Mendocino Submarine Fracture Zone," *J. Geophys. Res.*, vol. 64, 1959, p. 49.

PART 3

Applications

9

Lithology

One of the main applications of logs is the identification of rock types, or *lithology*. We shall first discuss methods for using known log responses to choose among several possible lithologies and then describe in some detail the characteristic response of three specific types of material: clay and shale, coal, and evaporites. The discussion will not be exhaustive, but it will be illustrative. We chose clay and shale for two reasons: (1) because shaly sands are found in many sedimentary sequences of interest to the petroleum industry and consequently a great deal of work has been done to use logs to find the fraction of clay present in shaly sands and (2) because the methods used to find this fraction are good illustrations of the pitfalls that can occur when one uses log data based on assumptions that are not always valid. We chose coal because it is so different from most other materials encountered in a borehole and evaporites because their characteristics too are different from those of most sediments and, in a way, are dissimilar to those of coal.

TABLE 9-1 Values of logging parameters from shales, coals, and evaporites. Typical values of parameters for sedimentary rocks also are shown. It is assumed that the neutron log is calibrated for a limestone matrix.

Material	γ-Ray (API)	Density, g/cm^3	Sonic, μs/ft	Neutron, I_H(LS)	Resistivity, Ω·m	Ref.*
Shale	80–150	2.4–2.6	70–150	0.3–0.4	2–10	R
Shale			60–180		1–100	P
Chlorite	1	2.6–2.96		0.34–0.51		E, F
Kaolinite	6	2.42–2.61		0.35		E, F
Illite	70	2.5–2.7		0.12–0.3		E, F
Smectite	3	2.1–2.7		0.13–0.4		E, F
Clay mixture 1	21			0.2–0.3		E
Clay mixture 2	85			0.1–0.2		E
Clay mixture 3	35			0.1–0.2		E
Anthracite	10–30	1.5–1.7	100–110	0.20–0.35	2–8	R
Anthracite		1.4–1.8				T
Anthracite	0	1.47		0.37		E
Bituminous	20–45	1.2–1.4	115–127	0.35–0.45	50–200	R

Bituminous		1.3–1.5	100–150			T
Bituminous	0	1.24		0.5–0.6		E
Subbituminous	20	1.4	128	0.62		T
Lignite	0–25	1.05–1.25	100–110	0.2–0.35	2–10,000	B, R
Lignite		0.7–1.5	140–170	0.47–0.52		E, T
Halite	0–10	2.0–2.2	40–67	−0.02–+0.02	500–1000	E, R, T
Trona	0	2.11	65	0.25–0.4	0	E, T
Anhydrite	0–10	2.8–3.0	40–55	−0.01–+0.03	500–1000	E, R, T
Gypsum	0–10	2.2–2.35	40–55	0.3–0.6	500–1000	R, E
Sylvite	500	1.98	74	−0.03–0		E, T
Carnallite	200	1.61	78	0.41–0.65		E, T
Polyhalite	180	2.78	58	0.15–0.25		E, T
Langbeinite	275	2.83	52	−0.02–0		E, T
Kainite	225	2.12		0.40–0.60		E, T
Quartz	0	2.65	50–55	−0.04		T
Dolomite	0	2.88	43.5	0.04		T
Limestone	0	2.71	47.5	0		T

[a]B, Brom and Driedonks (1981); E, Edmundson and Raymer (1979); F, Fertl and Frost (1980); P, Patchett (1975); R, Reeves (1981); T, Tixier and Alger (1970).

9-1 USE OF COMBINATIONS OF KNOWN RESPONSES TO DISTINGUISH LITHOLOGY

***M-N* plots.** Each pure rock type has a characteristic (but not necessarily unique) response on each of the three "porosity" logs: acoustic, density, and neutron. Table 9-1 lists the responses of many rock types on these and other logs. Burke et al. (1969) assume that, if density is plotted versus transit time (Δt) or neutron log response (ϕ_N) for a pure mineral whose pores are saturated with a single fluid, the result will be a straight line on the cross-plot (Sec. 10-2). They then define the variables M and N as the slopes of the lines of density versus transit time and density versus neutron log response, respectively:

$$M = \frac{\Delta t_f - \Delta t_{ma}}{\rho_{ma} - \rho_f} \times 0.01 \tag{9-1a}$$

$$N = \frac{\phi_{Nf} - \phi_{Nma}}{\rho_{ma} - \rho_f} \tag{9-1b}$$

where ϕ_{Nf} and ϕ_{Nma} are the porosities that would be obtained from a neutron log in the fluid and matrix, respectively. (Usually ϕ_N is expressed in units of porosity from a limestone calibration curve, Sec. 6-6b.) We see from Eqs. (9-1) that M and N depend on matrix and fluid properties but *not* on porosity. This follows from the assumption of linearity of the plots of ρ versus Δt and ρ versus ϕ_N. Consequently, any set of data from a saturated pure mineral with varying porosity maps to a single point in M-N space. If the measured values of Δt, ρ, and ϕ_N taken from logs in a formation composed of a given mineral and fluid are substituted for the matrix values, the slopes are

$$M = \frac{\Delta t_f - \Delta t}{\rho - \rho_f} \times 0.01 \tag{9-2a}$$

$$N = \frac{\phi_{Nf} - \phi_N}{\rho - \rho_f} \tag{9-2b}$$

If these values of M and N are plotted in M-N space, one may be able to identify the mineral from the position on the plot. If the lines of ρ versus Δt or ρ versus ϕ_N are not in fact straight, they are approximated by straight lines to obtain M and N.

Figure 9-1 shows an M-N plot for many common matrices, in which the fluid is fresh mud, in M-N space. Several things should be noted. There are two points for dolomite; they represent different matrix densities. Some M-N plots also show two different sandstone points; they represent different matrix velocities. Secondary porosity (Sec. 7-2a) causes M to increase but leaves N unchanged. Shale data will plot near

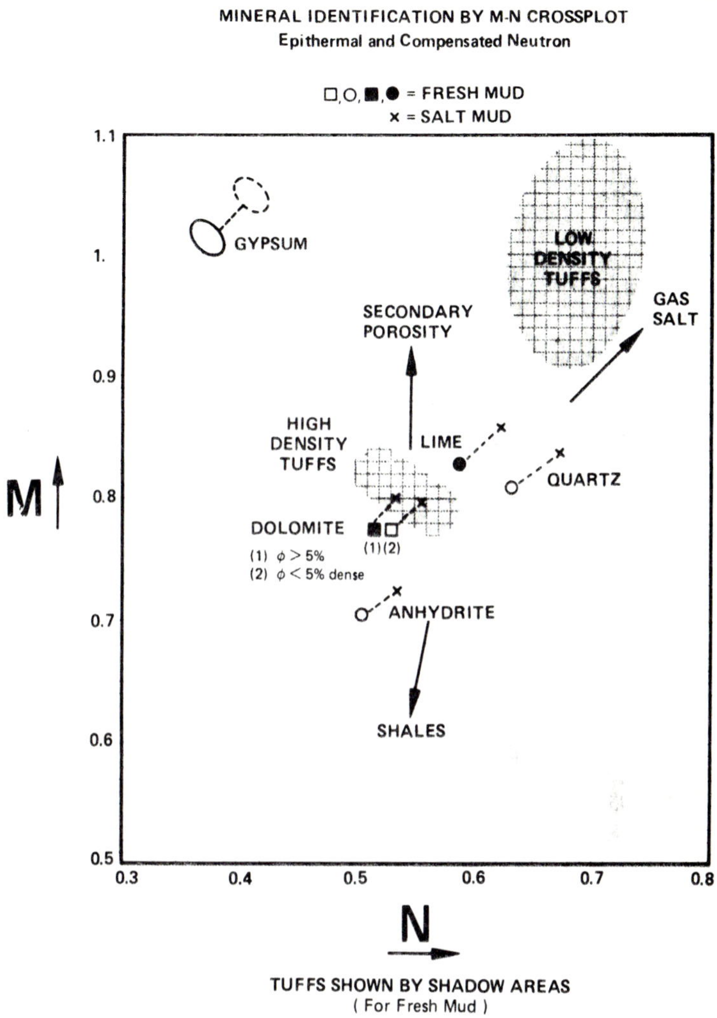

FIGURE 9-1 Standard *M-N* plot for mineral identification. (Dresser-Atlas, 1977.)

the end of the "shales" arrow. Gas in the rock shifts points up and to the right. In practice, one plots the measured values of M and N in the same manner as an ordinary cross-plot (Sec. 10-2). One then draws lines between three matrix points to make a triangle. If the triangle encloses the data, one concludes that the formation consists of the three matrices in question and calculates their proportions as described in Sec. 10-2. (One must, of course, assume that the matrix properties used to define M and N are correct.) Shale, gypsum, gas, secondary porosity, or measurement error can cause a point to lie outside the triangle. It may be possible to identify the cause of the displacement by the position of the point. Burke et al. (1969) give many examples.

Mineral identification (MID) plots. Another method (Clavier and Rust, 1976) uses plots of actual responses of logs in pure minerals. For example, a plot is made of ρ versus ϕ_N for each of several pure minerals by using specific logging sondes. It is then assumed that the differences in these plotted lines is a function only of ρ_{ma}. Consequently, it is possible to interpolate between the plotted response lines to obtain lines of constant ρ_{ma}. An example is shown in Fig. 9-2.

Next, the differences in ρ_{ma} between known lines (A and B in Fig. 9-2) are used to scale extrapolated lines on either side of the known

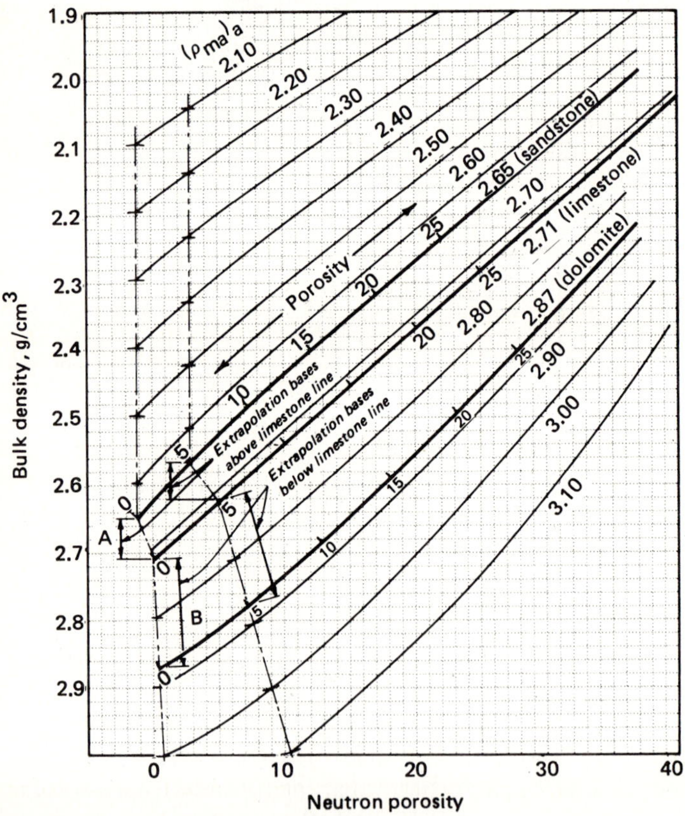

FIGURE 9-2 The grid used to determine apparent values of density $(\rho_{ma})_a$ for a mineral identification (MID) plot. (Clavier and Rust, 1976.)

lines. Ultimately a set of these lines is built up, as shown in Fig. 9-2. A similar chart is made for Δt versus ϕ_N. (Note that there is a different chart for each density and neutron sonde.)

When data from logs in a formation are obtained, each point is plotted on a grid like that of Fig. 9-2. There must be some value of ρ_{ma} corresponding to the observed combination of ρ and ϕ_N. It is designated $(\rho_{ma})_a$. In a similar manner, $(\Delta t_{ma})_a$ is obtained from a plot of Δt versus ϕ_N. The resulting values of $(\rho_{ma})_a$ and $(\Delta t_{ma})_a$ are plotted on a mineral identification (MID) plot (Fig. 9-3). As in the case of the *M-N* plot, each mineral has a characteristic position. Again, if three minerals are expected, data will lie within a triangle unless shale, gas, or measurement errors displace points.

Clavier and Rust (1976) stress that minerals other than those used to construct the grid may well not have values of $(\rho_{ma})_a$ and $(\Delta t_{ma})_a$ equal

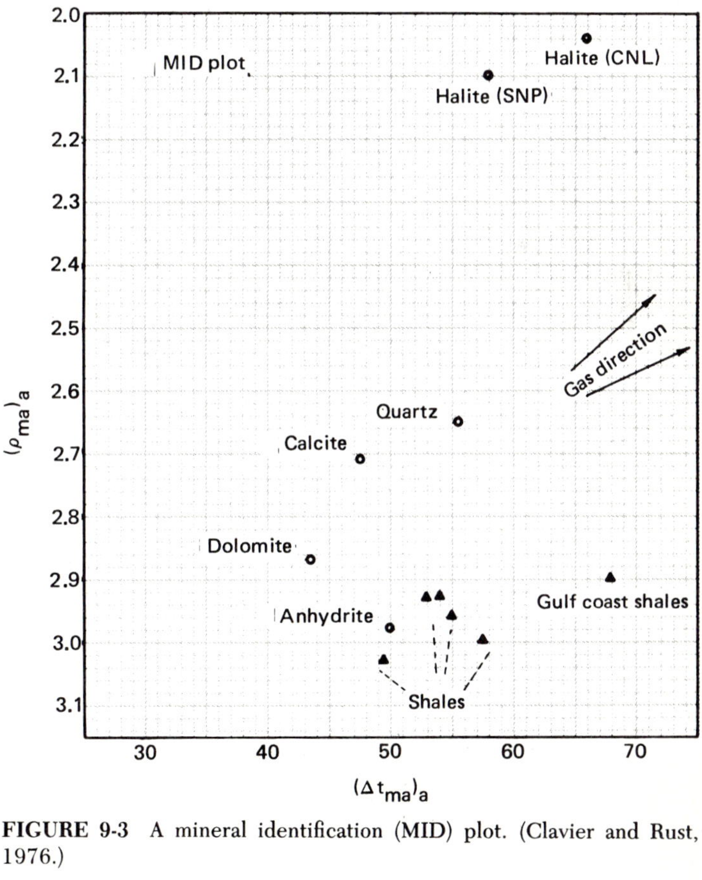

FIGURE 9-3 A mineral identification (MID) plot. (Clavier and Rust, 1976.)

to their true values of ρ_{ma} and Δt_{ma}. Indeed, the assumptions of extrapolability on which the MID plot technique rests are even less rigorous than those for the *M-N* plot.

The Litho-Density log. In a similar method, not yet in common practice, a selective γ-γ log is used (Sec. 6-4h). A quantity *U*, called the effective photoelectric cross section per unit volume, is defined as $P_e\rho_e$, where ρ_e is the electron density of the medium and $P_e = (Z/10)^{3.6}$ is an average photoelectric cross section per electron. An average value of P_e for a mixture of elements can be obtained by proper weighting of individual values (Bertozzi et al., 1981). Typical values are given in Table 6-1.

A value of $(U_{ma})_a$ is obtained in a manner similar to that for obtaining $(\rho_{ma})_a$ and is plotted against $(\rho_{ma})_a$ (Gardner and Dumanoir, 1980). *U*, of course, is really a representation of an effective atomic number, and it can therefore be used as a mineral identifier. Finally, a chart similar to Figs. 9-1 and 9-3 is used for matrix identification.

9-2 CLAY, SHALE, AND SHALY SANDS

The terms "shale" and "clay" seem sometimes to be used interchangeably. In fact, there are really many different clay minerals such as illite, montmorillonite, and kaolinite that, for our purposes, are of interest because they contain bound water and are conductive. Therefore, they give responses on some logs that can be confused with those of porous sedimentary rocks. Ruhovets and Fertl (1981) described the properties of clays and explained the nature of the high hydrogen content and low resistivity. Shale is a rock containing clays, water, very fine quartz (often referred to as silt) and perhaps other fine minerals as well. "Shaly sands" generally refers to sands or sandstone containing clay minerals and perhaps some silt. The differences between log responses of shaly sands and those of pure sandstone are caused by the clay minerals.

Shales per se are of little interest to most log users. Shale (or clay) introduced into a sandstone framework has been the object of much of modern log analysis because many methods of porosity determination require knowledge of the volume fraction of clay:

$$f_{cl} = \frac{\text{clay volume}}{\text{total rock volume}}$$

(See Chap. 10.) We shall describe some of the methods for determining f_{cl} in a later subsection. The mixtures of shale and sandstone have, for convenience, been divided into three types shown in Fig. 9-4 (Ransom, 1977).

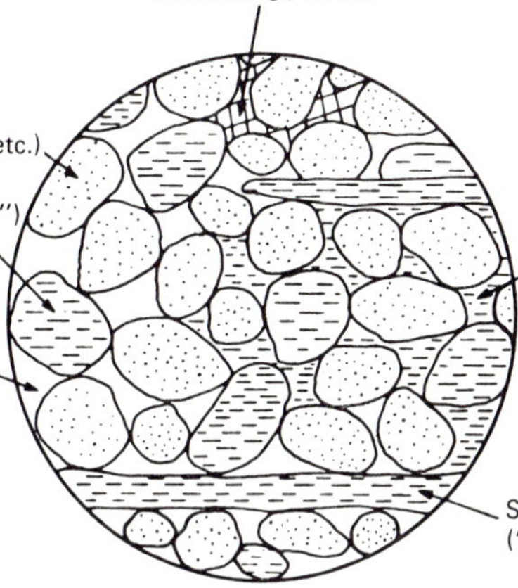

FIGURE 9-4 Idealization of the three ways in which shale can be mixed into a sandstone framework. (After Heslop, 1975.)

Dispersed clay. Clay minerals are deposited between the grains of the framework, and they fill some of the original pore spaces. Usually, clay fragments are found on grain surfaces. Dispersed clay can be either detrital (introduced into the voids as very fine grains) or diagenetic (formed after the framework was deposited). Some authors state that dispersed clay reduces porosity. They mean that, if there were no clay present, the total porosity would be that of the framework, but when dispersed clay is added, porosity is reduced below that which it would otherwise have been.

Structural clay. Clay particles are deposited as framework particles, or other particles are altered into clay by diagenetic processes. True porosity is not changed by the presence of structural clay.

Laminar clay. Clay is deposited in layers. It may be continuous, or it may occur as lenses. There is little or no porosity in the layers; that is, a layer has the same characteristics as a shale bed.

It is stated (Poupon and Gaymard, 1970) that

> It is generally assumed in the computations that, over limited intervals, the logging parameters of the clays remain constant. In formations containing many shale beds, a slow variation with depth of the well log response to clay is generally observed, reflecting the decrease in bound-water content with increasing depth. Over several hundred feet, the variation is usually negligible.

This idea is the basis for the common practice (Schlumberger, 1972) of assuming that shales in a sand framework have the same average properties as the shales in adjacent shale beds, although dispersed shales may have more bound water.

Almon (1979), however, finds that the properties of diagenetic (dispersed) clay minerals are not necessarily the same as those of associated shale beds. He sometimes finds radical differences, and he discusses the hazards of assuming that clays within sandstone have the properties of surrounding shales. Consequently, the user should be sure, before it is assumed that the shale of interest is the same as nearby shales, that anomalous diagenetic clays are not involved. This assurance is sometimes difficult to obtain.

Log responses. Clay minerals produce characteristic responses on several logs. These logs are often called "clay indicators" (Poupon and Gaymard, 1970), and they are commonly used to estimate the amount of clay present in a given sand-clay mixture. Quantitative data on the true values of log parameters versus clay content are sparse, because it is difficult to obtain reliable values of clay content by any method. Even core analysis is highly suspect (Heslop, 1975). Edmundson and Raymer (1979) have described the response of γ-ray, density, and neutron logs to many clay minerals and to three standard mixtures of clay minerals. There is a good deal of variation, but they seem to be the best data available. Table 9-1 lists responses of several logs to a number of different materials and to the mixtures described by Edmundson and Raymer. We shall discuss several *clay indicators* in turn.

Resistivity. Clay, because of the water bound to the surface of the grains, has a high *cation exchange capacity* (CEC) (Sec. 5-3), and it is therefore much more conductive than sand. Patchett (1975) discusses, at length, the effect of CEC on conductivity. He concludes that conductivity can be expressed as a function of CEC, bulk quartz concentration, confining pressure, and temperature, with CEC dominating. He finds a range of CEC of about an order of magnitude and a somewhat greater range of conductivity. Consequently, although it is a good indicator of clay, high conductivity is not very quantitative unless the CEC of the clay is known.

Self-potential. Clay minerals reduce self-potential (SP) deflection relative to that in a clay-free formation containing the same fluid (Sec. 5-5a,b). Define SP_{sh} as the reading of the SP log opposite shaly material (with borehole effects eliminated) and SP_{ss} as the SP reading opposite clean sand. Then (Poupon and Gaymard, 1970)

$$f_{cl} \leq 1 - \frac{SP_{sh}}{SP_{ss}} \tag{9-3}$$

Basically, the more clay, the more the SP reading is decreased compared with that opposite a clean formation with the same water content. Consequently, the estimate of f_{cl} is based wholly on the assumption that the SP values of the sand and shale do not change from formation to formation.

γ-Ray. Some clay minerals, in particular illite, contain potassium, and thorium tends to concentrate in a number of minerals including clays. Consequently, the γ-ray log is often used as an important clay indicator. In many cases, it is the only one used. The very simple formula for clay content is (Poupon and Gaymard, 1970)

$$f_{cl} = \frac{J - J_1}{J_2 - J_1} \tag{9-4}$$

where J is the γ-ray count rate in the formation of interest and J_1 and J_2 are the rates opposite clean sand intervals and clay beds, respectively. As usual, Eq. (9-4) assumes that the properties of clay are the same in the formation of interest as in the clay bed. Note, in Table 9-1, that some clay mixtures and shales have far greater γ-ray response than others. The γ-ray response depends on the specific clay minerals present. For example, montmorillonite often has little or no γ activity. Consequently, the γ log cannot be used to locate montmorillonite clay.

Kowalski and Asekun (1979) point out that while potassium and, generally, thorium are good clay indicators, uranium occurs in many minerals besides clay. Therefore, total γ-ray logs may give completely false indications of the presence of clay because they sense uranium from other sources. This is yet another situation in which the user should beware of the assumption that nearby shales can be used to calibrate a log in a formation of interest; uranium may be present in the formation in minerals other than shale. This problem can usually be eliminated by using a spectral γ-ray log and looking at the individual radioisotopes (Fig. 6-3).

Neutron and density. Clay contains bound water, which will give the same response on a neutron log as free water. But because the water is part of the crystal structure, the bulk density of the medium is not reduced as much as it would be if the water were free. In essence, the density is abnormally high for a medium with the amount of water indicated by the neutron log. Consequently, the combination of porosity readings from the neutron and density logs (usually displayed as a crossplot; Sec. 10-2) is inconsistent with the assumption that the water is free. This inconsistency is used as a clay indicator. Typical values of neutron and density log responses for clays are shown in Table 9-1.

If the response of the neutron log to clay ϕ_{Ncl} is known, then the ratio ϕ_{Ncl}/ϕ_N gives the maximum value of f_{cl}. The method of estimating the response of these logs to clay and porosity is discussed in more detail in Chap. 10.

Sonic. Because clay often has a lower sound speed than many other minerals, the sonic log can sometimes be used to estimate clay content. Table 9-1 lists sound speeds in several minerals. There is a wide variation in the values for clay.

Combinations. Fertl and Frost (1980) give expressions for f_{cl} in terms of readings of density and neutron, density and sonic, and neutron and sonic logs. They also discuss the conditions under which each individual log and each combination is most effective in determining f_{cl}. The use of the expressions, however, requires the knowledge of the response of each individual log to the clay (or shale) in question: ρ_{cl}, ϕ_{Ncl}, and Δt_{cl}. The expressions are then simple mixing laws for clay and matrix.

The wide variation in log responses to clay discussed above and shown in Table 9-1 show how very difficult it is to obtain accurate values of clay content from logs and known properties of clay. It is because of this variation that it is usually assumed that the clay in the formation of interest is the same as that in adjacent shales. The assumption may not be good, but the alternatives are often worse. All the user can do, quite often, is to be properly suspicious of the results.

9-3 COAL

Coal is a unique rock. It is a low-density hydrogenous medium with high carbon content. Coal is assigned several ranks depending on the degree of solidification from the original peat bog: anthracite, bituminous, subbituminous, and lignite. Low-rank coal such as lignite can have a volume fraction of water as large as 0.6 (Brom and Driedonks, 1981). Anthracite has a bulk density of the order of 1.5 g/cm^3, lignite of the order of 1.1 g/cm^3 or even as low as 0.7 g/cm^3 (Tixier and Alger, 1970), and values for bituminous coals fall in between. Coal often contains impurities, such as silica, alumina, or iron oxide. When the coal is burned, these impurities are left beind as "ash," and so the quantity of impurity present is usually referred to as the ash content. Logs are used (Peeters and Kempton, 1977; Liu and Harvey, 1982) both to identify and rank coal and to determine the ash content. Logs are also used (Chap. 15) to determine impurity content. Table 9-1 shows some log responses to different ranks of coal.

Resistivity. High resistivity has often been used as a coal identifier. Bituminous coals do indeed have high resistivity (Tixier and Alger, 1970), but lignite and anthracite can have quite low resistivity (Reeves, 1981). In fact, in the case of lignite, resistivity can decrease from 10^4 to 12 $\Omega \cdot m$ when moisture content changes from 0.1 to 0.6, and it can be used as a measure of moisture content (Brom and Driedonks, 1981). Consequently, resistivity should be used with caution.

γ-Ray. Most coals contain little or no potassium or thorium, and their γ-ray logs generally read very low, even lower than the log for "clean" sandstone. There are coals, however, that contain some uranium and will give anomalously high γ-ray readings. Consequently, although a very low γ-ray reading is a good coal indicator and an excellent way of distinguishing coal from interbedded shales, a higher reading does not necessarily indicate the absence of coal.

Neutron. Because coal is hydrogenous, even when it contains no water, it gives a very high I_H on a neutron log. This is interpreted, in most logs, as high porosity. Ordinarily, neutron logs are not very sensitive to changes in hydrogen index when the index is greater than about 0.4 to 0.5 (Fig. 6-27). Since the hydrogen index of coal is often higher than that, and it is desirable to distinguish among different values, a special short-spaced neutron log with a source-detector distance of about 25 cm has been built for coal evaluation (Brom and Driedonks, 1981). The hydrogen index can sometimes be correlated with the amount of *volatile matter* content, an index of coal quality. Bound water in shale gives similarly high values of I_H, but shale can usually be distinguished by its higher γ-ray reading.

Density. The best diagnostic of coal is its very low bulk density. Even with 50 percent ash, the bulk density of anthracite (the densest coal) will be less than 2.0 g/cm^3 (Reeves, 1981). It might be possible to confuse a coal having a great deal of ash with an extremely porous nonradioactive rock such as sandstone, limestone, or dolomite, but otherwise the density response for coal is unique. If the density of the impurity causing the ash, the rank of the coal, and therefore the bulk density of the pure coal are known and no moisture is present, the density log can be used to infer the ash content simply by (Liu and Harvey, 1982)

$$\text{Ash content} = \frac{\rho - \rho_{coal}}{\rho_{ash} - \rho_{coal}} \tag{9-5}$$

Sonic. Sonic velocity is very low in coal, and it can also be used to distinguish coal from other rocks. It may have the potential (Reeves, 1981) for use as a rank indicator.

9-4 EVAPORITES

Evaporite minerals such as halite (NaCl), anhydrite ($CaSO_4$), and gypsum ($CaSO_4 \cdot 2H_2O$) are characteristically very resistive and nonporous. Log responses are shown in Table 9-1. Responses are very different from those for coal, shale, or the common sedimentary rocks. It is desirable not only to locate evaporites but also to distinguish among them.

Resistivity. Resistivity is extremely high; it is a good identifier for evaporites in the absence of coal.

γ-Ray. Some evaporites, for example, sylvite and carnallite, contain primarily potash (potassium-bearing) minerals. These, of course, are quite radioactive. Their high radioactivity, coupled with high resistivity, is a good means of identification. Halite, trona, anhydrite, and gypsum are completely nonradioactive, and the very low γ-ray reading is another means of identifying them.

Sonic. Sound speeds in the common evaporites are not markedly different from those in porous sedimentary rocks.

Neutron and density. Halite and anhydrite contain no hydrogen, so the epithermal neutron log indicates zero porosity. The bulk density for each mineral is shown in Table 9-1. The density of anhydrite is particularly high, and it serves to distinguish anhydrite from halite. Gypsum contains large amounts of water of hydration, and the neutron log indicates high porosity. But the bulk density of gypsum is too high to be consistent with the high porosity; consequently, gypsum can be identified, in a manner similar to that for shale, by cross-plotting the responses of density and neutron logs (Sec. 10-2). Trona has a lower density and hydrogen index than gypsum, and it can be identified in a similar manner. The absence of a γ-ray response distinguishes gypsum and trona from many shales.

Potash minerals are often sought, but sylvite (KCl) is by far the most valuable. All can be located by radioactivity and then distinguished by neutron and density response. Sylvite and langbeinite are not hydrogenous, and they can be identified by low hydrogen index. Langbeinite is much denser than sylvite, so it can be distinguished with a density log. Carnallite and kainite have a high hydrogen index, which serves to distinguish them from sylvite. Polyhalite has a higher density and hydrogen index than sylvite, and it can be identified by the combination.

Often, however, evaporites occur in mixtures, and the simple analysis will not work. Tixier and Alger (1970) and Daniels et al. (1979) have used the individual responses to evaluate real data in complex evaporite sequences. They solve simultaneous equations using γ-ray, neutron, density, and sonic responses of the different evaporites to obtain fractions of each one. A general version of the method is discussed in Sec. 10-3. It is not as easy as it would seem.

REFERENCES

Almon, W. R.: "A Geologic Appreciation of Shaly Sands," *Proc. Soc. Prof. Well Log Analysts 20th Logging Symp.*, 1979, paper WW.

Bertozzi, W., D. V. Ellis, and J. S. Wahl: "The Physical Foundation of Formation Lithology Logging with Gamma Rays," Geophysics, vol. 46, 1981, p. 1439.

Brom, R. W. C., and F. Driedonks: "Applications of Petrophysical Logging in the Evaluation of Coal Deposits," *Proc. Soc. Prof. Well Log Analysts 22d Logging Symp.*, 1981, paper KK.

Burke, J. A., A. W. Schmidt, and R. L. Campbell: "The Litho-Porosity Crossplot," *Log Analyst*, vol. 10, November–December 1969, p. 25.

Clavier, C., and D. H. Rust: "MID Plot: A New Lithology Technique," *Log Analyst*, vol. 17, November–December 1976, p. 16.

Daniels, J. J., J. H. Scott, and R. J. Hite: "Analysis of Borehole Geophysical Data in an Evaporite Sequence at Salt Valley, Utah," *Proc. Soc. Prof. Well Log Analysts 20th Logging Symp.*, 1979, paper M.

Dresser Atlas: *Log Interpretation,* Dresser Atlas, Houston, Tex., 1977.

Edmundson, H., and L. L. Raymer: "Radioactive Logging Parameters for Common Minerals," *Log Analyst*, vol. 20, no. 5, 1979, p. 38.

Fertl, W., and E. Frost: "Evaluation of Shaly Clastic Reservoir Rocks," *J. Petr. Tech.*, September 1980, p. 1641.

Gardner, J. S., and J. L. Dumanoir: "Litho-density Log Interpretation," *Proc. Soc. Prof. Well Log Analysts 21st Logging Symp.*, 1980, paper N.

Heslop, A.: "Porosity in Shaly-Sands," *Proc. Soc. Prof. Well Log Analysts 16th Logging Symp.*, 1975, paper F.

Kowalski, J. J., and S. O. Asekun: "It May Not Be Shale," *Proc. Soc. Prof. Well Log Analysts 20th Logging Symp.*, 1979, paper P.

Liu, Jiajin, and A. Herbert Harvey: "Coal Logging in the People's Republic of China," *Log Analyst*, vol. 23, no. 2, 1982, p. 3.

Patchett, J.: "An Investigation of Shale Conductivity," *Log Analyst,* vol. 16, no. 6, 1975, p. 3.

Peeters, M., and N. H. Kempton: "Wireline Logging for Coal Exploration in Australia," *Log Analyst,* vol. 18, no. 3, 1977, p. 24.

Poupon, A., and R. Gaymard: "The Evaluation of Clay Content from Logs," *Proc. Soc. Prof. Well Log Analysts 11th Logging Symp.*, 1970, paper G.

Ransom, R. C.: "Methods Based on Density and Neutron Well-Logging Responses to Distinguish Characteristics of Shaly Sandstone Reservoir Rock," *Log Analyst,* vol. 18, no. 3, 1977, p. 47.

Reeves, D. R.: *Coal Interpretation Manual,* BPB Instruments, East Leake, England, 1981.

Ruhovets, N., and W. Fertl: *"Digital Shaly Sand Analysis Based on Waxman-Smits Model and Log-Derived Clay Typing,"* Dresser Atlas, Houston, Tex., 1981.

Schlumberger: *Log Interpretation: Principles,* Schlumberger, Houston, Tex., 1972.

————: *Log Interpretation: Applications,* Schlumberger, Houston, Tex., 1974.

Tixier, M. P., and R. P. Alger: "Log Evaluation in Nonmetallic Mineral Deposits," *Geophysics,* vol. 35, 1970, p. 124.

10

Porosity

Porosity, ϕ, defined as the ratio of the volume of the pores to the total volume, can take several forms: intergranular porosity (sometimes called primary porosity), the void space between grains; intragranular porosity, the void space within grains; fracture porosity; and vesicular (or vugular) porosity.

Intergranular porosity depends on the shape and sorting of the grains. The more uniform the grain size, and the more spherical the grains, the higher the porosity. McGeary (1961) showed experimentally that a packing of spheres of any constant diameter between 3 and 0.03 mm resulted in a porosity between 0.38 and 0.41, although theoretical calculations indicate that, with a rhombohedral packing, the porosity could be as low as 0.26. His results indicate that, for a given particle shape, porosity is independent of particle size. If two sizes of spheres, with a diameter ratio of at least 7, are used, porosity can be as low as 0.15; and if three sizes are used, porosity can decrease to 0.10. Beard and Weyl (1973) experimented with sand and found that well-sorted sand gave porosity as high as 0.43 and poorly sorted sand had a porosity as low as 0.25.

Fracture widths can vary from microscopic to several centimeters, and fracture porosity also can vary widely. If the fractures are large, the *scale* of the porosity, compared to the size of the sonde, is significant. One 5-cm-wide fracture per meter of depth will yield a total porosity of 0.05, but on the scale of a small logging sonde, the observed porosity could be as large as 0.5!

Vesicles, caused either by gas expanding as a volcanic rock cools or by solution in carbonate rocks, can have even larger size variations; think of a cavern in limestone. Again, scale will be important.

Cementation, shale deposition, diagenesis, and compression are among the causes that can reduce either intergranular or fracture porosity. Pressure can close fractures and reduce fracture porosity. Closed fractures are very common.

The range of porosity in earth media varies from as much as 0.80 in unconsolidated sediments on the ocean floor to less than 0.001 in some limestones and in igneous and metamorphic rocks. We have often found porosities up to 0.60 in desert alluvium and basalts. Schopper (1982) lists values between 0.004 and 0.53 in sandstones and between 0.001 and 0.365 in limestones.

There is no log that measures porosity, but log analysts use many different methods to calculate porosity from log data. If the assumptions on which these methods are based are satisfied, the results are often correct. But the correctness of the assumptions, and the results, can usually be verified only from core measurements of the porosity or of other medium properties, not from logs. There is evidence that, even with great care, log and core porosity measurements can disagree by large amounts (Mathews and LaDelfe, 1981). Furthermore, core measurements, if made on low-porosity rocks or on samples containing intergranular clays, are subject to considerable error (American Petroleum Institute, 1960).

Porosity is calculated from log data in three general ways: from single logs, from solutions of equations written in terms of variables whose values are obtained from different logs, and from statistical analysis of data from several logs. With one log we get only porosity; with more logs it becomes possible (at least conceptually) to solve for mineral fractions as well.

10-1 POROSITY FROM SINGLE LOGS

Porosity is sometimes obtained from resistivity logs, but ordinarily the three "porosity logs"—neutron, density, and acoustic—are used either singly or, more often, in combination. For each type of log there are analytical expressions for porosity that embody certain assumptions about the medium and different types of corrections to deal with some of the possible violations of the assumptions. Schlumberger (1972) gives a good description of these expressions.

Porosity from resistivity. The Archie equations for porosity (Sec. 5-2) are

$$F = \frac{R_o}{R_w} = \phi^{-m} \tag{5-54}$$

and

$$F = a\phi^{-m} \tag{5-56}$$

but this method is applicable only when the rock is completely water-saturated, when R_w is known, and when mineral and surface conduction contributions are negligible. These conditions are often satisfied within 5 to 7 cm of the borehole wall, where the formation fluid is entirely displaced by mud filtrate (the *flushed zone*). If there is no residual oil, the formation factor of the flushed zone will be equal to R_{xo}/R_{mf}, where R_{xo} is the measured resistivity of the flushed zone and R_{mf} is the resistivity of the mud filtrate. R_{xo} is obtained from a very short-spaced pad-type resistivity sonde. Then

$$\phi = \left(\frac{aR_{mf}}{R_{xo}}\right)^{1/m} \tag{10-1}$$

Common values of the constants are 0.62 for a and 2.15 for m (Winsauer et al., 1952).

Porosity from acoustic log. The time average, or Wyllie, equation (7-33) relates acoustic velocity in bulk material v_b to velocity in the matrix v_{ma} and the pore fluid v_f and the fractional porosity ϕ by:

$$\frac{1}{v_b} = \frac{\phi}{v_f} + \frac{1-\phi}{v_{ma}} \tag{7-28}$$

This equation applies only under certain conditions (Sec. 7-2a). In particular, it applies only to intergranular porosity, not to vesicular or fracture porosity, and only when the fluid is a liquid. It is assumed that the pressure is high enough that the velocity is not a function of pressure. When Eq. (7-28) is used in practice, the reciprocal velocity, or travel time Δt (microseconds per foot) is used instead of the velocity v. Then Eq. (7-28) becomes

$$\phi_S = \frac{\Delta t - \Delta t_{ma}}{\Delta t_f - \Delta t_{ma}} \tag{10-2}$$

for the *sonic porosity* ϕ_S.

Uncertainties in measured ϕ can arise from uncertainties in Δt, Δt_{ma}, and Δt_f.

$$\delta\phi = \frac{1}{\Delta t_f - \Delta t_{ma}} \delta\Delta t \simeq 0.007\ \delta\Delta t \tag{10-3a}$$

$$\delta\phi = \frac{\Delta t - \Delta t_f}{(\Delta t_f - \Delta t_{ma})^2} \delta\Delta t_{ma} \simeq -0.005\ \delta\Delta t_{ma} \tag{10-3b}$$

$$\delta\phi = \frac{\Delta t - \Delta t_{ma}}{(\Delta t_f - \Delta t_{ma})^2} \delta\Delta t_f \simeq 0.001\ \delta\Delta t_f \tag{10-3c}$$

where nominal values of 50, 75, and 200 μs/ft have been used for Δt_{ma}, Δt, and Δt_f, respectively.

Data on the range of acoustic velocities in pure minerals are scarce, but Δt_{ma} can vary at least as much as 4.3 μs/ft in sandstone or limestone (Gregory, 1977), and that alone would give rise to a porosity uncertainty of 0.0215. In addition, the matrix is rarely made up of a single pure mineral; consequently, Δt_{ma} will probably not be known as well as that for a pure mineral and the uncertainty in Δt_{ma} will likely be larger.

Shale also affects acoustic velocity, and different forms of shale affect velocity in different ways. Structural or laminar *shale* acts like solid material, but with a value of Δt_{ma} different from that for the rest of the matrix. The fraction of shale by volume f_{sh} is generally obtained by comparing "shale logs" (Sec. 9-2), for example, γ-ray or SP, in shales adjacent to those in the formation of interest and assuming that a fraction of shale equal to f_{sh} will result in a deflection on the log in the formation of interest equal to f_{sh} times the deflection in pure shale.

Then the apparent change in porosity caused by laminar and structural shale $\delta\phi_{shl}$ is obtained from

$$\delta\phi_{shl} = f_{shl} \frac{\Delta t_{shl} - \Delta t_{ma}}{\Delta t_f - \Delta t_{ma}} = f_{shl}\, \Delta\phi_{shl} \tag{10-4}$$

Equation (10-4) is based on the assumption that the shales occur in layers whose thickness compared to that of nonshales is the fraction f_{sh}. The value of Δt_{shl} is obtained from the Δt in adjacent shales, assuming that the velocity in the adjacent shales is the same as that in the shales of interest.

If the shale is dispersed, experiments show (Alger et al., 1963) that the sound speed in a water-shale slurry is close to that in water alone, so $\Delta t_{shd} = \Delta t_f$. Then

$$\delta\phi_{shd} = f_{shd}$$

Equation (10-2) is based on the assumption of compacted rock. If the medium is not fully compacted, Δt will be greater than that predicted by Eq. (7-28), and another correction for noncompaction must be applied. It is usually written as

$$\phi_S = C_p(\phi + \delta\phi_{shd}) \tag{10-5}$$

where C_p is always greater than 1.0. C_p is obtained by comparing ϕ from Eq. (10-2) with that measured by some other porosity log.

None of these corrections take *secondary porosity* into account. Secondary porosity is porosity in vugs or fractures that sometimes does not affect acoustic velocity at all. Equation (10-2) does not apply, and secondary porosity ϕ_2 must be obtained from other porosity logs.

When all of these corrections are put together, the porosity obtained from the acoustic (or sonic) log, in terms of the true total porosity ϕ, is [assuming Eq. (7-28) is satisfied in the competent rock]

$$\phi_S = C_p[(\phi - \phi_2) + f_{shd}] + f_{shl}\,\Delta\phi_{shl} \tag{10-6}$$

For a well-compacted clean sandstone the four correction terms in Eq. (10-6) are absent and $\phi = \phi_S$, as given by Eq. (10-2). When shale is present, the uncertainties in f_{shl} and f_{shd}, as well as the uncertainties in ϕ described by Eq. (10-3), will contribute to the total uncertainty. The total uncertainty in a sum of terms like those in Eq. (10-6) is equal to the square root of the sum of the squares of the individual uncertainties.

Porosity from density logs. The basic equation for calculating porosity from measured bulk density ρ is

$$\rho = \rho_{ma}(1 - \phi) + \rho_f\phi \tag{10-7}$$

Typical values of matrix density ρ_{ma} are given in Table 6-1. ρ_f is the density of the mud filtrate or pore fluid. From Eq. (10-7)

$$\phi_D = \frac{\rho_{ma} - \rho}{\rho_{ma} - \rho_f} \tag{10-8}$$

for the density porosity.

If the uncertainty in measured bulk density (Sec. 6-4f) is $\delta\rho$, then

$$\delta\phi = -\frac{1}{\rho_{ma} - \rho_f}\,\delta\rho \simeq -0.6\,\delta\rho \tag{10-9a}$$

The uncertainty in porosity caused by uncertainty in ρ_{ma} is

$$\delta\phi = \frac{\rho - \rho_f}{(\rho_{ma} - \rho_f)^2}\,\delta\rho_{ma} \simeq 0.4\,\delta\rho_{ma} \tag{10-9b}$$

where nominal values of 1.0, 2.0, and 2.6 g/cm^3 have been used for ρ_f, ρ, and ρ_{ma}, respectively.

Patchett and Coalson (1979) discuss quality control of density logs and state that very accurate values of ϕ can be obtained when ρ_{ma} is well known. ρ_{ma}, unlike Δt_{ma}, can be measured from cores or cuttings samples. Although both measurements and sampling are subject to error (American Petroleum Institute, 1960), the result is better than the assumption of a nominal value. But if a nominal value is used for ρ_{ma}, the uncertainty $\delta\phi$ [Eq. (10-9*b*)] can be nontrivial. If a value of 2.65 g/cm^3 were assumed, uncertainties as large as 0.035 in ϕ could result. Published mineral densities for quartz vary between 2.59 and 2.70 g/cm^3; those for calcite, between 2.70 and 2.82 g/cm^3; and those for dolomite, between 2.80 and 2.99 g/cm^3 (Wohlenberg, 1982). Also, Granberry et al. (1968) found ρ_{ma} of sandstone cores from the southeastern United States to vary between 2.58 and 2.70 g/cm^3, so uncertainties in ρ_{ma} of ± 0.05 (and therefore uncertainties in ϕ of ± 0.02) will not be uncommon. In a formation of varying mineralogy, such as alluvium, the situation will be much worse.

Shales introduce error only if their densities, usually lower than the density of crystalline rock, are not accounted for in ρ_{ma}. Then the apparent porosity caused by shale is

$$\delta\phi_{sh} = f_{sh}\frac{\rho_{ma} - \rho_{sh}}{\rho_{ma} - \rho_f} = f_{sh}\,\Delta\phi_{Dsh} \tag{10-10}$$

where ρ_{sh} is the shale bulk density obtained from a density log in adjacent shales, with the assumption that it is the same as in the formation of interest, and ρ_{ma} is the matrix density of that part of the formation which is not shale. Also,

$$\delta\phi = \frac{\rho_{ma} - \rho}{(\rho_{ma} - \rho_f)^2}\,\delta\rho_f \simeq 0.25\,\delta\rho_f \tag{10-9c}$$

Errors caused by wrong values of ρ_f can be very large if there is partial or total gas or air saturation and liquid saturation is assumed.

Hydrocarbons also affect the measured value of ϕ because ρ_f for hydrocarbons is not the same as ρ_f for the borehole fluid. Not only is the bulk density different, but the Z/A ratio [Eq. (6-32)] is not the same for water and hydrocarbon; and therefore the electron density is different even if the bulk density is the same. The density difference is, of course, largest if the fluid is gas. Figure 10-1 shows the true density and the observed response of a compensated density sonde for natural gas as a function of temperature and pressure (Schlumberger, 1972). The electron density for the gas is $\rho_e = 1.24\rho$, but the sonde is so calibrated that the observed density

$$\rho_o = 1.0704\rho_e - 0.1883$$

which gives a correct value of bulk density for a water-saturated limestone. Consequently, when ρ_e becomes small, the observed density can go negative. If the response of the density log to hydrocarbon is ρ_{Dh} and the hydrocarbon saturation is S_h, the change in apparent porosity caused by the hydrocarbon, following Eq. (10-9*c*), with

$$\delta\rho_f = S_h(\rho_{Dh} - \rho_w)$$

is

$$\delta\phi_h = S_h \frac{(\rho_{ma} - \rho)(\rho_{Dh} - \rho_w)}{(\rho_{ma} - \rho_w)^2} = \phi S_h \phi_{Dh} \tag{10-11}$$

We can now write an expression like Eq. (10-6) for the porosity as seen by the density log:

$$\phi_D = \phi S_w + \phi\phi_{Dh}(1 - S_w) + f_{sh}\,\Delta\phi_{Dsh} \tag{10-12}$$

The total uncertainty is again the square root of the sum of the squares of the uncertainties in the individual terms.

It is also possible, if both the density of the mud filtrate and the formation fluid are known, to obtain the porosity from the difference between density from a density log and that from a borehole gravimeter (Chap. 8).

Porosity from neutron logs. The neutron log, if all corrections have been properly made, measures the hydrogen index I_H, which is defined as the volume fraction of fresh water that would contain the observed amount of hydrogen (Sec. 6-6b). In practice, the neutron log is calibrated in porosity units, the assumption being that all hydrogen present is in fact in fresh water.

Several corrections are made before porosity is calculated (Sec. 6-6e). Among them are corrections for salinity of the borehole fluid and the response of the neutron log to the matrix material assumed to be present. The neutron log also responds to bulk density, and a correction for this (called the *excavation effect*) is made after the log has been recorded but before some of the other corrections are made. Since the porosity measured by the neutron log ϕ_N is ordinarily equal to I_H,

$$\delta\phi_N = \delta I_H$$

Of course, the neutron log cannot distinguish between bound water and water in pores. Shale contains a good deal of bound water (Sec. 9-2). The error in porosity due to shale is

$$\delta\phi_{sh} = f_{sh}\phi_{Nsh} \tag{10-13}$$

where ϕ_{Nsh} is the neutron log response to the specific shale present in the formation of interest. It is often determined from the neutron log response to adjacent shales, with the assumption that all shales in the same region have the same neutron log response per unit volume.

The I_H of some oils is the same as that of water; the I_H of others is not. There are several rules of thumb in the service company books that let one choose a value of I_H. Figure 10-1 shows how the I_H of a gas varies with temperature and pressure. Usually the response of the neutron log to hydrocarbon ϕ_{Nh} is taken as equal to the appropriate I_H.

The neutron log is usually calibrated for a specific matrix, generally limestone, sandstone, or dolomite, and for a specific fluid, often fresh water. The calibration algorithm is so designed that matrix response is equal to zero for the rock in which the log is calibrated, and the fluid response ϕ_{Nf} = 1.0. Real matrices are often mixtures. The response of

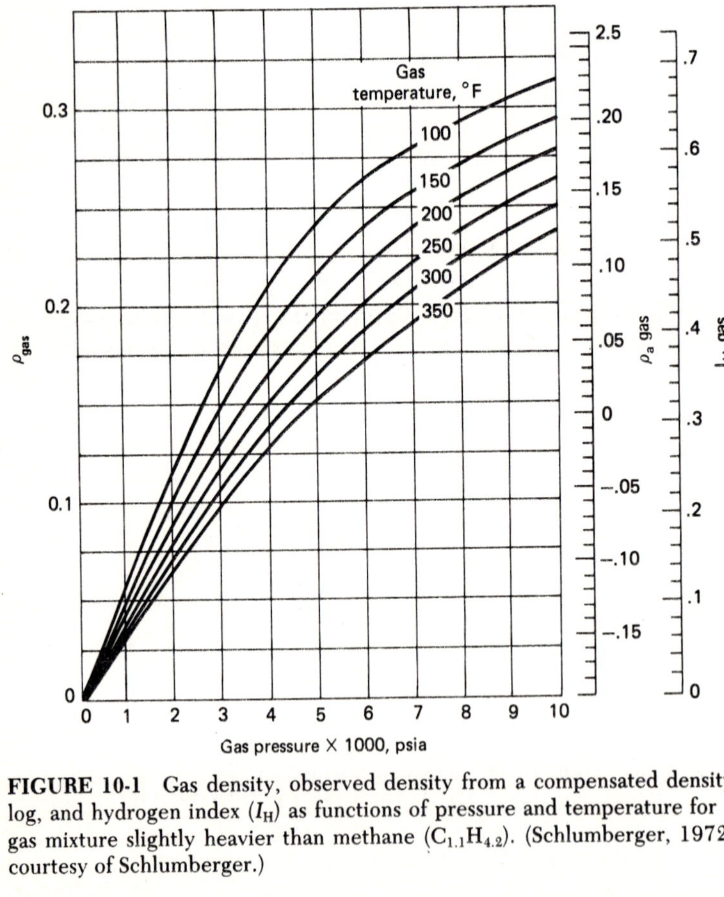

FIGURE 10-1 Gas density, observed density from a compensated density log, and hydrogen index (I_H) as functions of pressure and temperature for a gas mixture slightly heavier than methane ($C_{1.1}H_{4.2}$). (Schlumberger, 1972; courtesy of Schlumberger.)

a given matrix to a neutron log calibrated for a different matrix is ϕ_{Nma} (Fig. 6-27). For example, a neutron log run in sandstone of porosity 0.14 would register 0.10 on the limestone scale, in which case ϕ_{Nma} would be -0.04. (Unlike ρ_{ma}, ϕ_{Nma} cannot be measured on small samples.) And since not all borehole water is fresh, the response of the (corrected) neutron log to the water is ϕ_{Nw}.

When all of these effects are summed,

$$\phi_N = \phi S_w \phi_{Nw} + \phi S_h \phi_{Nh} + f_{sh}\phi_{Nsh} + (1 - f_{sh} - \phi)\phi_{Nma} \qquad (10\text{-}14a)$$

which simplifies to the expression

$$\phi_N = \phi_{Nw}\phi + \phi_{Nma}(1 - \phi) \qquad (10\text{-}14b)$$

for a log run in a water-saturated, shale-free rock other than that for which the sonde has been calibrated.

In summary, we have porosities from acoustic (sonic), density, and neutron logs. Each measured porosity involves uncertainties caused by assumptions about matrix and fluid properties. Each is affected differently by shale and hydrocarbons whose amount and properties must be determined somehow. Because of these effects and uncertainties, it is often desirable to find a way to combine some form of Eqs. (10-6), (10-12), and (10-14) to obtain more accurate values of porosiy.

10-2 POROSITY FROM TWO LOGS

Overlays. One common log interpretation procedure is to superpose density and neutron logs, both displayed in terms of porosity and both on the same scale. Figure 10-2 shows an idealization of such an overlay. If the true porosity is constant and gas is present, ϕ_D will be greater than ϕ and ϕ_N will be less than ϕ. The logs will appear to be mirror images. If shale is present (and ρ_{ma} is assumed to be that of the non-shale matrix), ϕ_D will be slightly less than ϕ and ϕ_N will be greater than ϕ. The logs will be mirror images, but they will have different porosity scales. The size of the effect will, of course, depend on ϕ. The effect will be greater for a compensated neutron log than for an epithermal neutron log (Truman et al., 1972). This overlay technique gives a quick indication of the presence of gas or shale. It is, however, not very quantitative.

Two-log cross-plots. Fertl (1981) gives a good overview of the cross-plot technique. The basic procedure is to plot log data for a given depth interval in a space defined by two log parameters. That is, one plots all pairs of observed values of ϕ_N versus Δt, ϕ_N versus ρ or ϕ_D, or ρ versus

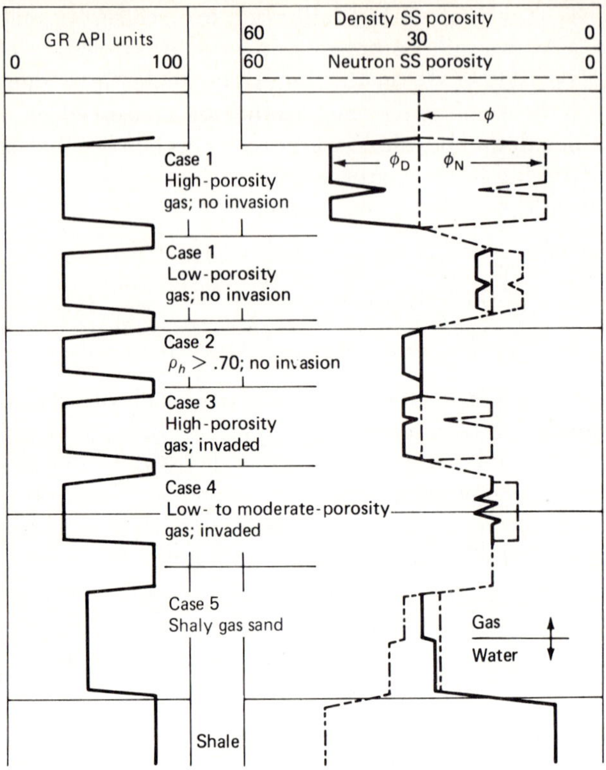

FIGURE 10-2 Schematic overlay of neutron (CNL) and density logs that shows the effect of gas and shale. (Truman et al., 1972.)

Δt. Often, at each point in the space corresponding to a measured pair of values, one plots a digit representing the number of times that pairs of values occur in the interval. Sometimes the value of a third variable, such as the γ-ray log, that is measured at the same depth as the pair of data being cross-plotted is plotted or is indicated by a color at the point in space. (This is called a *z* plot.) Figure 10-3 is a typical cross-plot.

For any pure matrix and specific pore fluid there is a single line in space that corresponds to all possible values of true porosity. If other minerals or pore fluids are present, the data will lie off the line. Cross-plots are constructed to treat cases where two or more minerals are present; lines corresponding to each mineral are plotted. Figure 10-4 shows the three cross-plots for sonic, density, and neutron data.

The sonic-density cross-plots in Fig. 10-4*a* can be constructed by

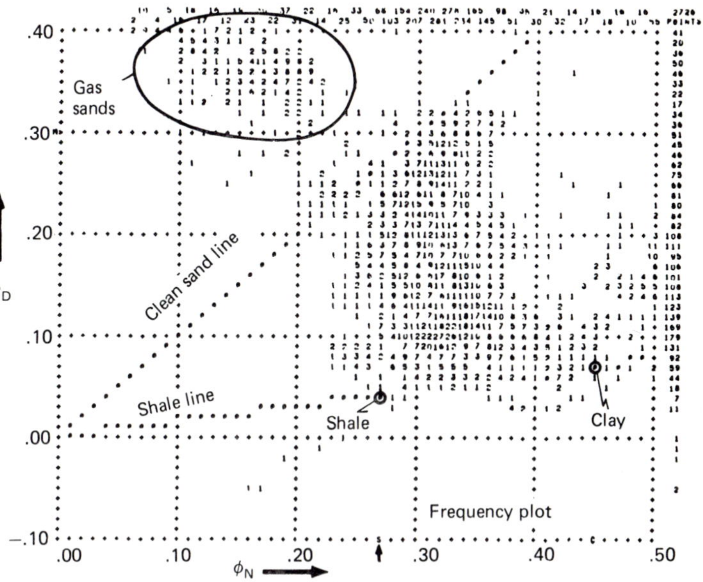

FIGURE 10-3 A typical density-neutron cross-plot of real data. (Schlumberger, 1972; courtesy of Schlumberger.)

equating Eqs. (10-2) and (10-8):

$$\frac{\Delta t - \Delta t_{ma}}{\Delta t_f - \Delta t_{ma}} = \frac{\rho_{ma} - \rho}{\rho_{ma} - \rho_f} \qquad (10\text{-}15)$$

Solving for the observed density yields the linear relation

$$\rho = \rho_{ma} - \frac{\Delta t - \Delta t_{ma}}{\Delta t_f - \Delta t_{ma}} (\rho_{ma} - \rho_f) \qquad (10\text{-}16)$$

Using the values of Δt_{ma} and ρ_{ma} appropriate for limestone, sandstone, or dolomite yields the corresponding straight lines in Fig. 10-4*a*.

Equations for neutron-sonic and neutron-density cross-plots can be generated directly from Eqs. (10-2) and (10-8) for the limestone scales in Fig. 10-4*b* and 10-4*c*, because the log gives the neutron porosity directly; $\phi = \phi_N$. Scales for sandstone and dolomite cannot be obtained directly, however, because the apparent porosities are functions of sonde design and are experimentally determined:

$$\phi_N = \phi_{Nw} + (1 - \phi)\phi_{Nma} \qquad (10\text{-}14b)$$

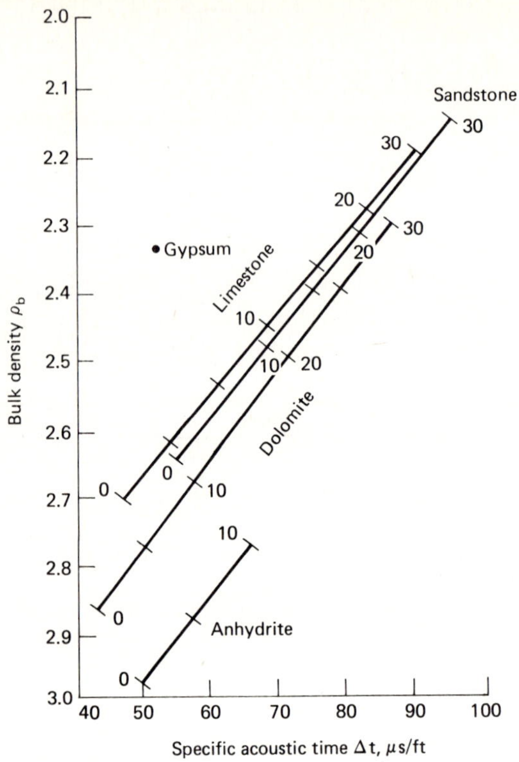

(a)

FIGURE 10-4 Standard cross-plot lines for (a) acoustic travel time and bulk density.(Fertl, 1981. Copyright, 1981, SPE–AIME.)

and ϕ_{Nma} is a function of porosity. The nonlinearity in the response to these minerals causes the curvature of the sandstone and dolomite lines in Fig. 10-4*b* and *c*.

Calculation of porosity. If no assumptions are violated and only two minerals are present, one can calculate not only ϕ but also the proportion of each mineral. If the neutron log is calibrated in terms of porosity in mineral m_1, ϕ_{N2} is defined as the apparent neutron porosity caused by the *matrix* of mineral m_2. If the fraction of m_2 is f_2, the measured values are

$$\rho = \rho_{m1}f_1 + \rho_{m2}f_2 + \rho_f\phi \tag{10-17a}$$

$$\Delta t = \Delta t_{m1}f_1 + \Delta t_{m2}f_2 + \Delta t_f\phi \tag{10-17b}$$

$$\phi_N = \phi_{N1}f_1 + \phi_{N2}f_2 + \phi_f\phi \tag{10-17c}$$

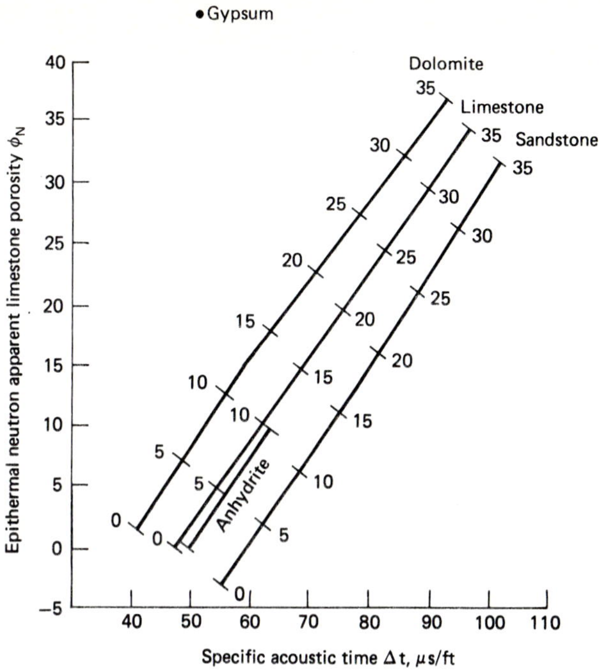

(b)

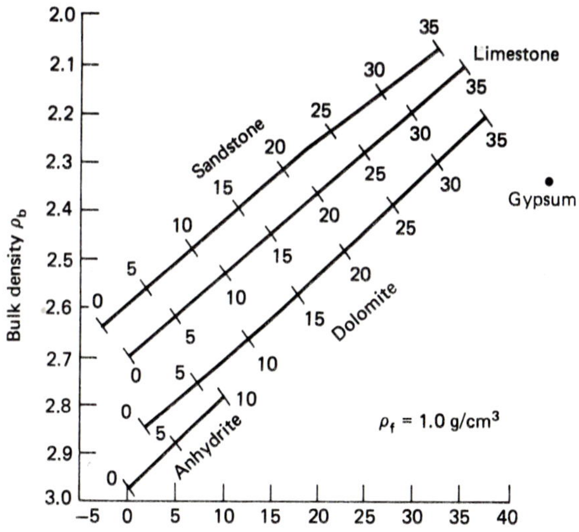

(c)

FIGURE 10-4 (*Continued*) Standard cross-plot lines for (*b*) Acoustic travel time and neutron porosity and (*c*) bulk density and neutron porosity. (Fertl, 1981. Copyright, 1981, SPE–AIME.)

But for matrix m_1, $\phi_{N1} = 0$ and $\phi_{Nf} = 1$, so

$$\phi_N = \phi_{N2} f_2 + \phi \qquad (10\text{-}17d)$$

Consequently, from density and neutron logs (Fertl, 1981),

$$\phi = \frac{(\rho - \rho_{m1})\phi_{N2} - \phi_N(\rho_{m2} - \rho_{m1})}{(\rho_f - \rho_{m1})\phi_{N2} - (\rho_{m2} - \rho_{m1})} \qquad (10\text{-}18a)$$

from acoustic and neutron logs

$$\phi = \frac{(\Delta t - \Delta t_{m1})\phi_{N2} - \phi_N(\Delta t_{m2} - \Delta t_{m1})}{(\Delta t_f - \Delta t_{m1})\phi_{N2} - (\Delta t_{m2} - \Delta t_{m1})} \qquad (10\text{-}18b)$$

and from density and acoustic logs

$$\phi = \frac{(\rho - \rho_{m1})(\Delta t_{m2} - \Delta t_{m1}) - (\rho_{m2} - \rho_{m1})(\Delta t - \Delta t_{m1})}{(\rho_f - \rho_{m1})(\Delta t_{m2} - \Delta t_{m1}) - (\rho_{m2} - \rho_{m1})(\Delta t_f - \Delta t_{m1})} \qquad (10\text{-}18c)$$

By reinserting ϕ into Eq. (10-17), the mineral fractions are found. The equivalent graphical solution is simply to plot the log values and interpolate along isoporosity lines drawn orthogonally to the matrix lines.

Calculation of uncertainty is complex; use of a statistics text is urged. The general form is

$$(\delta\phi)^2 = \sum \left(\frac{\partial \phi}{\partial x_i}\right)^2 (\delta x_i)^2 \qquad (10\text{-}19)$$

where δx_i is the uncertainty in an individual quantity. For example, from Eq.(10-18c),

$$(\delta\phi)^2 = \left(\frac{\partial\phi}{\partial\rho}\right)^2 (\delta\rho)^2 + \left(\frac{\partial\phi}{\partial\rho_{m1}}\right)^2 (\delta\rho_{m1})^2 + \left(\frac{\partial\phi}{\partial\rho_{m2}}\right)^2 (\delta\rho_{m2})^2$$
$$+ \left(\frac{\partial\phi}{\partial\Delta t}\right)^2 (\delta\Delta t)^2 + \left(\frac{\partial\phi}{\partial\Delta t_{m1}}\right)^2 (\delta\Delta t_{m1})^2 + \left(\frac{\partial\phi}{\partial\Delta t_{m2}}\right)^2 (\delta\Delta t_{m2})^2 \qquad (10\text{-}20)$$

10-3 USE OF CROSS-PLOTS TO SHOW VIOLATED ASSUMPTIONS

If the assumptions inherent in Eqs.(10-15) and (10-16) are violated, common practice is to "correct" the data. Cross-plots are used to show when assumptions are violated.

Density versus acoustic. Assume constant ϕ in a standard mineral (limestone, sandstone, or dolomite). For example, consider a limestone of water-saturated porosity equal to 0.10. Then in Fig. 10-4*a*, ρ versus Δt, light hydrocarbons, gypsum, and gas will give a lower density and shift the data up from the line. (The effect of these materials on velocity is not clear.) If some of the porosity is secondary porosity, velocity will be greater than if all of it were primary, so the data are shifted to the left (decreased Δt). Shales (replacing matrix) and poor compaction decrease density slightly and velocity substantially, so the data shift to the right. Borehole rugosity will decrease apparent density (if no correction has been made) and will shift data up.

Neutron versus acoustic. In Fig. 10-4*b*, ϕ_N versus Δt, shales replacing matrix increase both Δt and ϕ_N (shale contains bound water), and the data shift up and to the right. Gypsum increases ϕ_N and shifts the data upward. Poor compaction increases Δt but does not much change ϕ_N, so it shifts the data to the right. Gas decreases ϕ_N (since its I_H is much lower than that of water) and shifts the data down. Secondary porosity again shifts the data to the left.

Density versus neutron. In Fig. 10-4*c*, ρ versus ϕ_N (some users plot ϕ_D versus ϕ_N, but we prefer the more basic variable), the most commonly used cross-plot, shale (replacing matrix) and gypsum increase ϕ_N and decrease ρ slightly. Data shift to the right and slightly up. Gas decreases both ρ and ϕ_N; data shift up and to the left. If porosity is replaced by shale, there is an increase in ρ but not much of a decrease in ϕ_N. The *shale point* (actually a range, depending on the type of shale) denoting pure shale, is below and to the right of the lines.

Log analysts perform various manipulations of the kind described by Eqs. (10-6), (10-12), and (10-14) to correct the data and shift them back between the lines so that Eq. (10-18) can be used. Each logging company describes its method in its brochure. Schlumberger (1974) describes several methods.

10-4 CALCULATION OF POROSITY FROM THREE OR MORE LOGS

Savre (1963) uses a very straightforward method to solve for porosity and proportions of minerals once the three minerals have been chosen by a method such as the *M-N* plot. Let f_1, f_2, and f_3 be the volume fractions of the three minerals, ϕ_{N1}, ϕ_{N2}, and ϕ_{N3} be the response of the neutron sonde to the three matrices, and ρ_{m1}, ρ_{m2}, and ρ_{m3} be the matrix densities. Then

$$\phi_N = f_1\phi_{N1} + f_2\phi_{N2} + f_3\phi_{N3} + \phi \tag{10-21a}$$
$$\Delta t = f_1\,\Delta t_{m1} + f_2\,\Delta t_{m2} + f_3\,\Delta t_{m3} + \phi\,\Delta t_f \tag{10-21b}$$
$$\rho = f_1\rho_{m1} + f_2\rho_{m2} + f_3\rho_{m3} + \phi\rho_f \tag{10-21c}$$

and, of course,

$$1 = f_1 + f_2 + f_3 + \phi \tag{10-21d}$$

This gives us four equations in four unknowns. If Δt_{mi}, ρ_{mi}, and ϕ_{Ni} are well known, they can be readily solved for ϕ and f_i. Savre states that, if gas is present, it is often impossible to obtain a solution. It is also sometimes impossible to obtain a solution because of incorrect and inconsistent values of the constants ϕ_{Ni}, ρ_{mi}, and Δt_{mi}.

Equations (10-21) can be generalized for a system of $k = 1, \ldots,$ K logs and $i = 1, \ldots, n$ fractional components (Doveton and Cable, 1980). Porosity ϕ is now identified as one component f_i. The response of the kth log to the ith component is C_{ik}.

$$L_k = \sum_{i=1}^{n} C_{ik} f_i \tag{10-22}$$

For $n = 4$ and $f_4 = \phi$, Eq. (10-22) is the same as (10-21).

If the number of logs is greater than or equal to the number of components, Eqs.(10-22) are solved by least squares, thus optimizing the accuracy. If the number of logs is one less than the number of components, there is a unique solution.

If the number of logs is less than $n - 1$, the system of linear equations (10-22) is underdetermined and cannot be solved unless other conditions are stipulated. This is the usual situation of log analysis, as described in Chap. 1. We are generally faced with somehow selecting the "best" solution from an infinitely large number of possible solutions. Harris and McCammon (1971), among others, have formalized some of the possible approaches. As a start, we can require that all fractional components be nonnegative, that is,

$$f_i \geq 0 \qquad \text{for all } i$$

We might also require that a certain fractional component lie within bounds consistent with geological knowledge. For example,

$$\phi_{\min} \leq \phi \leq \phi_{\max}$$

where $\phi_{\min}$ might be 0.01 and $\phi_{\max}$ might be 0.5.

It is also common for one mineral component to be dominant within a rock. This condition can be expressed as

$$f_j \geq \sum_{i=1}^{n} f_i \qquad i \neq j \tag{10-23}$$

This condition can be satisfied by minimizing the *entropy*, which is defined as

$$H = -\frac{100}{\ln n} \sum_{i=1}^{n} f_i \ln f_i \tag{10-24}$$

or the *variance*

$$V = \frac{100n}{n-1} \sum_{i=1}^{n} f_i(1 - f_i) \tag{10-25}$$

Harris and McCammon state that V provides a better solution than H. Minimizing either V or H leads to one mineral fraction dominating the others.

Shales, gas, gypsum, and any other desired components may be included in the model if the appropriate C_{ik} are known. Of course, there is still uncertainty arising because the C_{ik} specified by the user are both poorly known and variable and because there are errors in measurements of Δt, ρ_b, and ϕ_N. The calculation of total uncertainty requires an equation like (10-19) for each of the L_k in Eq. (10-22). They must be solved simultaneously to obtain the uncertainties in the f_i.

These methods are effective in carbonates. Patchett and Coalson (1982) have found that methods using two or more logs do not give better results in shaly sands than a good density log alone would give. The other logs require too much information about the amount and type of clay minerals. They state the accuracy of porosity is a function only of the accuracy of the density log and the estimate of ρ_{ma}.

10-5 COMPREHENSIVE COMPUTER PROGRAMS USING BOTH POROSITY AND SATURATION

The ultimate refinement of log anaysis for porosity and saturation is an attempt to find ϕ and S_w from logs alone. In shaly sands, the procedure is to use three equations in four unknowns: ϕ, S_{xo}, ρ_{Dh}, and f_{sh}. In terms of Eq. (10-22), the f_i are ϕS_{xo}, ϕS_h, f_{sh}, and f_{ma}. The first three unknowns are obtained from logs; the fourth, f_{sh}, is obtained from "clay indicators" (Sec. 9-2). Poupon et al. (1970) developed a model of a "typical" sand-shale sequence on which the method is based. Many, many imaginative assumptions about the properties of the shales must be made. If these assumptions are consistent with reality, the program may give accurate results. No error analysis is described; it is likely to be very complex indeed.

Another method (Poupon et al., 1971) is used when minerals other than sand are present. A method like that described in Sec. 10-2 is used to obtain a first estimate of porosity, and a method like that described in Sec. 9-1 for a first estimate of mineral composition. Equations similar to (10-12) and (10-14a) are used to get a second estimate, but ρ_{Dh} must be supplied by the user. Then the resistivity is used to get S_{xo}.

In both methods, an interactive procedure is used to refine the values of the results. The references use figures to explain the process. We find the equations easier to understand.

Each service company has several versions of these and other kinds of computer programs to interpret logs in terms of porosity and saturation. The methods are too complex, and change too rapidly, to be discussed here. It is *vital* that the user understand the assumptions underlying the particular program that is being used and that he know (and believe!) the origin of all the input constants. Patchett and Coalson (1982) say: "The complexity of the methods lends undeserved credence to the results." There is an adage in computations: "Garbage in; garbage out." It is particularly relevant to log analysis programs.

REFERENCES

Alger, R. P., L. L. Raymer, Jr., W. R. Hoyle, and M. P. Tixier: "Formation Density Log Applications in Liquid-Filled Holes," *J. Petr. Tech.*, March 1963. Reprinted in SPE Reprint Series No. 1, *Well Logging.*

American Petroleum Institute. *API Recommended Practice for Core-Analysis Procedure,* Am. Petr. Inst., Dallas, Tex., 1960.

Beard, D. C., and P. K. Weyl: "Influence of Texture on Porosity and Permeability of Unconsolidated Sand," *Bull. Am. Assoc. Petr. Geol.*, vol. 57, 1973, p. 349.

Doveton, J. H., and H. W. Cable: "KOALA; Kansas On-Line Automated Log Analysis System," Kansas Geol. Survey, Lawrence, Kans., 1980.

Fertl, W. H.: "Openhole Crossplot Concepts—A Powerful Technique in Well Log Analysis," *J. Petr. Tech.*, vol. 33, March 1981, pp. 535–549.

Granberry, R. J., R. E. Jenkins, and D. C. Bush: "Grain Density Values for Some Gulf Coast Formations and Their Importance in Formation Evaluation," *Proc. Soc. Prof. Well Log Analysts 9th Logging Symp.*, 1968.

Gregory, A. R.: "Aspects of Rock Physics from Laboratory and Log Data That Are Important to Seismic Interpretation," Am. Assoc. Petr. Geol. Memoir 26, 1977, p. 416.

Harris, M. H., and R. B. McCammon: "A Computer-Oriented Generalized Porosity-Lithology Interpretation of Neutron, Density, and Sonic Logs," *J. Petr. Tech.*, vol. 23, 1971, p. 239.

Mathews, M., and C. LaDelfe: "Log Comparison and Quantification," *Proc. 8th Formation Eval. Symp. Can. Well Logging Soc.*, 1981.

McGeary, R. K.: "Mechanical Packing of Spherical Particles," *J. Am. Ceramic Soc.*, vol. 44, 1961, p. 513.

Patchett, J. G., and E. B. Coalson: "The Determination of Porosity in Sandstones and Shaly Sandstones. Part One—Quality Control," *Proc. Soc. Prof. Well Log Analysts 20th Logging Symp.*, 1979, paper QQ.

——— and ———: "The Determination of Porosity in Sandstone and Shaly Sandstone. Part Two—Effects of Complex Mineralogy and Hydrocarbons," *Proc. Soc. Prof. Well Log Analysts 23d Logging Symp.*, 1982, paper T.

Poupon, A., C. Clavier, J. Dumanoir, R. Gaymard, and A. Misk: "Log Analysis of Sand-Shale Sequences—A Systematic Approach," *J. Petr. Tech.*, vol. 22, 1970, p. 867.

———, W. R. Hoyle, and A. W. Schimdt: "Log Analysis in Formations with Complex Lithologies," *J. Petr. Tech.*, vol. 23, 1971, p. 995.

Ruhovets, N., and W. H. Fertl: *Digital Shaly Sand Analysis Based on Waxman-Smits Model and Log-Derived Clay Typing*, Dresser Atlas, Houston, Tex., 1981.

Savre, W. C.: "Determination of a More Accurate Porosity and Mineral Composition in Complex Lithologies with the Use of the Sonic, Neutron, and Density Surveys, *J. Petr. Tech.*, vol. 15, 1963, p. 915.

Schlumberger: *Log Interpretation: Principles*, Schlumberger, New York, 1972.

———: *Log Interpretation: Applications*, Schlumberger, New York, 1974.

Schopper, J. R.: "Porosity of Rocks," in G. Angenheister (ed.), *Landolt-Bornstein Numerical Data and Functional Relationships in Science and Technology*, group V, *Geophysics and Space Research*, vol. 1, *Physical Properties of Rocks*, Subvolume a, Springer-Verlag, New York, 1982.

Tixier, M. P., R. P. Alger, and C. A. Don: "Sonic Logging," *Petr. Trans. AIME*, vol. 216, 1959, p. 106.

Truman, R. B., R. P. Alger, J. G. Connell, and R. L. Smith: "Progress Report on Interpretation of the Dual-Spacing Neutron Log (CNL) in the U.S.," *Proc. Soc. Prof. Well Log Analysts 13th Logging Symp.*, 1972, paper U.

Winsauer, W. O., H. M. Shearin, P. H. Masson, and M. Williams: "Resistivity of Brine Saturated Sands in Relation to Pore Geometry," *AAPG Bull.*, vol. 36, no. 2, 1952.

Wohlenberg, J.: "Density of Rocks," in G. Angenheister (ed.), *Landolt-Bornstein Numerical Data and Functional Relationships in Science and Technology*, New Series, Vol. 1, *Physical Properties of Rocks*, Subvolume a, Springer-Verlag, Berlin, 1982.

Wyllie, M. R. J., A. R. Gregory, and L. W. Gardner: "Elastic Wave Velocities in Heterogeneous and Porous Media," *Geophysics*, vol. 21, 1956, p. 41.

11

Saturation

A major goal of log analysis is to establish the fraction of pore space occupied by hydrocarbons. Three methods of doing so with logging systems have been established thus far. The electrical resistivity and thermal neutron die-away methods measure the water saturation S_w, and the difference $1 - S_w$ is then the saturation of all other liquids and gases. Both the electrical resistivity and neutron die-away methods depend upon the presence of salts dissolved in the water and will be less effective in freshwater than in saltwater environments. The third method, neutron activation, can distinguish and measure fluid content by determining ratios of carbon, oxygen, and hydrogen atoms. Measurement of S_w from dielectric properties appears promising, but at this point the appropriate mixing laws (Sec. 5-7a) are not well enough established to warrant further discussion here.

The definitions of porosity and saturation deserve emphasis. Porosity is defined as

$$\phi = \frac{V_w + V_h}{V_b} \tag{11-1}$$

where V_w is the volume of water in the bulk volume of rock V_b and V_h represents all fluids other than water which fill the pore space. In most cases V_h represents oil plus gas, but above the water table it could represent air. For convenience, we shall refer to fluids other than water as "hydrocarbon" throughout this chapter. In similar fashion, we define the fractional water saturation

$$S_w = \frac{V_w}{V_w + V_h} \tag{11-2}$$

These definitions appear trivial but it is worth emphasizing that the fractions ϕ and S_w are based on volumetric ratios. We continue to use V to represent volume and the symbols S, ϕ, and f to represent fractional volumes. The distinctions become important when we discuss the various shale models for electrical conduction.

11-1 SATURATION FROM ELECTRICAL RESISTIVITY

Water saturation estimates are based upon the two empirical relations established by Archie and discussed in Sec. 5-3. Recall that R_w, R_o, and R_t refer respectively to the resistivity (ohm-meters) of water, rock saturated with water, and rock saturated with some mix of water and other fluids. From Eq. (5-53), $R_o/R_w = F$, where, for R_w sufficiently low, the formation factor F is a function only of porosity ϕ. Water saturation is related to the whole rock resistivity by Eq. (5-57), $R_t/R_o = S_w^{-n}$. Combination of these two empirical relations gives us

$$\frac{R_t}{R_w} = FS_w^{-n} \tag{11-3}$$

If the commonly used empirical relation $F = a\phi^{-m}$ is adopted, a practical equation for S_w is obtained:

$$S_w = \left(\frac{aR_w}{\phi^m R_t}\right)^{1/n} \tag{11-4}$$

In ideal circumstances, the coefficients a, m, and n are obtained from core analysis, R_w is obtained from measurement of an uncontaminated sample of formation water or else from an SP log (Sec. 5-6), and R_t is obtained from a resistivity measurement representative of the uninvaded zone. Porosity is obtained by one of the methods discussed in Chap. 10. It is also assumed that formation water is the only electrical conductor, that is, the rock is devoid of significant concentrations of clays or other conducting minerals. Rarely are all these conditions satisfied. The ingenuity of the log analyst is needed to supply an estimate of missing or poorly determined data and to compensate for the presence of shale.

Resistivity-porosity cross-plot. Graphical techniques are especially helpful when not all parameters required in Eq. (11-4) can be specified. By taking the logarithm of Eq. (11-4) and rearranging, we obtain

$$\log R_t = \log aR_w - n \log S_w - m \log \phi \qquad (11\text{-}5)$$

A plot based on Eq. (11-5) is shown in Fig. 11-1. One plots observed R_t and ϕ data on a log-log plot. The minimum values of R_t and ϕ should cluster on a straight line of slope m, which defines the $S_w = 1$ line. The user must supply an estimate of n in order to construct parallel lines, also of slope m, for lower values of S_w. The plot can be then used to estimate S_w without knowing a or R_w. Furthermore, the $S_w = 1$ line can be extrapolated to $\phi = 1$, giving the intercept log aR_w, so that if an estimate of either a or R_w is available, the other can be determined. Clusters of points on the graph can sometimes indicate changes in lithology or sonde malfunctions (Lang, 1972).

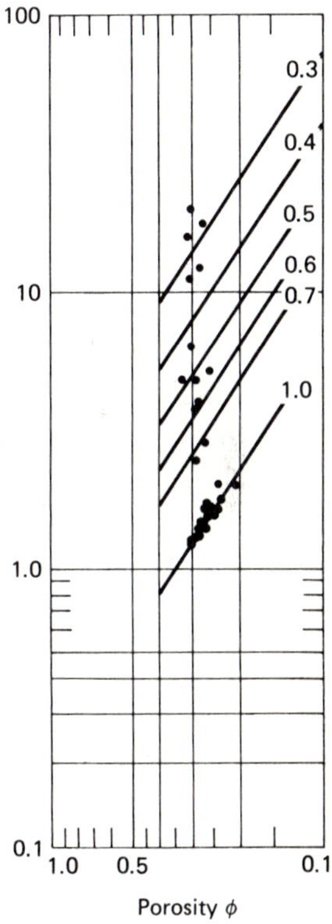

FIGURE 11-1 Resistivity-porosity cross-plot, with lines of constant S_w. (After Lang, 1972.)

R_{wa} indicator. The expression for S_w can be used to develop a qualitative indicator of the presence of hydrocarbons (Tixier et al., 1960). By rearranging Eq. (11-3) and using the expression for F, we have

$$R_w = \frac{R_t S_w^n}{F} = \frac{R_t S_w^n \phi^m}{a} \qquad (11\text{-}6)$$

An apparent water resistivity is defined by setting $S_w = 1$, as if the rock were fully water saturated:

$$R_{wa} = \frac{R_t \phi^m}{a} \qquad (11\text{-}7)$$

by which it can be seen that

$$R_{wa} = \frac{R_w}{S_w^n} \tag{11-8}$$

When the rock is fully water-saturated, R_{wa} and R_w are equal; but whenever S_w is less than 1.0, R_{wa} is greater than R_w. Hence, if R_w and ϕ are known in the formation of interest, a computed log of R_{wa} will indicate the presence of hydrocarbons. However, if clays are present, R_{wa} will be low, and the hydrocarbons will be overlooked. The R_{wa} method is only one of several semiquantitative techniques based upon manipulation of Eqs. (11-3) and (11-4) that have been introduced to provide a graphical "quick look" indicator of hydrocarbons (Schlumberger, 1974; Helander, 1983).

Clay-bearing sandstones. As described in Sec. 9-2, clays in sandstones can take several forms. In most formations clay reduces R_t and thereby erroneously increases the value of S_w that would be calculated from Eq. (11-4) if the presence of clay were ignored. Consequently, hydrocarbon saturations (given by $1 - S_w$) are underestimated. Much effort has been expended over the years to establish a method of compensating for the effect of clays. Summaries and a historical perspective of the "shaly sand" problem, along with an anthology of the literature, have been provided by the SPWLA (1982). Generally speaking, the problem has been addressed by attempting to measure the amount of clay or shale present (Sec. 9-2) and establishing a model and a method by which the correction can be computed.

A limiting form for most models is furnished by Eq. (11-3), which can be expressed in terms of conductivity:

$$C_t = \frac{S_w^n}{F} C_w \tag{11-9}$$

It is assumed that any general expression for S_w must reduce to this form as the concentration of clays approaches zero. The second requirement for the model is that an appropriate mixing rule for the clay and water components must be found. Different models rest upon different assumptions regarding the rock constituents. The volumetric assumptions of a few of the models are sketched in Fig. 11-2.

Laminated shale model. A series of sand-shale laminations is modeled by assuming that the two members behave electrically as parallel, independent conductors. We can extend Eq. (11-1) to write the porosity ϕ

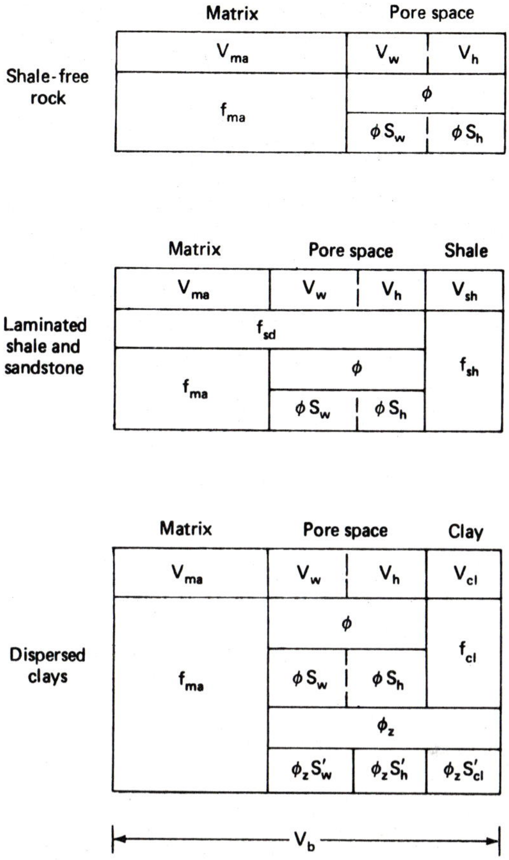

FIGURE 11-2 Nomenclature used in various shaly sand models. V is volume; f and ϕ represent fractions of total rock volume; and S (or S') designates fractions of ϕ (or ϕ_z).

of the sand-shale system as

$$\phi = \frac{V_w + V_h}{V_w + V_h + V_{ma} + V_{sh}} \tag{11-10}$$

However, as shown in Fig. 11-2*b*, the shale has zero porosity. Porosity is present only in the sandstone, and we can define ϕ_{sd} for the sandstone alone:

$$\phi_{sd} = \frac{V_w + V_h}{V_w + V_h + V_{ma}} = \frac{\phi}{f_{sd}} \tag{11-11}$$

As indicated in Fig. 11-2*b*, the sandstone fractional volume f_{sd} includes the porosity, so that

$$f_{sd} + f_{sh} = 1 \tag{11-12}$$

where f_{sh} is the shale fractional volume. Solving Eqs. (11-11) and (11-12) yields

$$\phi = \phi_{sd}(1 - f_{sh}) \tag{11-13}$$

The shale is assumed to have conductivity C_{sh}; the sandstone has conductivity C_{sd}. Because the two act in parallel, the conductivity of the formation is

$$C_t = (1 - f_{sh})C_{sd} + f_{sh}C_{sh} \tag{11-14}$$

Assuming that $m = 2$, the sandstone formation factor can then be written

$$F_{sd} = a\phi_{sd}^{-2} = a(1 - f_{sh})^2\phi^{-2} \tag{11-15}$$

Assuming that Eq. (11-9) is valid for the sandstone and that $n = 2$,

$$C_{sd} = \frac{S_w^2}{F_{sd}} C_w \tag{11-16}$$

Combining Eqs. (11-14) through (11-16) gives us

$$C_t = \frac{\phi^2 S_w^2}{a(1 - f_{sh})} + f_{sh}C_{sh} \tag{11-17}$$

The model often works for focused resistivity logs acquired in wells drilled normal to the bedding planes of a finely laminated sand-shale sequence. The shale fraction f_{sh} must be obtained from core or from a log capable of resolving the bedding. Shale conductivity C_{sh} is obtained from an electric log in a thick shale bed. Values of $m = n = 2$ have been assumed here in order to simplify the result given in Eq. (11-17).

Dispersed shale model. An approach adopted to treat dispersed clays (Alger et al., 1963) lumps together the clays and pore fluids to form a conceptual "slurry" which occupies a fractional (and fictional!) pore space (Fig. 11-2*c*):

$$\phi_z = \phi + f_{cl} \tag{11-18}$$

Each of the saturations of the components occupying the fictional pore space ϕ_z is denoted by a prime to distinguish it from conventional saturation. Thus

$$S_w' = \frac{V_w}{V_w + V_h + V_{cl}} = \frac{S_w\phi}{\phi_z} \tag{11-19}$$

and

$$S'_c = \frac{V_{cl}}{V_w + V_h + V_{cl}} = \frac{f_{cl}}{\phi_z} \tag{11-20}$$

and the three saturations S'_w, S'_c, and S'_h for water, clay, and hydrocarbon sum to 1.0. From Eqs. (11-2), (11-19), and (11-20) it follows that

$$S'_w = S_w(1 - S'_c) \tag{11-21}$$

The conductivity of the slurry is assumed to be the weighted arithmetic mean:

$$C_{sl} = \frac{S'_c C_c + S'_w C_w}{S'_c + S'_w} \tag{11-22}$$

Assuming that Eq. (11-9) applies to the slurry,

$$C_t = \frac{1}{F_z}(S'_c + S'_w)^n C_{sl} \tag{11-23}$$

where F_z is the formation factor for a clean sandstone with porosity ϕ_z, that is, $F_z = a\phi_z^{-m}$. Substitution of the slurry conductivity C_{sl} from Eq. (11-22) yields

$$C_t = \frac{1}{F_z}(S'_c + S'_w)^{n-1}(S'_c C_c + S'_w C_w) \tag{11-24}$$

Utilizing Eq. (11-21) and assuming $n = 2$,

$$C_t = \frac{1}{F_z}[S_w^2(1 - S'_c)^2 C_w + S_w S'_c(1 - S'_c)(C_c + C_w) + C_c S'^2_c] \tag{11-25}$$

Solution of the above quadratic expression for S_w gives us

$$S_w = \frac{1}{1 - S'_c}\left\{-\frac{S'_c(C_c + C_w)}{2C_w} + \left[\frac{F_z C_t}{C_w} + \frac{S'_c(C_c - C_w)^2}{2C_w}\right]\right\}^{1/2} \tag{11-26}$$

Alger et al. (1963) suggest that the variables for Eq. (11-26) be determined as follows. The sonic travel time is uninfluenced by the dispersed clay content, so that values of ϕ_z and F_z can be obtained from a sonic log. C_c is assumed to be the same as the conductivity of a nearby shale bed. A density log is used to get the fluid-filled porosity ϕ. From Eqs. (11-18) and (11-20), the clay fraction S'_c can then be determined as $S'_c = (\phi_z - \phi)/\phi_z$.

Waxman-Smits model. Development of a model in which the cation exchange capacity of a shaly sand is considered was described in Sec. 5-3b. The concentration of exchangeable ions is given by BQ_v, and the formation factor F^* is defined as the ratio C_o/C_w determined with C_w high enough to mask the contribution of exchange ions to C_o. The extended equation (5-62) for water-saturated rock is

$$C_o = \frac{C_w + BQ_v}{F^*} \tag{11-27}$$

Waxman and Smits (1968) extended the model to the partial saturation case by assuming that the exchange ion population remains unchanged as water is replaced by oil. The excess ions become more concentrated in the remaining water-filled pore space, which requires that Q_v be replaced by Q_v/S_w. Then, in accord with Eq. (11-9),

$$C_t = \frac{S_w^{n^*}}{F^*}\left(C_w + \frac{BQ_v}{S_w}\right) \tag{11-28}$$

or

$$C_t = \frac{S_w^{n^*}C_w}{F^*} + \frac{S_w^{n^*-1}BQ_v}{F^*} \tag{11-29}$$

The saturation exponent n^* differs from n slightly in that it strictly applies to a rock with pore structure identical with that of the actual rock but with its clays somehow rendered inactive. In practice, the distinction between n and n^* is usually ignored. Equation (11-29) can be solved as a quadratic if $n = 2$ or graphically or iteratively if n is other than 2. Waxman and Thomas (1974) confirmed the validity of Eq. (11-29) with experiments on samples saturated with a mix of oil and water. They also examined the effect of temperature on shaly sand conductivity and found that temperature affected not only the brine resistivity but also the equivalent conductivity of the exchange ions. They presented their results graphically and gave B as a function of R_w and temperature. Hence, determination of S_w with Eq. (11-29) by using log measurements of C_t requires temperature corrections of both C_w and B.

Limitations. These three models for shaly sands form only a partial list of all the models that have been proposed, but they serve to illustrate the assumptions used to correct for clay conduction in partially water-saturated rocks. Fundamental is the reliance on Archie's equation (11-3) as clay content drops to zero. It is worth emphasizing again that this equation is itself an empirical finding based on core measurements and that both m and n fluctuate statistically. In all the models it is assumed that some sort of parallel conduction mechanism applies; that although conduction in water and along clay surfaces are coupled in some fashion, the conduction pathways are ultimately independent. Moreover, the reservoir rock is taken to be water-wet.

The physical-chemical description of the clay-water interface is most detailed in the cation exchange models. However, application of Eq. (11-29) is not easy in practice, because direct determination of Q_v requires a laboratory measurement of cation exchange capacity. Attempts have been made to estimate Q_v by other means, but no one technique has been widely accepted. In searching for a completely log-based method for estimating S_w in shaly sands, other modifications of the basic cation exchange concepts have been proposed. In an introduction to a collection of the relevant papers, Brown (1982) summarizes the developments. It is clear that practitioners will welcome improved solutions to the shaly sand problem.

11-2 SATURATION USING THERMAL NEUTRON DIE-AWAY LOGS

Because it offers a way to monitor and measure S_w behind iron casing, the thermal neutron die-away method has enjoyed increasing popularity over the past decade. From Eq. (6-51), total cross section observed with a neutron die-away log is

$$\Sigma_t = \Sigma_{ma}(1 - \phi) + \Sigma_w S_w \phi + \Sigma_h(1 - S_w)\phi \qquad (11\text{-}30)$$

Rewriting in terms of S_w,

$$S_w = \frac{\Sigma_t - \Sigma_{ma} - \phi(\Sigma_h - \Sigma_{ma})}{\phi(\Sigma_w - \Sigma_h)} \qquad (11\text{-}31)$$

The method works if Σ_w differs sufficiently from Σ_h. This is indeed the case if the hydrocarbon is gas. Unfortunately, however, Σ_w of fresh water is quite close to Σ_h of oil, and only if the water contains enough chlorine will Σ_w be significantly different from Σ_h (Table 6-3). A value of Σ_w can be taken from charts, computed from chemical analysis, or measured directly with a thermal neutron absorption cell (Rinehart and Weber, 1975).

Σ_{ma} can be estimated from known values for rocks, or it can be obtained by plotting Σ_t against ϕ with data acquired in the formation of interest (Clavier et al., 1971*a*). On rewriting Eq. (11-30) with $S_w = 1$, we have

$$\Sigma_t = \Sigma_{ma} + \phi(\Sigma_w - \Sigma_{ma}) \qquad (11\text{-}32)$$

It can be seen that both water and matrix cross sections can be determined from the slope and intercept of such a plot if sufficient log data are available for fully water-saturated rock with a reasonable range of porosity.

The accuracy of Σ_t can be improved by logging more slowly or by obtaining repeat runs of the log and averaging the results. None of these improvements is to any avail if one of the parameters contains errors which overwhelm all the others, however. How much accuracy is required for each parameter? A sensitivity analysis of Eq. (11-31) is provided by Clavier et al. (1971*b*). They assume that the uncertainties are independent and can be expressed as partial derivatives, so that the total uncertainty for S_w would be

$$(\delta S_w)^2 = \left(\frac{\delta S_w}{\delta \Sigma_{ma}} \delta \Sigma_{ma}\right)^2 + \left(\frac{\delta S_w}{\delta \Sigma_w} \delta \Sigma_w\right)^2 + \cdots \tag{11-33}$$

The inverse of each partial derivative term, called the *tolerance factor*, is tabulated by the authors for several values of salinity and porosity. The user must estimate the uncertainty of each parameter, divide by the tolerance factor, and then use Eq. (11-33).

The presence of shale is not as troublesome with the neutron die-away method as it is with electrical resistivity because the mixing rule for capture cross section is known. The shale cross section Σ_{sh} is incorporated into Eq. (11-30) and gives us

$$\Sigma_t = \Sigma_{ma}(1 - \phi - f_{sh}) + \Sigma_{sh} f_{sh} + \Sigma_w S_w \phi + \Sigma_h (1 - S_w)\phi \tag{11-34}$$

which in terms of S_w is

$$S_w = \frac{\Sigma_t - \Sigma_{ma} - \phi(\Sigma_h - \Sigma_{ma}) - f_{sh}(\Sigma_{sh} - \Sigma_{ma})}{\phi(\Sigma_w - \Sigma_h)} \tag{11-35}$$

The value for Σ_{sh} can be measured at a nearby shale unit, provided that the shale where Σ_{sh} is measured has the same properties as the shale in the formation of interest. The methods (and frustrations) of determining f_{sh} have been described in Sec. 9-2. Other minerals also can be included explicitly in Eq. (11-34) or implicitly incorporated into Σ_{ma}. As the concentration of elements of high Σ (iron in pyrite, boron in shales) increases, the accuracy of S_w determinations progressively worsens.

Combined resistivity and thermal neutron die-away data. It can happen that Σ_w is poorly known, so that S_w cannot be determined from Eq. (11-31) alone. In an open hole, resistivity logs can be obtained as well as neutron die-away logs, and this problem can be attacked. The method requires the two governing equations (11-4) and (11-30):

$$S_w = \left(\frac{aR_w}{\phi^m R_t}\right)^{1/n} \tag{11-36}$$

and

$$\Sigma_t = \Sigma_{ma}(1 - \phi) + \Sigma_w S_w \phi + \Sigma_h(1 - S_w)\phi \qquad (11\text{-}37)$$

There are two approaches for combining Eqs. (11-36) and (11-37) so that R_t and Σ_t data can be cross-plotted and fluid saturations estimated. In the first approach (Clavier et al., 1971*a*), S_w is substituted from Eq. (11-36) into Eq. (11-37). Rearrangement then gives us

$$\Sigma_t = \Sigma_{ma} + \phi(\Sigma_h - \Sigma_{ma}) + (aR_w\phi^{n-m})^{1/n}(\Sigma_w - \Sigma_h)R_t^{-1/n} \qquad (11\text{-}38)$$

This equation simplifies considerably if one can assume $a = 1$ and $m = n = 2$:

$$\Sigma_t = \Sigma_{ma} + \phi(\Sigma_h - \Sigma_{ma}) + R_w^{1/2}(\Sigma_w - \Sigma_h)R_t^{-1/2} \qquad (11\text{-}39)$$

To apply Eq. (11-39), log values of $R_t^{-1/2}$ are plotted against values of Σ_t. Points from the water-saturated zone $S_w = 1$ will have high values of $R_t^{-1/2}$ and Σ_t. As S_w decreases, both values decrease, which defines a trend with slope $R_w^{1/2}(\Sigma_w - \Sigma_h)$. The intercept for $R_t^{-1/2} = 0$, corresponding to a hypothetical 100 percent oil-saturated zone of infinite resistivity, will yield a cross section value called Σ_o. From this value, the matrix cross section

$$\Sigma_{ma} = \frac{\Sigma_o - \phi\Sigma_h}{1 - \phi} \qquad (11\text{-}40)$$

can be determined. A chart for which Σ_{ma} and ϕ are known is shown in Fig. 11-3. Each of the four solid lines corresponds to fluid of a particular salinity with specific values of R_w and Σ_w. They originate at Σ_0, where $S_w = 0$, and terminate on a dashed line where $S_w = 1$. Intermediate saturation values are obtained by linear interpolation between the two end points. A change in porosity requires a shift in the intercept, as can be seen from Eq. (11-39), but the shift may not be of much consequence for porosity ranges encountered in practice (Clavier et al., 1971).

The second approach to cross-plotting (Aguilera, 1979) is to eliminate ϕ from Eqs. (11-36) and (11-37). This is done by rewriting the equations in terms of ϕ:

$$\phi = \left(\frac{aR_w}{S_w^n R_t}\right)^{1/m} \qquad (11\text{-}41)$$

and

$$\phi = \frac{\Sigma_t - \Sigma_{ma}}{(1 - S_w)\Sigma_h + S_w\Sigma_w - \Sigma_{ma}} \qquad (11\text{-}42)$$

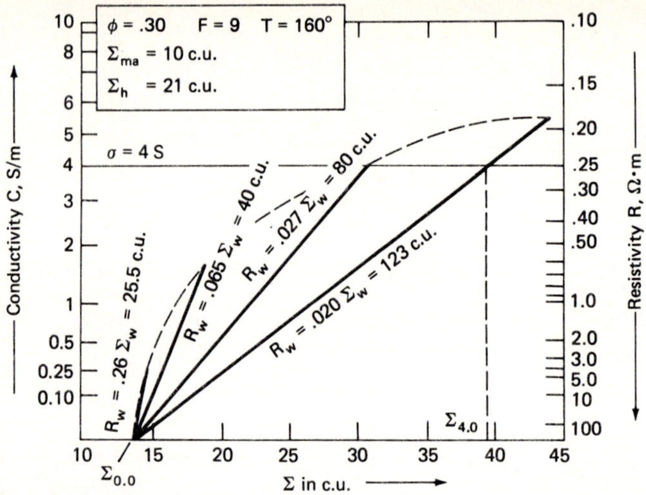

FIGURE 11-3 Plot of Σ_t versus R_t, based on Eq. (11-39), for different values of R_w and σ_w. (From Clavier et al., 1971*a*. Copyright 1971, SPE–AIME.)

Equating Eqs. (11-41) and (11-42) and taking the logarithm results in

$$m \log (\Sigma_t - \Sigma_{ma}) = \log aR_w - n \log S_w + m \log \Sigma' - \log R_t \quad (11\text{-}43)$$

where $\Sigma' = (1 - S_w)\Sigma_h + S_w\Sigma_w - \Sigma_{ma}$. Construction of the cross-plot based on Eq. (11-43) is an iterative process that requires successive estimates of Σ_{ma} until a straight line is obtained on a plot of $\log (\Sigma_t - \Sigma_{ma})$ against $\log R_t$. Only data for $S_w = 1$ can be used to determine Σ_{ma}, and a large spread in porosity is desirable. The slope then gives the cementation exponent m. Lines of constant ϕ and S_w can then be added to complete the analysis for zones with S_w less than 1.

11-3 SATURATION FROM CARBON/OXYGEN LOGS

The carbon/oxygen (C/O) log, described in Sec. 6-8b, can be used to measure oil saturation S_o by responding to the carbon content of oil and oxygen content of water. The method assumes that carbon and oxygen contributions from rock and wellbore are either known or can be ignored. If the logging situation is similar enough to the environment in which the sonde was calibrated (Fig. 6-40*a*), then those calibrations can be applied directly.

Hertzog (1980) gives a description of the C/O calibration plots which incorporates borehole effects. It states that the carbon peak is proportional to the carbon concentration in rock and fluid plus a background term B_c which represents the borehole contribution. The oxygen peak is treated similarly, so the ratio becomes

$$\frac{[C]}{[O]} = A \frac{[C]_{ma}(1 - \phi) + [C]_o \phi S_o + B_c}{[O]_{ma}(1 - \phi) + [O]_w \phi S_w + B_o} \tag{11-44}$$

where [C] and [O] are carbon and oxygen concentrations of matrix (*ma*), oil (*o*), and water (*w*). Equation (11-44) must be regarded as an approximation because it ignores the occurrence of γ's in each spectral peak caused by interactions other than those from the desired element (Sec. 6-8). However, the equation provides a useful framework with which to treat the contribution of borehole effects (Schweitzer et al., 1982).

Because the C/O method detects high-energy γ's due to inelastic scattering, it is not affected by the absorption of thermal neutrons and consequently, unlike the die-away log, it is not sensitive to small amounts of elements with high absorption cross section. However, it is quite sensitive to lithological changes which result in enhancements or declines in carbon or oxygen abundance. In sediments these can be compensated by consideration of the Ca/Si ratio (Sec. 6-8b), and various algorithms have been suggested to carry out the adjustment for varying mixes of carbonate material with sandstone (Lawrence, 1979). Another shortcoming of both the C/O log and the thermal die-away log is limited depth of penetration, which is of the order of 20 cm measured radially from the wall of the hole (Smith and Schultz, 1974). This shallow penetration reduces the utility of these logs where invasion of drilling fluids is significant. Because of this, and because they can measure in the presence of casing, both the C/O and thermal neutron decay methods are more used in cased hole than in open hole logging.

11-4 ERROR ASSESSMENT

The accuracy with which S_w can be determined rests upon the accuracy of the measured parameters, whether one is using resistivity, neutron die-away, or carbon/oxygen measurements. Consider Eq. (11-4), where six parameters must be determined: a, R_w, ϕ, R_t, m, and n. Recall that some of these parameters must be estimated or indirectly determined. For example, R_w may be determined from the SP log by using a measured value for mud filtrate R_{mf} [Eq. (5-133)]. Moore and Kaufman (1981) have discussed the problems in getting a good determination of R_{mf} and R_w. Both the resistivity determination itself and the extrapolation of a room temperature value to a formation temperature value are subject to error. They compared fluid resistivity measurements and found that a 25.7 percent difference existed in the R_w value determined by two service companies. This error translated into a 15.5 percent error is S_w.

A full error analysis provides interesting insight into the worth of a computed result. Walstrom et al. (1967) used a Monte Carlo simulation to evaluate the error associated with Eqs. (11-3) and (11-4). They assumed a range of values for each parameter. For example, R_w may have any value between 0.055 and 0.075 Ω·m, and the exponent n may range between 1.8 and 2.2. A uniform or triangular probability distribution function is assigned to each variable. For each computational case, an actual value for each parameter is chosen randomly from within the parameter's range. Many cases are run (of the order of 500) in order to define the distribution of the computed result. Values of porosity were determined from a density log; so in order to determine the probability distribution of ϕ, distributions of bulk density and

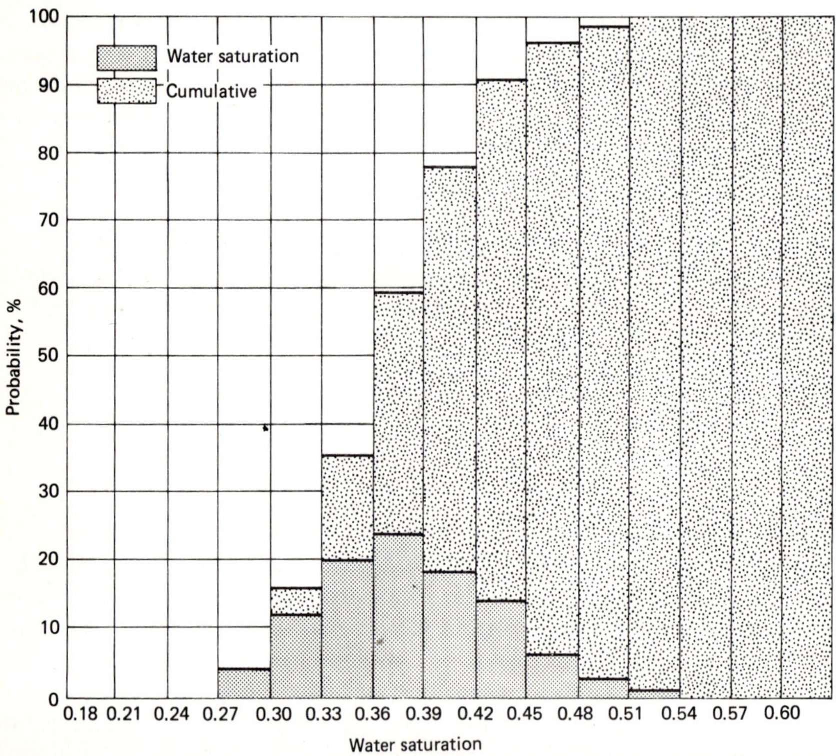

FIGURE 11-4 Probability distribution for S_w based on Monte Carlo simulation. (From Walstrom et al., 1967. Copyright 1967, SPE–AIME.)

matrix density were assumed as input to a series of Monte Carlo runs. Likewise, the distribution of formation factor F was determined from a second Monte Carlo simulation by using porosity as a random variable with its range and distribution defined by the previous simulation. Finally, S_w was determined in a similar manner from Eq. (11-3) by using values of R_t obtained from a log. The result is shown as Fig. 11-4. The probability of S_w attaining a value in any 0.03 interval is shown by the histogram. S_w is between 0.274 and 0.54, with its most probable value between 0.36 and 0.39. From the cumulative probability distribution one can state that there is about a 90 percent chance that S_w is equal to or less than 0.45.

The Monte Carlo technique is quite general. It can also be applied to neutron die-away and carbon/oxygen logging methods. It is only necessary to establish ranges and probability distributions for each parameter before applying the proper response equation.

REFERENCES

Aguilera, R.: "A New Approach for Log Analysis of Pulsed Neutron and Resistivity Log Combinations," *J. Petr. Tech.*, vol. 31, no. 4, 1979.

Alger, R. P., L. L. Raymer, Jr., W. R. Hoyle, and M. P. Tixier: "Formation Density Log Applications in Liquid-Filled Holes," *J. Petr. Tech,* March 1963. Reprinted in SPE Reprint Series No. 1, *Well Logging.*

Brown, A. A.: "Cation Exchange Models," in SPWLA reprint vol. *Shaly Sand,* Soc. Prof. Well Log Analysts, Houston, Tex., 1982.

Clavier, C., W. R. Hoyle, and D. Meunier: "Quantitative Interpretation of Thermal Neutron Decay Time Logs, Part I. Fundamentals and Techniques," *J. Petr. Tech,* vol. 23, June 1971*a*, pp. 743–755.

———, W. R. Hoyle, and D. Meunier: "Quantitative Interpretation of Thermal Neutron Decay Time Logs, Part II. Interpretation Example, Interpretation Accuracy, and Time-Lapse Technique," *J. Petr. Tech.*, vol. 23, June 1971*b*.

Coates, G., R. P. Schulze, and W. H. Throop: "VOLAN—An Advanced Computational Log Analysis," *Trans. 23d Ann. Logging Symp. SPWLA*, 1982, paper A.

Helander, D. P.: *Fundamentals of Formation Evaluation,* Oil & Gas Consultants, Tulsa, Okla., 1983.

Hertzog, R. C.: "Laboratory and Field Evaluation of an Inelastic Neutron Scattering and Capture Gamma Ray Spectrometry Tool," *Soc. Petr. Eng. J.*, October 1980.

Lang, W. H., Jr.: "Porosity-Resistivity Cross-Plotting," *Trans. 13th Annual Logging Symp. SPWLA*, 1972, paper F.

Lawrence, T. D.: "Continuous Carbon/Oxygen Log Interpretation Techniques," 54th Annual Fall Tech Conf. SPE Paper 8366, 1979.

Moore, C. V., and R. L. Kaufman: "Your Unsuspected Problems: Fluid Resistivity and Water Analysis," *Trans. 22d Ann. Logging Symp. SPWLA*, 1981, pp. C1–C16.

Rinehart, C. E., and H. J. Weber: "Measuring Thermal Neutron Absorption Cross Sections of Formation Brines," *Trans. 16th Annual Logging Symp. SPWLA*, 1975, pp. G1–G14.

Schlumberger: *Log Interpretation*, Vol. II: *Applications*, Schlumberger, New York, 1974.

Schweitzer, J. S., R. A. Manente, and R. C. Hertzog: "Gamma-Ray Spectroscopy Tool Environmental Effects," 57th Ann. SPE Fall Tech. Conf. Paper 11144, 1982.

Smith, H. D., Jr., and W. E. Schultz: "Field Experience in Determining Oil Saturations from Continuous C/O and Ca/Si Logs Independent of Salinity Shaliness," *Log Analyst*, November–December, 1974.

SPWLA Reprint Volume, *Shaly Sand*, Soc. Prof. Well Log Analysts, Houston, Tex., 1982.

Tixier, M. P., R. P. Alger, and D. R. Tanguy: "New Developments in Induction and Sonic Logging," *J. Petr. Tech.*, May 1960. Reprinted in SPE Reprint Series No. 1, *Well Logging*.

Walstrom, J. E., T. D. Mueller, and R. C. McFarlane: "Evaluating Uncertainty in Engineering Calculations," *J. Petr. Tech.*, December 1967, pp. 1595–1603.

Waxman, M. H., and L. J. M. Smits: "Electrical Conductivities in Oil-Bearing Shaly Sands," *Soc. Petr. Eng. J.*, June 1968, pp. 107–122. Reprinted in *Shaly Sand*, SPWLA, 1982.

——— and E. C. Thomas: "Electrical Conductivities in Oil-Bearing Shaly Sands. I. The Relation between Hydrocarbon Saturation and Resistivity Index. II. The Temperature Coefficient of Electrical Conductivity," *Soc. Petr. Eng. J.*, no. 14, February 1974, pp. 213–225. Reprinted in *Shaly Sand*, SPWLA, 1982.

12

Permeability

Wherever a hydraulic gradient exists within interconnected porous or fractured rock, fluid migration will occur. Permeability is a measure of the magnitude of fluid flow per unit of hydraulic gradient. If permeability is known, the producibility of an aquifer or reservoir under varying conditions of pressure and flow can then be predicted by using the methods of reservoir engineering. Because most wells are drilled to recover water or hydrocarbons, permeability determination is obviously of prime economic significance. We would like to be able to run one or more logs from which we can use physical principles to derive permeability. Despite the incentive to measure it, however, rock permeability has proved to be one of the most elusive of all rock parameters, and at this stage logging methods for permeability are far less satisfactorily developed than for porosity.

Most logging methods for permeability measurement are based on considerations of pore space, pore structure, and water saturation. That is, permeability is assessed from rock properties under static, no-flow conditions.

12-1 DARCY'S LAW

Experimental work has shown that, at low flow rates in porous media, the volumetric rate of flow Q across an area A normal to the direction of flow is in direct proportion to the hydraulic head applied along the direction of flow. The ratio Q/A is the apparent (macroscopic) velocity v. Darcy's law in a homogeneous isotropic porous medium can be written

$$v = -K\nabla\left(\frac{P}{\rho g} - z\right) \tag{12-1}$$

where z is taken positive downward. The gradient term, which is dimensionless in Eq. (12-1), is referred to as the *hydraulic head.* It depends upon pressure P, fluid density ρ, gravitational attraction g, and vertical separation z. K is the hydraulic conductivity, which is a function of the properties of both the fluid and the porous medium. It has the dimensions of velocity, and it is convenient for hydrological applications where the properties of water at standard temperature and pressure are regarded as constant. For reservoir conditions at depth, the physical properties of rock and fluid are separated by writing

$$v = -\frac{k}{\mu}\nabla(P - \rho g z) \tag{12-2}$$

where μ is the dynamic viscosity of the fluid. In Eq. (12-2) the gradient term is referred to as the *hydraulic gradient.* The *intrinsic* or *effective permeability* k reflects the rock properties only. Because μ is the ratio of stress to velocity gradient, it can be seen from Eq. (12-2) that k has the dimensions of length squared (m^2). This is consistent with one's intuitive notion of k, in that it must be proportional to the effective area presented by the pore space to fluid flow. From Eqs. (12-1) and (12-2), hydraulic conductivity and permeability are related by

$$K = \frac{\rho g k}{\mu} \tag{12-3}$$

Equations (12-1) and (12-2) are satisfactory in any self-consistent set of units. However, most hydrological and petroleum usage is in mixed units, so that conversion factors are required. Most common is the darcy, which is the permeability of a porous medium filled with a single-phase fluid of one centipoise viscosity flowing at a rate of one cubic centimeter per second per cross-sectional area of one square centimeter under a gradient of one atmosphere pressure per centimeter (API, 1956). One darcy converts to 0.9869×10^{-12} m^2, which, rather conveniently, is approximately one square micron. Conversions between K and k must be applied with care because both fluid density ρ and viscosity μ are temperature- and pressure-dependent. Conversion factors are usually given for water at room temperature and pressure (Narasimhan, 1982).

The terms incorporating depth z in Eqs. (12-1) and (12-2) account for the presence of a continuous hydrostatic column. The pressure P_2 at depth z_2 is greater than P_1 at some shallower depth z_1, but no flow occurs because the column is in hydrostatic equilibrium such that $P_2 - P_1 = \rho g(z_2 - z_1)$. Special cases involving the flow of compressible or incompressible fluids in simple geometries can be readily derived from Eq. (12-1) or (12-2) (API, 1956; Amyx et al., 1960). For example, if a pressure ΔP is applied across a cylindrical rock sample of length L positioned horizontally, Eq. (12-2) becomes

$$v = \frac{Q}{A} = -\left(\frac{k}{\mu}\right)\left(\frac{\Delta P}{L}\right) \tag{12-4}$$

and k can be determined from measurement of flow rate Q and sample cross-sectional area A.

12-2 RANGE OF PERMEABILITY IN ROCKS

Permeability of porous rock is determined in the laboratory by using core or small plugs cut from core. Values of k are generally high enough that steady-state methods based on Eq. (12-4), or a modification of it if gas is the fluid, can be used. Permeability values in rocks and unconsolidated deposits range over many orders of magnitude, as shown in Fig. 12-1. Good water aquifers have permeabilities in excess of 1 μm^2 (1 darcy). Good petroleum reservoir permeabilities are greater than one millidarcy, although in tight gas reservoirs permeability can drop much lower.

Determination of permeability in nonporous rocks poses special problems. That is because permeability of low-porosity carbonate, igneous, metamorphic, and argillaceous rocks is often controlled by fractures rather than by the matrix. In situ measurements show that permeability in crystalline rock varies by more than six orders of magnitude even within a single site (Brace, 1980). Shales form permeability barriers for many reservoirs because their permeability is in the microdarcy and submicrodarcy range. The permeability of such rocks can be determined in the laboratory by using transient pressure techniques, but the values obtained are valid only for the matrix and usually rep-

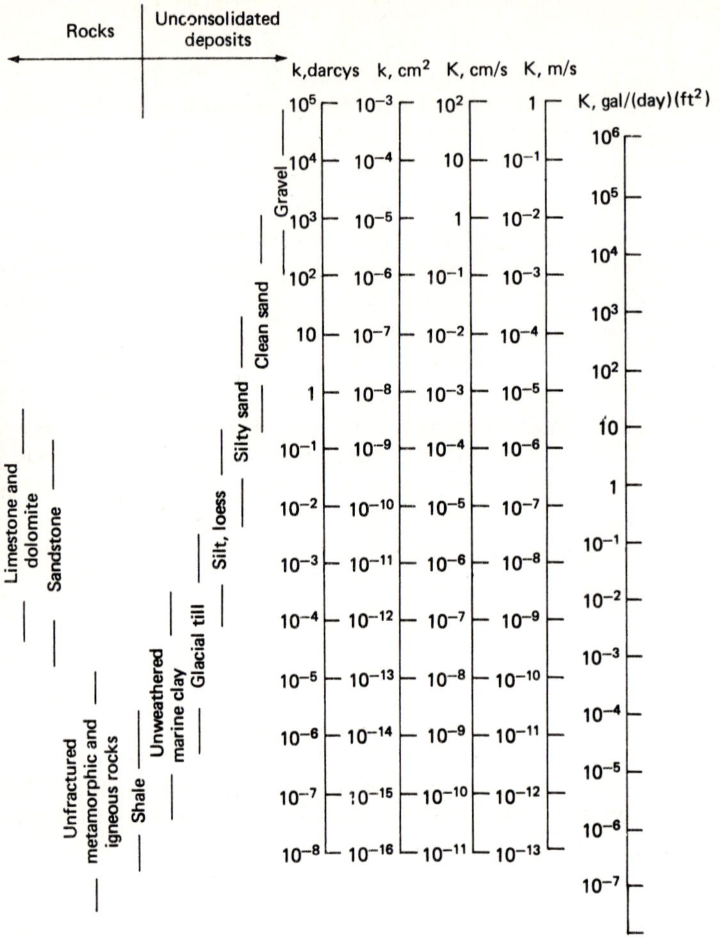

FIGURE 12-1 Range of permeability and hydraulic conductivity values in rocks. (From Freeze and Cherry, 1979.)

resent a lower limit to permeability in situ. Fractures cannot be sampled with standard core recovery methods, so special field and laboratory techniques are required (Gale, 1982).

The effects of pressure and temperature upon permeability depend greatly upon rock type. In well-cemented porous rock, the variations caused by pressure and temperature are not large in comparison with the variation in permeability from sample to sample, although they can be important for reservoir engineering calculations. Permeability can decrease by 20 percent if hydrostatic pressure is increased from 0 to 40 MPa (Somerton, 1982). Experimental evidence suggests that permeability also decreases with increasing temperature; however, it is difficult to distinguish the flow effects caused by thermal expansion from effects which depend on fluid-rock interaction. The latter include temperature-dependent chemical reactions on grain surfaces and plugging due to dissolution and reprecipitation.

12-3 PERMEABILITY MODELS FOR ROCK

We would like to assume a pore structure for a rock, calculate the flow as a function of applied pressure, and thereby derive an expression for k in terms of other rock parameters which characterize the pore structure. That has been done only for geometries so simple that they lack realism. The model for porous rock which was investigated years ago (Amyx et al., 1960, provide a summary and references) might be described as the "mangled tube" model. Flow from one end to another of a porous rock sample is assumed to occur in a bundle of tubes of different radii. Within each tube the flow rate is low enough so as not to be turbulent. Poiseuille's law gives the average flow velocity in a cylindrical tube of radius r and length L:

$$v_i = \frac{r^2 \, \Delta P}{8\mu L} \tag{12-5}$$

We will assume here that all tubes are of equal radius r. The macroscopic (Darcy) flow velocity Q/A will be less than the velocity in an individual tube by the ratio of all tube cross-sectional areas to the total area A. But this area ratio is simply the porosity ϕ, in which case the macroscopic velocity $v = \phi v_i$ and

$$v = \phi \frac{r^2 \, \Delta P}{8\mu L} \tag{12-6}$$

so that, by comparison with Eq. (12-4), the permeability $k = \phi r^2/8$; that is, the permeability should be directly proportional to porosity and to the square of the radius of the fluid flow path. To accommodate flow paths which are not circular in cross section, Wyllie and Spangler (1952) introduced the mean hydraulic radius r_h, which is defined as the volume of the tube divided by the wetted surface area. For a tube of circular cross section, $r_h = r^2$. The flow equation is modified by adopting a dimensionless shape factor f:

$$v = \frac{r_h^2 \, \Delta P}{f\mu L} \tag{12-7}$$

Because fluid obviously does not flow in straight paths through a porous rock, the tubes are given a length L_a greater than the sample length L, so that each flow channel comprises a twisted, tortuous, yet independent path from one end of the sample to the other. To compute the macroscopic (Darcy) velocity for this conceptual model, the extra path length must be accounted for. By substitution and rearrangement, Wyllie and Spangler show that Eq. (12-6) becomes

$$v = \frac{\phi r_h^2 L^2 \, \Delta P}{f \mu L_a^2 L} \tag{12-8}$$

so the permeability can be written

$$k = \frac{\phi r_h^2}{f \tau} \tag{12-9}$$

where $\tau = (L_a/L)^2$ is called the *tortuosity*. The product $f\tau$ is called the Kozeny-Carman coefficient.

The Kozeny equation (12-9) has been cast in a variety of equivalent forms in attempts to express k in terms of other rock properties (Wyllie and Spangler, 1952; Brace, 1977). This is done by writing r_h in terms of specific surface area, which is the surface area of the pore space divided by a characteristic volume. Specific surface area will be designated by a subscripted Λ (units of inverse length) to distinguish it from area A (units of length squared). From the definition of hydraulic radius, $r_h = \phi/\Lambda_v$, where the specific surface area Λ_v is the ratio of pore surface area (cm^2) to rock volume (cm^3) and has the dimensions of inverse length. With this substitution,

$$k = \frac{\phi^3}{f \tau \Lambda_v^2} \tag{12-10a}$$

If specific surface area is instead expressed as Λ_ϕ, defined as the ratio of pore surface area to pore volume, $r_h = 1/\Lambda_\phi$ and

$$k = \frac{\phi}{f \tau \Lambda_\phi^2} \tag{12-10b}$$

Another choice is Λ_g, defined as the ratio of pore surface area to grain volume, in which case $r_h = \phi/(1 - \phi)\Lambda_g$ and

$$k = \frac{\phi^3}{f \tau \Lambda_g^2 (1 - \phi)^2} \tag{12-10c}$$

Equations (12-9) and (12-10) have been evaluated in a variety of ways (Wyllie and Spangler, 1952; Carman, 1956; Brace 1977) by estimating $f\tau$ and r_h (or specific area) independently and comparing with measured permeability of laboratory samples. For example, f can be estimated from geometric considerations, τ can be evaluated from the electrical formation factor since $\tau = F^2\phi^2$ for the bundle-of-tubes model, and r_h can be estimated from pore size estimates based on scanning electron micrographs or from capillary pressure measurements.

It is easy to criticize the bundle-of-tubes model because the geometry is obviously a poor representation of a porous rock. Moreover, it appears that Poiseuille flow is a special case of Darcy flow and hence a poor basis upon which to derive a Darcy permeability (Hubbert, 1956). In light of these objections, the experimental agreement obtained between measured and calculated permeabilities is surprisingly good, especially in unconsolidated samples, as shown by data presented by the writers cited above. But even simple models have limited applicability because detailed knowledge of the pore structure is required. Measurement of specific surface area or average pore radius appears to be just as troublesome as measurement of permeability itself.

12-4 DEPENDENCE OF PERMEABILITY ON WATER SATURATION

Despite its limitations, Eq. (12-10) has been used as a starting point for acquiring values of k from well log data alone. The key assumption is that residual (irreducible) water saturation S_{wi} is proportional to specific surface area. Timur (1968) tested a general power law relation

$$k = \frac{a\phi^b}{S_{wi}^2} \qquad (12\text{-}11)$$

by using a data base of 155 sandstone samples from three oil fields (Fig. 12-2). The three sandstone types exhibit varying degrees of sorting, consolidation, and ranges of porosity. If the best statistical fit for the exponent b were 3, then Eq. (12-10*a*) would hold; forms of Eqs. (12-10*b*) and (12-10*c*) could be tested in a similar manner. An exponent of 6, advocated by Schlumberger (1972) was also tested. Timur's statistical results show that the power law model is reasonable for values of b between 3 and 6. The insensitivity to the porosity exponent may seem surprising, but it results from the scatter inherent in k and the variation in k over more than four decades (0.1 to 4000 millidarcies), corresponding to a porosity range of only a single decade (0.03 to 0.3). Results for $b = 4.4$, which produced a fit somewhat better than other values, are shown in Fig. 12-3; the fit was obtained by taking the logarithm of both sides of Eq. (12-11) and testing the correlation coeffi-

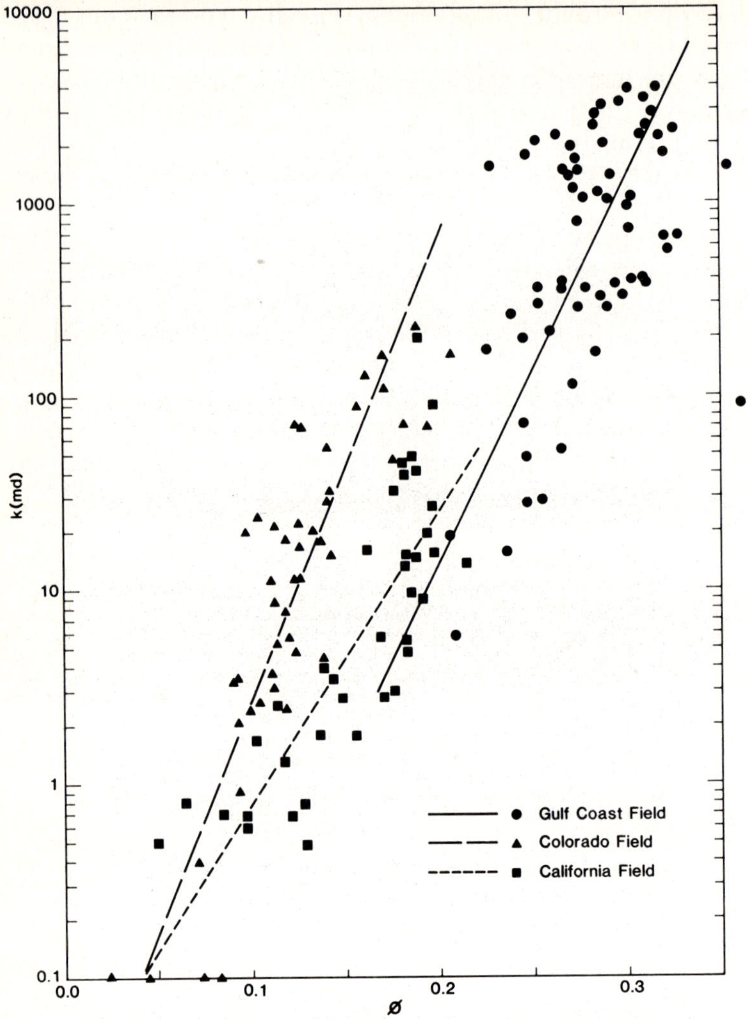

FIGURE 12-2 Permeability and porosity of core samples from three sandstone reservoirs. (From Timur, 1968.)

cient with respect to ϕ^b/S_{wi}^2. The empirical nature of the result deserves emphasis: although in form it is similar to the Kozeny relation, there is no theoretical basis for the substitution of S_{wi} for specific surface area.

There are some practical considerations which further complicate the application of Eq. (12-11) to well logs from a petroleum reservoir. The S_{wi} data on which Eq. (12-11) is based were obtained for a fixed value of capillary pressure P_c. But in a reservoir P_c varies with height;

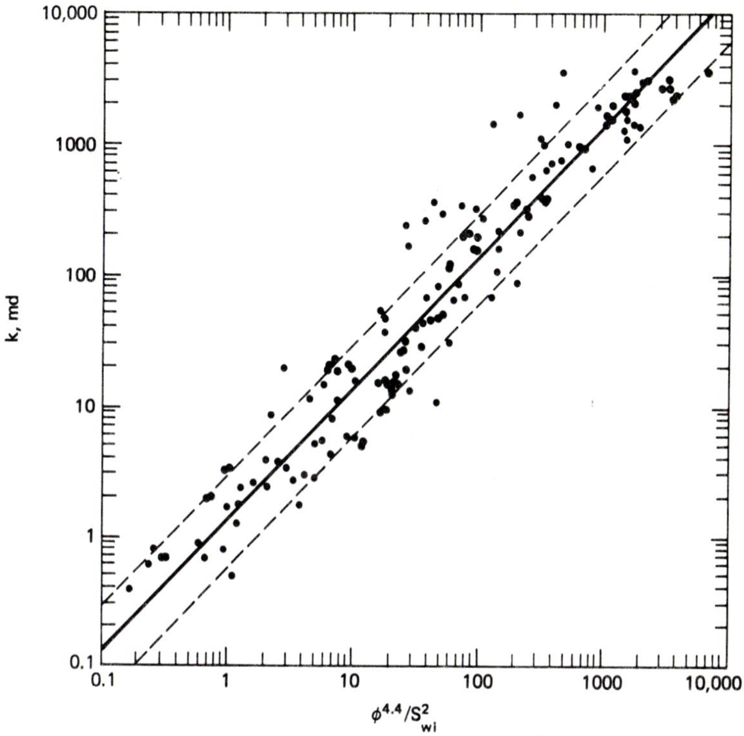

FIGURE 12-3 Fit of permeability to porosity and irreducible water saturation, based on 155 core samples from three fields. (From Timur, 1968.)

and because S_{wi} varies with P_c, it is necessary to assume a functional dependence of S_{wi} upon P_c (Raymer, 1981). Another problem arises within the *transition zone* immediately above the oil-water contact. Above the transition zone all water can be assumed to be irreducible and is attributed to S_{wi}. However, within the transition zone only part of the water is irreducible; the remainder is said to be free (that is, movable). Thus a log-based evaluation of saturation immediately above the oil-water contact will overestimate S_{wi}.

12-5 APPLICATION OF NUCLEAR MAGNETISM FOR PERMEABILITY DETERMINATION

Because the nuclear magnetic measurement (Sec. 5-8d) is insensitive to water close to grain surfaces, it has received special attention as a method for permeability assessment. Recall from Eq. (5-160) that the free fluid index (FFI) is taken to be equal to ϕ_f, the pore space containing free water, and is obtained by extrapolating the decay curve back to zero time. Seevers (1966) describes the relation between two measurable time constants, T_2, the observed decay time for a saturated sample, and T_f, the decay time of a fluid sample, and the porosity by

$$\frac{1}{T_2} - \frac{1}{T_f} = \frac{\phi - \phi_f}{\phi_f T_s} \tag{12-12}$$

where $\phi - \phi_f$ is the fraction of the volume of the rock near the grain surfaces and T_s is a time constant characteristic of the grain-liquid interface. ϕ_f and T_2 are obtained from the intercept and slope of the nuclear magnetic decay curve.

Seevers argues that the surface water coats grains as layers of thickness h. He combines Eq. (12-12) with a version of the Kozeny equation (12-10c) to obtain

$$k = \frac{\phi}{\tau}\frac{h^2}{T_s^2}\left(\frac{T_2 T_f}{T_f - T_2}\right)^2 \tag{12-13}$$

Seevers established experimentally that Eq. (12-13) is valid for packs of sized quartz sand. Further, his experimental work with consolidated sandstones led him to propose the relation

$$k = c\phi_f \frac{(T_2 T_f)^2}{(T_f - T_2)^2} \tag{12-14}$$

The factor c incorporates the complexities of pore structure and surface chemistry. Hence, c varies with rock type and must be independently established by core measurement for each formation of interest.

Another method for converting nuclear magnetic data to permeability is based on Eq. (12-11). If the free fluid in the pore space, given by ϕ_f, and the residual water saturation comprise the total pore space,

$$\phi_f + \phi S_{wi} = \phi \tag{12-15}$$

then by substitution into Eq. (12-11)

$$k = a\frac{\phi^b}{(1 - \phi_f/\phi)^2} \tag{12-16}$$

Timur (1969) developed a more elaborate form of Eq. (12-16) and, by using laboratory determinations of k and ϕ_f, found that k could be satisfactorily predicted from the magnetic decay measurements. Nuclear magnetism logging is so new that neither Eq. (12-14) nor Eq. (12-16) has been adequately tested with field data. It is noteworthy that both Eqs. (12-14) and (12-16) can be traced, directly or indirectly, to the Kozeny equation.

12-6 OTHER METHODS

Resistivity. It often happens that permeability can be inferred, albeit qualitatively, from the manner in which fluids have redistributed themselves in response to some pressure perturbation. An excellent example is the invasion of drilling fluid into the formation as drilling progresses. As discussed in Chap. 2, the mud cake builds to a greater thickness on zones which are highly permeable. Because the mud cake resistivity R_{mc} is less than R_{xo}, electrical resistivity can be used to indicate mud cake thickness. A microlog pad-type device contains electrodes which provide two resistivity measurements, one penetrating deeper than the other. If mud cake is thin or nonexistent, the logs will read the same but the shallow probe will record lower resistivity values in zones with mud buildup and thus reveal the more permeable zones.

Acoustic. There is considerable evidence (Biot, 1956; Berryman et al., 1983) that attenuation of acoustic waves depends on permeability. Unfortunately, attenuation also depends on matrix properties, porosity, saturation, and fracturing (Sec. 7-2b). Consequently, it seems unlikely that attenuation can be used to infer permeability except under very restricted conditions.

12-7 SUMMARY

We have seen that there is no logging sonde which measures permeability directly. This is understandable, because a direct measurement requires that fluid flows through the rock while pressure and flow are measured. The methods which provide the best promise of determining permeability are nuclear magnetism and (perhaps) acoustic attenuation. Both are recent developments and are not yet fully tested. Both methods, and any others which come along, will benefit from improved understanding of what the microscopic controls on permeability really are and how they relate to a specific physical measurement technique. Until then, the most common approach to permeability determination will remain the construction of a plot of core data such as those in Fig. 12-2. If the permeability-porosity relation is known for the formation of interest, and if the scatter is not too great, permeability can be estimated from porosity logs. The clever interpreter may be able to improve the estimation by utilizing water saturation data or other information on grain size.

REFERENCES

American Petroleum Institute: "Recommended Practice for Determining Permeability of Porous Media," API RP 27, Dallas, Tex., 1956.

Amyx, J. W., D. M. Bass, Jr., and R. L. Whiting: *Petroleum Reservoir Engineering*, McGraw-Hill, New York, 1960.

Berryman, J. G., B. P. Bonner, and R. C. Y. Chin: "Correlation of Ultrasonic Attenuation and Fluid Permeability in Very Low Porosity Water-Saturated Rocks," *Geophys. Res. Letters*, 1983.

Biot, M. A.: "Theory of Propagation of Elastic Waves in a Fluid-Saturated Porous Solid," *J. Acous. Soc. Am.*, vol. 28, 1956, p. 168.

Brace, W. F.: "Permeability from Resistivity and Pore Shape," *J. Geophys. Res.*, vol. 82, no. 23, Aug. 10, 1977.

———: "Permeability of Crystalline and Argillaceous Rocks," *Int. J. Rock Mech. Min. Sci. & Geomech. Abstr.*, vol. 17, 1980, pp. 241–251.

Carman, P. C.: *Flow of Gases through Porous Media*, Academic, London, 1956.

Freeze, R. A., and J. A. Cherry: *Groundwater*, Prentice-Hall, Englewood Cliffs, N.J., 1979.

Gale, J. E.: "Assessing the Permeability Characteristics of Fractured Rock," in T. N. Narasimhan (ed.), *Recent Trends in Hydrogeology*, Spec. Paper 189 Geol. Soc. Am., Boulder, Colo., 1982, pp. 163–182.

Hubbert, M. K.: "Darcy's Law and the Field Equations of the Flow of Underground Fluids," *Trans. AIME*, vol. 207, 1956, pp. 222–239.

Narasimhan, T. N.: "Conversion Tables," in *Recent Trends in Hydrogeology*, Spec. Paper 189 Geol. Soc. Am., Boulder, Colo., 1982.

Raymer, L. L.: "Elevation and Hydrocarbon Density Correction for Log-Derived Permeability Relationships," *Log Analyst*, vol. 22, no. 3, May–June 1981, pp. 3–7.

Schlumberger: *Log Interpretation: Principles*, Schlumberger, Houston, Tex., 1972.

Seevers, D. O.: "A Nuclear Magnetic Method for Determining the Permeability of Sandstones," *SPWLA Trans.*, 1966, pp. L1–L14.

Somerton, W. H.: "Porous Rock—Fluid Systems at Elevated Temperatures and Pressures," in T. N. Narasimhan (ed.), *Recent Trends in Hydrogeology*, Spec. Paper 189 Geol. Soc. Am., Boulder, Colo., 1982, pp. 183–198.

Timur, A.: "An Investigation of Permeability, Porosity, and Residual Water Saturation Relationships for Sandstone Reservoirs," *Log Analyst*, July–August 1968, pp. 8–17.

————: "Pulsed Nuclear Magnetic Resonance Studies of Porosity, Movable Fluid, and Permeability of Sandstones," *J. Petr. Tech.*, 1969. Reprinted in SPE Reprint Vol. 1, *Well Logging*.

Wyllie, M. R. J., and M. B. Spangler: "Application of Electrical Resistivity Measurement to Problem of Fluid Flow in Porous Media," *Bull. AAPG*, vol. 36, no. 2, 1952, pp. 359–403.

13

Fluid Movement

In the preceding chapter we discussed the use of logs to measure permeability and indicated that this remains an unsolved problem. In many cases detecting the movement of fluids in the subsurface is more important than the measurement of permeability itself. Sometimes it is necessary to detect fluid movement within or around a wellbore. Wells designed to produce or inject fluids often do not function as designed because of a problem such as a poor cement bond, plugged perforations, or corrosion of casing. If an oil well is producing too much water along with oil, it is necessary to determine which zone or zones is producing the water so that remedial action can be taken. In the petroleum industry, the diagnosis of this class of problems by logging in cased wells has become specialized to the point that it is termed *production logging* to distinguish it from open-hole logging.

In other cases it may be desirable to determine how fluids have migrated through rock. As an example, consider the leakage of fluid from an underground storage facility, the commingling or contamination of water in a good-quality aquifer with water from an aquifer poor in quality, or the progress of injected fluids across an oil reservoir during secondary or tertiary recovery. Logs sensitive to a particular fluid can be run sequentially to map fluid movement as a function of time.

13-1 FLUID IDENTIFICATION

When multiphase flow within a wellbore is recognized, it is often desirable to know the fraction of each fluid type present.

Electrical properties. Salinity can be roughly determined by resistivity, as described in Chap. 5, and hence it can be used to distinguish between different sources of groundwater. A dielectric measurement has been used as the basis for a sonde which can determine the ratio of water to oil in a mixture (Carlson and Roesner, 1982).

Differential pressure. Another method is the differential pressure probe commonly employed in petroleum production logging and referred to as the gradiomanometer (Schlumberger, 1973). The sonde measures the pressure drop between two transducers spaced a distance h apart in the wellbore fluid. The pressure drop is the sum of the hydrostatic pressure ρgh and frictional pressure losses. Although the frictional losses are usually negligible, corrections for various pipe diameters and flow rates are available. If the hole is inclined, correction must be made for the reduced vertical separation between the two transducers. The log is scaled in units of density; and when a single phase is present, it can readily distinguish among gas, oil, and water.

Fractional volumes can be determined if two phases such as water and oil are present and their downhole densities are known. In the wellbore, the gradiomanometer measures the fluid density

$$\rho = f_w \rho_w + (1 - f_w)\rho_o \tag{13-1}$$

so the water fraction is

$$f_w = \frac{\rho - \rho_o}{\rho_w - \rho_o} \tag{13-2}$$

Thermal neutron absorption. Both steady-state and neutron die-away measurements of the thermal neutron population can readily distinguish gas because the absorption cross section of gas is very low relative to oil or water. McGhee et al. (1974) present a number of field examples of the use of the thermal neutron die-away sonde in which gas is identified both within and behind casing.

13-2 GAIN OR LOSS OF FLUID FROM WELLBORE

Zones of fluid entry or exit from a wellbore can be located either by imprinting or by flow profiling. By "imprinting" we mean that some record of fluid entry remains within the formation itself for a time long enough to be recorded. One method of imprinting (not commonly used) is first to run a γ-ray base log and then inject a radioactive material in the form of particles or a liquid into the formation. After the fluid remaining in the wellbore is circulated out, a second γ-ray log is run to locate the injected fluid. The same procedure can be followed by using neutron sondes after injection of a neutron-absorbing material such as a chemical compound of boron (Matiisen, 1973).

Thermal imprinting. The principle of thermal imprinting was discussed in Chap. 4: the injection of fluid colder than the formation will cool the loss zone so that the resulting temperature perturbation and its recovery can be recognized on subsequent temperature logs. Figure 13-1 shows three such cooling features where surface waters entered a formation at depths greater than 3000 ft (915 m). Flow rates cannot ordinarily be "backed out" from analysis of such anomalies; fortunately, more accurate methods are available for quantitative flow estimates.

Tracers. When the linear rate of flow within a wellbore is measured and the volumetric rate is computed and plotted as a function of depth, the resulting plot is called a flow profile. There are many methods of sensing flow within a pipe, but only a few have been successfully adapted to logging. The radioactive tracer method is one of three which we will discuss here. The probe incorporates a chamber filled with a radioactive fluid, ejection ports, and a pair of γ-ray detectors separated some distance from the ejector. One of several isotopes used as the tracer material is ^{131}I, which has an 8-day half-life. With the sonde held at a fixed depth, the tracer is ejected into the well bore and entrained by the flowing wellbore fluid. The passage of the radioactive fluid past the pair of γ-ray detectors is then timed and converted to flow velocity.

The measured flow velocities are converted directly to volumetric flow rates after corrections for variations in wellbore diameter are made. Such corrections, obtained from caliper measurements, are required in open wells and may be necessary in cased wells if the casing is corroded or if scale is present. With a number of such measurements at different depths, a flow profile of the type shown in Fig. 13-1 can be produced. Tracer logs were run at three stages during the drilling of the well. The abrupt offsets in flow rate demarcate three major zones of exodus in the final flow profile. The tracer log shows how much flow was lost to each interval; the corresponding temperature logs show where loss occurred, but quantitative information cannot be extracted from the temperature excursions.

The advantage of tracer logging is that quite low flow rates can be determined. If the rate is so low that dispersion obscures the arrival of

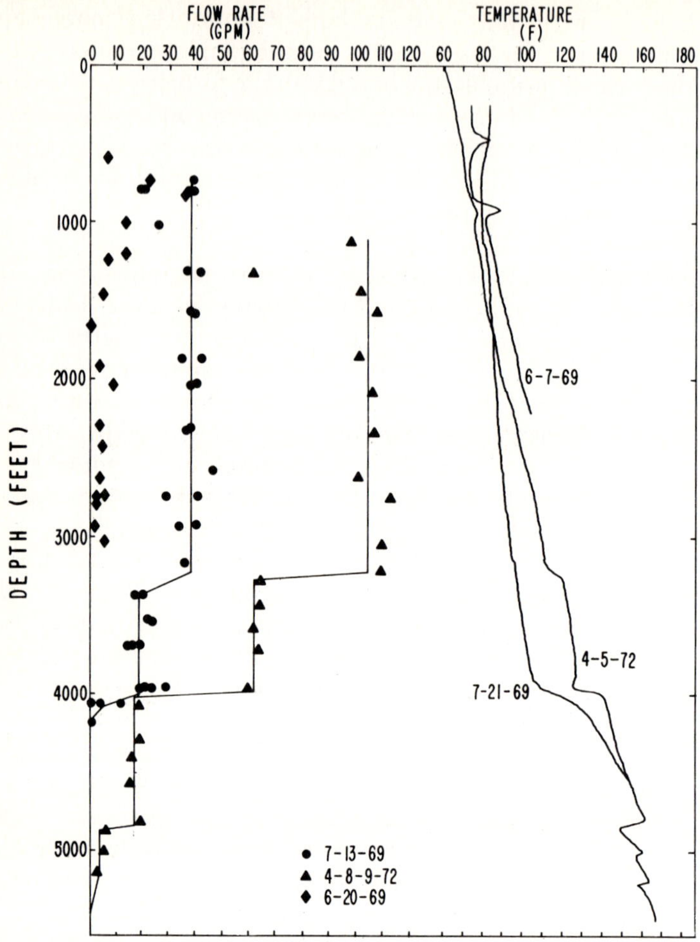

FIGURE 13-1 Temperature logs and injection flow profiles taken in a layered basalt sequence (Apps et al., 1979). Flow profiles are based on radioactive tracer logs; each point represents a separate tracer slug. Logs were acquired on three separate dates as the well was drilled deeper.

the tracer slug, the operator has the option of moving the sonde in the wellbore to locate the slug. As illustrated by the γ-ray logs shown in Fig. 13-2, a sequence of such logs can reveal the dispersion and migration of a single pulse with time, and an apparent velocity can then be computed. The reduction in amplitude and change of shape of the third log is interpreted as loss of fluid to a fracture at 544 m.

It may happen, however, that the tracer does not move at all, and the operator must decide if there is a zone of no flow or if the tracer

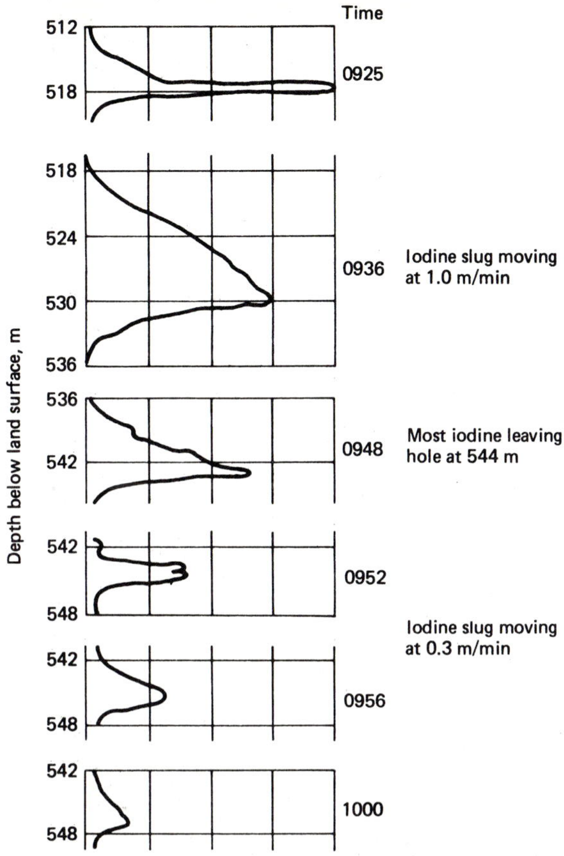

FIGURE 13-2 γ-Ray logs of iodine-131 tracer slug shortly after it is ejected at time 0925 and on five subsequent passes as it migrates downhole. (Marine, 1980.)

has adhered to the wall. If the ejected fluid is immiscible with the well-bore fluid, the ejected slug may simply stick to the wall next to the ejector port and leave only a small fraction entrained in the flow (Roesner et al., 1982). A cure for this problem may be to alter the ejection rate or to choose a tracer fluid which will be miscible with the flow stream.

Spinner. The spinner sonde senses the relative velocity of an impeller vane with respect to the fluid column. Some sondes incorporate a packer which expands against the wall of the hole and forces all fluid to flow past the spinner. Only station measurements can be obtained;

but if made properly, they are accurate and no caliper correction is necessary.

In another design use is made of a spinner with blades which fold up to allow the sonde to pass through small-diameter tubing. Once below the tubing, the vanes unfold and subtend a greater cross-sectional area than is possible with fixed vanes. The sensor is referred to by Leach et al. (1974) as a "rotating vane type velocimeter." Vane pitch is so designed that the vector sum of fluid velocity and vane angular rotation is as parallel as possible to the blade face. Spin rate is directly proportional to relative velocity for an ideal vane, although viscous and frictional effects produce departures from linear response and reduce sensitivity at low spin rates. To overcome this problem, the operator can generate high spin rates by logging up or down at a uniform rate. Figure 13-3 shows a sequence of runs at different pull speeds. Water flows upward from the three perforated intervals where it is being produced. The three downward logging runs show a decrease with depth as the spinner encounters steadily decreasing upward flow. The spin rate of the three upward runs is necessarily less than that of the downward runs and reverses direction where the logging speed is equal to the flow rate.

The fraction of total flow at a given depth can be simply defined on the log by appending a scale with end points at zero and maximum flow, as shown in Fig. 13-3 for the logging run done at 115 ft/min (35 m/min). However, the multiple runs can be used to construct a graph which permits quantitative flow rate measurements. This is done by picking spin rates at depths A through D (Fig. 13-3) away from the perforations where the flow is judged to be free of turbulence. The threshold speed at which the spinner does not spin can be calculated from the reversals. Then a response curve of spin rate to sonde velocity (and equivalently, to volumetric flow rate) can be drawn (Schlumberger, 1973).

Noise logs. Flow can also be sensed with the sibillation (noise) log, which consists of an acoustic transducer sensitive to energy in the 200- to 2000-Hz frequency range (Robinson, 1976). Noise generated at a constriction by flow at a rate Q is proportional to the energy lost due to pressure drop ΔP and hence to the product $\rho Q \, \Delta P$ (McKinley and Bower, 1979). In turbulent flow, pressure drop ΔP is proportional to $(Q/A)^2$, so that the noise intensity

$$N = \frac{C\rho Q^3}{A^2} \tag{13-3}$$

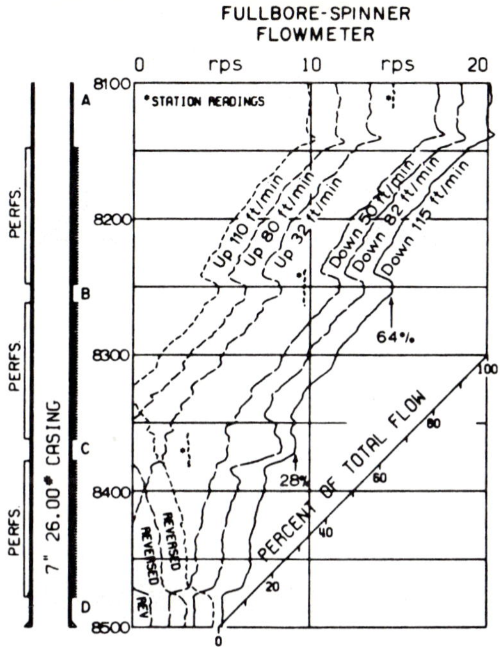

FIGURE 13-3 Multiple runs of a spinner sonde. (Courtesy of Schlumberger, 1973.)

where A is the cross-sectional area open to flow, ρ is fluid density either estimated or measured with a gradiomanometer, and C is a constant for Reynold's numbers in excess of 6000.

Like the tracer and spinner logs, the noise log can be used to obtain a flow profile by measuring noise intensity at selected depth intervals. The area A is the cross-sectional area between sonde and casing. The noise intensity is converted to flow rate Q by using Eq. (13-3), and the change in computed Q between stations gives the loss from the wellbore.

McKinley and Bower estimate the lower detection limit to be about 0.0023 m^3/s (1250 bbl/day) for single-phase flow. They point out that the dependence on the cube of Q makes the noise log sensitive to small changes in Q. They also examine the characteristics of noise generated by liquids and gases flowing through perforations; single-phase flow through a perforation requires modification of Eq. (13-3). Their test data show that the empirical relation

$$N = B\left(\frac{\rho Q}{\mu d}\right)^{2.86} \tag{13-4}$$

is a useful approximation, where μ is viscosity and d is the perforation diameter. Because of the exponential dependence on Q, a few high-flow perforations can generate more noise than a lot of low-flow perforations.

Multiphase flow. Schlumberger (1973) describes the different flow regimes which can exist in a producing well where gas, oil, and water are present. We will confine our discussion to the case of *bubble flow*, in which oil flows as bubbles in a continuous water phase. The fraction of water present f_w is called the *water holdup*. The oil bubbles flow faster than the water by a velocity v_s called the *slip velocity*. Slip velocity is a function of f_w and of the density difference $\rho_w - \rho_o$. Because of the slip effect, the flow rate of oil Q_o can be substantially greater than that of water Q_w. The two are related by

$$\begin{aligned} Q_w &= f_w A v_w \\ Q_o &= (1 - f_w) A (v_w + v_s) \end{aligned} \tag{13-5}$$

where A is the cross-sectional area open to flow and v_w is the linear flow rate of water. Schlumberger (1973) gives a graphical method for solving Eqs. (13-5) based on gradiomanometer and flowmeter measurements. Once Q_o and Q_w are known at several depths in a producing well, action can be taken to eliminate the zones which contribute excessive quantities of water.

13-3 CHANNELING OF FLUID BEHIND CASING

Logs can be used to detect flow through leaks within the cemented annulus between casing and the formation. Two methods discussed in Sec. 13-2 are applicable here. The noise log can sometimes be used to detect changes in the noise level generated as the fluid flows past restrictions in the annulus. A radioactive tracer released next to perforations during injection will flow through the perforations and either enter the formation or flow into channels in the cemented annulus. If it goes into channels, its movement can be detected by running a series of γ-ray logs.

Temperature. Temperature measurements also can be used to search for flow behind casing. A device described by Cooke (1979) presses a temperature sensor against the casing wall and rotates it around the circumference at a speed of one revolution every 4 min. The difference between the casing temperature and the fluid temperature is recorded as a function of angle. Small temperature anomalies are then indicative of fluid movement behind the casing. The sonde can then be moved and reset at different depths to map the vertical extent of a fluid channel. This application illustrates the extreme sensitivity of differential temperature measurements: changes of 0.005°C are significant.

Neutron activation. An interesting application of neutron activation is described by Arnold and Paap (1979). A 14-MeV neutron source produces the reaction $^{16}O(n, p)^{16}N$ in water. If the water is flowing, the activated ^{16}N moves with it, decaying as it moves with a decay constant λ of 0.0972 s^{-1}. The decay γ's are detected by a pair of detectors located in the sonde either above or below (as appropriate for the flow direction) the neutron source. Because the decay of the activated nitrogen is exponential with time (Sec. 6-1b), the count rate J_2 at the farther detector will be less than J_1 at the nearer detector. The expression for flow velocity is given by

$$v = \frac{\lambda L}{\ln (J_1/J_2)} \tag{13-6}$$

where L is the detector spacing. The Arnold and Papp analysis indicates that the method is sensitive to flow velocities in the range 0.01 to 0.25 m/s. Besides measuring flow rate, the authors claim that the distance of the flow channel from the detectors can be established. Gammas of different energies attenuate with different attenuation coefficients (Sec. 6-2b), so spectral ratio measurements of different-energy γ's can be used to determine the distance from a common source.

13-4 MOVEMENT OF FLUIDS WITHIN THE FORMATION

We now turn our attention to the problem of determining how fluids have redistributed within a formation once production or injection has commenced. Monitoring is important in conventional reservoir depletion and even more so when secondary or tertiary recovery methods are used. The approach is to measure some physical property of the replacing fluid which differs from that of the original fluid. If production is from an open hole, resistivity can be used to monitor changes in S_w (Chap. 11). Usually, however, the presence of casing or well screen in the wellbore precludes running the resistivity logs. In special circumstances, nonconducting fiberglass casing may be installed to permit use of the induction sonde. But in steel casing, a nuclear method such as thermal neutron die-away must be used.

Neutron die-away. The thermal neutron die-away log is well suited to monitor changes in S_w by obtaining a series of runs spread throughout the production life of the reservoir. Such a process is referred to as *time-lapse logging.* Youmans et al. (1970) recommend that the base log be obtained days or weeks after cementing in the casing. This allows time for the filtrate to diffuse away so the fluids near the wellbore are more nearly representative of the original fluid saturations. Next it is assumed that no new fluids are introduced between runs; that is, water and hydrocarbon cross sections Σ_w and Σ_h remain unchanged and only S_w changes as water replaces the produced oil. If Σ_1 represents the cross section measured from the first of two runs (it might be the base log) and Σ_2 is from the second run, then, following Eq. (6-51),

$$\Sigma_1 = S_{w1}\phi\Sigma_w + (1 - S_{w1})\phi\Sigma_h + (1 - \phi)\Sigma_{ma} \qquad (13\text{-}7a)$$

$$\Sigma_2 = S_{w2}\phi\Sigma_w + (1 - S_{w2})\phi\Sigma_h + (1 - \phi)\Sigma_{ma} \qquad (13\text{-}7b)$$

By taking the difference between Eqs. (13-7*a*) and (13-7*b*) and defining $\Delta S_w = S_{w2} - S_{w1}$, we get

$$\Delta S_w = \frac{\Sigma_2 - \Sigma_1}{\phi(\Sigma_w - \Sigma_h)} \qquad (13\text{-}8)$$

Porosity can be estimated from an open hole log; Σ_w and Σ_h must be measured or estimated. Clavier et al. (1971) give a table of tolerance factors (Sec. 11-2) for Σ and ϕ; the factors indicate how accurately these parameters must be known for ΔS_w to be within a specified range of error.

Residual oil saturation from neutron die-away. The residual oil saturation S_{or} is the fraction of oil remaining in pore space after all the movable oil has been produced during waterflood recovery. Hence, it is a key parameter in assessing the economics of a tertiary recovery operation. The neutron die-away log is used to measure S_{or} in what is referred to as the *log-inject-log method,* so named because an injection stage is used to sweep out the formation waters near the wellbore and replace it with fluid of contrasting Σ (Murphy and Owens, 1973; Richardson et al., 1973). The logs run before the injection measure the total cross section of the formation containing fluid with cross section Σ_{w1}; the post-injection log responds to the formation containing fluid with cross section Σ_{w2}. It is assumed that the oil saturation was already at S_{or} when injection began and that injection leaves S_{or} unchanged, so S_w is the same for both logs. Hence, the capture cross sections measured by the before-and-after runs are

$$\Sigma_1 = S_w\phi\Sigma_{w1} + (1 - S_w)\phi\Sigma_h + (1 - \phi)\Sigma_{ma} \qquad (13\text{-}9a)$$

$$\Sigma_2 = S_w\phi\Sigma_{w2} + (1 - S_w)\phi\Sigma_h + (1 - \phi)\Sigma_{ma} \qquad (13\text{-}9b)$$

By taking the difference between Eqs. (13-9*a*) and (13-9*b*) and rearranging, we get

$$S_w = \frac{\Sigma_1 - \Sigma_2}{\phi(\Sigma_{w1} - \Sigma_{w2})} \tag{13-10}$$

Or, in terms of the residual oil saturation S_{or},

$$S_{or} = 1 - \frac{\Sigma_1 - \Sigma_2}{\phi(\Sigma_{w1} - \Sigma_{w2})} \tag{13-11}$$

The authors cited above describe the use of Eq. (13-11) to estimate S_{or} in wells drilled specifically for the purpose. Porosity can be obtained from open hole logs and core. S_{or} values thus obtained should be compared with values measured from core or from resistivity logs obtained before casing was installed. The authors also describe extensions of the above sequence in which the volume near the wellbore is chemically flooded to remove all the residual hydrocarbon. Other variations on the theme also are possible.

Measurement of flow past observation wells. It is common practice to inject water into the downdip zone of a reservoir in order to maintain adequate fluid pressure to produce the oil. The operation is referred to as a *waterflood.* As oil is produced, the injected water migrates from the injection wells toward the producers. Under ideal circumstances it will move through the formation as a uniform front and displace some fraction of the oil and push it ahead as a "bank" of increased oil saturation. There are many complications which can upset this simple picture. The water may channel directly through the more permeable layers, bypass a large fraction of the oil in the less permeable layers, and produce a high water-to-oil ratio in the producing wells. Or flow may be blocked by discontinuous shale members or by lateral gradations in permeability within the reservoir rock.

To monitor the progressive movement of injection fluid across a field, observation wells can be drilled at suitable locations between the injectors and producers. Logs run in the observation wells at different times then give a series of pictures of fluid infiltration as time progresses. The time duration may be of the order of days to observe "breakthrough" of the fluid front in a pilot injection test, or it may be of the order of years to observe the encroachment of flood waters displacing an oil column.

The examples shown in Fig. 13-4 are taken from a surfactant flood pilot test which lasted 17 months. Continuous injection of surfactant mobilizes oil and forms an "oil bank," which is pushed away from the

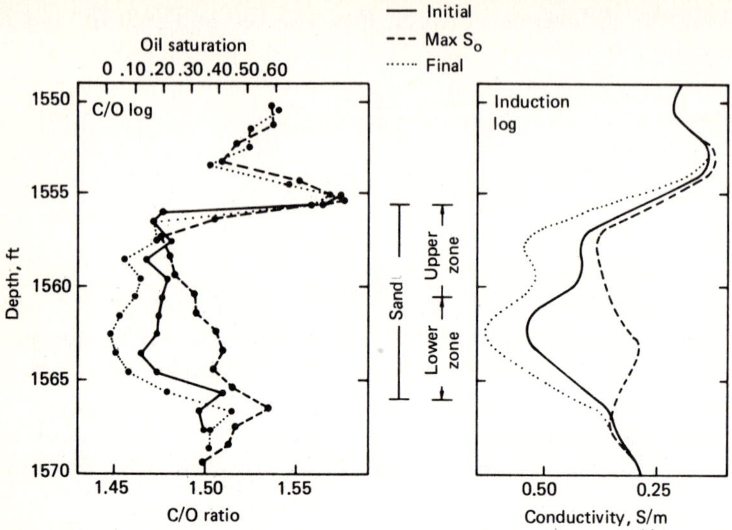

FIGURE 13-4 Logs in observation well monitoring a surfactant flood. Only three of the logs acquired over months of testing are shown: base, maximum oil saturation as oil bank passed by, and final. (From Felder and Hoyer, 1982. Copyright 1982, SPE–AIME.)

injection well. One of several observation wells, lined with fiberglass casing over the test interval, was located 14 m from the nearest injection well. Carbon/oxygen (C/O) and induction logs were run to monitor changes in oil saturation as the surfactant flood progressed. Three key logs shown in Fig. 13-4 are the base log, the log obtained as the oil bank passed and oil saturation was maximum, and a final log obtained after the oil bank had passed. Conductivity decreases and the C/O ratio increases as the oil bank passes the observation well.

By knowing the distance between injector and observation wells and measuring the transit time for the injected fluid to arrive at the observation well, the macroscopic flow velocity for the formation can be established. If some physical property such as R_w of the two fluids involved is sufficiently different, it is possible to measure the change in saturation as a function of time and thereby gain information on sweep efficiency. Under circumstances in which flow is rapid enough and there is a source of water hotter or colder than the formation, temperature measurements can be used (Keys and Brown, 1978). Usually either salinity or hydrocarbon content is the contrasting property and the suitable logs from which to choose are the induction resistivity, carbon/oxygen, and neutron die-away logs.

In principle, logging in observation wells does not differ greatly from logging in production wells. Case histories discussed by Felder and Hoyer (1984), Youngblood (1980), and Richardson (1979) illustrate some of the unique aspects. Because the observation wells are drilled specifically for logging, well completion can be designed to suit the logging program. Fiberglass casing can be used to line the observation wells if induction logs are to be used. Induction logs have a greater radius of investigation than the neutron lifetime or activation logs. If open hole completion is possible, the wellbore can be filled with heavy oil which will invade little into the formation and will produce minimal response on the induction or neutron logs.

All practitioners emphasize the need for quality control in sequential logging operations. Depth control is especially important, and logs are usually adjusted to a common depth at which no invasion changes

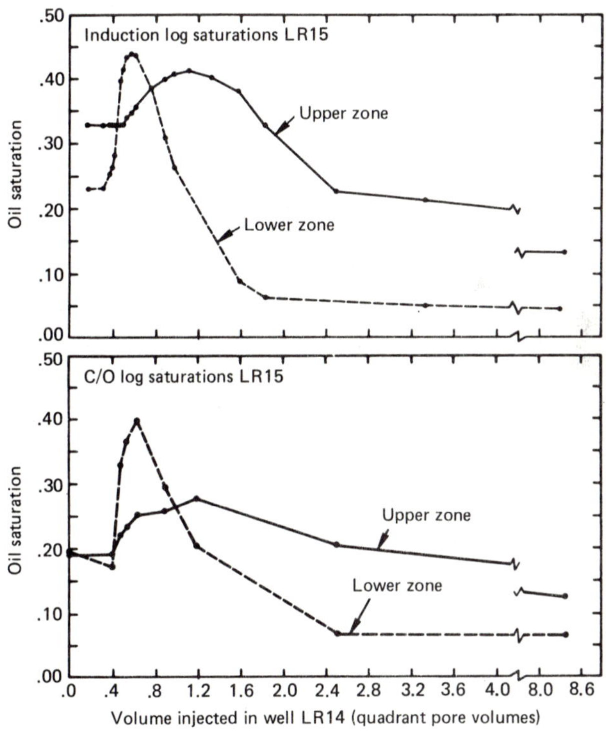

FIGURE 13-5 History of log response, converted to oil saturation, throughout surfactant flood for the well shown in Fig. 13-4. Response is given with respect to volume injected into a nearby well, rather than time. Each point represents the response of an individual log. (From Felder and Hoyer, 1984. Copyright 1982, SPE–AIME.)

occur in a formation above or below the formation in which flow is occurring. If possible, the same sondes should be used throughout the measurement sequence, and they should be calibrated regularly at the wellsite by following the same procedure each time. Logs should be compared with previous ones, and they should be studied to be sure there are no changes in formations where none are expected.

If sufficient care is used in running the logs and if core data are available to provide ϕ and S_w calibrations, the sequence of time-lapse logs (Fig. 13-4) can be converted to fluid saturation changes as a function of time or of volume injected (Fig. 13-5). Saturations for the induction log are obtained by using Eq. (11-2) and for the C/O log by using Eq. (11-44). Averages in this example are taken over two zones designated "upper" and "lower." Agreement between the two methods is good in the lower zone but poor in the upper zone, where thin beds may have affected the induction log values. An error analysis by Felder and Hoyer (1984) shows that both logs have an error bound of ± 0.06 for S_w. Relative errors in S_w expected between different runs of the same type of log in the same well are about ± 0.03.

REFERENCES

Apps, J., and 12 others: "Geohydrological Studies for Nuclear Waste Isolation at the Hanford Reservation," Lawrence Berkeley Lab. Rept. LBL-8764, 1979.

Arnold, D. M., and H. J. Paap: "Quantitative Monitoring of Water Flow behind and in Wellbore Casing," *J. Petr. Tech.*, vol. 31, no. 1, January 1979.

Carlson, N. R., and R. E. Roesner: "Water-Oil Flow Surveys with Basket Fluid Capacitance Tool," *Trans. 23d Annual Logging Symp. SPWLA*, 1982, paper F.

Cooke, C. E., Jr.: "Radial Differential Temperature (RDT) Logging—A New Tool for Detecting and Treating Flow behind Casing," *J. Petr. Tech.*, vol. 31, no. 6, June 1979.

Clavier, C., W. R. Hoyle, and D. Meunier: "Quantitative Interpretation of Thermal Neutron Decay Time Logs," Part II, *J. Petr. Tech.* vol. 23, June 1971.

Felder, R. D., and W. A. Hoyer: "The Use of Well Logs to Monitor a Surfactant Flood Pilot Test," *J. Petr. Tech.*, vol. 36, no. 9, 1984, pp. 1379–1391.

Keys, W. S., and R. F. Brown: "The Use of Temperature Logs to Trace the Movement of Injected Water," *Ground Water*, vol. 16, no. 1, January–February 1978.

Leach, B. C., J. B. Jameson, J. J. Smolen, and Y. Nicolas: "The Full Bore Flowmeter," SPE paper 5089, 1974.

Marine, I. W.: "Determination of the Location and Connectivity of Fractures in Metamorphic Rock with In-Hole Tracers, *Ground Water*, vol. 18, no. 3, May–June 1980.

Matiisen, A.: "Defining Injection Profiles Using the Neutron Lifetime Log and High Sigma Fluids, *Trans. 14th Ann. Logging Symp. SPWLA*, 1973, paper V.

McGhee, B. F., J. A. McGuire, and H. L. Vacca: "Examples of Dual Spacing Thermal Neutron Decay Time Logs in Texas Coast Oil and Gas Reservoirs," *Trans. 15th Ann. Logging Symp. SPWLA*, 1974, paper R.

McKinley, R. M., and F. M. Bower: "Specialized Applications of Noise Logging," *J. Petr. Tech.*, vol. 31, no. 11, November 1979.

Murphy, R. P., and W. W. Owens: "The Use of Special Coring and Logging Procedures for Defining Reservoir Residual Oil Saturations," *J. Petr. Tech.*, vol. 25, July 1973.

Richardson, J. E.: "Monitoring Flood Profiles with Induction Logs," *J. Petr. Tech.*, vol. 31, 1979, p. 19.

————, R. E. Wyman, J. R. Jordan, and F. R. Mitchell: "Methods for Determining Residual Oil with Pulsed Neutron Capture Logs," *J. Petr. Tech.*, vol. 25, May 1973.

Robinson, W. S.: "Recent Application of the Noise Log," *Trans. 17th Annual Logging Symp. SPWLA*, 1976, paper Y.

Roesner, R. E., A. J. LeBlanc, J. E. Strassner, and J. R. Bragg: "Visual Flow Loop Investigation of Nuclear Flolog Performance in Non-Newtonian Fluids," *Trans. 23d Annual Logging Symp. SPWLA*, 1982, paper JJ.

Schlumberger: *Production Log Interpretation*, Schlumberger, New York, 1973.

Youmans, A. H., D. W. Bishop, and P. A. Wichman: "Application of the Neutron Lifetime Log in New Wells," *Trans. 11th Ann. Logging Symp. SPWLA*, 1970, paper N.

Youngblood, W. E.: "Application of Pulsed Neutron Decay Time Logs to Monitor Waterfloods with Changing Salinity," *J. Petr. Tech*, vol. 32, no. 6, June 1980.

14

Fractures

In many rocks, fractures provide the permeable pathways for fluid flow and limit the mechanical strength of the rock mass. Often, stimulation of fractured zones will improve or initiate production of oil and gas wells. Consequently, one desires to be able to locate individual fractures or fractured zones so that one can choose the depth intervals for stimulation or production. In ore bodies, the amount of explosive required to produce fragments of a given size depends on the number of fractures present. Knowledge of the location of fractures is therefore useful for the design of blast patterns. The presence of fractures is one of the major considerations in the design of repositories for high-level radioactive wastes. One wishes to locate sites without fractures if possible. For all of these reasons, there is considerable motivation to be able to use logs to locate fractures.

14-1 INTRODUCTION

A major problem in the study of fracture detection is the establishment of "ground truth." If logging methods are to be tested, the quantities measured in the test well must be known. For variables like porosity and density we can rely on core for truth, but for fracturing it is more difficult. Fractures can be induced in core by drilling, so core is not necessarily reliable. Sometimes an impression packer can be used to establish truth.

One of the difficulties is the nature of fractures. The fracture aperture (the perpendicular distance across the fracture) can vary from micrometers to centimeters either in an individual fracture or from one fracture to another. Some fractures may be closed by pressure or filled with minerals and therefore no longer have any opening at all. Some may be fault zones and be filled with rubble. (They are easy to detect.) Consequently, methods that work well for one type of fracture are quite likely to be useless for another.

In some situations (for example, geotechnical engineering) there is greater interest in the *quality* of the rock than in individual fractures. A term called the *rock quality designation* (RQD) is defined as the ratio of the recovered core consisting of pieces greater than 10 cm long to the total length drilled in a core hole. Then the logging problem becomes one of relating whatever is measured by the log to the rock quality.

The publication of a review paper listing, and briefly describing, different methods of using logs to locate fractures appears to be an approximately annual ritual in the logging community (Shanks et al., 1976; Heflin, 1979; Suau and Gartner, 1980; Julian, 1982). Austin chalk and the surrounding formations seem to be the formations of choice for many of these publications. Only one of these (Paillet, 1981) compares the ability of different logging methods to locate the same fractures.

Most fractures of interest in the petroleum industry are more or less vertical *(subvertical)*. It is assumed that this is the case because horizontal fractures at depth tend to be closed by overburden pressure and therefore are not very permeable. The planes of fractures caused, either accidentally or deliberately, by pressure in the borehole *(hydraulic fractures)* are perpendicular to the least principal stress. The greatest principal stress is generally vertical, so hydraulic fractures will generally be vertical. Consequently, many of the methods described are designed to locate vertical fractures. In this book we shall consider vertical fractures to be those whose planes are approximately parallel with the borehole in which measurements are made.

In other fields, such as hydrology, waste isolation, and engineering geology, horizontal fractures are as important as vertical fractures. At the relatively shallow depths of interest in these technologies, horizontal fractures can open, and they are often very significant.

14-2 ACOUSTIC METHODS

As we saw in Sec. 7-2a, the equation commonly used to obtain porosity from acoustic velocity [Eq. (7-27)] often does not work if the porosity is in fractures. In some methods of fracture location it is assumed that the difference between porosity from an acoustic log and that from, for example, a neutron log is due to fractures and that such a difference at a given depth is evidence for fractures at that depth. We shall not consider methods such as this, which infer the presence of fractures from otherwise unexplained anomalies in acoustic logs; we shall consider only methods that look for direct influence of individual fractures or fracture zones.

It should be noted that acoustic logs generally work only in liquid-filled holes. Consequently, they cannot be used for fracture analysis in dry media unless either special sondes are used or the holes are filled with liquid.

Wave propagation in fractured media. Morris et al. (1964) describe three ways in which fractures reduce the amplitude of compressional and shear arrivals. First, liquid-filled fractures act as thin beds having acoustic impedance different from that of the rest of the medium. Consequently, the transmission coefficient [Eqs. (7-26), (7-27)] can be small if Z_1 is significantly different from Z_2. Second, fractures can cause increased borehole rugosity and thereby decrease the coupling of the signal to the formation. Third, one side of the fracture can move with respect to the other, parallel with the plane of the fracture, and absorb energy in the process.

They cite Knopoff et al.'s (1957) calculation of the transmission coefficient as a function of the dip angle of the fracture for plane compressional and "dip-polarized" shear waves crossing an infinitely thin, liquid-filled fracture, with no contact between the faces, in an infinite medium. (They do not describe the direction of the waves; it appears to be vertically downward.) The compressional wave was attenuated significantly for dips between 33 and 78°, whereas the shear wave was very severely attenuated at angles outside that range. There was no shear transmission at all when the fractures were either horizontal or vertical (parallel and perpendicular to the wavefront).

Morris et al. (1964) also performed experiments with scale models in limestone, with fractures that had smoothly ground surfaces. They obtained the results shown as the points on Fig. 14-1, which are in reasonable agreement with Knopoff's calculations. They therefore concluded that the difference between compressional and shear attenuation would be diagnostic of the angle of the fracture. Dzeban (1970),

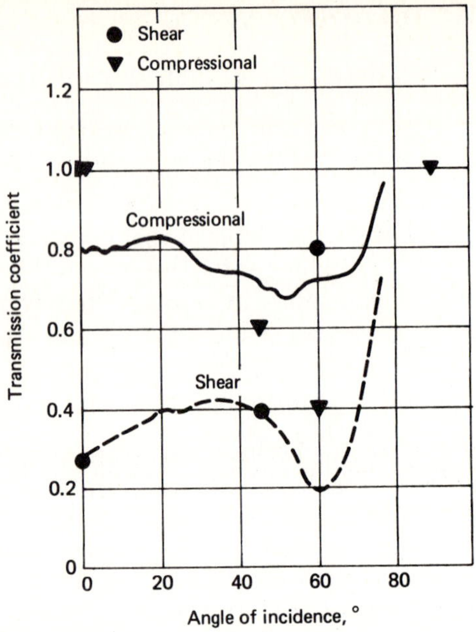

FIGURE 14-1 Amplitude transmission coefficient of compressional and shear waves crossing a fracture as a function of angle of incidence. (Curves from Dzeban, 1970; points from Morris et al., 1964.)

however, performed similar experiments with "real" fractures, the surfaces of which were rough, and obtained the results shown by the curves in Fig. 14-1; they disagree markedly with those of Morris et al. The transmission of compressional waves is approximately constant below an angle of incidence of about 70°, and then it increases. Shear transmission is finite across a horizontal fracture, increases slowly with angle, decreases at about 60°, and then becomes large for a vertical fracture.

Anderson and Walker (1961) calibrated an amplitude log in a test pit with several horizontal fractures. They found that the logarithm of the compressional amplitude was inversely proportional to the number of fractures m, so that the amplitude A could be expressed as

$$A = 0.77^m \tag{14-1}$$

Paillet (1980) studied the amplitude of shear and tube waves crossing near-horizontal fractures by determining the total energy arriving in the part of the record attributable to those waves. Both waves decreased in amplitude when the fracture was between the source and detector. He did not find normal waves to be useful; they apparently had less energy than tube waves.

From these results we can see that compressional wave amplitudes are sometimes attenuated by horizontal fractures and sometimes not, that they are generally attenuated by fractures with intermediate angles, and that they are not attenuated by vertical fractures. Shear amplitudes are always attenuated by horizontal fractures, sometimes less so by fractures with intermediate angles, and sometimes by vertical fractures.

Field experiments. Morris et al. (1964) compared amplitude logs with observations of core fractures in limestone. In general, they found shear amplitude reduced when the fractures were horizontal and compressional amplitude reduced when the fractures were subvertical, in agreement with their laboratory data.

Anderson and Walker (1961) used the calibration described by Eq. (14-1) to study fractures with an unspecified dip in limestone. In general, they found a high *fracture index,* that is, large attenuation, when the log was in a fractured formation.

Walker and Kessler (no date) used a variable-intensity log (Sec. 7-3c) to investigate vertical fractures in Austin chalk. They found that both shale laminations and fractures produced anomalies in the log. They distinguished fractures from laminations by noting that the W pattern characteristic of a horizontal boundary (Fig. 7-19) indicated laminations, which were horizontal, but not vertical fractures. There was usually no reduction in the amplitude of the first arrival when the anomaly was caused by fractures, whereas there often was when the anomaly was caused by a lamination. They also correlated the anomalies with dip logs (Sec. 14-3a). It is often difficult for the uninitiated to distinguish fractures from laminations on their logs.

Myung and Baltosser (1972) also used a variable-intensity log. They found that vertical fractures had little effect on either compressional or shear waves, whereas horizontal fractures completely suppressed the shear signal. Figure 14-2 shows a log in granite. In zone B there are no fractures. In zone C there is a horizontal fracture that eliminates the shear signal but not the compressional signal and causes reflections (*diagonal energy transmission;* Fig. 7-18) below the zone. In zone A both compressional and shear signals are removed; the authors attribute this to an oblique fracture for which they have no other evidence.

Shanks et al. (1976) used a sonde with source and detector very close together on a pad forced against the wall of the borehole, and they used a variable-intensity display to locate fracture anomalies on a

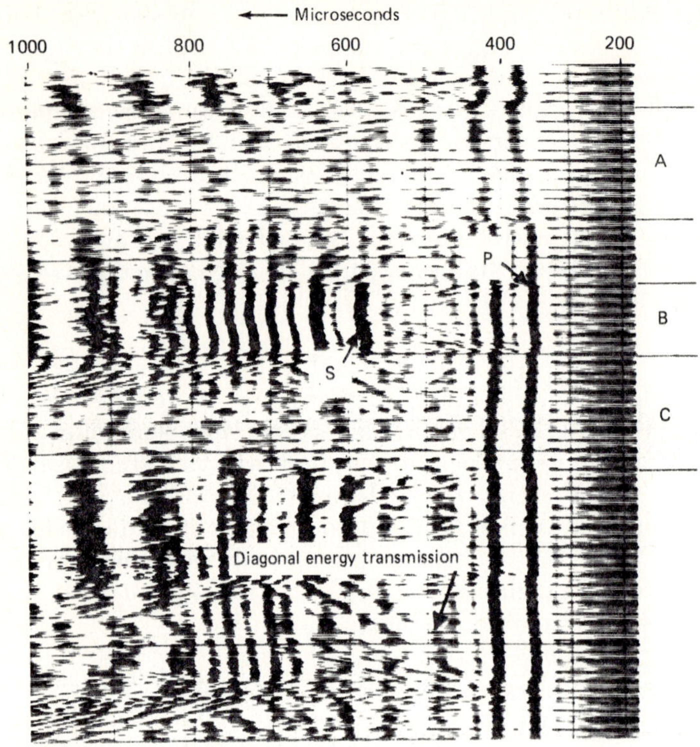

FIGURE 14-2 Variable-intensity log in fractured granite. (Myung and Baltosser, 1972.)

fine depth scale. Koerperich (1978) was unable to reliably locate vertical fractures with either an amplitude log or a variable-intensity log in tight sandstone. Hearst (1980) observed that the coda (the signal observed after the passage of the tube waves) of a variable-intensity log was greatly reduced in amplitude when the sonde was in a zone in which a sibilation log (Sec. 14-3c) indicated the presence of (probably vertical) fractures.

We see that the users of variable-intensity logs disagree as to the effect of vertical fractures. We suspect the effect depends on the type of formation, the structure of the fractures, the number of contact points, and the details of the equipment.

King et al. (1978) used a dry hole sonde with expanding pistons to locate open fractures that intersected a borehole. They found that the ratio of compressional velocity measured by the log to that measured on intact cores in the laboratory varied inversely with measured fracture frequency (or RQD). They also found that individual fractures decreased compressional wave amplitude.

Paillet (1980) compared shear and tube wave amplitudes as fracture diagnostics in a granite batholith. The fractures appear to be betweeen 60 and 90° with respect to the axis of the hole. Both amplitudes were reduced in heavily fractured zones, but shear waves were more effective in locating small fractures. He states, "The tube wave amplitude anomalies have been especially effective in identifying major open fractures within intervals containing other closed or only partially closed fractures." This implies that the tube wave anomalies did *not* identify the closed fractures.

We can conclude from these reports that both compressional and shear amplitudes are often affected by fractures. Some users can distinguish vertical fractures from horizontal ones by comparing compressional and shear amplitudes; some cannot. Some observers use the early part of the wave; some a later part. There are no consistent results, partly because factors other than fractures also affect the amplitudes. We can draw no clear conclusions, except that fractures often reduce amplitudes.

The Borehole Televiewer. The Televiewer is described briefly in Sec. 7-4a. The sonde contains a transducer that acts as both a transmitter and receiver. The transducer is rotated by a motor, and it emits bursts of acoustic energy as a directional beam. These travel through the borehole fluid, are reflected from the borehole wall, and are detected at the transducer. The received amplitude modulates the intensity of the signal on a raster, so that a picture of the reflectivity of the borehole wall is displayed (Fig. 14-3). A smooth wall reflects more energy than a rough wall, so fractures and other rough parts of the wall are displayed as dark patches.

Davison et al. (1982) found that the largest discrepancy between core-fracture data and Televiewer logs involves the detection of closed fractures. Many of the fractures described as completely closed on the core log were not detected by the Televiewer.

If the borehole is not circular, or the sonde is not centered, the signal from a smooth wall may be reflected away from the sonde so that a low amplitude will be recorded. This low amplitude may be interpreted as a fractured zone. A washed-out portion of a borehole wall may give a similar low-amplitude signal. Taylor (1983) has developed a method for dealing with these problems. He recorded not only a picture of amplitude versus azimuth and depth but also, in parallel, a picture showing the arrival time of the reflected signal (Fig. 14-3*b*).

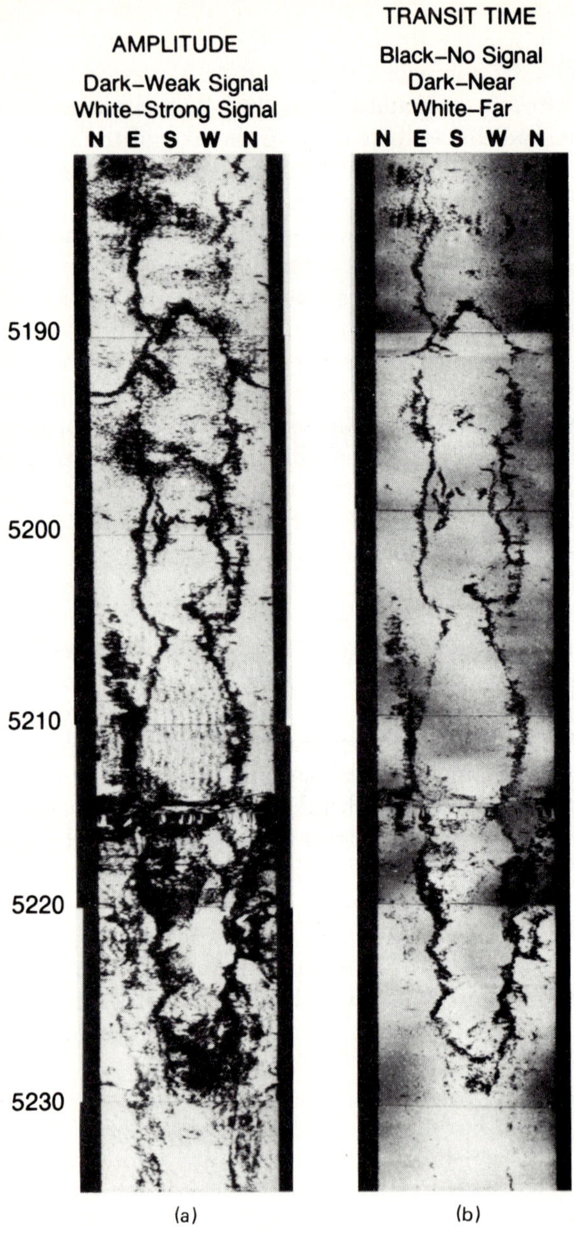

FIGURE 14-3 (*a*) Amplitude picture and (*b*) transit time picture from a Borehole Televiewer in a hole with open vertical fractures. (Taylor, 1983.)

On the amplitude picture (Fig. 14-3*a*), white represents a strong signal and dark represents weak or no signal. On the time picture (Fig. 14-3*b*), white represents long time (wall far from the sonde), dark represents short time (near the sonde), and black represents no signal at all. A noncircular borehole will give a characteristic pattern on the transit time record: dark stripes for the part of the hole near the sonde, light stripes for the part far from the sonde, and black for parts that reflect the signal away from the sonde (Fig. 14-4). This situation can be distinguished from signals caused by fractures.

The transit time picture can also be used to distinguish filled fractures from open fractures. Both will cause dark images on the amplitude picture, but the transit time picture will not record "no signal" from the filled fractures; it will record "no signal" only from the open ones. Consequently, the open fractures will be observed on both pictures, but the filled fractures only on the amplitude picture. Figure 14-5 shows pictures of partly filled fractures.

Most authors consider the Televiewer to be the standard of comparison for other fracture logs. Only core is considered more reliable, and this point is debatable. Of course, to be detected, the fractures must intersect the hole, but that is true of most fracture logging methods. Keys and Sullivan (1979) made extensive use of the Televiewer to characterize fractures in a geothermal well, and with excellent results. Wiley (1980) developed improved signal processing and image enhancement for the Televiewer, and he was able to see fine detail of both vertical and horizontal fractures. He reports no difficulties in detailed evaluation of fractures. Paillet (1981) finds that thick mud can sometimes attenuate the signal so that not all fractures are seen. He also finds that the fractures appear wider than they really are. But he too uses the Televiewer as a standard for comparison.

It is reasonable to ask why one does not simply use the Televiewer routinely and be confident in the results. It is a very costly log to run, and its displays require detailed visual interpretation, so it is also costly to analyze. It does not always detect closed fractures. It cannot be run in holes in which the fluid contains much particulate matter. Nonetheless, if one really wants to know about fractures, it appears that the Televiewer will often do the job.

If the hole is dry or contains a clear fluid, television or photography can be used. Each gives excellent results if good equipment and plenty of light are used.

Circumferential acoustic logging. Many fractures are vertical, and we have seen from the above discussion of wave propagation in fractured

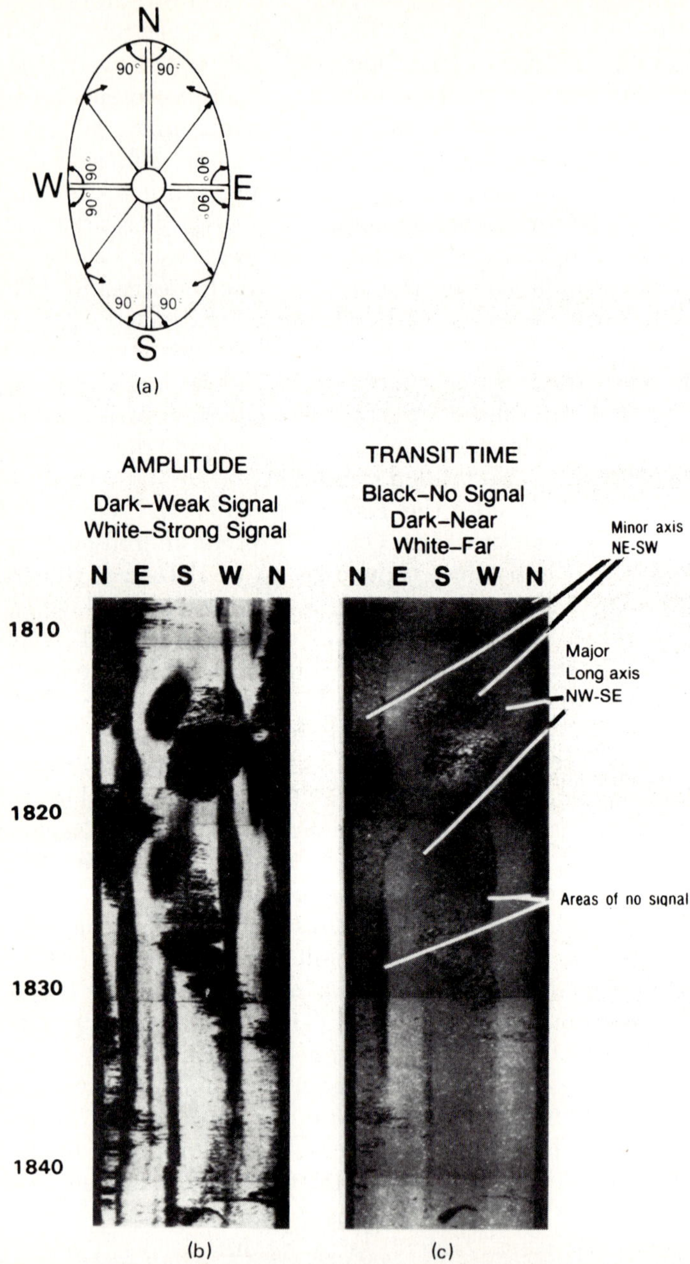

FIGURE 14-4 (*a*) Sketch of the direction of reflected signals in an elliptical borehole, (*b*) amplitude picture, and (*c*) transit time picture from a Borehole Televiewer in an elliptical borehole. (Taylor, 1983.)

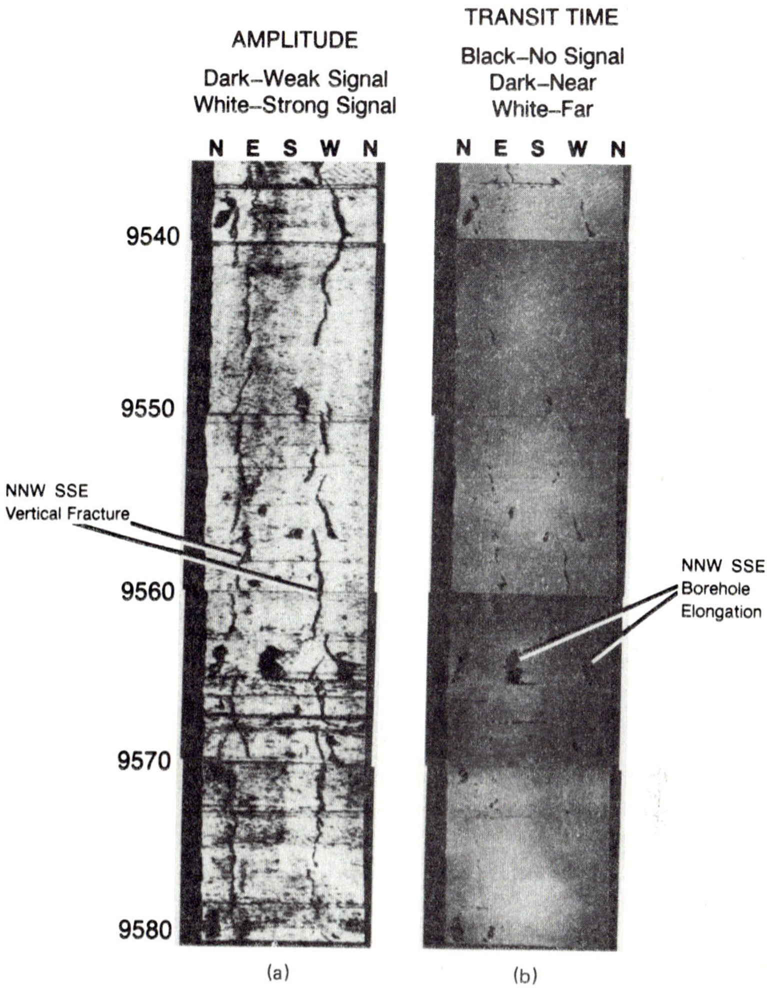

FIGURE 14-5 (*a*) Amplitude picture and (*b*) transit time picture from a Borehole Televiewer in a hole with partly filled vertical fractures. (Taylor, 1983.)

media that vertically traveling waves are often unreliable for detecting vertical fractures. Consequently, a device for inducing waves to travel around the circumference of the borehole has been developed (Vogel and Herolz, 1977). It has four arms positioned 90° apart around the circumference of the hole at the same depth. Each arm carries a transmitter or receiver, and the transmitters, long PZT cylinders, are said to be directional. It is claimed that the signal is restricted to a thin horizontal band of rock around the circumference of the hole. It is reported that only shear waves and water waves are observed. Figure 7-15 is a sketch of the principle of the device.

Koerperich (1975) studied such a logging system. He considered strong first-arrival signals in two quadrants to be a prerequisite for interpretation. Such strong signals indicate that the formation will support a circumferential wave, and so weak first-arrival signals in other quadrants indicate the presence of a fracture. If there are no strong signals, the attenuation could have been caused by intact formation material. In a well with cored intervals of sandstone, shale, and carbonate he found that the circumferential log responded to 21 of 24 fractures observed in the core.

Unfortunately, it also responded to 71 zones without fractures. He attributes this to roughness of the hole. He dealt with the problem by use of auxiliary logs that indicated shaliness, and he states that the rough parts of the hole were in shale. A reduced amplitude in nonshaly formation was considered to be a good fracture indication. This result indicates that the circumferential log by itself is not a reliable fracture indicator.

Kleinberg et al. (1982) used an "offset acoustic path" method in which the sonde is rotated through small angles and the semblance (Sec. 7-4b) between pairs of received signals is recorded. Let $x_1(n)$ and $x_2(n)$ be corresponding samples of two waveforms. For a section of waveform centered at sample n, and with $2m + 1$ samples in the section, they express the semblance as

$$S(n) = \frac{\sum_{k=-m}^{m} \tfrac{1}{4}[x_1(n+k) + x_2(n+k)]^2}{\sum_{k=-m}^{m} \tfrac{1}{2}[x_1^2(n+k) + x_2^2(n+k)]} \tag{14-2}$$

S is unity if the waveforms are identical, equal to 0.5 if they are uncorrelated, and zero if they are negatives of each other. In practice, Kleinberg et al. (1982) use an average of $S(n)$ over sliding windows, such that

$$Y(n_1, n_2) = \frac{1}{n_2 - n_1 + 1} \sum_{n=n_1}^{n_2} S(n) \tag{14-3}$$

The quantity Y is the measure used for comparison of the waveforms. If the pair of signals crosses a fracture, semblance is low; if not, it is high. Kleinberg et al. compare the energy and the semblance of signals propagating along a fractured surface. Both decrease sharply when a fracture is present, but the semblance is much less sensitive to lithological changes such as porosity increases. The semblance method works less well than the energy method when the surface is rough.

Botter (1982) studied the circumferential sonde in a formation in which the acoustic velocity of the formation was so low that the first arrival at the detectors was a direct fluid wave. He used the energy in the part of the signal starting at a time late enough that the circumferential wave would be analyzed, rather than the first-arriving fluid wave. He overlaid amplitude logs from the adjacent quadrants which have a transmitter in common; the resulting log emphasizes the difference between the quadrants. If the energy in the signals in the two adjacent quadrants is the same, there is no fracture in either quadrant (vertical fractures in both quadrants are quite unlikely) and the two signals will overlay. If the relative energy is different, the difference may be caused by a fracture in one quadrant and not in the other. The curves will not overlay. The shaded area in the log shows the difference between the curves.

Figure 14-6 shows some of Botter's results compared to the fracture pattern observed in core. The log shows some, but not all, of the fractures. Response is good above 1635 but poor below 1635. Botter attributes the inability to see some fractures to high porosity in the formation causing all amplitudes to be low. The log also sometimes shows energy differences where there are no fractures. These are not explained. Botter states that a circumferential semblance log run in the same hole missed several of the fractures in the cored interval.

From these results we may conclude that the circumferential log, like the ordinary amplitude log, sees some of the fractures some of the time.

Problems. Several things besides fractures can reduce the amplitude of the signal and make an acoustic log respond as if fractures were present. Morris et al. (1964) point out several.

1 If the sonde is not centered in the hole, different parts of the returning signal can add out of phase and reduce the amplitude. They find that an eccentricity of 6 mm can reduce the amplitude by 50 percent and an eccentricity of 5 cm by a factor of 10.

2 Hole size changes will cause an increase or decrease in the length of the mud path. This in turn will lead to an increased or decreased attenuation in the mud. The type of mud in the hole can affect the amplitude. If gas or air bubbles are present, the amplitude is greatly reduced. A change in mud impedance can change the mismatch at the wall and therefore change the energy coupled to the formation.

3 Changes in borehole rugosity will change the amplitude of the signal entering the formation from the fluid and the amplitude of the signal entering the fluid from the formation.

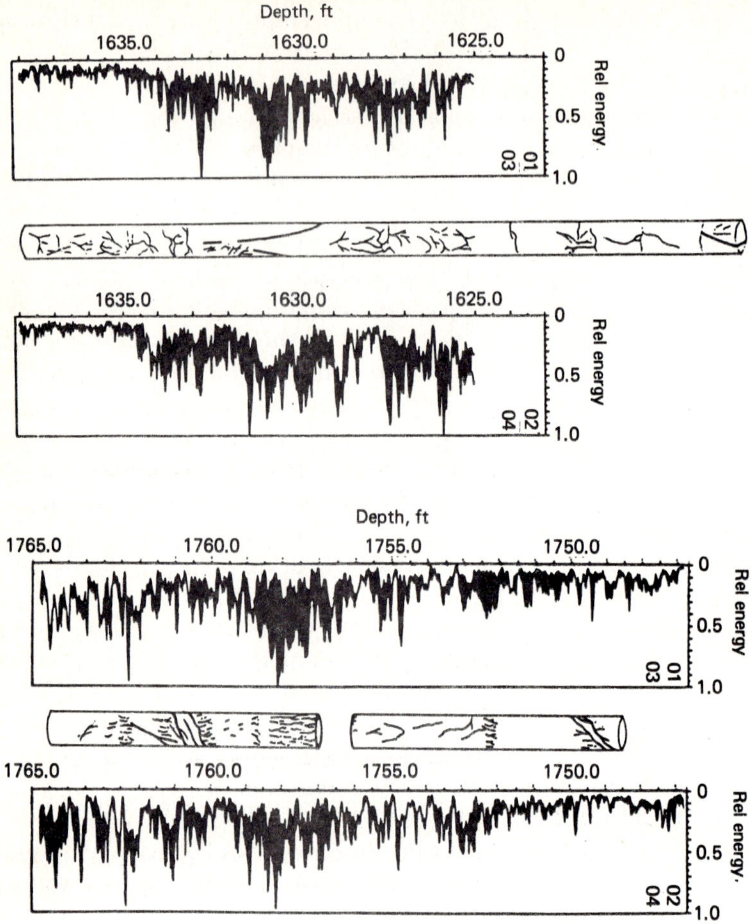

FIGURE 14-6 Comparison of a circumferential acoustic log with core fractures. Traces from adjacent quadrants are overlaid. Shading shows energy difference between traces. (Botter, 1982.)

4 Lithology changes will change the mud-formation mismatch. Also, the attenuation in different formations can be different. Both of these phenomena can change the amplitude.

5 Of course, bed boundaries cause both amplitude decreases and reflections and often give rise to spurious fracture signals.

Beck et al. (1977) point out that a variable-intensity log, as well as an amplitude log, can be affected by changes in centralization, hole size, and borehole rugosity. Lithology changes and bed boundaries will have very visible effects. In fact, the variable-intensity log is deliberately so designed that almost any change in conditions will be visible. Consequently, it is important, but not always possible, to recognize the changes caused by fractures.

14-3 RESISTIVITY METHODS

Measurements. Various semiempirical methods have been used for some time to infer fracture locations from electric logs. Spikes on shallow-reading resistivity logs can sometimes be used as indicators of fractures. The differential resistance sonde is so constructed that impedance variations are controlled by the mud in the annulus between the sonde and the wall of the borehole. It can be used in small-diameter holes in hard rock to detect fractures that have caused very slight enlargements in borehole diameter or that are themselves more conductive than the host rock (Keys and MacCary, 1971; Nelson et al., 1983). Aguilera (1976) developed a method for deducing the presence of fracture porosity; he compared the log-measured total porosity and saturation exponent [Eq. (11-4)] to the laboratory-measured values of the same quantities in intact rock. The difference between log and laboratory values (in saturated rock) is taken to be caused by fracture porosity. Julian (1982) finds that the difference in resistivity on shallow and deep induction logs can be used. He states, "Induction log resistivity anomalies sometimes indicate fracturing. . . . The existence of these anomalies depends on the depth of invasion of a non-conductive fluid into the fractures."

Beck et al. (1977) describe a method of fracture location by using a modification of a four-arm dipmeter (Chap. 18). Each arm is a shallow-reading resistivity device. If an arm passes over a fracture filled with conductive fluid, a low-resistivity spike will occur on the signal from that arm. The authors use two presentations. One, using signals from all four arms, shows spikes on all four signals when a horizontal or near-horizontal fracture is present. The other is an overlay of signals from adjacent arms. If a vertical fracture is present, low-resistivity anomalies are likely to appear on opposite arms. Therefore, the overlay will show a difference between adjacent arms (Fig. 14-7). The difference is shown on the log by shading. The authors found the results described above for vertical and permeable horizontal fractures in Austin chalk. Healed fractures and asphalt-filled fractures gave no response.

Comparison of several acoustic and resistivity methods. Paillet (1981) studied the response of a simple short-range resistivity sonde (R_{xo}), a dipmeter-type sonde (designated FIL), a Borehole Televiewer (ATV),

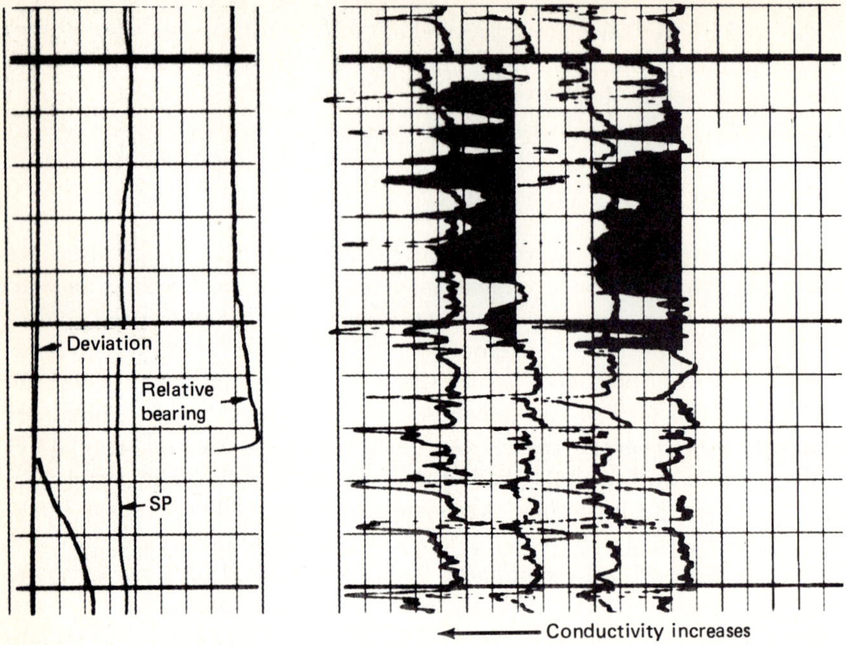

FIGURE 14-7 Dipmeter-type fracture log showing overlay of traces from adjacent pads. Shading shows difference between signals in adjacent pads. (Schlumberger, no date.)

his own amplitude log (AWF) (Paillet, 1980), and two versions of a circumferential log (CAD, CMT). He compared the responses of the logs with data from core in an 80-m section of a limestone with a low matrix porosity. The mud filtrate and formation water had a low resistivity, so that resistivity logs would respond well to fractures. The fractures were vertical. Table 14-1 presents his results. Zone B was the only one in which all of the logs responded to the fractures. The circumferential logs showed only a small response. Paillet states that the responses from all of the logs except ATV and AWF would be difficult to recognize as fracture responses in formations with more complex lithology than the limestone he used for this experiment.

None of the logs but the ATV gave clear responses in the other zones, despite the presence of several fractures. In some zones it was difficult to identify fractures seen on the ATV with those in the core. In zone F, where the largest fractures were seen in the core, none of the logs responded. Paillet attributes the lack of ATV response to high-density mud, and he has no explanation for the lack of response of the other logs. He finds that, in general, the response of the ATV is inconsistent.

TABLE 14-1 Summary of fracture characterization results in a 7.5-cm-diameter hole in limestone. After Paillet (1981).

Core zone	Fracture description	Fracture response present?					
		ATV	R_{xo}	FIL	CAD	CMT	AWF
A	One, closed	Yes	No	No	No	No	No
B	Several, near vertical; as much as 1 mm	Yes	Yes	Yes	Yes	Yes	Yes
C	One, near vertical, 0.5 mm wide	Yes	No	Yes?	No	No	No
D	One, near vertical, 2 mm wide	Yes?	No	No	No	No	No
E	Several, as much as 0.75 to 1 mm wide	Yes?	No	No	No	No	No
F	As many as five, each 1 mm wide	No	No	No	No	No	No

It responded to a sealed fracture in one zone much more strongly than to a much larger fracture in another.

We must conclude that all methods are unreliable for vertical fractures of width 0.5 to 2 mm in limestone.

14-4 DETECTION OF FLOW

Fractures can sometimes be located by detection of flow into or near boreholes. These methods are described in detail in Chaps. 2 and 13. Heflin (1979) and Suau and Gartner (1980) state that the SP log (Sec. 5-6) sometimes gives characteristic response opposite fractures. This response can be caused either by flow (Sec. 5-5), by pyrite in fractures, or by electrochemical effects.

The injection of mud in a fracture can sometimes reduce the temperature in the wellbore enough to be visible on a temperature log (Suau and Gartner, 1980) (Sec. 4-6). When gas at formation temperature enters the borehole through a fracture, it expands and cools. This cool gas reduces the borehole temperature below the geothermal gradient (Sec. 4-5), and the reduction can be observed on a temperature log. If gas enters a borehole through a small orifice, it can produce high-frequency sounds. A sibilation log (Chap. 13) containing piezoelectric transducers with peak sensitivity at about 40 kHz can sometimes be used to detect those sounds (Myung, 1976). It does not detect the lower-frequency noise caused by motion of the sonde in the hole or by gas flowing in the hole. It works, however, only for small fractures.

14-5 DETECTION OF DEPOSITS LEFT BY FLOW

γ-Ray. It has been found that in some rocks (for example, Austin chalk) which exhibit low total γ-ray or K, Th, and U readings on a spectral γ-ray log (Sec. 6-3) precipitation of uranium can occur in fractures. Wichmann et al. (1975) used a spectral γ-ray log to locate fractures in "basement rocks" by high uranium content. West and Laughlin (1976) located fractures in crystalline Precambrian rocks by γ-ray spectroscopy. Fertl et al. (1980) found that fractures could be located by high uranium in Austin chalk even when acoustic logs failed to locate them. Fertl and Reike (1980) used the presence of very high uranium readings in shales in which all three natural radioactive elements were present to find natural fractures in calcerous, cherty, or silty zones in organic-rich shales. The authors do not directly confirm the presence of fractures; rather, they infer the presence of fractures by stating that oil or gas production comes from fractured zones and that the log locates zones that are good producers. Since the kerogens that are the sources of production in many of these shales are high in uranium, it may be that the log simply locates zones of high kerogen content.

Heflin (1979) points out that other sources of uranium can cause confusion in the location of fractures by this method. Of course, the technique works only if the fractures contain precipitated uranium.

Other methods. The *correction curve* $\Delta\rho$, on a compensated density log (Sec 6-4d) can exhibit peaks caused by fractures (Suau and Gartner, 1980; Julian, 1982). In low-porosity granite, fractures that are filled with hydrogenous materials such as hydrated clay can be detected as low count rate spikes on a neutron log (Daniels et al., 1983). Julian (1982) compares the reading of Σ from a neutron die-away log (Sec. 6-7) with total γ-ray count to distinguish hydrocarbon-productive fractures (that contain uranium) from nonproductive fractures (that do not). He does not explain why Σ is sensitive to fractures.

All these methods are sometimes useful; none is always useful. At present there is no reliable way to log for fractures in an arbitrarily chosen formation. But we really do not know how to define a fracture. It may be that the apparent failures of detection are a result of our inability to define; the fractures may be so thin or so remineralized as to be the same as the rock mass. Fracture logging has a long way to go.

REFERENCES

Aguilera, R.: "Analysis of Naturally Fractured Reservoirs from Conventional Well Logs," *J. Petr. Tech.*, vol. 28, 1976, p. 764.

Anderson, W. L., and T. Walker: "Application of Open Hole Acoustic Amplitude Measurements," Soc. Petr. Eng. Paper SPE 122, 1961.

Beck, J., A. Schultz, and D. Fitzgerald: "Reservoir Evaluation of Fractured Cretaceous Carbonates in South Texas," *Proc. Soc. Prof. Well Log Analysts 18th Logging Symp.*, 1977, paper M.

Botter, B. J.: "Circumferential Acoustic Waves in Boreholes for the Delineation of Vertical Fractures," *Proc. Soc. Prof. Well Log Analysts 23d Logging Symp.*, 1982, paper T.

Daniels, J. J., G. R. Olhoeft, and J. H. Scott: "Interpretation of Core and Well Log Physical Property Data from Drill Hole UPH-3, Stephenson County, Illinois," *J. Geophys. Res.*, vol. 88B, 1983, p. 7346.

Davison, C. C., W. S. Keys, and F. L. Paillet: "Use of Borehole-Geophysical Logs and Hydrologic Tests to Characterize Crystalline Rock for Nuclear-Waste Storage, Whiteshell Nuclear Research Establishment, Manitoba, and Chalk River Nuclear Laboratory, Ontario, Canada," Office Nuc. Waste Isolation Rept. ONWI-418, 1982.

Dzeban, I. P.: "Elastic Wave Propagation in Fractured and Vuggy Media," *Izv. Earth Physics* (translation), no. 10, 1970, p. 629.

Fertl, W. H., and H. H. Rieke: "Gamma Ray Spectral Evaluation Techniques Identify Fractured Shale Reservoirs and Source-Rock Characteristics," *J. Petr. Tech.*, vol. 32, 1980, p. 2053.

————, W. L. Stapp, D. B. Vaello, and W. C. Vercellino: "Spectral Gamma-Ray Logging in the Texas Austin Chalk Trend," *J. Petr. Tech.*, vol. 32, 1980, p. 481.

Hearst, J. R: "Acoustic Logging for Fractures," *Second Eastern Gas Shales Symp.*, 1980.

Heflin, J. D.: "Fracture Detection in West Coast Reservoirs Using Well Logs," Soc. Petr. Eng. Paper SPE 7976, 1979.

Julian, P. J.: "Fracture Detection Techniques in the Georgetown and Austin Chalk Formations," Soc. Petr. Eng. Paper SPE 11035, 1982.

Keys, W. S., and L. M. MacCary: "Application of Borehole Geophysics in Water Resources Investigations," in *Techniques of Water-Resources Investigation of the U.S. Geological Survey,* 1971.

———— and J. K. Sullivan: "Role of Borehole Geophysics in Defining the Physical Characteristics of the Raft River Geothermal Reservoir, Idaho," *Geophysics,* vol. 44, 1979, p. 1116.

King, M. S., M. R. Stauffer, and B. I. Pandit: "Quality of Rock Masses by Acoustic Logging," *Proc. 3d Internal. Cong. IAEG* sec. IV, vol. 1, 1978, p. 156.

Kleinberg, R. L., E. Y. Chow, T. J. Plona, M. Orton, and W. J. Canaday: "Sensitivity and Reliability of Fracture Detection Techniques for Borehole Application," Soc. Petr. Eng. Paper SPE 11036, 1982.

Knopoff, L., et al., *Second Annual Report Seismic Scattering Project,* Inst. Geophys. UCLA, April 1957, chap. 12.

Koerperich, E. A.: "Evaluation of the Circumferential Microsonic Log—A Fracture Detection Device," *Proc. Soc. Prof. Well Log Analysts 16th Logging Symp.*, 1975, paper JJ.

————: "Investigation of Acoustic Boundary Waves and Interferance Patterns as Techniques for Detecting Fractures," *J. Petr. Tech.*, vol. 30, 1978, p. 1199.

Morris, R. L., D. R. Grine, and T. E. Arkfeld: "Using Compressional and Shear Acoustic Amplitudes for the Location of Fractures," *J. Petr. Tech.*, vol. 16, 1964, p. 623.

Myung, J. I.: "Fracture Investigation of the Devonian Shale Using Geophysical Well Logging Techniques," *Proc. 7th Ann. Appalachian Petr. Geol. Symp.*, 1976.

———— and R. W. Baltosser: "Fracture Evaluation by the Borehole Logging Method," *Proc. 13th Symp. Rock Mech. ASCE*, 1972, p. 31.

Nelson, P. H, K. A. Magnusson, and R. Rachiele: "Applications of Borehole Geophysics at an Experimental Waste Storage Site," *Geophys. Pros.*, vol. 30, 1983, p. 910.

Paillet, F. L.: "Acoustic Propagation in the Vicinity of Fractures Which Intersect a Fluid-Filled Borehole," *Proc. Soc. Prof. Well Log Analysts 21st Logging Symp.*, 1980.

————: "A Comparison of Fracture Characterization Techniques Applied to Near-Vertical Fractures in a Limestone Reservoir," *Proc. Soc. Prof. Well Log Analysts 22d Logging Symp.*, 1981, paper XX.

Schlumberger: *Schlumberger's Fracture Identification Log,* Schlumberger, Houston, Tex., no date.

Setser, G. G.: "Fracture Detection by Circumferential Propagation of Acoustic Energy," Soc. Petr. Eng. Paper SPE 10204, 1981.

Shanks, R. T., B. S. Kwon, M. R. DeVries, and P. A. Wichmann: "A Review of Fracture Detection with Well Logs," Soc. Petr. Eng. Paper SPE 6159, 1976.

Suau, J., and J. Gartner: "Fracture Detection from Well Logs," *Log Analyst,* vol. 21, no. 2, 1980, p. 3.

Taylor, T. J.: "Interpretation and Application of Borehole Televiewer Surveys," *Proc. Soc. Prof. Well Log Analysts 24th Logging Symp.* 1983, paper QQ.

Vogel, C. B., and R. A. Herolz: "The CAD, a Circumferential Acoustical Device for Well Logging," Soc. Petr. Eng. Paper SPE 6819, 1977.

Walker, T., and C. Kessler: *"Detection of Natural Fractures with Well Logs,"* Welex, Houston, Tex., no date.

West, F. G., and A. W. Laughlin: "Spectral Gamma Logging in Crystalline Basement Rocks," *Geology,* vol. 4, 1976, p. 617.

Wichmann, P. A., V. C. McWhirter, and E. C. Hopkinson: "Field Results of the Natural Gamma Ray Spectralog," *Proc. Soc. Prof. Well Log Analysts 16th Logging Symp.*, 1975, paper O.

Wiley, R.: "Borehole Televiewer—Revisited," *Proc. Soc. Prof. Well Log Analysts 21st Logging Symp.*, 1980, paper HH.

15

Elemental Analysis

15-1 INTRODUCTION

Many log users wish to measure the concentration of fluids, minerals, and elements in rock. The quantitative evaluation of water and petroleum in sedimentary rock has been more or less successful for many years (Chap. 11). The evaluation of specific minerals has met with some isolated successes. Coal can be inferred from resistivity and density (Chap. 9); magnetite can be quantified from magnetic susceptibility (Zablocki, 1974); native sulfur can be determined by several methods (Tixier and Alger, 1970); and sulfide minerals can be roughly estimated by using induced polarization (Nelson and Van Voorhis, 1983). The empirical relations established for iron ore and sulfides, however, depend on mineral and rock type and cannot be expected to be applied in different locales without modification.

Nuclear methods are now making it possible to quantify the concentration of specific elements. Logging is expected to become more economical than the combination of core sampling and laboratory analysis. Moreover, there is both expectation and evidence (Nargolwalla et al., 1977*a*) that because a log sees a larger volume of rock than is usually obtained from a core sample, it gives a more accurate estimate of elemental concentration.

Three groups have been in the forefront of development of borehole methods for elemental analysis over the years. Each has produced a major review article in the fairly recent past (Nargolwalla et al., 1977*a;* Czubek, 1979; Senftle, 1980). We shall use those articles as the basis for this chapter.

Many of the methods used for elemental analysis were first developed for analysis of laboratory, surface, or marine samples. Their use in boreholes is just beginning. We shall describe some of those methods (using Chap. 6 as background) and then discuss some of the elements for which they are used. We shall, when possible, discuss accuracy. But information on accuracy is sparse in the literature. Often a *sensitivity limit,* or the smallest amount the author believes to be detectable by the method, is given instead. Different authors give different values for both expected accuracy and sensitivity limit.

15-2 METHODS

While almost all the methods to be discussed involve detection of photons, they can be conveniently sorted into those using photon sources and those using neutron sources. Because so many methods are used to assay uranium, a separate subsection is devoted to that element.

15-2a Photon Sources

γ-Ray backscattering methods. Both density logging (Sec. 6-4b–g) and selective γ-γ logging (Sec. 6-4h) are used for quantitative assay of the grade (or fraction of the desired mineral) of ores whose general composition is known. Usually, a two-material mixture (such as zinc sulfide in limestone) is studied (Donhoffer, 1969). One exploits the known difference in density or atomic number Z between the material of interest (the zinc sulfide) and that of no interest (the limestone). One chooses the source-detector spacing and γ-ray energy to optimize the resolution for the density or Z in question. One can estimate Z from the ratio of high- to low-energy γ-rays as in Bertozzi et al. (1981) (Sec. 6-4a). Some accuracies for γ-ray methods are shown in Table 15-1.

The result can be affected by grain size (Charbucinski and Umiastowski, 1977). Czubek (1979) says "For some range of γ-ray energies, the concurrent, reciprocal influence of grade and grain size on scattered radiation becomes so strong that grade determination is impossible unless the grain size, or heterogeneity, is fixed or known."

Photonuclear reactions. It is possible to create neutrons by the (γ, n) reaction, in which a γ-ray causes a nucleus to produce a neutron. Only beryllium has a threshold energy for this reaction (1.67 MeV) that is low enough to be accessible with convenient radioactive sources such as ^{124}Sb. The emitted neutrons can be detected with BF_3 or ^{3}He counters.

x-Ray fluorescence (XRF). When a photon interacts with an orbital electron of an element by means of the photoelectric effect, and the electron returns to its orbit, it emits an x-ray whose energy is characteristic of the element. (Recall that the difference between an x-ray and a γ-ray is only in name; the name distinguishes the manner in which the ray is produced.) Clayton (1969) reviews the method in some detail. The photon source is either a low-energy γ-ray from a radioisotope source, x-rays generated by "bremsstrahlung"—the slowing down of a beta particle—from a radioisotope beta source, or an x-ray tube. In some cases, monochromatic x-rays, with energy slightly above that of the x-ray from the element sought, are obtained from radioactive sources that have decay modes that emit x-rays (for example, electron capture) or from x-ray tubes with targets whose x-ray emissions have the proper energy.

The energy of x-rays is low compared with that of most γ-rays used in logging, and the range of the x-rays is very short compared with that of γ-rays: a few micrometers to as much as a few centimeters, depending on the element being investigated. (Consequently, the detector must have very thin windows.) Table 15-2 lists energies and ranges of some typical x-rays. Smooth, clean, dry holes are required, although measurements have been performed in a liquid-filled hole with a very tight fit between sonde and hole. Because the range of the x-ray can often be small compared with the size of a particle, the method works best in "fine-grained, evenly disseminated, homogeneous deposits" (Clayton, 1969). Of course, the volume sampled is less than that with a core.

Czubek (1979) considers an accuracy of 25 percent of the true value to be the limit in borehole measurements. Several authors (Wogman, 1977; Clayton, 1969; Clayton, 1977) give accuracies and detection limits as shown in Table 15-1.

Natural γ-rays. Of course, radioactive elements (potassium, thorium, uranium) can be detected by their own emissions (Sec. 6-3).

Infrared. It is possible to excite molecular energy levels by bombardment with infrared radiation, either artificial or solar. Infrared radiation

TABLE 15-1 Detectability and accuracy of measurements of selected elements with x-ray fluorescence (columns A to E), γ-γ(F), neutron activation (G to I), and neutron capture (J and K).

		x-Ray. fluorescence					γ-γ	Neutron-γ				
								Activation			Capture	
	A or C°	Det., ppm			Acc., %		Acc., abs	Det., ppm		Acc., %	Det., ppm	Acc., %
Element		A	B	C	D	E	F	G	H	I	J	K
Be							10 ppm					
F								100		10–20		
Na	A									4		
Mg								100	1000			
Al	A							1000	900	5	43,000	
Si	C										29,000	3
S	C											5–7
K	C									18		
Ca	C										1700	
Sc								1				

Ti	C	400	10		0.2			10			1500	
V	A		10					10	30			
Cr	C							40,000		17		
Mn	A	5						100	10	3.8		
Fe	C	7000	5		2.6		1%			9.2	14,000	2–4
Co	A		5					1000		7		
Ni	C	800	5	1400		14		10,000				0.09–0.38
Cu	A	250	5	1000	0.1	12		1000	200			5–20
Zn	C	150	5		0.15	18	1%			23		
As	A		5	1000				1000		20		
Mo	A	100	5	100	0.01		0.01%	100		24		
Ag	A	20								0.3 ppm		
Sn	A		10	500	0.02			30,000				
Sb	A	100	10					10,000		28		
Ba	C		20	1000			0.03%	1000		14		
W	A		10	1000				1000				
Au			5					10				
Hg	C										800	
Pb	C	200	5		0.1		0.14%					

°A or C indicates the recommendation of Senftle et al. (1972) for use of activation or capture γ-rays for analysis.

TABLE 15-2 Energy of emitted x-rays from selected elements and thickness of three different rocks for which the intensity of the x-ray is halved. After Clark et al. (1977).

Element	x-Ray energy, keV	Half-value thickness, μm		
		Quartz	Rhyolite	Calcite
Mg	1.25	1.5	1.2	1.0
Al	1.49	2.4	1.9	1.7
Si	1.74	3.6	2.2	2.5
K	3.31	5.9	5.5	15.2
Ca	3.69	8.0	11.3	20.6
Ti	4.51	14.1	24.0	7.0
Fe	6.4	38.9	30.5	18.3
Cu	8.9*			
Sr	14.16	423.7	284.4	178.6
^{109}Cd	22.16	1,414.2	930.2	569.6
W	69.51*			
Au	80.7*			
Pb	15.87*	100*		
	88.0*	30,000*		

*Clayton, 1977.

emitted from these levels is characteristic of the molecules, and spectroscopy can be used for quantitative evaluation. Although no borehole instrument has yet been developed, the method has been used for analysis of minerals at the surface, and borehole methods may some day be feasible. The range of investigation can be micrometers to millimeters, depending on the molecules present. Vanasse et al. (1971) provide a good introduction to the field.

15-2b Neutron Sources

Capture and activation γ's. In Sec. 6-8a we briefly described logging with neutron-induced γ-rays. Recall that γ's can be produced either by inelastic scattering of fast neutrons, capture of a neutron (usually thermal) and immediate emission of a γ, or activation by capture of a neutron and slower decay (Fig. 6-1). Senftle et al. (1972) compare the efficiencies of logging with capture γ's (where the γ is detected during irradiation) and with activation γ's (where the source is removed and the γ is detected after irradiation) for determination of quantities of many different elements by using a ^{252}Cf source and a Ge(Li) detector. (They refer to the activation γ's as "delayed" γ's.) They consider the γ-ray yield and energy, detector efficiency for the γ energy, and irradiation time. Although they discuss it, they do not take into account the fact that the activation γ's, generally being lower-energy, are attenuated more rapidly in the medium and therefore do not sample as large a volume. In Table 15-1, an A after the symbol for the element indicates that they find activation the method of choice, whereas a C indicates the choice of capture. Nargolwalla et al. (1977*a*) have compared the advantages and disadvantages of capture and activation logging, and they strongly favor capture. It is usual now to use Ge(Li) or high-purity Ge as a detector. In practice, the method is used routinely for only a few elements at present (Senftle, 1980). Clayton et al. (1983) are using capture γ's to analyze impurities in coal. Table 15-1 shows some accuracy data for elemental analysis with both capture and activation γ's, and Table 6-2 shows typical γ-ray energies.

As in the situation of neutron die-away (Sec. 6-7b) diffusion effects are significant. All calibrations assume capture of thermal neutrons. Senftle et al. (1977) study the effects of the energy distribution of the neutrons in hydrogenous and nonhydrogenous media. If enough hydrogen (quantity not specified) is present, all the neutrons are thermalized, and one can compensate for the effect by normalizing to the number of γ's in the peak corresponding to emission of γ-rays from capture in hydrogen. The intensity of the peak of interest can be corrected by multiplying it by the square root of the intensity of the hydrogen peak. The method does not work in rock with very low hydrogen content because both the intensity and energy of the γ's are affected in dry material. Senftle (1980) states

> When the spatial distribution of thermal neutrons is nonuniform, calibration can vary even within a given formation. . . . Compensating for the variable distribution of thermal neutrons is a problem that has not yet been completely solved but that can be assisted by appropriate design of the sonde.

The source intensity is limited to avoid radiation damage to the detector, to minimize dead time in the counting electronics, and, for NaI detectors, to reduce pileup in the detector (Sec. 6-2c). Pileup is a particularly severe problem if digitization is performed at the surface.

Calibration is often done in the material being studied. A matrix method such as spectral fitting [Sec. 6-3b; Eq. (6-22)] is used to determine the quantity of each element represented in the spectrum. Tanner et al. (1972) describe a somewhat simpler method [similar to Eq. (6-20)] used when a detailed spectrum is not available. If the concentration of one element is known, then that of all others can be obtained. Otherwise, only ratios can be found.

Inelastic scattering. A few elements such as C, Ca, Si, O (Westaway et al., 1980) have been analyzed with inelastic scattering of neutrons

(Sec. 6-8b). In some cases both inelastic and capture methods can be used and the analysis thereby improved. Generally a pulsed neutron source is used. Sowerby (1979), however, proposes use of inelastic scattering of neutrons from a Pu-Be source for analysis of many elements. He considers this method best suited to samples from 10 to 100 kg because of interference from capture γ's, but it may be adaptable to the borehole.

15-3 INDIVIDUAL ELEMENTS

15-3a Table 15-1

Table 15-1 shows many of the published values of accuracy and limits of detectability for different elements. The elements are listed in order of atomic number, with A or C designating the Senftle et al. (1972) recommendation for use of activation or capture γ-ray for analysis.

Columns A to E refer to x-ray fluorescence; A to C are limits of detectability in parts per million; D and E are accuracy. Column A is from Clayton's (1969) review paper, and the data are from several studies on powdered ore samples, not from logging. The accuracy data from that paper are given in column D, and they are in units of percent by weight of the element. Column B is from Wogman (1977), and it refers to results with a system designed for sea bottom use. The experiments were done with very thin samples, so absorption in the sample evidently did not limit the sensitivity. Even so, it is a great improvement over the results quoted in Clayton (1969). Column C, from Czubek (1979), refers to detectability limits for actual borehole data. Therefore, it is much more realistic for our purposes. Column E shows accuracies for a few elements in stream sediments (Clayton and Packer, 1977). The x-ray energies were chosen to be as suitable as possible for the elements in question. The data were taken in the laboratory, so they are not directly applicable to boreholes.

Column F consists of accuracy data from Czubek (1971, 1979); it is taken primarily from γ-γ logging. In the case of Be, it is (γ, n). The other numbers refer to the *best* results he quotes. For Fe, Zn, Mo, and Ba the accuracy is in percent by weight, for Pb it is 14 percent of the value of the quantity of Pb.

Columns G to I are data for γ's from neutron activation. G is from Wogman (1977), and the data are his calculated sensitivity limits for activation from irradiation with ^{252}Cf and a 15 percent efficient Ge(Li) detector. Note that, for many of the elements he discusses, capture rather than activation would actually be the method of choice. Column

H is from the work of Moxham et al. (1976) on in situ analysis of sediments; the data are the estimated limits of detectability using ^{252}Cf and Ge(Li). They refer to the real case including the diffusion discussed in the preceding section, rather than theoretical calculations, and they are therefore more applicable to field use. But they are still not for a borehole. Column I, giving accuracy limits, is mostly from Clayton (1977), who quotes Plant et al. (1976) on laboratory analysis of samples of sediments.

Columns J and K refer to γ-rays from neutron capture. Column J, like column H, is from the work of Moxham et al. (1976). Column K, mostly from Nargolwalla et al. (1977*b*), describes the results of real field work with a logging sonde. Both columns I and K are in units of percent of the true value, often for values of less than 1 percent by weight of the element.

The table is, perhaps, more confusing than one would desire, but it seems to be a reasonable way to summarize the available data. Before using any of the results, the reader should refer to the original source for details.

15-3b Specific Elements

We shall now discuss some specific elements for which some interesting work has been done. We shall confine ourselves primarily to results from borehole work. Czubek (1971, 1979) describes much of the Soviet work. We shall refer to his papers rather than to the original work in Russian. Although there are many references to it, x-ray fluorescence is in fact quite difficult to use in a borehole. It is still an experimental logging technique (Rhodes et al., 1968). Only high-Z materials such as Hg or Pb give x-rays with enough penetration to really be useful.

H. Senftle et al. (1978) have used the 2.22-MeV γ from capture of ^{252}Cf neutrons to analyze the amount of H in coal. Coal beds could be located by making a log of the number of γ's in a window centered on 2.22 MeV; a distinct peak was found. Even with the widest plausible variation in the amount of oxygen, their results agree very well with ordinary chemical analysis. They used a ratio method that depends on knowing the amount of oxygen present and did not estimate accuracy. Flaum and Pirie (1981) used capture after irradiation with 14-MeV neutrons, and they inferred salinity from the ratio of H to Cl.

Be. The (γ, n) method discussed in Sec. 15-2a has evidently proved very satisfactory for routine analysis for Be (Czubek, 1979).

C. The carbon/oxygen log (Sec. 6-8b) is routinely used to determine the amount of carbon. The uncertainty of the measurement with standard methods appears to be about 4 atomic percent of carbon (Hertzog, 1980). Senftle et al. (1978) obtained an accuracy of about 1% C in coal, and Hearst (1978) obtained a similar accuracy in a laboratory simulation of a field experiment.

F. Several measurements have been reported (Czubek, 1971; Senftle, 1980), but accuracy is poor.

Al. One of the first elements to be investigated by neutron activation was aluminum, because it can be used as a clay indicator. Senftle (1980) points out that aluminum, because of its high neutron cross section, is particularly easy to activate and is one of the few elements for which the activation method is really practical. Czubek (1979) states that "Aluminum is the main element detected by nuclear methods." Wichmann and Webb (1970) used the ratio of the total number of γ-rays that decayed with the 9.5-min half-life characteristic of ^{27}Mg, the product of ^{27}Al (n, p) ^{27}Mg, to that of the γ-rays that decayed with the 2.3-min half-life characteristic of ^{28}Al, the product of ^{28}Si (n, p) ^{28}Al, to determine the Si/Al ratio. Scott and Smith (1973) developed a log that used the number of counts in a window appropriate to the γ energy, and they tested it in a number of holes to obtain shale content. But its use as a shale indicator has not been reported in recent years.

Senftle et al. (1978) used the *absence* of the activation γ's to infer the presence of coal beds in shales—if no γ's, no shale and, therefore, coal.

Si. There is really very little interest in determining the amount of Si per se; there's plenty of it available. But it occurs as an impurity in coal, and the analysis leads to the ash content (Senftle et al., 1978). The Ca/Si ratio is used with the carbon/oxygen log to learn what fraction of the carbon is in $CaCo_3$ rather than in oil. The H/Si (Westaway et al., 1980) and Al/Si (Wichmann and Webb, 1970) ratios are used to determine lithology. Nargolwalla et al. (1977a) obtained accuracies of 3 percent of the true value of Si.

S. Crusciel et al. (1977) used three methods to obtain the amount of sulfur in limestone. Their main indicator was the ratio of intensities of S and Ca peaks, with a Po-Be source. The standard error was 5 to 7 percent by weight.

Cl. The amount of chlorine is of interest in correcting ordinary neutron logs, and it can be obtained from capture data (Westaway et al.,

1980). Chlorine has a higher thermal capture cross section than most other materials commonly found while logging. Consequently, it is usually measured by obtaining the total cross section from a thermal neutron die-away log (Sec. 6-7c) and attributing the difference between measured cross section and that of the matrix to chlorine.

K. Potassium is generally investigated with natural γ-rays and a spectral γ-ray log (Sec. 6-3).

Ca. The Ca/Si ratio is of interest, when a carbon/oxygen log is used, to distinguish the carbon in $CaCo_3$ from that in oil. It can be obtained from both capture and inelastic scattering (Westaway et al., 1980).

Mn. Many methods—x-ray fluorescence, activation, and selective γ-γ logs—have been used to investigate Mn (Czubek, 1971). No details are given.

Fe. Iron is perhaps the element for which the most work in in situ analysis has been done. Practically every possible method has been tried; the one used in a particular case depends on the geochemical situation. Both density and selective γ-γ logging have been used extensively (Czubek, 1971, 1979). Eisler et al. (1977) obtained good results—0.05 percent by weight iron for mean hole grade, 2 percent by weight for a specific location—with neutron capture. The rock was quite dry, so the diffusion effect discussed in Sec. 14-1 was significant. They measured the thermal and epithermal neutron flux and then calibrated by regression with the neutron fluxes, the intensity of the iron γ-ray, and the intensity of the hydrogen γ-ray. This method appears to work in the dry situation.

Nargolwalla et al. (1977*b*), while using capture γ's, obtained accuracies of 3.8 percent of the amount of the iron present in a lateritic nickel deposit.

Ni. Senftle et al. (1971) estimated that nickel in concentrations of 1 percent could be analyzed with capture γ's. Nargolwalla et al. (1977*b*) obtained an accuracy of 4 percent of the nickel present in concentrations greater than 10 percent in the lateritic deposit discussed above. They detected capture γ's from ^{252}Cf with a NaI detector.

Cu. Copper is another element for which neutron capture is used routinely. Czubek (1971) reports a standard deviation of 20 percent of the true value in disseminated ores and 5 to 6 percent in massive ores, when concentrations are less than 1% Cu and activation is used. Nargolwalla et al. (1977*b*) report a standard error of less than 10 percent of the true value for concentrations less than 0.5 percent; they used capture of neutrons from ^{252}Cf and a NaI detector. Czubek (1979) discusses the use of x-ray fluorescence, selective γ-γ, and epithermal neutron logs.

Zn. Donhoffer (1969) routinely used γ-γ backscatter density logs to obtain the concentration of zinc with an uncertainty of 1% zinc. The presence of lead in the ore causes confusion. Czubek (1979) reports that x-ray fluorescence is also a satisfactory method.

Ag. Senftle et al. (1970) compared the use of 3-MeV neutrons from an accelerator with neutrons from ^{252}Cf for activation analysis of silver. Silver is one of the two elements (Al is the other) for which use of activation γ's is practical, according to Senftle (1980). They preferred ^{252}Cf, and they obtained accuracies of 0.1 oz per ton and a detection limit of 0.01 oz/ton, which is 0.4 part per million!

Sn. Czubek reports use of x-ray fluorescence for tin and the use of the ratio of two x-ray energies to determine the grade.

Sb. Czubek (1971) reports accuracies of 15 percent of the true value at concentrations of 0.2 to 0.3 percent with x-ray fluorescence.

W. x-Ray fluorescence, activation, and selective γ-γ logging can all be used for tungsten, with a detection level of 0.2 to 0.4 percent (Czubek, 1979).

Hg. Czubek (1979) describes the use of neutron-neutron methods similar to those used by Eisler et al. (1977) for determining mercury concentrations. Use of capture γ's is difficult because the γ from iron in the equipment has an energy similar to that of the γ from mercury, and it interferes with the results. This problem is solved by using a ratio of two γ windows 4 to 5 MeV and 6.1 to 8 MeV to achieve a sensitivity of 0.06 to 0.08% mercury. Czubek (1979) also discusses borehole use of x-ray fluorescence to determine mercury with about 20 percent accuracy.

Pb. Czubek (1971) reports use of selective γ-γ logs for lead with a detection limit of 0.2 to 0.3% lead and an accuracy of 14 percent of the value. x-Ray fluorescence also is used. Evidently, selective γ-γ logging for lead has been routine in the Soviet Union since 1971. The high Z of lead makes this method convenient.

15-3 Uranium

Killeen (1983) has reviewed the large literature on γ-ray logging for uranium in considerable detail. He described calibration methods needed for good accuracy and depth resolution, and he discussed some of the pitfalls. He did not discuss sensitivity limits. Only a brief description of the methods is given here.

Total γ-ray logging. For many years the normal method for quantitative analysis for uranium in place was the grade-thickness product method (Scott et al., 1961) described in Sec. 6-3a. The total γ-ray count was used, either from a Geiger counter or a scintillator. Dodd and Eschliman (1972) give a detailed discussion of the practical problems in using the method. One must correct for dead time, hole diameter and fluid, casing, and effect of the Z of the formation (Sec. 6-2b).

Sondes are calibrated in special uranium test pits. Calibration models at Grand Junction, Colorado contain various grades of uranium with different bulk density, porosity, and neutron cross sections (Koizumi, 1979).

It is recognized that total γ-ray logs (or even γ-ray spectral logs with broad windows) sample γ's from many daughters of uranium (Fig. 6-3). Consequently, the method works for quantitative analysis only if both the calibration models and the ore body are in the same state of equilibrium; preferably secular equilibrium (Sec. 6-1). It is known that many natural uranium ores were not in secular equilibrium, so methods have been developed to deal with the problem. It is also assumed that uranium dominates thorium and potassium.

Accurate spectral γ-ray logging. In one method of spectral γ-ray logging, described by Tanner et al. (1977), an intrinsic germanium detector is used to obtain very accurate γ spectra. Unless changes have occurred in the past few months, ^{238}U, with a half-life of 4.5×10^9 years, is in equilibrium with its first daughters, ^{234}Th and ^{234}Pa, with half-lives of 24 days and 1.17 min, respectively. ^{234}Th emits a 63.3-keV γ that can be used for analysis with a germanium detector. ^{234}Pa emits a 1001-keV γ that also can be used. Tanner et al. (1977) used two detectors of different resolutions to obtain the number of γ's of each of these energies, and showed that each is proportional to the amount of uranium present. Consequently, they could calibrate for uranium without worry about equilibrium. They obtained a sensitivity limit of 80 ppm.

Coles et al. (1976) used the 63.3-keV γ to analyze for uranium in soils that were *not* uranium ores. They found that the emission rate of the 1001-keV γ is too low to give a good count rate, partly because the detector is, of course, less efficient at the higher energy (Sec. 6-2c). They corrected the count rate for self-absorption in the soil. One would also expect that corrections similar to those discussed above would be required. Their sensitivity is 10 ppm.

They find uranium levels of 17 ppm in soils in which only 2 ppm had been found by total γ-ray logging. They attributed this to deposition of uranium from fertilizer over the past 30 years, which therefore could not be in equilibrium.

Delayed fission neutrons. When neutrons interact with uranium, fission occurs; the atom splits into fission products and emits neutrons. In 2 to 3 percent of the cases, the product is unstable, β-decays to another nuclide, and emits another neutron. These are called delayed neutrons. Different fractions of the delayed neutrons are emitted with different half-lives, between 0.2 and 50 s.

One can make a quantitative analysis for uranium by irradiating with neutrons, removing the source, waiting for the source neutrons to disappear, and then counting the delayed neutrons. This method has the advantage that the neutrons indicate the presence of uranium itself, not its daughters, and can therefore be used in any state of equilibrium. Both pulsed and radioactive neutron sources have been used.

Givens et al. (1976) used 14-MeV neutrons from a neutron generator. They pulsed the generator, turned it off, waited for the neutrons to die away, and then counted the delayed neutrons. Of course, as in any other method of analysis using neutrons, they had to compensate for the effect of diffusion in the medium (Sec. 6-7b). Their count rate depended not only on the uranium content, but also on density, water content, and neutron cross section of the other components of the medium.

They used the reaction

$$^{17}\mathrm{O}\ (n, p)\ ^{17}\mathrm{N} \rightarrow {}^{16}\mathrm{O} + n$$

which is usually considered part of the background in logging for uranium, to obtain neutrons with a delay and energy similar to those from uranium. They argue that they can calibrate their system for the effects of diffusion by assuming that, since most media have about the same oxygen content, the delayed oxygen neutron fluxes would be the same in all media if the diffusion effects were removed. Then the change in the number of delayed oxygen neutrons caused by the medium would be the same as the change in the number of delayed fission neutrons. They also assume that the count from an ordinary neutron-neutron log would not be affected by the amount of uranium present. Therefore, the neutron log count rate would be a measure of the diffusion properties of the medium.

They use these ideas to develop a set of uranium calibration curves for different neutron log count rates; the different calibration curves take the diffusion properties into account. They also have to subtract the background of delayed oxygen neutrons from their data. Their accuracy is approximately 0.05 percent by weight U_3O_8 for small amounts of uranium.

The difficulty with using a radioactive source is timing. The half-life of the most common delayed neutrons is too short to get a good count rate when the source and detector are moved together as the logging sonde moves up the hole. Steinman et al. (1980) solve the problem by shuttling a ^{252}Cf source inside the moving sonde. They move it back and forth between two detectors. While one detector is counting delayed neutrons, the source is irradiating the medium next to the other detector. There is no ^{17}O reaction with the lower-energy ^{252}Cf neutrons, so no background subtraction is required. They use the count in other neutron detectors to obtain diffusion corrections. Their accuracy is 0.04 percent by weight for low values of uranium.

Prompt fission neutrons. Barnard (1978) and Humphreys et al. (1983) used the neutrons emitted immediately after fission to obtain the uranium content. Since the time for neutron emission is finite, they found that they could gate an epithermal neutron detector off during a neutron pulse, and then count epithermal neutrons immediately after the pulse, while thermal neutrons were still present and causing fission. The ratio of epithermal neutrons to thermal neutrons or to thermal neutron die-away time is used to determine uranium content. It is necessary to correct for hole diameter, bulk density, scattering and absorption cross sections, and logarithmic energy decrement (Sec. 6-5b). The first experiments agreed with core to within a factor of 2. At typical logging speed the detection limit is 500 ppm.

Comparison of the methods. Wilson et al. (1980) compared four sondes: one using accurate spectral γ-ray logging, one using a pulsed neutron source and delayed fission neutrons, one using a radioactive source and delayed fission neutrons, and one using prompt fission neutrons. They first calibrated the sondes at the Grand Junction test models and then ran them in 30 holes in the field. They obtained grade-thickness products versus depth from each sonde in each hole and compared the results with core data. The range of uranium content is 40 to 2300 ppm.

They found a great deal of scatter in all the comparisons. The radioactive-source delayed-fission sonde was systematically low. The prompt fission sonde is best, especially above 1100 ppm. In general, they found that, after all corrections, the log results were within 50 percent of the core results 75 percent of the time and within 30 percent of the core results 50 percent of the time.

REFERENCES

Barnard, R. W.: "Interpretation of Prompt Fission Neutron Uranium Borehole Logs;" *Trans. Am. Nuc. Soc.*, vol. 28, 1978, p. 107.

Bertozzi, W., D. V. Ellis, and J. S. Wahl: "The Physical Foundation of Formation Lithology Logging with Gamma Rays," *Geophysics*, vol. 46, 1981, p. 1439.

Charbucinski, J., and K. Umiastowski: "Some Factors Affecting Accuracy in Determinations of Heavy Element Concentrations in the Selective Gamma-Gamma Method," *Proc. IAEA Symp. Nuc. Tech. Min. Res.*, 1977, p. 281.

Clark, B. C., A. K. Baird, P. Toulmin, H. J. Rose, and K. Keil: "x-Ray Fluorescence Geochemical Analysis on the Surface of Mars," in J. G. Morse (ed.), *Nuclear Methods in Mineral Exploration and Production*, Elsevier, Amsterdam, 1977.

Clayton, C. G.: "Review Paper: Applications of Radioisotope x-Ray Fluorescence Analysis in Geological Assay, Mining and Mineral Processing," *Proc. IAEA Symp. Nuc. Tech. Min. Res.*, 1969.

———: "Some Recent Applications of Nuclear Techniques in the Exploration and Mining of Metalliferous Minerals," *Proc. IAEA Symp. Nuc. Tech. Min. Res.*, 1977.

——— and T. W. Packer: "Analysis of Stream Sediment and Soil Samples Using a High Resolution Energy-Dispersive x-Ray Fluorescence Spectrometer," *Proc. IAEA Symp. Nuc. Tech. Min. Res.*, 1977.

———, A. M. Hassan, and M. R. Wormald: "Multi-element analysis of Coal During Borehole Logging by Measurement of Prompt γ-Rays from Thermal Neutron Capture," *Int. J. Appl. Rad. Isotopes*, vol. 34, 1983, p. 83.

Coles, D. G., J. W. T. Meadows, and C. L. Lindeken: "The Direct Measurement of PPM Levels of Uranium in Soils Using High-Resolution Ge(Li) Gamma-Ray Spectroscopy," *Sci. Total Env.*, vol. 5, 1976, p. 171.

Crusciel, E., J. Niewodniczanski, and K. W. Palka: "Determination of Sulphur Content in Boreholes by Neutron Capture," *Proc. IAEA Symp. Nuc. Tech. Min. Res.*, 1977.

Czubek, J. A.: "Recent Russian and European Developments in Nuclear Geophysics Applied to Mineral Exploration and Mining," *Log Analyst*, vol. 12, no. 6, 1971, p. 20.

———: "Modern Trends in Mining Geophysics and Nuclear Borehole Logging Methods for Mineral Exploration," in P. J. Hood (ed.), *Geophysics and Geochemistry in the Search for Metallic Ores*, Geol. Survey Can. Econ. Geol. Rept. 31, 1979, p. 231.

———, J. Loskiewicz, J. Gyurcsak, A. Lenda, K. Umiastowski, and T. Zorsk: "Geostatistical Method of Interpretation of Nuclear Well Logs," *Proc. IAEA Symp. Nuc. Tech. Min. Res.*, 1977.

Dodd, P. H., and D. H. Eschliman: "Borehole Logging Techniques for Uranium Exploration and Evaluation," in S. H. U. Bowie et al. (eds.), *Uranium Prospecting Handbook,* Inst. Min. Metall., London, 1972.

Donhoffer, D. K.: "Determination of Ore Concentration in Exploration Holes by the Gamma-Ray Backscattering Technique," *Proc. IAEA Symp. Nuc. Tech. Min. Res.*, 1969, p. 23.

Eisler, P. L., P. Huppert, P. J. Mathew, A. W. Wylie, and S. F. Youl: "Use of Neutron Capture Gamma Radiation for Determining Grade of Iron Ore in Blast Holes and Exploration Holes," *Proc. IAEA Symp. Nuc. Tech. Min. Res.*, 1977.

Flaum, C., and G. Pirie: "Determination of Lithology from Induced Gamma-Ray Spectroscopy," *Trans. Soc. Prof. Well Log Analysts 22d Logging Symp.*, 1981, paper H.

Givens, W. W., W. R. Mills, C. L. Dennis, and R. L. Caldwell: "Uranium Assay Logging Using a Pulsed 14-MeV Neutron Source and Detection of Delayed Fission Neutrons," *Geophysics*, vol. 41, 1976, p. 468.

Hearst, J. R: "Laboratory Studies of an Improved Carbon/Oxygen Log," *Log Analyst*, vol. 19, November–December 1978, p. 26.

Hertzog, R. C.: "Laboratory and Field Evaluation of an Inelastic Neutron Scattering and Capture Gamma Ray Spectrometry Tool," *Soc. Petr. Eng. J.*, vol. 20, 1980, p. 327.

Humphreys, D. R., R. W. Barnard, H. M. Bivens, D. H. Jensen, W. A. Stephenson, and J. H. Weinlein: "Uranium Logging with Prompt Fission Neutrons," *Int. J. Appl. Rad. Isotopes*, vol. 34, 1983, p. 261.

Killeen, P. G.: "Borehole Logging for Uranium by Measurement of Natural γ-Radiation," *Int. J. Appl. Rad. Isotopes*, vol. 34, p. 231, 1983.

Koizumi, C. K.: "Logging Calibration Models for Fission Neutron Sondes," Bendix Field Engineering Corp., Grand Junction, Colo., 1979.

Moxham, R. M., A. B. Tanner, and F. E. Senftle: "In Situ Neutron Activation Analysis of Bottom Sediments, Anacostia River, Washington," *MTS J.*, vol. 11, 1976, p. 14.

Nargolwalla, S. S., N. D. Nguyen, and A. U. Rehman: "Radioactive Methods," in J. G. Morse (ed.), *Nuclear Methods in Mineral Exploration and Production*, Elsevier, Amsterdam, 1977*a*.

———, A. Kung, O. J. Legrady, J. Strever, A. Csillag, and H. O. Seigal: "Nuclear Metallog Grade Logging in Mineral Deposits," *Proc. IAEA Symp. Nuc. Tech. Min. Res.*, 1977*b*.

Nelson, P. H., and G. D. Van Voorhis: "Estimation of Sulfide Content from Induced Polarization Data," *Geophysics*, vol. 48, 1983, pp. 62–75.

Plant, J., G. C. Goode, and J. Herrington: "An Instrumental Neutron Activation

Method for Multi-element Geochemical Mapping," *J. Geochem. Explor.*, vol. 6, 1976, p. 299.

Rhodes, J. D., T. Furata, and P. F. Berry: "A Radioisotope x-Ray Fluorescence Drill-Hole Probe," *Proc. IAEA Symp. Nuc. Tech. Min. Res.*, 1968.

Scott, H. D., and M. P. Smith: "The Aluminum Activation Log," *Trans. Soc. Prof. Well Log Analysts 14th Logging Symp.*, 1973, paper F.

Scott, J. H., P. H. Dodd, R. F. Droullard, and P. J. Mudra: "Quantitative Evaluation of Gamma-Ray Logs," *Geophysics*, vol. 26, p. 182, 1961.

Senftle, F. E.: "Application of Gamma-Ray Spectral Analysis to Subsurface Mineral Exploration," in G. K. Muecke (ed.), *Short Course in Neutron Activation Analysis in the Geosciences*, Mineralog. Assoc. Can., Halifax, 1980.

———, P. F. Wiggins, D. Duffey, and P. Philbin: "Nickel Exploration by Neutron Capture Gamma Rays," *Economic Geol.*, vol. 66, 1971, p. 583.

———, R. M. Moxham, and A. B. Tanner: "A Comparison of Radiative Capture with Decay Gamma-Ray Method in Bore Hole Logging for Economic Minerals," *Nuc. Instr. Meth.*, vol. 104, 1972, p. 485.

———, R. M. Moxham, A. B. Tanner, P. W. Philbin, G. R. Boynton, and R. E. Wager: "Importance of Neutron Energy Distribution in Borehole Activation Analysis in Relatively Dry, Low-Porosity Rocks," *Geoexploration*, vol. 15, 1977, p. 121.

———, A. B. Tanner, P. W. Philbin, G. R. Boynton, and C. W. Schram: "In Situ Analysis of Coal Using a ^{252}Cf-Ge(Li) Borehole Sonde," *Mining Eng. (AIME)*, vol. 30, p. 666, 1978.

Sowerby, B. D.: "Elemental Analysis by Neutron Inelastic Scatter Gamma Rays with a Radioisotope Neutron Source," *Nuc. Inst. Meth.*, vol. 166, 1979, p. 571.

Steinman, D. K., J. Stokes, J. A. Adams, R. Voigt, C. S. Pepper, K. V. Laubscher, W. J. Smith, and J. John: "^{252}Cf-Based Delayed-Neutron Logging System for In Situ Assay of Uranium Ore," *Trans. Soc. Prof. Well Log Analysts 21st Logging Symp.*, 1980, paper I.

Tanner, A. B., R. C. Bhargava, F. E. Senftle, and J. M. Brinkerhoff: "Radiative-Neutron-Capture Gamma-Ray Analysis by a Linear Combination Technique," *Nuc. Inst. Meth.*, vol. 102, 1972, p. 61.

———, R. M. Moxham, and F. E. Senftle: "Assay for Uranium and Determination of Disequilibrium by Means of In Situ High-Resolution Gamma-Ray Spectrometry," U.S. Geol. Survey Open-File Rept. 77-571, 1977.

Tixier, M. P., and R. P. Alger: "Log Evaluation of Nonmetallic Mineral Deposits," *Geophysics*, vol. 35, 1970, p. 124.

Vanasse, G. A., A. T. Stair, Jr., and D. J. Baker: "Aspen International Conference on Fourier Spectroscopy, 1970," Air Force Cambridge Res. Lab. Rept. AFCRL-71-0019, 1971.

Westaway, P., R. Hertzog, and R. E. Plasek: "The Gamma Spectrometer Tool Inelastic and Capture Gamma-Ray Spectroscopy for Reservoir Analysis," Soc. Petr. Eng. Paper SPE 9461, 1980.

Wichmann, P. A., and R. W. Webb: "Neutron Activation Logging for Silicon to Aluminum Ratios," *J. Petr. Tech.*, vol. 22, 1970, p. 201.

Wilson, R. D., M. S. Cosby, and J. M. Stone: "Field Evaluation of Direct Uranium Borehole Logging Methods," *Trans. Soc. Prof. Well Log Analysts 21st Logging Symp.*, 1980, paper S.

Wogman, N. A.: "In Situ x-Ray Fluorescence and Californium-252 Neutron Activation Analysis for Marine and Terrestrial Mineral Exploration," *Proc. IAEA Symp. Nuc. Tech. Min. Res.*, 1977.

Zablocki, C. J.: "Magnetite Assays from Magnetic Susceptibility Measurements in Taconite Production Blast Holes, Northern Minnesota," *Geophysics*, vol. 39, 1974, p. 174.

16

Cement

16-1 INTRODUCTION

Cement between casing and the formation serves two purposes: it supports the casing and it provides a seal that prevents fluid from flowing behind the casing from one part of the formation to another (Chap. 2). Cement powder can be mixed with water in various proportions; the properties of the product depend on the proportion. Compressive strength of fully cured cement, for example, varies from 50 MPa for a ratio of water to cement of 0.3, to 35 MPa for a ratio of 0.5, to 20 MPa for a ratio of 0.75, and to 12 MPa for a ratio of 1.0. The compressive strength increases as the cement sets. A typical bulk density is 1.9 g/cm^3, but again it depends on the proportion of water and cement. Sound speed varies between 2.7 and 4.8 km/s.

Cement does not always set as desired. Sometimes a thin *microannulus* exists between the casing and the cement; this can sometimes be a leakage path. Sometimes only part of the cement sets, and channels of borehole fluid are present in the cement. It is desirable to know several things about the cement: whether it is in contact with the casing, its quality, and whether channels are present in it. The setting of cement is exothermic. Consequently, the top of cement can be located by running a temperature log shortly after installation of the cement. [Figure 13-1 shows two such responses above 1000 ft (305 m).] Once the cement is set, it can be studied with either acoustic or nuclear logs. The methods used are interesting examples of the use of the physical principles behind those types of logs.

16-2 ACOUSTIC METHODS

When acoustic logs were run in cased holes, it was sometimes found that strong signals, with velocity corresponding to thin-plate velocity (Sec. 7-1b) in steel (5300 m/s), arrived at the detector. It was soon realized that these signals were caused by ringing in free casing. If the casing is not in good mechanical contact with a solid, much of the energy of the signal is retained in the casing, and a strong signal propagates. If there is good coupling, the energy enters the solid, and the signal in the casing is suppressed. This phenomenon permits using acoustic amplitude as a *cement bond log.*

Amplitude logging. A time gate is set at the time corresponding to the expected arrival of the casing signal, and the amplitude of the signal in that gate is recorded. A high amplitude indicates a large casing signal, and therefore a poor cement bond; a low amplitude indicates a good bond.

If full-wave or variable-intensity displays (Sec. 7-3c) are used, more information can be obtained. If there is no bond, an arrival is seen at the time corresponding to the plate velocity. If there is thick cement bonded to the casing but not to the formation, a very weak signal is seen until the arrival of the fluid wave; the thicker the cement, the weaker the casing signal. If there is a good bond to both the cement and the formation, there will be a strong signal at formation velocity. Often both the P and S waves in the formation can be seen, and one can get information about formation properties. Figure 16-1 shows a comparison of amplitude, full-wave, and variable-intensity displays for each of these situations.

Pardue et al. (1963) performed a calculation in which they simulated the casing by a thin steel plate in contact with an infinite solid and obtained the logarithmic energy decrement per unit distance as

$$\xi = \frac{\text{const}}{x} \frac{\rho_2/\rho_1}{[(v_c/v_p)^2 - 1]^{-1/2} + [(v_c/v_s)^2 - 1]^{-1/2}} \tag{16-1}$$

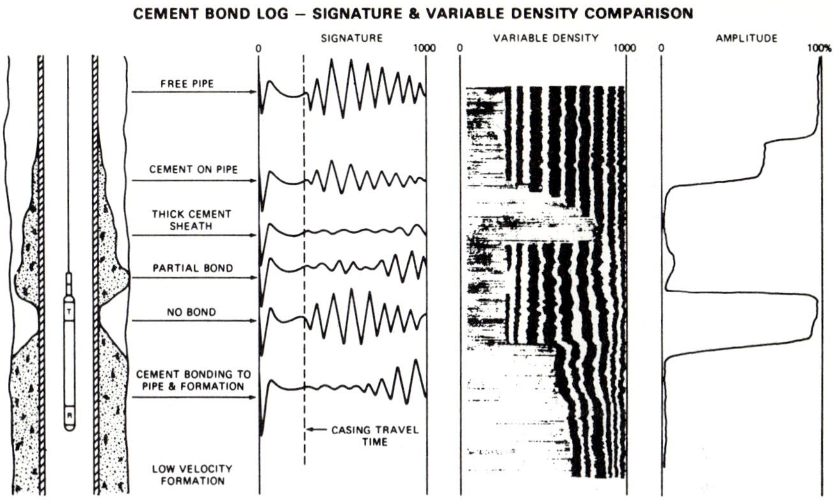

FIGURE 16-1 Comparison of amplitude, full-wave, and variable-density cement bond logs for different situations. (Rollman, no date.)

where ρ_1 and ρ_2 are the densities of the steel and solid, respectively, x is casing thickness, v_c is the plate velocity in the steel, and v_p and v_s are the longitudinal and shear velocities in the solid. Note that Eq. (16-1) blows up for a fluid for which v_s is zero, and therefore it does not apply to free casing.

The attenuation is inversely proportional to the casing thickness and directly proportional to solid density. The term in v_s is considerably greater than the term in v_p, so v_s is the most important cement property. The authors state that, since the shear velocity is important, the shear modulus, and therefore the strength, can be related to attenuation. They studied attenuation as a function of time after pouring cement. As the cement cured, the attenuation became greater. They plotted strength versus time for the cement, and they showed that attenuation did indeed increase with increasing compressive strength (Fig. 16-2). Furthermore, they confirmed that attenuation was inversely proportional to casing thickness for bonded casing but not for free casing. (Consequently, casing collars can be detected when the casing is bonded.) They also showed that attenuation rate was proportional to the fraction of the circumference in contact with cement.

There are many things that can cause incorrect amplitudes to be observed on a cement bond amplitude log (Bade, 1963; Pilkington and Fertl, 1975). If the gate is set at the wrong time, an amplitude that is either too low (if the gate is set too early and a low-amplitude *forerun-*

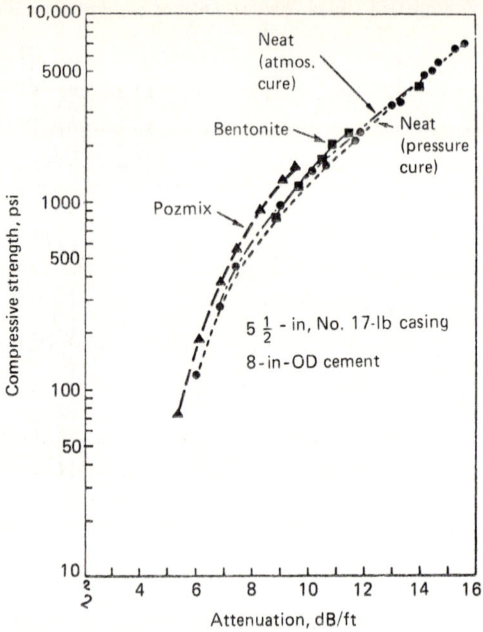

FIGURE 16-2 Compressive strength of cement versus attenuation of the casing signal for several cements. (After Pardue et al., 1963. Copyright 1963, SPE–AIME.)

ner is recorded) or too high (if the gate is set too late and a later cycle is recorded) is observed. A change in borehole fluid can change the arrival time of the casing signal and have the same effect as an incorrect gate time. A change in fluid can also change amplitude because different fluids attenuate signals differently. If the sonde is not well centralized, destructive interference (Sec. 7-3b) can cause low amplitudes. Gas in the borehole fluid will reduce amplitude. If the compressional velocity in the formation is greater than the plate velocity in the casing, the signal transmitted through the formation will arrive early enough to be mistaken for casing signal and give a spuriously high amplitude.

A *microannulus*, or very thin fluid layer between casing and cement, will give a high-amplitude signal corresponding to no contact. If the casing is pressurized, the microannulus will close and a low amplitude will result. Whether this effect is a defect or virtue depends on the user.

Many of these effects can be recognized if the displays discussed above are shown along with the amplitude log.

Recently (Gollwitzer and Masson, 1982) a two-detector version of the bond log has been developed. With it, one can observe the amplitude difference between the receivers and therefore can record attenuation rather than amplitude. This eliminates the effects of changing fluid attenuation on amplitude. It is also said to be less sensitive to poor centralization.

Some perceived disadvantages of the ordinary bond log are:

1 It requires good centralization.

2 One cannot tell high-strength cement with channels from low-strength cement without channels.

3 A microannulus gives low amplitudes, indicating a good bond.

4 A formation with high velocity gives spuriously high amplitude.

Broding (1984) has applied his improved method of analyzing Borehole Televiewer data (Sec. 19-4) to visualize the details of the condition of cement behind casing. He can see details of channeling in the cement. He can also relate the strength of the cement to its acoustic velocity.

Thickness mode. A new type of system (Froelich et al., 1981) has been built to overcome the difficulties described above. It uses frequencies so high (about 450 kHz) that the wavelength (about 1 mm) is less than the thickness of the casing. It has eight transducers mounted at different azimuths. Each transducer is both a transmitter and receiver. The transducers are pulsed in turn. The signal goes through the fluid and then reverberates in the casing, having been reflected from the cement and fluid in turn. The casing resonates in the thickness mode.

Each time the signal makes a round trip through the casing ($t = 2x/v$, where v is the bulk sound speed in the casing and x is the thickness) another pulse is returned to the transducer. The amplitude of the reflected signals depends on the impedances of the borehole fluid and of the material behind the casing. The reflection coefficients at the inner and outer surfaces of the casing are (Sec. 7-1c)

$$R_1 = \frac{Z_s - Z_f}{Z_s + Z_f} \qquad R_2 = \frac{Z_c - Z_s}{Z_c + Z_s} \tag{16-2}$$

where Z_s is the impedance of the steel, Z_f is that of the borehole fluid, and Z_c is that of the cement (or fluid) behind the casing. Havira (1982) has analyzed the situation and shows that, since Z_s and Z_f are fairly constant, the amplitude of the returned signal depends primarily on Z_c. If the material behind casing is cement, Z_c is large, R_2 is small, and amplitude decreases rapidly with time; if it is water (so the bond is bad), Z_c

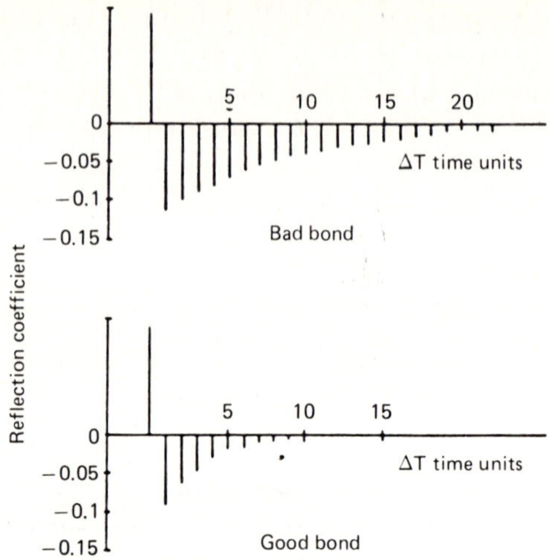

FIGURE 16-3 Amplitude versus time response to a unit impulse for a thickness mode cement log. (Havira, 1982.)

is small and amplitude decreases slowly. Figure 16-3 illustrates his result.

To measure both the value of the amplitude and its rate of decay, a segment of the amplitude versus time response is integrated as an indicator of Z_c (the first reverberation is not included). A large integrated amplitude response indicates small Z_c, and vice versa. Eight amplitude traces and eight adjacent variable-intensity records are displayed, giving a visual image of possible channels in the cement.

Transmission through a microannulus is a function of wavelength. If the microannulus is thin compared to the wavelength, it will not affect the signal. Havira has found that small microannuli (less than 0.08 mm) are not visible on the log. He states (without confirmation) that such a microannulus is not thick enough to affect the hydraulic seal. Of course, thicker microannuli have effects that depend on thickness, and there must be some thickness that can cause problems.

16-3 NUCLEAR METHODS

Tracers in cement. Nuclear methods have been used for cement investigation for many years. The mixture of radioactive tracers with the cement has been used in the Soviet Union to investigate the presence of cement as a function of both depth and azimuth. Some rather complex instrumentation (Barsukov et al., 1965) with a rotating collimated γ-ray detector can be used to estimate thickness and resolve channels in the cement.

γ-Ray backscattering. The most satisfactory nuclear method is γ-ray backscattering. One can use either a standard density log (usually decentralized) or a special centralized sonde made of a γ-ray source, a shield that fits fairly closely inside the casing, and a detector on the other side of the shield. This latter device has been used for at least 20 years (Caldwell and Sippel, 1958). A centralized sonde is very effective for investigating the amount and quality of gravel behind the casing. When used for this purpose, it is called the *photon log* (Chudy, 1981).

Gulin et al. (1960) discussed the use of a sonde made of three decentralized γ-γ sensors spaced symmetrically around the circumference of the hole. The source-detector spacing was chosen to be such that the calibration curve was single-valued. The sonde was slowly rotated as it came up the hole, so that the entire circumference could be examined.

Not only did Gulin et al. observe high count rates for water in the azimuth and low count rates for cement, but they could also study the decentralization of the casing (if the annulus was thin enough). The formation density is generally larger than cement density. As one of the sensors comes close to the formation, during rotation of the sonde, the count rate will be lower than that corresponding to the material in the annulus. Consequently, a shape like a half cycle of a cosine will be observed for a decentralized casing, with the lowest counts at the azimuth where the casing touches the formation. Figure 16-4 shows records for a decentralized casing with water and cement in the annulus. Note that, for the case with water in the annulus and a decentralized casing (*b*), the count at 180° is greater than that with a centralized casing (*a*). Evidently the annulus is so thin that γ-rays interact with the formation when the casing is centralized, and so the count rate does not correspond to pure water. Apparently, even if the system is well calibrated, it is possible to get inaccurate values of cement density if the annulus is thin enough that the formation affects the count rate.

A similar system has been used in large (2- to 3-m-diameter) holes at the Nevada Test Site. The accuracy of measurement of cement density for a system calibrated with the real casing is limited by the tolerance on the casing thickness. The thickness can vary by 0.25 cm, which can cause a change of 10 percent in apparent cement density.

γ-Ray backscattering methods can be powerful tools for investigating conditions behind casing. Standard sondes are satisfactory to locate cement and learn if channels are present. If, however, quantitative val-

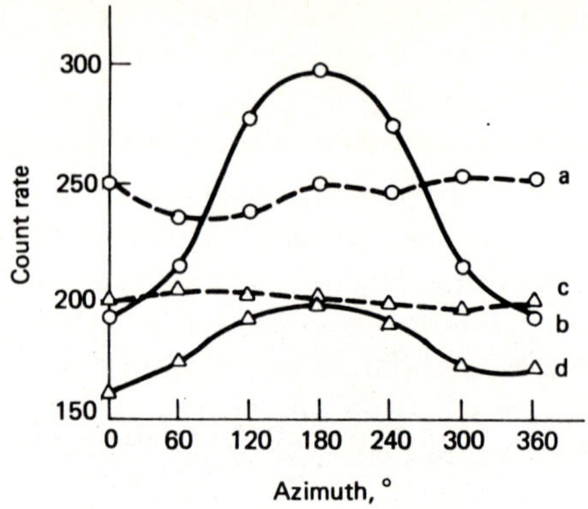

FIGURE 16-4 Count rate versus azimuth for a γ-ray backscattering cement density sonde. (*a*) Water in the annulus, centralized casing; (*b*) water in the annulus, casing decentralized; (*c*) cement in the annulus, centralized casing; (*d*) cement in the annulus, casing decentralized. (After Gulin et al., 1970.)

ues of cement density are required, it is imperative that calibration be done with sections of the actual casing to be investigated and with the materials that are expected to be found behind the casing. The shield and source-detector spacing should be designed for the specific casing diameter. The spacing should be optimized for maximum sensitivity and to assure that the count rate is a single-valued function of density (or water content). This is most easily done experimentally. In fact, a sonde of this nature is sometimes improvised from a conventional γ-ray sonde.

Neutron methods. As we saw in Sec. 16-1, cement can contain a good deal of water—more than in many formations but less than in borehole fluid. Consequently, a neutron log (Sec. 6-6) can be used to distinguish among cement, rock, air, and borehole fluid behind casing. It can locate the top of cement in either liquid-filled or dry holes.

REFERENCES

Bade, J. F.: "Cement Bond Logging Techniques—How They Compare and Some Variables Affecting Interpretation," *J. Petr. Tech.*, vol. 15, 1963, p. 17.

Barsukov, O. A., N. M. Blinova, S. F. Vybornykh, Yu. A. Gulin, V. N. Dakhnov, V. V. Larionov, and A. I. Kholin: *Radioactive Investigations of Oil and Gas Wells*, Macmillan, New York, 1965.

Broding, R. A.: "Application of the Sonic Volumetric Scan Log to Cement Evaluation," *Proc. Soc. Prof. Well Log Analysts 25th Logging Symp.*, 1984, paper JJ.

Birdwell Company: personal communication, 1970.

Caldwell, R. C., and R. F. Sippel: "New Developments in Radioactive Well-Logging Research," *Bull. Am. Assoc. Petr. Geol.*, vol. 42, 1958, p. 159.

Chudy, S.: "Photon Log," *Proc. Soc. Prof. Well Log Analysts 22d Logging Symp.*, 1981, paper GG.

Froelich, B., D. Pittman, and B. Seeman: "Cement Evaluation Tool—A New Approach to Cement Evaluation," Soc. Petr. Eng. Paper SPE 10207, 1981.

Gollwitzer, L. H., and J. P. Masson: "The Cement Bond Tool," *Proc. Soc. Prof. Well Log Analysts 23d Logging Symp.*, 1982, paper Y.

Gulin, Yu. A., D. A. Bernshtein, and Yu. I. Solokov: "New Techniques and Apparatus for Studies of the Distribution of Cement in the Annulus in Cased Wells," in N. N. Petrov, L. I. Petrenko, and P. S. Savitskii (eds.), *The Uses of Radioactive Isotopes and Nuclear Radiation in the U.S.S.R.*, Proc. All-Union Conf., 1960, vol. IV.

Havira, R. M.: "Ultrasonic Cement Bond Evaluation," *Proc. Soc. Prof. Well Log Analysts 23d Logging Symp.*, 1982, paper N.

Pardue, G. H., R. L. Morris, L. H. Gollwitzer, and J. H. Moran: "Cement Bond Log—A Study of Cement and Casing Variables," *J. Petr. Tech.*, vol. 15, 1963, pp. 545–554.

Pilkington, P. E., and W. H. Fertl: "Field Tests of Cement Bond Logging Tools," *Log Analyst*, vol. 16, no. 4, 1975, p. 13.

Rollman, E. E.: *"Acoustic Cement Bond Log, CBL,"* Dresser Atlas, Houston, Tex., no date.

17

Directional Surveying

Once a well is drilled, its course must be determined before any log and core information can be located and related to information from other wells. If conditions are such that the well is known to be straight enough for practical purposes, no survey is necessary. However, if rock and drilling conditions are such as to cause the hole to wander from its desired course, a survey is required. If a hole is intentionally deviated by employing directional drilling methods, it is necessary to survey during drilling in order to keep the hole on course. In such cases the user must determine if the survey made during drilling is accurate enough for his purposes or if another survey needs to be done when drilling is completed. Postdrilling surveys are typically run after casing is installed and the rig is removed from the site.

Directional surveys produce measurements of inclination φ, azimuth θ, and distance along hole ℓ (Fig. 17-1), typically at stations spaced 30 m apart. The distance between the sequential measurement stations must now be computed in x, y, z coordinates. Starting at some known point (usually the kelly bushing), these incremental distances are then summed sequentially to determine the course of the well.

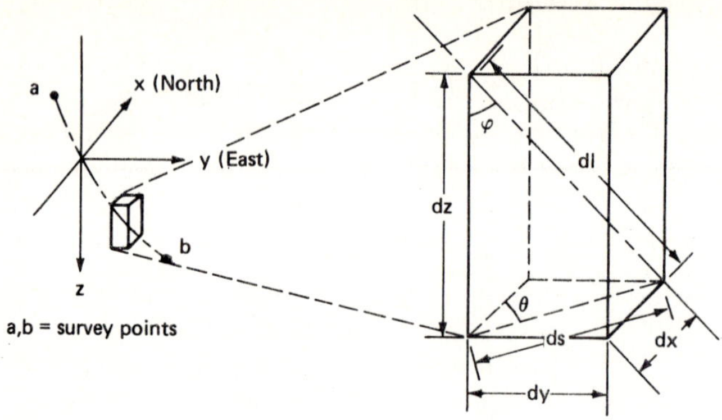

FIGURE 17-1 Geometry and nomenclature for directional surveying.

Once this conversion is done, any point intermediate between survey stations can be interpolated in x, y, z coordinates. Because the actual shape of the well course is unknown, some shape must be assumed in order to connect the survey points. No one computational method is the "right" method, and the conversion method to be used can be chosen from one of several in the literature.

Location of marker horizons. Survey data and well logs must be adjusted to a common surface elevation. The measured depths of geological and fluid marker horizons picked from the well logs can then be converted to x, y, z coordinates by using the interpolation scheme appropriate to the computational method adopted for the survey data. If the beds are horizontal, since the z coordinate gives the true vertical depth below surface, the stratigraphic thicknesses can be readily computed. Well logs from neighboring wells can be compared once the measured depths on the logs are converted to true vertical depths. It is best to replot all well logs to the same vertical scale if some wells are highly deviated and some are straight. In this way records obtained along slant sections in deviated holes are compressed to the same vertical scale as those acquired in straight holes.

In areas where beds are dipping, it is sometimes advantageous to scale well logs to projections other than vertical (Holt et al., 1977; Bateman and Hepp, 1981). Such projections are useful to compare logs from different wells on the basis of the true stratigraphic thickness. However, such projections cannot be accomplished unless dip information is available from a dipmeter or from geological interpretations.

17-1 COMPUTATIONAL METHODS

As illustrated in Fig. 17-1, the inclination φ is taken to be the angle of the well course from the vertical, azimuth θ is taken clockwise from geographic north, and measured depth increases downward from some reference point near the surface. In order to maintain a right-handed coordinate system, coordinate x increases to the north and y to the east. The differential relations between measured and rectangular coordinates are:

$$\begin{aligned} dx &= \sin\varphi \cos\theta \, d\ell = \alpha \, d\ell \\ dy &= \sin\varphi \sin\theta \, d\ell = \beta \, d\ell \\ dz &= \cos\varphi \, d\ell = \gamma \, d\ell \end{aligned} \tag{17-1}$$

The direction cosines α, β, and γ defined in Eqs. (17-1) represent the cosines of the angles between the well course and the x, y, z directions. The simplest method of computing a well course, called the tangential method, is to fit a series of straight-line segments to the well course and establish the direction of each segment by the lower of each pair of stations. The method is an extension of Eq. (17-1) applied to data from two sequential measurement stations labeled a and b, so that

$$\begin{aligned} \Delta x &= \sin\varphi_b \cos\theta_b \, \Delta\ell = \alpha_b \, \Delta\ell \\ \Delta y &= \sin\varphi_b \sin\theta_b \, \Delta\ell = \beta_b \, \Delta\ell \\ \Delta z &= \cos\varphi_b \, \Delta\ell = \gamma_b \, \Delta\ell \end{aligned} \tag{17-2}$$

where the incremental distances are the differences between values taken at a and b, for example, $\Delta x = x_b - x_a$. It is generally held to be less accurate than other methods, and its use is not recommended (Wilson, 1968). In more satisfactory methods, some means of averaging the measurements taken at two adjacent stations is utilized.

The tangential method can be improved by averaging the directional cosines of stations a and b. The result is referred to as the balanced tangential method

$$\begin{aligned} \Delta x &= \tfrac{1}{2}(\alpha_a + \alpha_b) \, \Delta\ell \\ \Delta y &= \tfrac{1}{2}(\beta_a + \beta_b) \, \Delta\ell \\ \Delta z &= \tfrac{1}{2}(\gamma_a + \gamma_b) \, \Delta\ell \end{aligned} \tag{17-3}$$

The interpolation procedure used for the balanced tangential method (Taylor and Mason, 1971) is to use the direction cosines in this manner: If the new point lies less than halfway to b, it is positioned on a tangent through point a; if more than halfway, it is positioned on a tangent through point b. Consequently, the continuous well course consists of segments joined together halfway between survey stations. Despite its simplicity, the balanced tangential method is reported to give results which agree quite well with those obtained with more sophisticated schemes (Walstrom et al., 1972).

The radius-of-curvature method first assumes that the well course increment lies in a vertical plane such that a smooth arc of length $\Delta\ell = \ell_b - \ell_a$ connects a and b. This arc has a radius of curvature r given by $\Delta\ell/(\varphi_b - \varphi_a)$. Thus in the vertical plane the vertical separation Δz between a and b is $r(\sin\varphi_b - \sin\varphi_a)$ and the horizontal separation $\Delta\lambda$ is $r(\cos\varphi_a - \cos\varphi_b)$. It can then be argued that $\Delta\lambda$ is an arc lying in the horizontal plane with radius of curvature r' given by $\Delta\lambda/(\theta_b - \theta_a)$. This effectively bends the vertical plane in a circular arc about a vertical axis. North-south and east-west increments are then given by $r'(\sin\theta_b - \sin\theta_a)$ and $r'(\cos\theta_a - \cos\theta_b)$. Rivero (1971) gives the resulting incremental distances as

$$\begin{aligned}\Delta x &= \frac{(\cos\varphi_a - \cos\varphi_b)(\sin\theta_b - \sin\theta_a)}{(\varphi_b - \varphi_a)(\theta_b - \theta_a)}\,\Delta\ell\\ \Delta y &= \frac{(\cos\varphi_a - \cos\varphi_b)(\cos\theta_a - \cos\theta_b)}{(\varphi_b - \varphi_a)(\theta_b - \theta_a)}\,\Delta\ell \qquad (17\text{-}4)\\ \Delta z &= \frac{\sin\varphi_b - \sin\varphi_a}{\varphi_b - \varphi_a}\,\Delta\ell\end{aligned}$$

The similarity between Eqs. (17-4) and (17-2) becomes clear when the differential relations $d(\sin\varphi)/d\varphi = \cos\varphi$ and $d(\cos\varphi)/d\varphi = -\sin\varphi$ are recalled. As the difference in angles becomes small, the radius-of-curvature method becomes equivalent to the tangential method, as it should for the case of a straight hole. Rivero utilizes the differential equivalences to get the versions of Eq. (17-4) required when either φ or θ is constant from one station to the next.

In the minimum curvature method a circular arc in three-dimensional space with curvature minimized is generated by applying the variational calculus (Taylor and Mason, 1971). The result can be expressed as

$$\begin{aligned}\Delta x &= \tfrac{1}{2}(\alpha_a + \alpha_b)g\,\Delta\ell\\ \Delta y &= \tfrac{1}{2}(\beta_a + \beta_b)g\,\Delta\ell \qquad (17\text{-}5)\\ \Delta z &= \tfrac{1}{2}(\gamma_a + \gamma_b)g\,\Delta\ell\end{aligned}$$

where $g = (2/\psi)\,[(1 - \cos\psi)/\sin\psi]$ for $\psi \neq 0$

$g = 1$ for $\psi = 0$

$\cos\psi = \alpha_a\alpha_b + \beta_a\beta_b + \gamma_a\gamma_b$

The angle ψ, sometimes referred to as the *dogleg*, is the angle between the tangent at a and the tangent at b. It can be seen from Eq. (17-5) that expressions based on minimum curvature revert to the balanced tangential method as the angular measurements at a and b become more nearly equal. The minimum curvature method appears to be advantageous in terms of the rigor of its derivation, although the interpolational computations are more complex than in most other methods. Taylor and Mason (1971) give the expressions needed to interpolate between stations.

We have by no means exhausted the list of computational methods possible by using data from two sequential stations. Both Walstrom et al. (1972) and Taylor and Mason (1971) have presented and compared several models by adopting mathematical frameworks which emphasize how the averaging schemes differ. Calculations by different methods of sample wells indicate that the different methods (except the tangential method) generally agree to within 0.3 m vertically and 3 to 6 m horizontally for a 3000-m well.

It is possible to compute position by using data from more than two stations, but lack of knowledge about the true nature of wellbore curvature discourages the use of such a technique. A better approach is to increase the station density if the hole neither is straight nor is changing course smoothly. As station density increases, the assumptions regarding shape become less important.

Dogleg severity. Wellbore tortuosity is best defined as the rate of change of total wellbore curvature (Wilson, 1968). In practice, wellbore tortuosity is estimated by differencing the dogleg angle between two consecutive stations. From Eqs. (17-1) and (17-5) we have

$$\cos \psi_{ab} = \sin \varphi_a \sin \varphi_b \cos (\theta_b - \theta_a) + \cos \varphi_a \cos \varphi_b \tag{17-6}$$

If data are not taken between stations spaced 30 m apart, the value for ψ_{ab} is normalized and expressed in degrees per 30 m. In this form it is referred to as *dogleg severity*. Values greater than 5° per 30 m are usually judged excessive. During a survey with a surface-readout system, the dogleg severity can be computed and displayed as data are acquired. If the value is excessive, fill-in stations can be obtained at tighter spacing, so that greater accuracy can be had in zones where wellbore curvature is changing rapidly.

17-2 SURVEY INSTRUMENTS

Directional survey instrumentation has undergone three successive stages of evolution. In the first stage, a gyroscopic or magnetic determination of azimuth and a plumb bob determination of inclination were recorded at each station by photographing the compass and plumb bob or by using some other recording scheme which could be read only after retrieval from the hole. In the second stage (now in general use), transducer readings are sensed electronically downhole, telemetered to surface, and read out after digital processing. The third stage is now underway as technology developed for inertial guidance and control systems is adapted for downhole survey use. Ease of operation and survey accuracy improve with each evolutionary stage.

Distance along the hole ℓ is determined from a tally of drill pipe or from measurement of cable length expended, depending upon whether the survey instrument is lodged in a drill collar or suspended from a wireline. Incorrect pipe tallies can arise from a bent or damaged measuring tape, omission of a joint from the tally, or failure to correct for distance from the bit to the sonde. Wireline depths can be incorrect if the magnetic cable marks are off or if the cable-measuring wheel sticks (Sec. 3-2). The wireline unit used for directional surveying will usually not be the unit used to acquire well logs, so any systematic differences in depth control will result in mislocation of depths. However, a casing collar log run in conjunction with the directional survey can be used to tie depths back to open-hole logs if a log with both CCL and γ-ray has also been run.

Azimuth is measured either with a magnetic compass (before a well is cased) or with a gyroscopic sensor. A set of flux gate magnetometers can substitute for a magnetic compass. If a magnetic sensor is run in drill pipe, nonmagnetic sections must be inserted into the drill string to house the survey instrument. Azimuth obtained magnetically must be corrected for magnetic field declination. Because declination varies smoothly over the earth's surface, no errors will result as long as the same declination is applied to all wells in a field. At high geomagnetic latitudes the inclination of the earth's field results in a smaller horizontal component to align the compass. Consequently, local field distortions due to residual and induced magnetization in the drill pipe and drill bit will affect the compass more at high latitudes than at low latitudes. The usual response to this problem is to add more nonmagnetic sections to the drill bit assembly. Significant disturbances can also arise if magnetic minerals are present in the formation. At high latitudes, temporal fluctuations of several degrees in declination can occur during magnetic storms. These fluctuations attenuate with depth, and so they can be expected to influence shallow readings more than deep ones.

Conventional single-axis gyroscopes substitute for a magnetic compass by providing a nearly stable reference orientation in the horizontal plane. A reference azimuth is then established on the surface by aligning the case with a known survey point. A gyroscope will drift throughout the survey; consequently, an error in initial alignment cannot be recovered or checked. The drift rate is checked every 10 or 20 min during the survey by holding the sonde at a survey station for several minutes. When the sonde is returned to surface, a final sighting of the survey stake is taken. Drift is regarded as a piecewise linear function of time. The total cumulative drift computed from the drift measurements should check with the difference between stake sightings taken before and after the survey. Bumping of the sonde is to be avoided because it can alter the drift rate, sometimes drastically. Drift rates will range from a few to more than 10°/h, but it is the drift check (closure) rather than the drift rate that matters. Wolff and de Wardt (1981) state that a closure equivalent to a drift of a few degrees per hour constitutes a good survey.

Inclination is measured by sensing the earth's gravitational attraction. A plumb bob will align with the local vertical. A disk, scribed with the angle with respect to the sonde axis, is placed next to the plumb bob, and both are photographed. Better resolution is achieved with servoed systems which keep a suspended mass aligned with the tool axis. The force required to hold the mass long the axis and away from vertical is proportional to wellbore inclination. Resolutions of 0.01° appear to be possible.

In most of the newer systems use is made of accelerometer and gyroscopic sensors developed for inertial guidance which have been adapted to the borehole situation. Different gyroscopic and accelerometer packages are being used by different service companies, so the operational procedures and the accuracies of each package will be different. The gyroscopic units can sense the direction of the earth's spin axis and thereby directly determine azimuth. There is no need to reference the system to surface surveys, so that surface alignment errors are eliminated. Drift errors also are eliminated because the gyro essentially realigns itself at each measurement station. There is one adverse consideration. At high latitudes the horizontal component of the earth's angular velocity diminishes, so the north-seeking ability of some of these systems progressively declines as the geographic poles are approached.

The systems described thus far have been developed for surveying wells drilled for petroleum. For small-diameter boreholes common in geotechnical and mineral exploration, an optical system which operates to depths of several hundred meters has been developed. The system consists of three circular rings mounted concentrically within the sonde and spaced 1.5 or 3 m apart. If the borehole is straight, the three rings appear on a common center. As the borehole course deflects, the rings will be progressively offset. A photograph is taken of the rings at each station, and angular deflection is reckoned from the ring offset. Vertical is tracked from the position of a bubble in the photograph. In nearly vertical holes, sonde orientation is maintained at the drill collar by orienting a scribed mark which is keyed through all tool and extension sections.

17-3 ANALYSIS OF ERRORS

Survey errors fall into two classes: random and systematic. Random errors vary from station to station within a single survey. A source of random error is the resolution with which an angle can be read from a film record or from a transducer. By contrast, systematic or station errors persist throughout an entire survey of a single well. They might be due to an incorrect surface orientation reading, to an improperly calibrated compass or inclinometer, or to a small fractional error in cable length measurement. Systematic errors will vary from survey to survey, and hence from well to well, assuming that different equipment is used on different wells. Bad positioning of the sonde in the hole can produce both random and systematic errors. Small attitudinal changes of the sonde with respect to the hole axis as the sonde is moved from station to station will produce random errors; on the other hand, if the sonde is cocked at some angle because of poor centralizers, the resulting error will be systematic. Systematic errors are generally more detrimental to overall survey accuracy than random errors are.

As discussed previously, the different computational methods will give somewhat different results in x, y, z determination. These differences are not due to measurement errors; instead they are due to our lack of knowledge concerning the geometric shape of the wellbore.

There are three ways to assess survey errors. If a number of surveys have been run on a particular well, the variations in x, y, z of a deep geological marker can be established relative to the mean value of all surveys. The statistical base for vertical depth can be extended to nearby wells by tying to a horizontal fluid contact or to a geological marker of known dip. Broadening the data base in this manner does not establish a "true" depth for a given marker; it simply increases confidence in the mean value against which any single survey can be compared.

The other two methods require an estimate of the various measurement errors. A base set of survey data from a well is adopted. By using the Monte Carlo technique, the well course is then recomputed many times. Each computational run adds to the measurements a different set of errors selected by a random process from the assumed error distribution. Ideally this technique simulates an actual sequence of repeat surveys. The results can be presented statistically or graphically as shown in the horizontal projection of Fig. 17-2.

The third technique is to compute a range of uncertainty as exemplified by the ellipse drawn around the individual Monte Carlo runs shown in Fig. 17-2. To establish the ellipse, Walstrom et al. (1969) assumed random errors in inclination of $\pm\frac{1}{4}°$ and in azimuth of $\pm 2°$ for inclination angles greater than 1°. The random errors were distributed uniformly over those ranges. Systematic errors were not incorporated into the model. It can be seen that the horizontal projection of the uncertainty ellipsoid closely bounds the locus of the Monte Carlo results.

Treatment of systematic errors has been incorporated into a deterministic error model developed by Wolff and de Wardt (1981). Each measurement produces an error vector

$$\Delta u = \sum m_i w_i(\ell) x_i(\ell) \tag{17-7}$$

comprised of i sources of error of magnitude m_i. The three orthogonal unit vectors x_i are oriented along the well course, horizontal and nor-

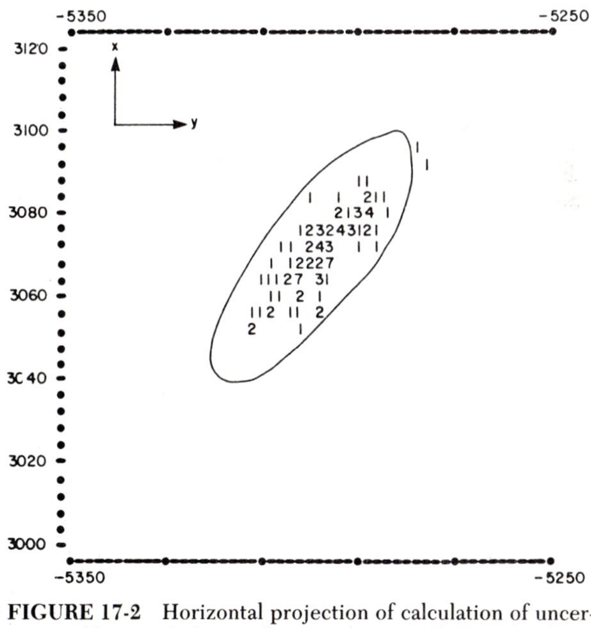

FIGURE 17-2 Horizontal projection of calculation of uncertainty of bottom-hole coordinates of an actual directional survey with maximum φ of 59° and total ℓ of 3350 m. Each digit represents the number of Monte Carlo simulations arriving at the grid point. The solid curve is a .99 probability ellipse. (From Walstrom et al., 1969. Copyright 1969, SPE–AIME.)

mal to the well course, and vertical and normal to the well course. The weights w_i account for the dependence of some of the errors upon φ and θ. Because both φ and θ change with ℓ, the terms w_i and x_i are shown as varying with ℓ. Error magnitudes m_i established for magnetic and gyroscope surveys run in the North Sea are shown in Table 17-1. Errors for a poor-quality survey are substantially greater than for a good survey. Also given in Table 17-1 are the functional weights w_i assigned to the six sources of error.

The total error accumulated at some measurement depth ℓ_o is given as

$$\Delta r_e = \int_0^{\ell_0} u(\ell)\,\Delta\ell \tag{17-8}$$

The matrix algebra required to convert the total error vectors to error ellipsoid parameters is too complex to give here. The error ellipsoid is unique to an individual hole; however, errors for a straight (but inclined) hole can be calculated by using three relatively simple algebraic expressions: one for vertical error, one for radial (horizontal plane, in the direction of the well) error, and one for lateral (horizontal plane, normal to well direction) error. Errors in these three orthogonal directions are shown in Fig. 17-3, which is based on the m_i and w_i values given in Table 17-1. It can be seen that lateral errors increase sharply with φ. For example, at an inclination of 50°, the uncertainty in the lateral direction from a good gyro survey of a 3000-m well would be 30 m; for a poor-quality magnetic survey the uncertainty would exceed 300 m. Vertical and radial errors are considerably less than lateral errors.

TABLE 17-1 Magnitude of measurement errors m_i and weighting factors w_i for four survey types. From Wolff and de Wardt (1981).

Survey type	Depth error $\Delta\ell/\ell$	Error in φ, °		Ref. error	Error in θ, °	
		Misalignment	True inclination		Drill-string magnetization	Gyro error
Good gyro	0.5	0.03	0.2	0.1		0.5
Poor gyro	2.0	0.2	0.5	1.0		2.5
Good magnetic	1.0	0.1	0.5	1.5	0.25	
Poor magnetic	2.0	0.3	1.0	1.5	5.0 + 5.0	
Weighting factor w_i	1	1	$\sin\varphi$	1	$\sin\varphi \sin\theta$	$1/\cos\theta$

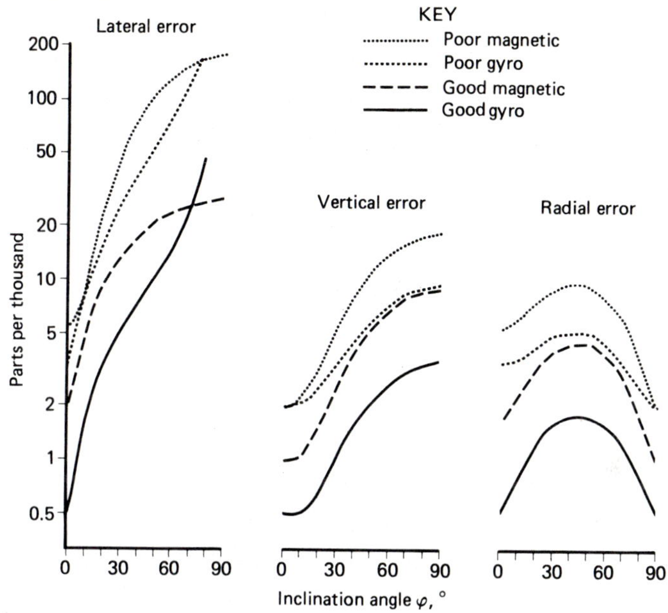

FIGURE 17-3 Expected survey errors in three orthogonal directions for an inclined straight well at varying incination angle. Errors are given in meters per 1000 m of distance along the well. (After Wolff and de Wardt, 1981.)

REFERENCES

Bateman, R. M., and V. R. Hepp: "Application of True Vertical Depth, True Stratigraphic Thickness and True Vertical Thickness Log Displays," *Trans. 22d Ann. Logging Symp. SPWLA*, 1981, pp. YY1–YY14.

Holt, O. R., L. G. Schoonover, and P. A. Wichman: "True Vertical Depth, True Vertical Thickness and True Stratigraphic Thickness Logs," *Trans. 18th Ann. Logging Symp. SPWLA*, 1977, pp. Y1–Y19.

Rivero, R. T.: "Use of the Curvature Method to Determine True Vertical Reservoir Thickness," *J. Petr. Tech.*, April 1971, pp. 491–496.

Taylor, H. L., and C. M. Mason: "A Systematic Approach to Well Surveying Calculations," SPE Paper 3362, 1971.

Walstrom, J. E., A. A. Brown, and R. P. Harvey: "An Analysis of Uncertainty in Directional Surveying," *J. Petr. Tech.*, April 1969, pp. 515–523.

———, R. P. Harvey, and H. D. Eddy: "A Comparison of Various Directional Survey Models and an Approach to Model Error Analysis," *J. Petr. Tech*, vol. 21, August 1972, pp. 935–943.

Wilson, G. J.: "An Improved Method for Computing Directional Surveys, *J. Petr. Tech.*, August 1968, pp. 871–876.

Wolff, C. J. M., and J. P. de Wardt: "Borehole Position Uncertainty—Analysis of Measuring Methods and Derivation of Systematic Error Model," *J. Petr. Tech.*, vol. 33, December 1981, pp. 2339–2350.

18

Dipmeter

The dipmeter is the most specialized and probably the most complex of all logging sondes. Its purpose is to measure the strike and dip of planar features intercepted by the borehole. The interpretation of dipmeter data is best done by someone familiar with the concepts of structural geology and stratigraphy, and the interpretation is usually separate from that of other logs. A dipmeter log is sometimes referred to as the "geologist's log."

In order to provide strike and dip measurements, a dipmeter must sense the location and orientation of (presumably) planar features which intersect the hole. Conceivably the sensing could be done by imaging the entire wall of the borehole, as can be done with a Televiewer (Sec. 14-2). From a full image, the continuity (or lack of continuity) of individual planar features could be noted and the corresponding angles measured. However, rather than image the bore wall, commercial sondes use three or more pads pressed against the bore wall to detect small-scale physical property changes. Any sensor with good resolution (such as an acoustic transducer) could be mounted on the pads, but the electrical method has been found to give good results in sedimentary sequences. Consequently, present-day sondes incorporate small electrode arrays mounted on pads which are pressed against the wall by articulated arms. A four-arm system is shown schematically in Fig. 18-1. Each pad produces a signal proportional to either conductivity or resistivity, depending upon the electronic design. The trend in pad design has been to enhance resolution in order to provide as much detail for the interpreter as possible (Allaud and Ringot, 1969). High resolution requires high sampling rates ranging from 2 to 4 samples per centimeter and large data handling capacity.

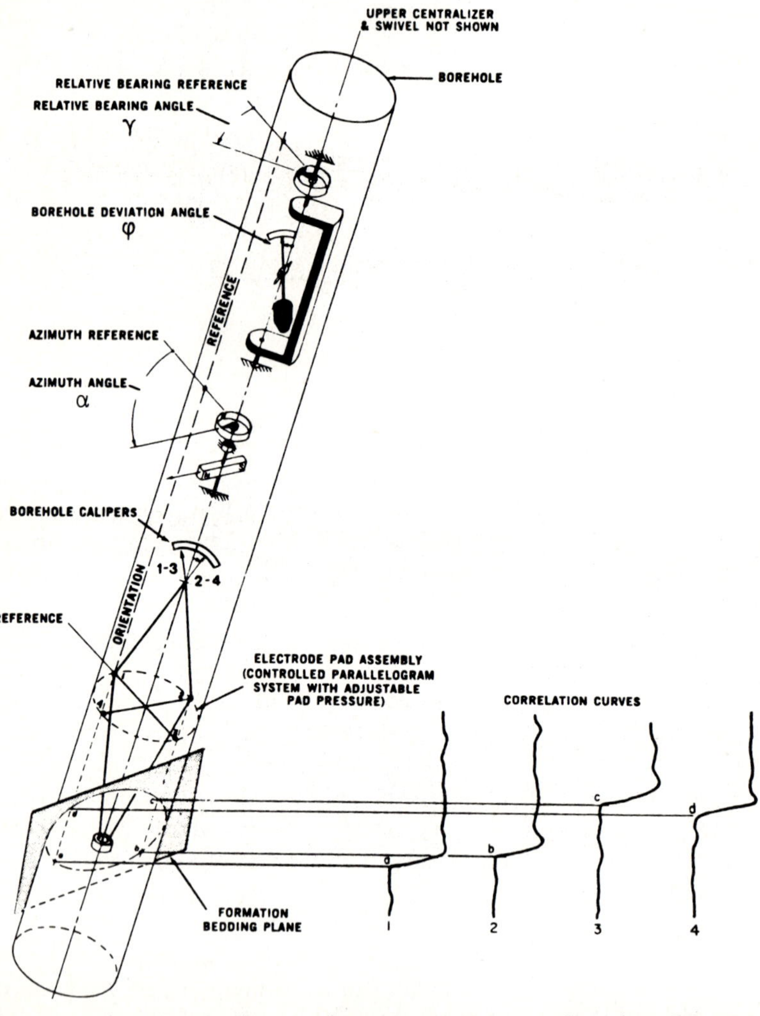

FIGURE 18-1 Schematic of a four-arm dipmeter intersecting a planar boundary separating two rock units of differing resistivity. Pad 1 crosses the plane at point *a*. (After Dresser Atlas, no date.)

The fundamental problem of dipmeter measurement is the identification and correlation of the individual features which cause the electrical responses. As illustrated in Fig. 18-1, the traces produced by each of the pads will be offset one from another because of the dip of the beds and the angle of the hole. Hence, the traces must be shifted to correlate individual features. It is also necessary to measure borehole diameter, because wellbore rugosity causes erroneous computation of apparent dip. Once a planar feature has been identified and corrected for borehole rugosity, the orientation of that feature can be found with respect to a coordinate system fixed in the wellbore.

It then remains to determine the true strike and dip of a feature with respect to geographic coordinates. This can be done only if the borehole orientation is known. Hence, a dipmeter must incorporate a directional survey device capable of continually measuring inclination φ and azimuth θ. In addition, it must measure a rotational angle γ, which specifies the angle between the high side of the hole and the number 1 resistivity pad. [As a matter of interest, these three orientation measurements are also required for survey control of directional drilling (Chap. 2). To drill directionally, the orientation of the bit face within the hole, as well as hole inclination and azimuth, must be monitored. Hence, some aspects of the following development also apply to directional drilling control.]

We see, then, that a dipmeter actually requires three kinds of measurements: (1) sonde orientation with respect to a geographic frame of reference, (2) the detection of geological features intersecting the wall of the hole, and (3) a caliper measurement of wellbore diameter. In the following two sections we shall show how these measurements are used to determine strike and dip of geological features.

18-1 IDENTIFICATION OF PLANAR FEATURES

The central problem of dipmeter design is the identification of individual features which intersect the borehole and determination of their angles of intersection. Bear in mind that a geological contact or bedding plane intersected by a well can be idealized as a plane cut by a cylinder, the locus of intersection being an ellipse.

Let ℓ be the measured position of the center of the sonde in a borehole. Suppose there is one feature in the hole, such as that with the step-like response shown in Fig. 18-1. If it intersects the hole at some angle ψ, then each pad will register a response at a different position ℓ. Let ℓ_{ab} be the shift, or the difference in depth between the response of pads a and b to the same feature. For four pads, there are a total of six possible shifts (ℓ_{12}, ℓ_{13}, ℓ_{14}, ℓ_{23}, ℓ_{24}, ℓ_{34}) to determine. Kemp (1980) shows that, for a planar surface, the shifts are related by

$$\ell_{12} = -\ell_{34} = \frac{\ell_{13} - \ell_{24}}{2} \qquad \ell_{14} = \ell_{23} = \frac{\ell_{13} + \ell_{24}}{2} \tag{18-1}$$

as long as the pads hold their diamond-shaped pattern coming up the hole and the sonde does not rotate as it passes through the section of interest. Under these conditions, Eq. (18-1) shows that the shifts ℓ_{13} and ℓ_{24} obtained along orthogonal diameters are the only two of six shifts which are independent. However, if one pad fails to contact the wall, then the other three pads must be used. Extraneous signals due to small-scale geological irregularities will also contaminate the response. Consequently, there are many practical considerations for selecting a planar surface, and the algorithms to do so will be as varied as the ingenuity of the programmer. Suffice it to say that somehow the shift required to overlay a given pair of traces *a* and *b* must be determined in order to establish ℓ_{ab}. We next describe two approaches to determining the shifts: cross-correlation and pattern recognition.

Cross-correlation. Moran et al. (1962) consider the conductivity traces from two different pads to be two "vectors" A_i and B_i ($i = 1, \ldots, n$), where the number of points *n* defines the *search length.* Each point in the vector is a sample of the conductivity curve. A measure of the distance between the two vectors is given as

$$D = \left[\sum (A_i - B_{i+j})^2\right]^{1/2} \tag{18-2}$$

The value of *j* which minimizes *D* determines the optimum correlation ℓ (ℓ is *j* multiplied by the spacing between samples). A closely related approach is to first normalize the vectors so they have the same average value. Then a measure of their correlation is the scalar product

$$\Phi_{ij} = \sum A_i B_{i+j} \tag{18-3}$$

To use Eq. (18-3), the curves are sequentially shifted by *j* and the summed product is determined. Maximum correlation occurs when Φ_{ij} is greatest, and ℓ_{ab} is then known from the shift.

This treatment of the correlation procedure is somewhat simplified. Kemp (1980) gives a complete formulation for calculating the correlation coefficient $r(\ell)$. The closer $r(\ell)$ is to a magnitude of 1.0, the closer A_i will be to B_{i+j} in the least-squares sense and the better the determination of ℓ_{ab} will be. Cross-correlation requires time-consuming computations which can be performed faster if specialized computer peripherals called array processors are available (Holt and Schoonover, 1973).

An important aspect of cross-correlation methods is that all features within a given search interval receive the same shift. This may not be desirable in many cases because it means that structural features (having one strike and dip) will be treated the same as stratigraphic features (having another strike and dip).

Pattern recognition. The second approach to curve correlation is more akin to the procedure used when lining up traces by eye. The techniques of pattern recognition (Vincent et al., 1979) are applied so that a hierarchy of correlatable features is established within a specified search interval. This is done by describing the curves as a series of elements which have parameters describable in terms of shape and derivatives. Each element is defined by its boundaries and an associated pattern vector V_i which serves to characterize its properties. Just as a human interpreter would select certain features as more significant than others, so the elements can be ranked as, say, large, medium, or small. The computerized search procedure is to first search out all large elements of a particular type, as exemplified by the trial correlations for positive peaks in Fig. 18-2*a*. The correlation choice is based on the likeness distance

$$Y = \sum [V(e)_i - V(e')_i]^2 \tag{18-4}$$

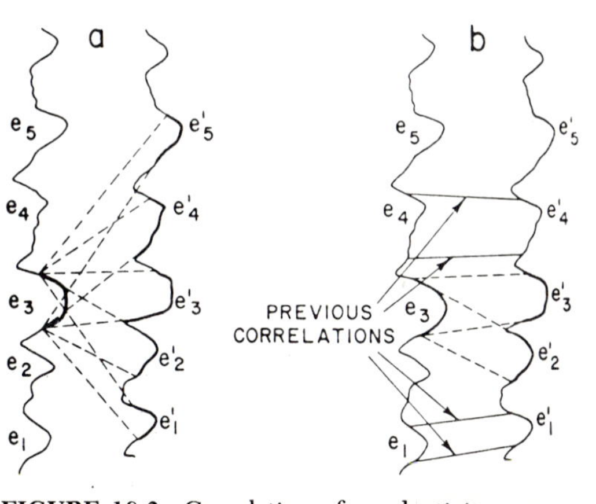

FIGURE 18-2 Correlation of conductivity curves by using a pattern recognition technique. (From Vincent et al., 1979. Copyright 1979, SPE–AIME.)

where e and e' are elements of the same type from two different curves and V_i are the values of the pattern vector. A Y value is computed for all element pairs of the same rank and type. Elements are then correlated for the minimum value of Y. The search sequence then proceeds for other types and ranks of correlatable elements. Once a correlation has been made, no subsequent correlation can cross it, as illustrated in Fig. 18-2*b*. Here correlations $e_4 - e'_4$ and $e_1 - e'_1$ are established before e_3 is examined. Hence e_3 cannot be correlated with e'_4 or e'_1, nor can it be correlated with e'_5 because that would require crossing an existing correlation. Once two elements have been correlated, ℓ_{ab} is the difference between their respective depths.

Cone of silence. Suppose the major axis of the ellipse characterized by the wellbore and the plane lies in the direction of the number 1 pad. Then the angle of the ellipse relative to a plane normal to the wellbore is

$$\tan \psi = \frac{\ell_{13}}{d_{13}} \tag{18-5}$$

where d_{13} is the wellbore diameter between pads 1 and 3. Regardless of how correlations are established, it is impractical to search the entire record of two traces for a match. Instead, a finite search length ℓ_m is chosen as a computational parameter and no attempt is made to correlate features separated by more than that length. With ℓ_m defined as the maximum displacement permitted in the correlation procedure, and d the average borehole diameter, by analogy with Eq. (18-5) we can define

$$\tan \psi_m = \frac{\ell_m}{d} \tag{18-6}$$

so that ψ_m, sometimes called the *search angle*, is the maximum angle of a plane which can be detected by the data reduction process. Rotation of this plane around the wellbore axis generates a *cone of silence* within which no planar features will be detected. An example will be discussed in Sec. 18-4.

18-2 DETERMINATION OF STRIKE AND DIP

Once the shifts ℓ_{13} and ℓ_{24} and the corresponding borehole diameters d_{13} and d_{24} are determined, the attitude of a planar surface can be specified with respect to the dipmeter axis and its pads. Let the center of the dipmeter sonde be the center of the (u_3, v_3, w_3) coordinate system.

The w_3 component points downward along the axis of the wellbore; and if the sonde has four pads, the u_3 and v_3 components point to the number 1 and number 2 pads, respectively. The components of the unit vector normal to the planar surface are given as (Kemp, 1980; Holt and Schoonover, 1973)

$$\begin{aligned} u_3 &= \frac{\ell_{13}/d_{13}}{\Upsilon} \\ v_3 &= \frac{\ell_{24}/d_{24}}{\Upsilon} \\ w_3 &= \frac{1}{\Upsilon} \end{aligned} \tag{18-7}$$

where

$$\Upsilon = \left[1 + \left(\frac{\ell_{13}}{d_{13}}\right)^2 + \left(\frac{\ell_{24}}{d_{24}}\right)^2\right]^{1/2}$$

Moran et al. (1962) give a similar expression for the unit normal vector determined by a three-arm dipmeter.

The unit normal vector can be expressed in terms of a geographic coordinate system by performing a series of three coordinate rotations. The first such rotation produces components u_2, v_2, and w_2, with w_2 kept oriented along the wellbore axis and unchanged from w_3. However, the u and v axes are rotated through an angle γ so that u_2 points to the high side of the well bore. The required rotation can be written

$$\begin{vmatrix} u_2 \\ v_2 \\ w_2 \end{vmatrix} = \begin{vmatrix} \cos\gamma & -\sin\gamma & 0 \\ \sin\gamma & \cos\gamma & 0 \\ 0 & 0 & 1 \end{vmatrix} \begin{vmatrix} u_3 \\ v_3 \\ w_3 \end{vmatrix} \tag{18-8}$$

In the next rotation the v axis is held fixed so that $v_1 = v_2$. The u and w axes are rotated through an angle φ which is defined as the wellbore inclination (Fig. 17-1). After this rotation, the w axis points vertically downward and the u axis points in the direction of wellbore azimuth.

$$\begin{vmatrix} u_1 \\ v_1 \\ w_1 \end{vmatrix} = \begin{vmatrix} \cos\varphi & 0 & \sin\varphi \\ 0 & 1 & 0 \\ -\sin\varphi & 0 & \cos\varphi \end{vmatrix} \begin{vmatrix} u_2 \\ v_2 \\ w_2 \end{vmatrix} \tag{18-9}$$

The final rotation holds the w axis fixed vertically and rotates the u and v axes through an angle θ, which is defined as the wellbore azimuth (Fig. 17-1). After this rotation the u axis points north (u_0), the v axis points east (v_0), and the unit vector (u_0, v_0, w_0) is now expressed in a geographic frame of reference or geomagnetic frame of reference, depending on the convention adopted by the user. (If geomagnetic, it is assumed that the user will correct for magnetic declination.)

$$\begin{vmatrix} u_0 \\ v_0 \\ w_0 \end{vmatrix} = \begin{vmatrix} \cos\theta & -\sin\theta & 0 \\ \sin\theta & \cos\theta & 0 \\ 0 & 0 & 1 \end{vmatrix} \begin{vmatrix} u_1 \\ v_1 \\ w_1 \end{vmatrix} \tag{18-10}$$

In the geographic coordinate system, the unit normal vector can also be defined in terms of the orientation of the planar feature.

$$\begin{aligned} u_0 &= \sin\varphi_p \sin\theta_p \\ v_0 &= \sin\varphi_p \cos\theta_p \\ w_0 &= \cos\varphi_p \end{aligned} \tag{18-11}$$

The angle θ_p is orthogonal to the strike of the plane, and the angle φ_p is the same as the dip of the plane.

To solve for strike and dip in terms of dipmeter measurements, we can carry out the sequence of matrix operations indicated by Eqs. (18-7) through (18-10) and equate the result to Eq. (18-11). These operations can be summarized as

$$\begin{vmatrix} u_3 \\ v_3 \\ w_3 \end{vmatrix} = \begin{vmatrix} R_\gamma \end{vmatrix} \begin{vmatrix} R_\phi \end{vmatrix} \begin{vmatrix} R_\theta \end{vmatrix} \begin{vmatrix} u_0 \\ v_0 \\ w_0 \end{vmatrix} \tag{18-12}$$

where R_γ, R_ϕ, and R_θ represent the 3 by 3 matrices of Eqs. (18-8), (18-9), and (18-10). From Eq. (18-12), expressions for θ_p and φ_p can be found.

In this development it is assumed that the directional survey package of the dipmeter measures γ, θ, and φ, as in the system described by Johnson and Angehrn (no date). Instead of θ, in some directional (inclinometer) systems α is measured; α is the angle in the horizontal plane between magnetic north and the number 1 pad. Such a system is illustrated schematically in Fig. 18-1. Moran et al. (1962) and Bateman and Konen (1978) give relations equivalent to Eq. (18-12) which incorporate the angle α.

18-3 SPECIAL PROBLEMS AND ERRORS

Because of the statistical nature of geological processes, any good dipmeter survey will produce data with a certain amount of *scatter*. Bedding and structural features are not truly planar, may not extend all the way across the wellbore, and may have different widths or different electrical properties on different sides of the wellbore. Further, the measurement may be adversely affected by borehole enlargement. As a consequence, individual data points do not repeat precisely on subsequent logging runs. These statistical vagaries, which we might call "geological noise," produce scatter. In addition to geological noise, the correlation of traces also produces scatter. Individual features may not repeat on subsequent processing runs of the same data if different search parameters are applied to the two runs. Scatter of this sort is not really due to measurement error; instead, it is due to imperfect correlation processes. Whatever the origin of scatter, interpretive errors can result if too much significance is assigned to individual points. A realistic interpretation must take into account the statistical nature of the data.

Measurement errors can introduce both random and systematic errors into the determination of bed attitudes. Some of the known sources of dipmeter error have been corrected by design improvements. An interesting example is the error caused by the tendency of the sonde to stick and slide; so that even if the cable pull appears to be uniform at surface, the pad motion is actually erratic. This causes errors because the digitized samples are not gathered at uniform spacings. A cure for this problem is to measure the sonde velocity by inserting an extra sensor called a "speed button" on one of the pads immediately above the existing one (Allaud and Ringot, 1969). The two identical traces are shifted by a constant amount if the velocity is constant. If the velocity is not constant, a stretch factor is derived from the two curves and applied to the traces from each pad.

The accuracy of survey systems has improved markedly over the years. Some older systems are reported to produce errors in azimuth θ of $\pm 10°$ or $\pm 20°$ under unfavorable circumstances (Johnson and Angehrn, no date). Even worse, Fitzgerald et al. (1980) state that, in one unit, the azimuth determination could be off by as much as $\pm 40°$! Newer systems incorporating magnetometers have improved these figures to about one degree at low geomagnetic latitudes and better than 5° at high latitudes. Errors in inclination φ appear to be of the order of $\pm 0.5°$ to $\pm 1.0°$, but they may become as large as 2 to 4°. The error in older systems in which a weighted cylinder is used to measure the pad rotation γ may be no better than $\pm 10°$.

The impact of measurement error (θ, φ, γ) on computed dip and strike (φ_p, $\theta_p + 90°$) can be assessed by using the same computational program as was used for the field survey to simulate runs with specified error in the measured parameters. Fitzgerald et al. present a series of such error runs which show the impact of θ and φ errors on artificial and real dipmeter runs. They also show examples for a well inclined at 45° in which a 10° error in γ can produce a 7° error in dip or a 45° error in strike, depending upon the azimuth of the well.

An error in the caliper measurement of diameter d directly affects the estimate of the normal vector (u_0, v_0, w_0) as shown by Eq. (18-7). Fitzgerald et al. (1980) state that the caliper is one of the most critical sources of error of a dipmeter. They show examples in which a caliper error of 5 cm in a well inclined 45° causes a dip error of 7° and strike errors of 45°.

More difficult to assess are the effects of the finite depth of investigation of the conductivity pad. Because the current penetrates into the formation, the electrical reading sees deeper than the mechanical caliper. If not compensated, the apparent dip angles will be slightly in error. One method of compensation is to add an estimate of the average electrical depth of investigation to the mechanical diameter. There is another effect of the finite depth of current penetration: the reading is influenced to some extent by beds immediately above or below the pad. The effect is asymmetric, so that the apparent shift ℓ_{ab} can be lengthened or shortened depending on the position of pads and bed boundaries.

The direction of dip can be of extreme importance in delineating prospective petroleum traps. If the true dip is to the east, but the dipmeter indicates it to be to the west, then the consequences are obvious! For this reason, as Fitzgerald et al. point out, errors in a highly inclined hole penetrating a flat formation can be more serious than errors in a vertical hole penetrating steeply dipping formations. In the former case a small error in borehole φ can cause the computed dip direction to be incorrect.

18-4 DISPLAYS AND INTERPRETATION

Dipmeter interpretation is intimately tied to the presentation of the processed data. Several graphical techniques conveying three-dimensional information on a two-dimensional plot have been developed to portray dipmeter results. One of the more commonly used plots is the "tadpole" plot (Fig. 18-3). The head of a tadpole is plotted at the magnitude of dip angle and its tail points toward the direction of dip. For example, beds above the unconformity in Fig. 18-3 dip to the southeast, whereas beds below the unconformity dip to the southwest. In this way strike and dip information are efficiently conveyed as a function of depth.

Another method of presentation (Bengtson, 1981) is to not present strike angle but instead present the apparent dip on two orthogonal vertical planes, one transverse to the dominant structural trend and the second parallel to it. These plots are easier to interpret where signifi-

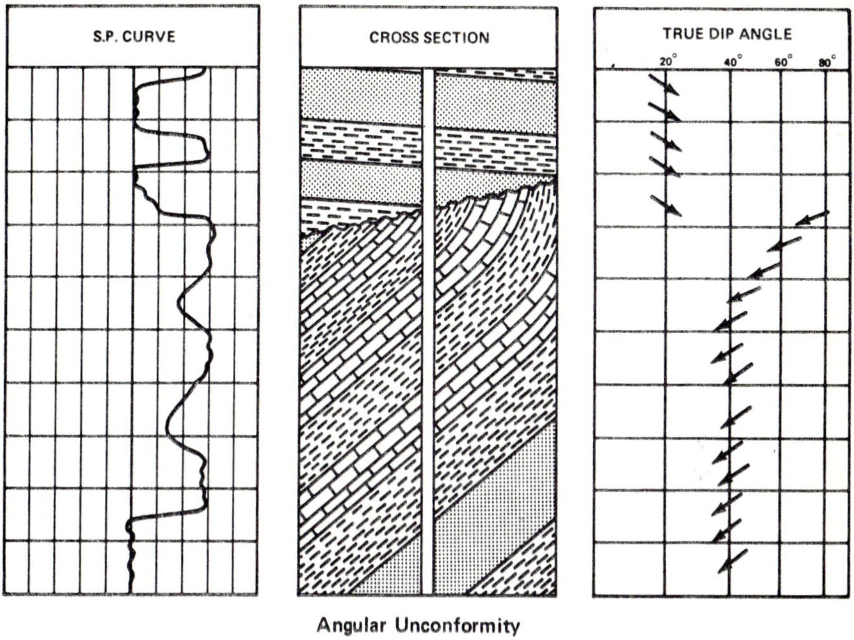

FIGURE 18-3 Hypothetical geological cross section and resulting SP and diameter plot. The dipmeter results are shown on a "tadpole" plot in which the orientation of arrow gives the strike angle. (From Dresser Atlas.)

cant structural curvature exists, as in the case of plunging folds or drag folds produced by faulting. By viewing dip in the two orthogonal directions, the interpreter remains aware of the three-dimensional aspects of the data. In contrast, by viewing a tadpole plot, an interpreter may forget that he is not always viewing the transverse plane.

Dipmeter data need not be displayed as a function of depth. It is often advantageous to group strike and dip determinations from a zone of particular interest which is likely to represent a consistent structural or sedimentological environment. Polar plots of azimuth-frequency, Schmidt stereonet projections (Schlumberger, 1981), and half-tangent stereonet methods (Bengtson, 1983) can be used to summarize data from a vertical interval or even from several wells intersecting the same structural zone. Bengtson (1981) shows that plots of dip versus azimuth can be used to diagnose basic categories of structural curvature.

Dip-azimuth plots also display the effect of the cone of silence (Fig. 18-4). Consider a vertical well for which the search angle ψ_m is chosen to be 80° (Fig. 18-4, A). Any plane dipping at an angle of 80° or more will lie within the cone and cannot be detected by the dipmeter. This

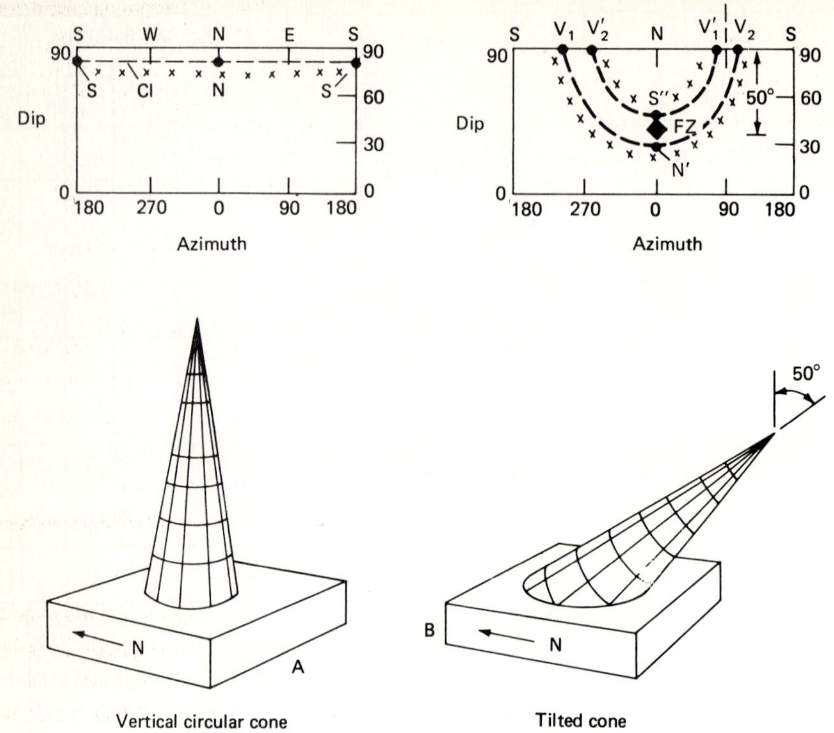

FIGURE 18-4 Presentation of dipmeter data on a dip-azimuth plot. (From Bengtson, 1983.)

cone of silence produces a forbidden zone above the dashed line in the corresponding dip-azimuth plot. If the well is not vertical, but is inclined at an angle of 50°, the cone of silence maps as an arcuate band (shown as *FZ* in Fig. 18-4, B). The solid diamond shows the dip and azimuth of the wellbore axis: obviously the dipmeter cannot detect any plane coincident with the wellbore axis nor can it detect any plane with dip and azimuth lying in the cone of silence. Bengtson notes that some computerized search algorithms produce data points immediately outside and adjacent to the forbidden zone. These spurious points are shown as x's in Fig. 18-4. It is easy to see that the cone of silence can eliminate meaningful data in inclined wells.

REFERENCES

Allaud, L. A., and J. Ringot: "The High Resolution Dipmeter Tool," *Log Analyst*, vol. 10, no. 3, 1969, pp. 3–11.

Bateman, R. M., and C. E. Konen: "The Log Analyst and the Programmable Pocket Calculator, Part III: Dipmeter Computation," *Log Analyst*, January–February 1978, pp. 3–9.

Bengtson, C. A.: "Statistical Curvature Analysis Techniques for Structural Interpretation of Dipmeter Data," *Am. Assoc. Petr. Geol. Bull.*, vol. 65, no. 2, 1981, pp. 312–332.

————: "Easy Solutions to Stereo-Net Rotation Problems," *Am. Assoc. Petr. Geol. Bull.*, April 1983.

Dresser Atlas: "Relating Diplogs to Practical Geology," Dresser Atlas, Houston, Tex., no date.

Fitzgerald, D. D., J. C. Theriot, and P. I. York: "Dipmeter Validity in Deviated Boreholes," *Log Analyst*, May–June 1980.

Holt, O. R., and L. G. Schoonover: "Computer Methods of Diplog Correlation," *Soc. Petr. Eng. J.*, 1973, pp. 31–38.

Johnson, W. M., and J. Angehrn: *Advances in Diplog Instrumentation*, Dresser Atlas, Houston, Tex., no date.

Kemp, F.: "An Algorithm for Automatic Dip Computation," *Computers Geosci.*, vol. 6, 1980, pp. 193–209.

Moran, J. H., M. A. Coufleau, G. K. Miller, and J. P. Timms: "Automatic Computation of Dipmeter Logs Digitally Recorded on Magnetic Tapes," *J. Petr. Tech.*, July 1962.

Schlumberger: *Dipmeter Interpretation*, Vol. I, *Fundamentals*, Schlumberger, New York, 1981.

Vincent, P., J. Gartner, and G. Attali: "An Approach to Detailed Dip Determination Using Correlation by Pattern Recognition," *J. Petr. Tech.*, vol. 31, February 1979, pp. 232–240.

19

Some Other Interesting Applications

In this chapter we describe some unusual applications of the logging methods described in preceding chapters. In each case, unusual circumstances or problems were encountered; they required the use of methods that depended on an understanding of the basic principles of the logging technique rather than routine "cookbook" methods.

19-1 COMPARISON OF A DENSITY LOG AND A GRAVITY LOG TO INFER STRUCTURE

When one wishes to understand the difference between the γ-γ density and gravimetric density, as described in Sec. 8-7, sometimes the geologic structure is fairly complex. Then it cannot be described by simple shapes such as cylinders, spheres, or truncated beds. A more complicated structure can be represented by a two-dimensional geological cross section. If the cross section can be described by a set of polygons (in two dimensions), a computer program developed by Hearst (1981), using a formula due to Talwani et al. (1959), can be used to calculate gravity as a function of depth. The program compares the calculated gravimetric density with the measured gravimetric and γ-γ densities. This calculation is useful in deciding which of two possible cross sections is the more plausible.

For example, Fig. 19-1a and b shows two postulated cross sections for a region near a proposed nuclear test at the Nevada Test Site. Mea-

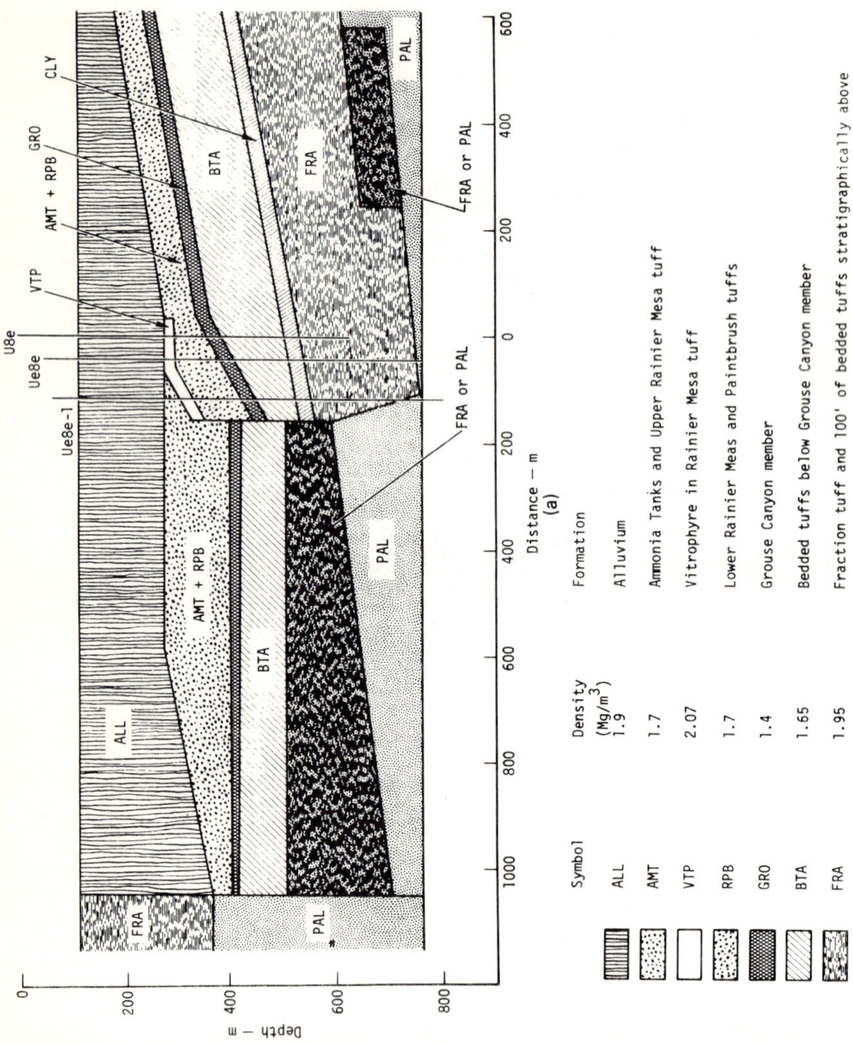

sured and calculated gravimetric density in two holes shown on the figures are compared in Fig. 19-2*a* and *b*. The high-density Paleozoic rock near hole Ue8e in Fig. 19-1*a* causes the density calculated below 600 m in that hole to be too high in Fig. 19-2*a*. The Paleozoic is farther away from the hole in Fig. 19-1*b*, so the calculated density is more cor-

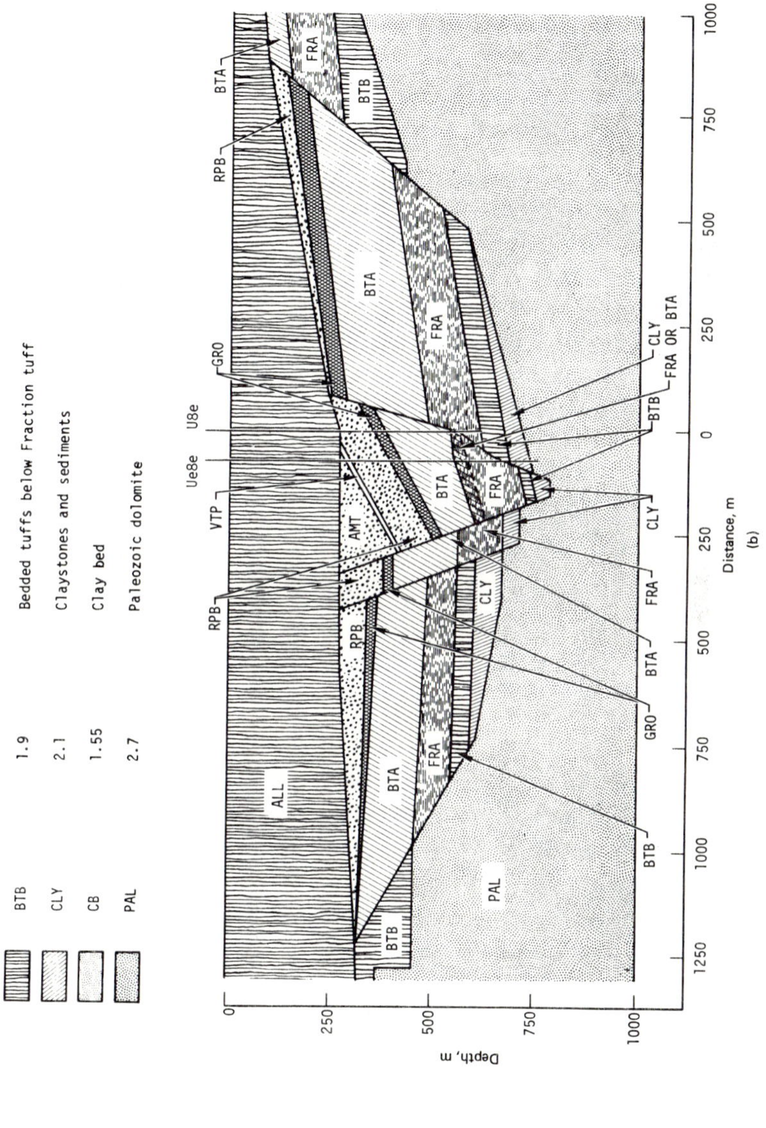

FIGURE 19-1 (*a*) One-scarp cross section. Gravity is measured and calculated in wells marked U8e and Ue8e. (*b*) Two-scarp cross section (Hearst and Carlson, 1982.)

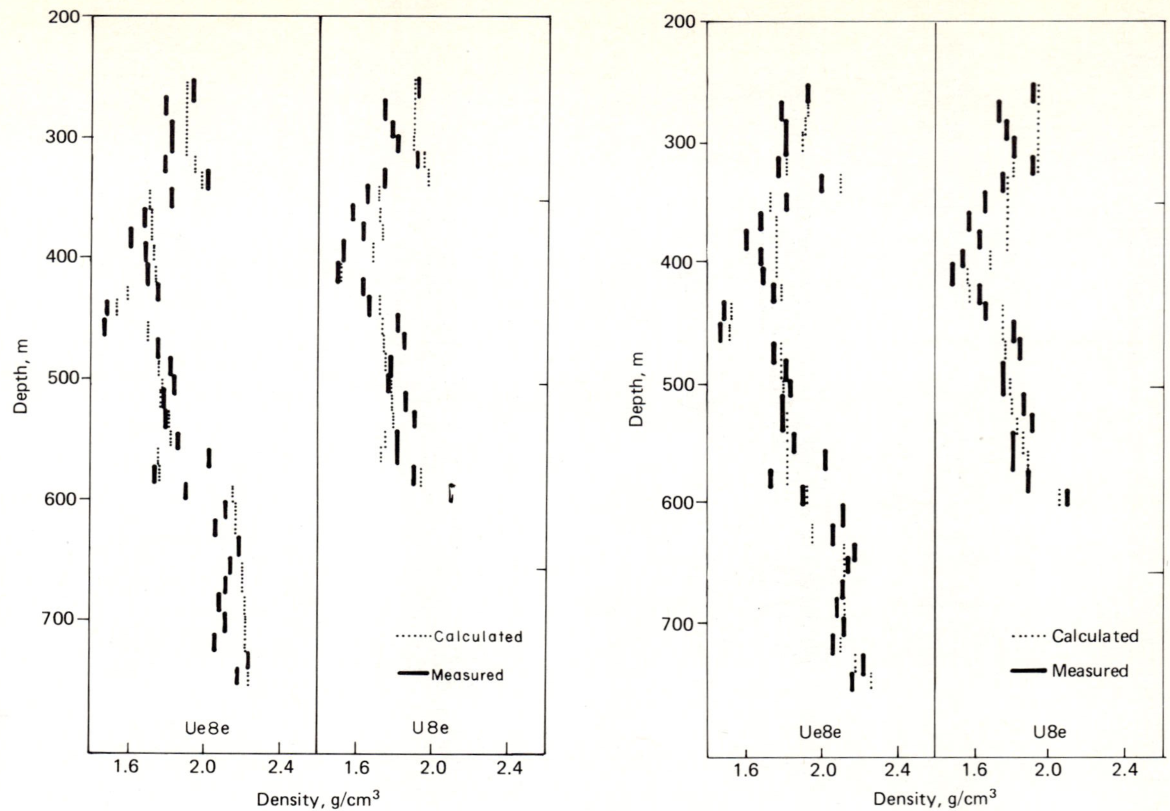

FIGURE 19-2 (*a*) Calculated gravimetric density versus depth for cross section in Fig. 19-1*a* compared with gravimetric density measured in holes U8e and Ue8e (Hearst and Carlson, 1982.) (*b*) Calculated gravimetric density versus depth for cross section in Fig. 19-1*b* compared with measured gravimetric density in two holes, U8e and Ue8e.

rect. Analysis of the rms difference between measured and calculated gravimetric density shows the calculated density in Fig. 19-2*b* to be much closer to the measured density than that in Fig. 19-2*a*, and therefore the cross section in Fig. 19-1*b* is more plausible than that in *a*.

19-2 SPECIFIC USES OF γ-γ DENSITY LOGS

Fragmentation problems in coal overburden. Certain portions of the overburden of the Callide strip coal mine in Queensland, Australia were much more difficult to fragment than the remainder of the overburden. Geologists (Flood and Wilson, 1980) examined the highwall of the mine and found that the hard-to-fragment bodies were lenticular sandstones. The rest of the overburden consisted of shales and poorly sorted sandstones. The lenses were characterized as "channel sand deposits associated with a fluvial environment," and "well-sorted clay-free sands." The lenses evidently were difficult to fracture both because they were more lithified than the rest of the overburden, and therefore stronger, and because, being well sorted, their porosity was higher than that of the rest of the overburden. High-porosity rock will absorb more of the energy of an explosion than hard, dense rock. This absorption can reduce fracturing.

It was desirable to be able to use logs in blast holes to locate the lenses so that the explosive loading could be better designed. Since the sands were characterized as "well-sorted," one could assume the density was lower than that of the surrounding medium (Sec. 10-1). Therefore, a density log could be used to identify the lenses. It was found that there were several low-density regions visible on logs in the blast holes, but some were associated with clay and had high γ-ray readings. When the combination of low density and low γ-ray was used, correlation with observations of lenses in core was very good. Evidently that combination can be used to identify the lenses.

The volume of a petroleum storage facility. Baembitov (1969) describes the use of γ-ray backscattering in monitoring the underground storage of petroleum products in rock salt. The shape of the solution-mined reservoir was not well known, so it was important to locate the brine-product surface. It was necessary to measure through a steel tube inserted in the reservoir, so resistivity could not be used. The brine density was 1.05 to 1.20 g/cm^3 and that of the product was 0.7 to 0.9 g/cm^3. It was desirable that even at this low density the count rate decrease with increasing density (Fig. 6-15). This would require a large distance between source and detector, so a 90-cm-long sonde was used. Depth of investigation was found to be 30 to 40 cm for these low densities. A calibration curve was developed, and it was easy to locate the product-brine interface and also the top and bottom of the reservoir.

It was, of course, desirable to learn the shape and volume of the reservoir. As the reservoir was filled with gasoline, the density log was used to monitor the height of the gasoline-brine interface as a function of time; and each time the interface was located, the volume of gasoline that had been inserted was recorded. Thus a curve of interface height versus volume could be plotted. Figure 19-3 is a composite of all the logs. At the top of the figure, the count increases as the sonde enters the gasoline. In the middle, the count decreases at the interface at a different depth for each log. The hatched line is then the height of the interface for each log. The results showed that the reservoir was roughly spherical, as specified.

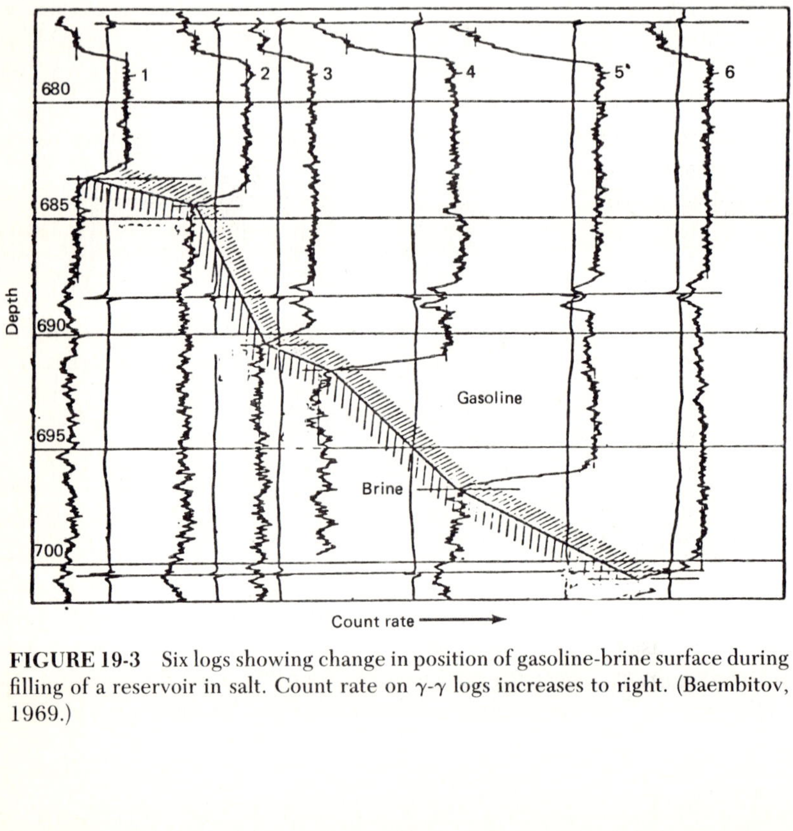

FIGURE 19-3 Six logs showing change in position of gasoline-brine surface during filling of a reservoir in salt. Count rate on γ-γ logs increases to right. (Baembitov, 1969.)

19-3 MAGNETIC LOGGING FOR STRATIGRAPHY IN TUFFS

The presence of remanent and induced magnetism (Sec. 5-8) can be used to identify rock units. Magnetic anomalies observed in a borehole are opposite in sign to those in the formation, because the borehole represents the removal of a magnetic dipole from a slab of formation. The void left by the hole can be thought of as a path by which magnetic lines of force go from the bottom to the top of the slab—sort of a short circuit. The lines of force have a direction opposite to those in the formation and to the ambient field. Consequently, an induced or normal remanent field causes a negative anomaly in a magnetic field log, and a reversed remnant field causes a positive anomaly.

Douglas and Millet (1978) found that many tuffs at the Nevada Test Site have characteristic magnetic "signatures." Figure 19-4 shows a typical magnetic log through units exhibiting positive and negative remanent magnetism. Some of the other tuffs in the column have significant susceptibility and give rise to negative anomalies. The shape of each anomaly is consistent from hole to hole and therefore can be used to identify the unit causing it.

This logging method is now used routinely at the Nevada Test Site to identify tuff units.

19-4 USE OF LOGS TO LOCATE ONE HOLE FROM ANOTHER

In several circumstances—a blowout in an oil well, for example—the drilling of a ventilation hole to intersect a tunnel, or preparation of connecting holes for in situ coal gasification, it is desirable to drill a hole to intersect an already existing hole. Directional surveys (Chap. 17) are often not accurate enough for the task. If the first hole is accessible, an acoustic, electric, or magnetic transmitter can be inserted, and the transmitter is fairly easy to locate from the second hole. In the case of a blowout, however, this method is impossible. Fortunately, the target hole is usually cased. Three methods have been used to locate the casing from another hole: resistivity, magnetic, and acoustic.

Resistivity. Obviously, a conductive casing in a hole will affect the resistivity measured in another hole nearby. But the size of the effect will depend on the distance between holes, the formation resistivity, and the electrode spacing. Mitchell et al. (1972) found experimentally

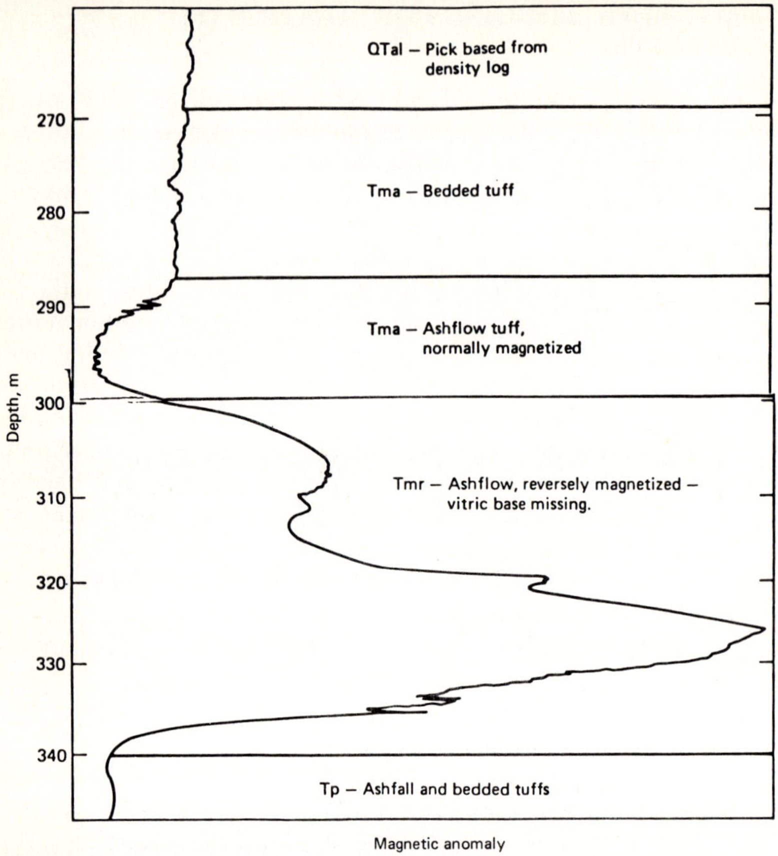

FIGURE 19-4 Typical total magnetic intensity log through normally and reverse-magnetized ashflow tuffs. (Douglas and Millett, 1978.)

that in a formation with a resistivity of the order of 10 Ω·m or less, ordinary "64-inch" resistivity logs could detect casing at a distance of about 3 m. They also found that if long-spaced resistivity logs (25- to 50-m *AM* spacing, 200-m *AN* spacing, Fig. 5-19) were used, they could detect casing at distances of more than 15 m. The method used was to assume that the resistivity of the medium near the depth of interest was approximately the same as that at the depth of interest and then compare the measured resistivity with the "normal" resistivity. They used laboratory experiments to obtain curves of the ratio of measured to normal resistivity versus distance to the casing. The accuracy of the method depends on the ability to estimate normal resistivity at the depth of interest.

Field results showed that casing could be detected 20 m away and distance could be estimated within a range of 15 m. They suggest 10 percent accuracy at long range, with improved accuracy at closer range. Warren (1981) found that the method gave good qualitative evidence that a casing was nearby but distance resolution was poor: a distance of 4 to 5 m was calculated when the true value was 9 m.

Magnetic. Resistivity devices do not give the direction to the target. Robinson and Vogiatzis (1972) proposed the use of a magnetometer to locate the casing. They calculated that, if the casing had no remanent magnetization, a magnetometer would not be useful because the effect of the casing, at useful distances, would be small compared to the earth's field. But if the casing had been inspected for flaws by a magnetic device (a common practice), remanent magnetization would exist and be detectable.

Remanent magnetism caused by casing inspection is ordinarily periodic in depth because of the individual casing joints. When the magnetism was periodic, Robinson and Vogiatzis could obtain the distance to the hole from the depth difference between maxima of the observed field in a second hole. When, because of stresses in the casing, the field was no longer periodic, they mathematically modeled the magnetic field caused by the casing based on an assumed remanent magnetization and fitted the measured response in the second hole to the model for various distances and directions to the casing. They concluded that the distance and direction giving the best fit between the model and the data were correct.

Morris et al. (1977) developed a simple magnetic gradiometer to eliminate the effect of the earth's field and permit detection of unmagnetized casing. The method works because the spatial gradient of the earth's field is very much less than that caused by casing. Morris et al. mounted two vertical magnetometers "several feet" apart vertically in a sonde and used the difference in their readings to detect casing. They also installed two orthogonal horizontal magnetometers to determine direction. They claimed a range of 30 m, but they did not state accuracy.

Warren (1981), however, compared the above method to the resistivity method. He concluded the range was about 15 m. He found the accuracy "nearly correct" for distance and either perfect or within 20° for direction. Evidently the distance was within 1 or 2 m of that determined by other methods, and there was no way of telling which method was truly correct.

Acoustic. Broding (1981) developed an improved method of analysis of Borehole Televiewer data (Sec. 7-4a) which, among other things,

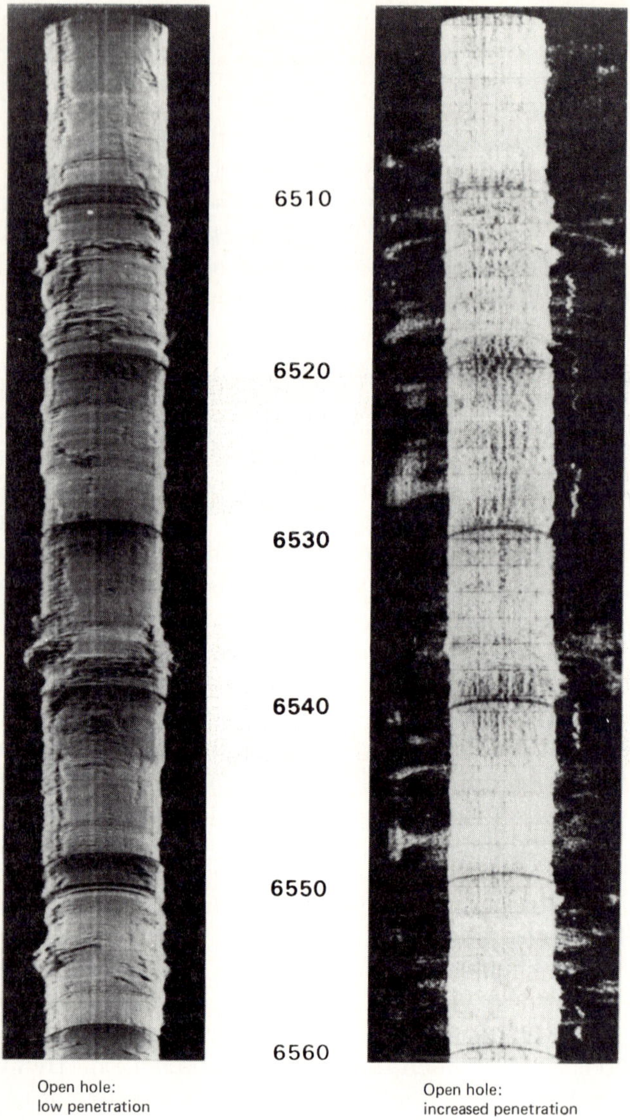

FIGURE 19-5 Open-hole polar displays of "volumetric scan log" at two different penetrations. (Broding, 1981.)

permits detection of nearby casing. Instead of just recording the amplitude of the first arrival on film, 512 data points of the entire signal are recorded on magnetic tape. Consequently, signals reflected from distances of a meter or more beyond the wall of the hole are recorded.

Once the recording is made, many different playbacks are possible. The simplest, besides the ordinary Televiewer presentation, is one in which the intensity is proportional to the time of arrival of the first high-amplitude signal. This becomes an intensity-modulated sonic caliper. Another presentation is a *polar plot,* a horizontal slice through the hole and its surroundings showing arrivals versus azimuth. This plot shows any reflections occurring in that slice as bright lines.

The most spectacular plot is a tilted polar image. Figure 19-5 shows the same hole at two different penetrations (that is, two different signal lengths). Vertical and horizontal fractures and bedding planes can be seen. Figure 19-6 shows this method used to locate a blowout well from a relief well. The blowout well can be seen as a slanted line on the right of the figure. Although the range is short compared to that of electrical and magnetic methods, distance and directions can be obtained to very high accuracy indeed, and they can be determined in real time. As of this writing, Broding had not developed an easy method for obtaining all possible information from these displays, but it is clear that in the future this technique will be of immense value.

19-5 DISTINCTION BETWEEN MAFIC AND FELSIC ROCKS IN GEOTHERMAL WELLS

A new frontier in logging is the study of geothermal reservoirs. Special sondes have been developed to withstand the temperatures, and special

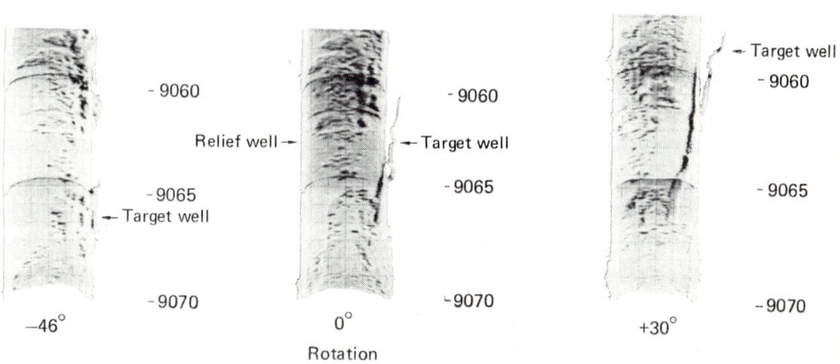

FIGURE 19-6 The intercept of well 1 from well 2 by polar scanning with rotation. (Broding, 1981.)

interpretation techniques are being used (Keys and Sullivan, 1979; Keys, 1979; Daniels et al., 1981; Glenn et al., 1982). Many geothermal wells are drilled in igneous and metamorphic rocks, and one of the objectives of logging is distinguishing between mafic (high Mg, Fe, Al) and felsic rocks. The authors listed above have found that many mafic rocks exhibit the seemingly impossible combination of high bulk density and high neutron porosity (or low neutron count rate). Figure 19-7 shows a set of logs on which a biotite-hornblende quartz monzonite gneiss has a bulk density of at least 3.0 g/cm^3 and yet a neutron porosity of almost 0.1. These values appear inconsistent. The authors use a density-neutron cross-plot to locate these apparently anomalous materials and identify them as mafic. The high density is understandable because mafics, with their high content of heavy elements such as iron and magnesium, have a high grain density. There are two plausible reasons for the neutron response:

1. Many of the mafic minerals, such as biotite and hornblende, contain some hydrogen in their crystal structure. This gives rise to an apparent neutron porosity even when no free water is present (Nelson and Glenn, 1975). But there is not enough hydrogen to explain the observed neutron porosity.
2. Iron has a thermal neutron absorption cross section that is high compared with that of most other elements in rocks (Table 6-2).

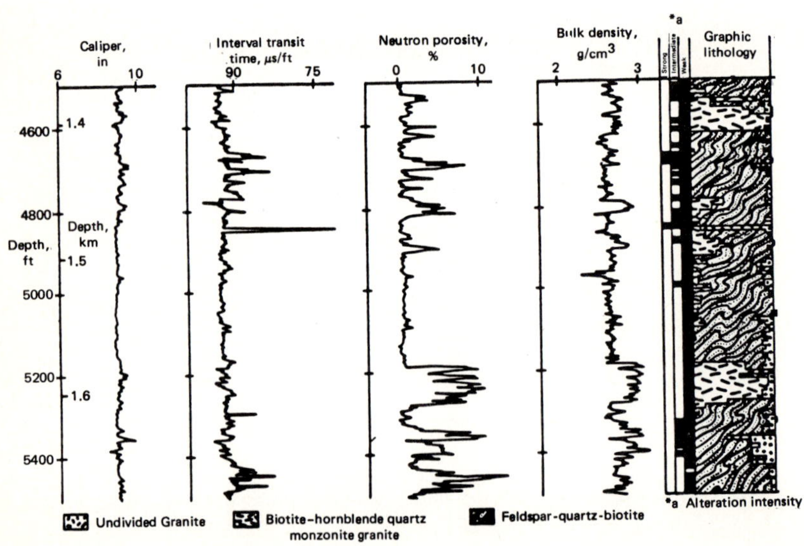

FIGURE 19-7 Logs from a geothermal well (Utah State Well 52-21) showing high density and neutron porosity in biotite-hornblende quartz monzonite gneiss. (Glenn et al., 1982. Copyright 1982, SPE–AIME.)

Although the authors do not emphasize the fact, *thermal* neutron logs are generally used for this work. It is quite likely that the high iron content contributes to the high apparent porosity (Nelson et al., 1983*a*).

All of the above authors state that high density and high neutron porosity are characteristic of mafics. An unwary user might run a density log and an *epithermal* neutron log and be quite unable to locate the mafics. He could reach a totally wrong conclusion. This situation illustrates the importance of understanding the characteristics and behavior of specific logging equipment rather than simply following a recipe. One can find mafics with a thermal but not an epithermal neutron log. If, however, the user wanted to locate water in the formations, an epithermal neutron log would be much better than a thermal one.

19-6 USE OF A TEMPERATURE LOG TO ESTIMATE POROSITY IN PERMAFROST

The geothermal gradient (Sec. 4-5) is smaller in ice-saturated permafrost than in the water-saturated sediments below the permafrost. Lachenbruch et al. (1982) use this difference in gradient, measured with a temperature log, to estimate porosity. Figure 19-8 shows tem-

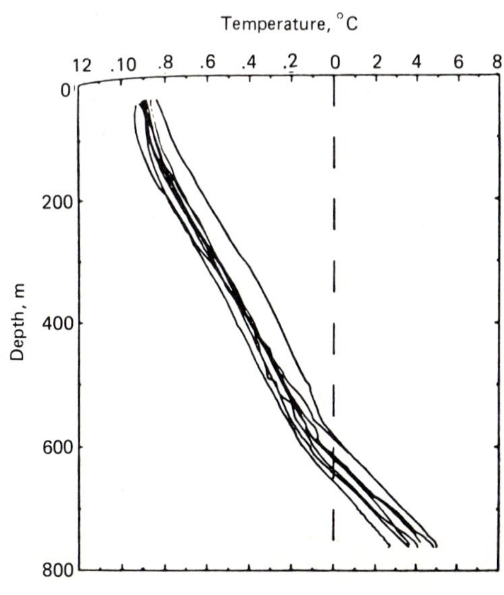

FIGURE 19-8 Temperature profiles from nine holes in the Prudhoe Bay field. (Lachenbruch et al., 1982.)

perature measured in nine holes drilled through permafrost in the Prudhoe Bay field in Alaska. The boundary between ice and water saturation is at 0°C. For a steady-state condition, the heat flow across the boundary must be constant,

$$Q = G_{th}k_{th} = G_{fr}k_{fr} \tag{19-1}$$

where G is the geothermal gradient, k is the conductivity, and the subscripts *th* and *fr* refer to thawed (or water-saturated) and frozen (or ice-saturated), respectively. If we assume a geometric mean mixing law for thermal conductivity, then, from Eq. (4-11),

$$k_{fr} = k_{ic}^{\phi}k_{ma}^{1-\phi} \tag{19-2a}$$
$$k_{th} = k_{w}^{\phi}k'^{1-\phi}_{ma} \tag{19-2b}$$

where $k'_{ma} = \gamma k_{ma}$ and γ is the constant of proportionality between conductivity of the matrix at the temperature of the frozen material and that at the temperature of the thawed material. Equation (19-2) can be solved to yield

$$\phi = \frac{\ln (k_{fr}/k_{th}) + \ln \gamma}{\ln (k_{ic}/k_{w}) + \ln \gamma} \tag{19-3a}$$

k_{ic}, k_{w}, and γ are known from laboratory data, and the measured ratio of gradients from Eq. (19-1) can be substituted for the unknown ratio of conductivities to give

$$\phi = \frac{\ln (G_{th}/G_{fr}) + \ln \gamma}{\ln (k_{ic}/k_{w}) + \ln \gamma} \tag{19-3b}$$

Substitution of known and measured values yielded a porosity of about 0.38. Porosities measured on core varied between 0.39 and 0.43.

19-7 APPLICATIONS OF THE CHANGE OF VALUE OF A PROPERTY WITH DEPTH

Use of velocity logs to reconstruct geologic history. It is often desirable to reconstruct past geologic events. In sedimentary basins it is sometimes possible to infer the history from velocity logs. Several authors (Athy, 1930; Hedberg, 1936) have shown that density of sediments increases with depth of burial. From this idea, Lang (1978) assumed that velocity also increased with depth of burial and did not subsequently decrease if uplift or erosion decreased the depth of burial. He assumed that all sediments in a given basin are sufficiently similar that, if they had been buried the same amount, they would have had the same compaction and therefore the same velocity. Consequently, the velocity, or transit time, can be used as an indicator of maximum depth of burial.

Lang plotted transit time Δt versus current depth of the sediments in different parts of the basin, and he obtained a gradient from the slope of the plot. He attributed this gradient to compaction, and he assumed that all holes in the basin had the same "compaction gradient." Figure 19-9 is an example of such plots in two holes. Paleontological evidence indicates that hole 1 is 500 m structurally higher than hole 2. The compaction gradient is fitted to each well, and points of equal transit time are connected. Lang reconstructed the history:

> The "F-marker" and "F-2" zones were deposited and buried to an equivalent depth. After burial and compaction, well 1 was uplifted 580 m more than well 2. The structural difference in this example is due entirely to uplift rather than uplift and subsidence.

Methods such as this require assumptions, but not the same assumptions as those used in the inference of porosity from velocity. This shows that the utility of logs can depend on the assumptions that can be made about the medium.

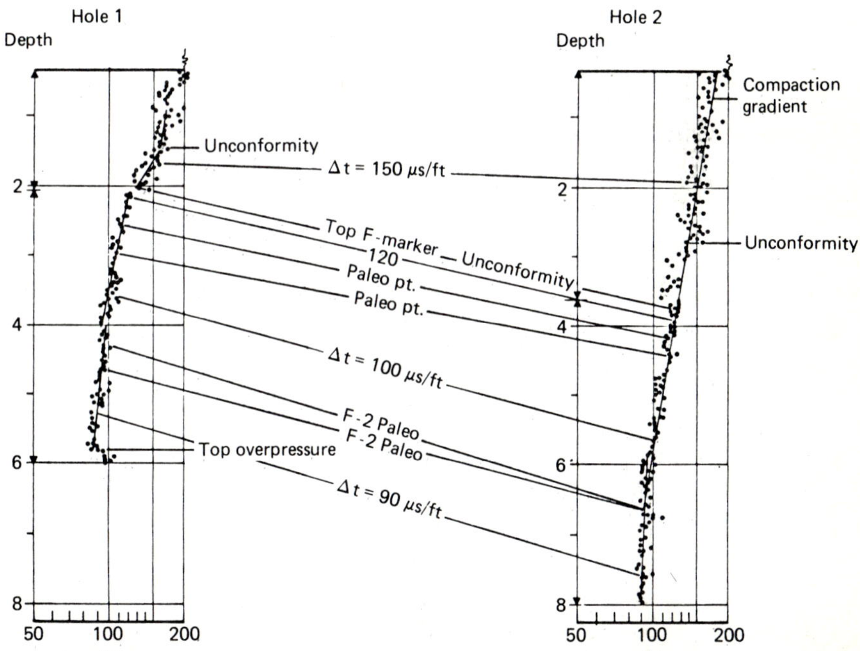

FIGURE 19-9 Transit time versus depth plots and compaction gradients for two wells in a sedimentary basin showing depths of equal transit time. (Lang, 1978.)

Use of logs to estimate unusual formation pressure. Fertl (1976) described a similar method for using several logs to estimate pore pressure in "overpressured zones," zones whose pore pressure is greater than that which would be expected from overburden (Sec. 2-1) alone. He plotted trend lines for measured properties (density, transit time, resistivity, macroscopic cross section) in shale. He cited evidence that each of these properties is a function of pressure in the shale. He then used the values of these properties in two ways to infer pressure.

There are empirical data for some regions that give the difference between the measured values of the properties and those which would be inferred from the trend line at the depth in question as a function of the overpressure. He simply used that function to infer the overpressure.

It is also possible to use the "equivalent depth method." For any actual depth of interest z_a Fertl found an equivalent depth z_e on the trend line for which the value of the property was equal to the observed value. Then the pore pressure

$$P_f = G_l z_a - (G_l - G_h) z_e \tag{19-4}$$

where G_h is the hydrostatic pressure gradient and G_l is the lithostatic gradient (Sec. 2-1). For fresh water G_h is 9.8 kPa/m (0.433 psi/ft) and a typical value of G_l is 22.6 kPa/m (1.0 psi/ft), so

$$P_f = z_a - 12.8 z_e$$

If the point is on the trend line, of course, $P_f = G_h z$.

19-8 USE OF A γ-RAY LOG TO INFER RATE OF ENTRY OF FLUID

At an underground site at Stripa, Sweden the granite contains an unusually large concentration of natural radioelements, particularly uranium. The groundwater seeping into open boreholes is consequently high in radon. The result is unusual; the γ-emitting daughters of radon often dominate the γ-ray log.

Nelson et al. (1983*b*) showed that, if the radon-containing water does not mix significantly with the rest of the water in the hole, the concentration of the radon is

$$C(z) = C_0 e^{-\lambda A z/q} \tag{19-5}$$

where z is the distance from the point of entry to the measurement point, A is the cross-sectional area of the hole, λ is the decay constant of the radon and q is the rate of flow. Because the half-life of radon is

3.8 days and entry of liquid into the borehole occurs continuously, the method is sensitive to flow rates as low as several liters per day.

Nelson et al. used a total γ-ray log (no spectral log was available) to estimate the flow in several holes at the site. In one hole they found that a change in flow from 3.0 to 11.5 L/day at a certain depth was correlated in time with the packing off of a nearby "copiously flowing hole" 35 m away. They found an outflow of 13.5 L/day at the top of the hole by direct measurement at the time they observed inflow of 11.5 L/day with γ-rays. This gives some confidence in the method.

REFERENCES

Athy, L. F.: "Density, Porosity, and Compaction of Sedimentary Rocks," *AAPG Bull.*, vol. 14, 1930, p. 1.

Baembitov, F. G.: "Use of Gamma-Gamma Logging in the Operation of Underground Reservoirs for Gas and Petroleum Products in Rock Salt Deposits," *Trudy Vsesoyuznogo Nauchno-issledovatel'skogo Inst. Yadernoi Geofiz. I Geok.*, no. 7, 1969, p. 178. UCRL-Trans-11314 (1977).

Broding, R. A.: "Volumetric Scanning Well Logging," *Proc. Soc. Prof. Well Log Analysts 22d Logging Symp.*, 1981, paper B.

Daniels, J. J., J. H. Scott, and G. Olhoeft: "Interpreting Precambrian Metamorphic Geology from Geophysical Well Logs," *Proc. Soc. Prof. Well Log Analysts 22d Logging Symp.*, 1981, paper T.

Douglas, A. C., and M. R. Millett: "Total Intensity Magnetometer Logging as a Stratigraphic Tool in Tertiary Volcanic Rock," Lawrence Livermore Lab. Rept. UCRL-52617, 1978.

Fertl, W. H.: *Abnormal Formation Pressure,* Elsevier, Amsterdam, 1976.

Flood, P. G., and M. M. Wilson: *Progress Report, 1979, 3-D Modelling and Paleoenvironmental Reconstruction of the Conditions of Coal Measure Sedimentation, Callide and Moura-Kianga Mines,* Julius Kruttschnitt Min. Res. Centre, Brisbane, 1980.

Glenn, W. E., H. P. Ross, and J. W. Atwood: "Review of Well Logging in the Basin and Range Known Geothermal Resource Areas," *J. Petr. Tech.*, vol. 34, May 1982, pp. 1104–1118.

Hearst, J. R.: "BIFUR III, a Program for Calculating Borehole and Surface Gravity Caused by Two-Dimensional Structure in Cartesian or Cylindrical Coordinates," Lawrence Livermore Natl. Lab. Rept. UCID-19030, 1981.

——— and H. L. McKague: "Structure Elucidation with Borehole Gravimetry," *Geophysics,* vol. 41, 1976, p. 491.

——— and R. C. Carlson: "Measurement and Analysis of Gravity in Boreholes," in Fitch (ed.), *Developments in Geophysical Exploration Methods—3,*" Applied Science, London, 1982.

Hedberg, H. G.: "Gravitational Compaction of Clays and Shales," *Am. J. Sci.*, vol. 31, 1936, p. 241.

Keys, W. S.: "Borehole Geophysics in Igneous and Metamorphic Rocks," *Proc. Soc. Prof. Well Log Analysts 20th Logging Symp.*, 1979, paper OO.

——— and J. K. Sullivan: "Role of Borehole Geophysics in Defining the Physical Characteristics of the Raft River Geothermal Reservoir, Idaho," *Geophysics*, vol. 44, 1979, p. 1116.

Lachenbruch, A. H., J. H. Sass, B. V. Marshall, and T. H. Moses: "Permafrost, Heat Flow, and the Geothermal Regime at Prudhoe Bay, Alaska," *J. Geophys. Res.*, vol. 87B, 1982, p. 9301.

Lang, W. H.: "The Determination of Prior Depth of Burial (Uplift and Erosion) Using Interval Transit Time," *Proc. Soc. Prof. Well Log Analysts 19th Logging Symp.*, 1978, paper B.

Mitchell, F. R., J. D. Robinson, J. P. Vogiatzis, F. Pehoushek, J. H. Moran, B. E. Ausburn, and L. N. Berry: "Using Resistivity Measurements to Determine Distance between Wells," *J. Petr. Tech.*, vol. 24, 1972, p. 723.

Morris, F. J., R. L. Waters, G. F. Roberts, and J. P. Costa: "A New Method of Determining Range and Direction from a Relief Well to a Blowout Well," Soc. Petr. Eng. Paper SPE 6781, 1977.

Nelson, P. H., and W. E. Glenn: "Influence of Bound Water on the Neutron Log in Mineralized Igneous Rocks," *Proc. Soc. Prof. Well Log Analysts 16th Logging Symp.*, 1975.

———, K. A. Magnusson, and R. Rachiele: "Application of Borehole Geophysics at an Experimental Waste Storage Site," *Geophys. Pros.*, vol. 30, 1983*a*, p. 910.

———, P. Rachiele, and A. Smith: "Transport of Radon in Flowing Boreholes at Stripa, Sweden," *J. Geophys. Res.*, vol. 88B, 1983*b*, pp. 2395–2405.

Robinson, J. D., and J. P. Vogiatzis: "Magnetostatic Methods for Estimating Distance and Direction from a Relief Well to a Cased Wellbore," *J. Petr. Tech.*, vol. 24, 1972, p. 741.

Talwani, M., J. L. Worzel, and M. Landisman: "Rapid Gravity Computations for Two-Dimensional Bodies with Application to the Mendocino Submarine Fracture Zone," *J. Geophys. Res.*, vol. 64, 1959, p. 49.

Warren, T. M.: "Directional Survey and Proximity Log Analysis of a Downhole Well Intersection," *J. Petr. Tech.*, vol. 33, 1981, p. 2351.

Appendix

SYMBOLS

A	Constant
A	Area (including cross-sectional area)
A	Atomic mass
A	Amplitude (acoustic)
A	Area under grade-thickness curve
A	Elevation of terrain compartment
A_i	Conductivity trace of dipmeter pad
A_{ik}	Constant relating count in *i*th window to concentration of *k*th element
A_n	Fourier coefficient
a	Constant
a	Chemical activity
a	Gravimeter leg length
a	Radius of spherical shell
a	Amplitude of one of two signals
a	Radius of induction coil
a_i	Activity of *i*th species of nuclide
a_i	Constant (with index)
B	Constant
B	γ-Ray buildup factor
B	Waxman-Smit parameter
B	Background count rate
B	Formation γ-ray spectrum vector
B	Terrain datum elevation
B	Magnetic field or induction
$B(G, \rho)$	Change in density log count rate caused by sonde-wall gap
$B(G, I_H)$	Change in neutron log count rate caused by sonde-wall gap
B_C	Background in carbon channel
B_O	Background in oxygen channel

B_i	Number of counts in channel i of n channels
B_i	Conductivity trace of dipmeter pad
B_n	Fourier coefficient
b	Constant
b	Exponent
b	Gravimeter lever arm length
b	Amplitude of one of two signals
b	Radius of spherical shell
C	Conductivity (electrical)
C	Constant
C	Concentration
C	Gravity station elevation
C	Heat capacity per unit mass
$\mathcal{C}$	Capacitance
C_a	Apparent conductivity
C_c	Conductivity of clay component of fictional clay-fluid slurry
C_e	Conductivity of exchangeable cations
C_i	Concentration of ith element in the field
C_{ij}	Concentration of ith radioelement in jth calibration model
C_{ik}	Response of kth log to ith material
C_{ma}	Matrix conductivity
C_o	Conductivity of water-saturated formation
C_p	"Compaction" constant
$C_{r,i}$	Real and imaginary parts of conductivity
C_{sd}	Sandstone conductivity
C_{sh}	Shale conductivity
C_{sl}	Conductivity of fictional clay-fluid slurry
C_t	True formation conductivity
C_w	Water conductivity
C/O	Ratio of count rates in carbon and oxygen channels
$[\mathrm{C}]_{ma}$	Carbon concentration in the matrix
$[\mathrm{C}]_o$	Carbon concentration in oil
$(\mathrm{C/O})_{os}$	Ratio of count rates for full oil saturation
$(\mathrm{C/O})_{ws}$	Ratio of count rates for full water saturation
CEC	Cation exchange capacity
c	Constant
c_i	Ionic concentration
D	Diffusion coefficient
D	Distance between dipmeter pad vectors
D'	Second diffusion coefficient
d	Diameter
d	Gravimeter lever arm
d	Perforation diameter
d	Rate of change of wellbore curvature (dogleg)
d_c	Cable diameter
d_h	Hole diameter
d_{ij}	Hole diameter between dipmeter pads
d_s	Sheave diameter
Δd	Change in hole diameter
E	Energy
E	Electric field strength
E	Elastic (Young's) modulus
E	Environmental factor (nuclear magnetic resonance)
E_γ	γ-Ray energy
E_n	Neutron energy
e	Base of natural logarithms
e	Electronic charge
e_i	Element of dipmeter curve
F	Force
F	Formation factor
$\mathcal{F}$	Faraday's constant
$F(\ )$	Function
F_{sd}	Formation factor for sandstone only
F_z	Formation factor for clean sandstone with porosity z
F^*	Formation factor (Waxman-Smit model)
FFI	Free fluid index (NMR)
f	Frequency
f	Shape factor
f	Vector of fractions
$f(\ )$	Function
f_{cl}	Volume fraction of clay
f_i	Fraction of ith component
f_{sd}	Fraction of sandstone or sand
f_{sh}	Fraction of shale

f_{shd}	Fraction of dispersed shale
f_{shl}	Fraction of laminar shale
f_w	Fraction of water (by either weight or volume)
G	Hydrostatic pressure gradient
G	Gravitational constant
G	Geothermal gradient
G	Geometric spreading factor
G	Gap
G	Geometric factor
G	Complex admittance
G'	Integrated geometric factor
$G(\ )$	Function
$G'(\ )$	Function
G_l	Lithostatic pressure gradient
G_h	Hydrostatic pressure gradient
G_{fr}	Geothermal gradient in frozen permafrost
G_{th}	Geothermal gradient in thawed permafrost
G_γ	Ore grade
g	An arbitrary formation property
$\mathbf{g}$	Gravitational acceleration
g	Vertical component of gravitational acceleration
g_a	Apparent value of an arbitrary formation property
$g_{c,d}$	Value of arbitrary property in formations surrounding the formation of interest
g_m	Value of arbitrary property in mud
g_t	True value of an arbitrary formation property
g_t	Total gravitational acceleration
g_{te}	Gravity terrain effect
δg	Smallest measurable gravity difference
Δg	Gravity anomaly
Δg_m	Maximum gravity anomaly
H	Magnetic field strength (intensity)
H	Entropy (for several components)
$H(\ )$	Function
H_z	Component of magnetic field strength perpendicular to coil area
h	Arbitrary formation property
h	Average thickness of grain-liquid interface
h	Heat transfer coefficient
h	Annular thickness of volume of γ-emitter
h	Piezoelectric coefficient
I	Electric current
I_H	Hydrogen index
I_i	Current entering capacitor
I_o	Current leaving capacitor
I_0	Current amplitude
i	Index
i	$\sqrt{-1}$
J	Count rate
J	Electric current density
J_0	Initial count rate in a beam
J_a	Measured count rate
J_c	Corrected count rate
J_i	Fractional current carried by a species
J_{ij}	Count rate in ith energy window in jth calibration model
J_{max}	Maximum possible count rate
J_o	Observed count rate
J_t	True count rate
J_t	That part of measured count rate not caused by the formation
J_t	Total current density
ΔJ	Uncertainty in ratemeter output
j	Index
j	Number of γ's flowing across unit area normal to flow per unit time
j	Number of collisions to slow a neutron from energy E_1 to E_2
K	Bulk modulus
K	Hydraulic conductivity
K	Environmental factor for nuclear magnetism
K	Spring constant
K_i	Constant
k	Thermal conductivity
k	Index
k	Constant

k	Intrinsic permeability
k	Wavenumber
k	Propagation factor
k_c	Permeability from core
k_{fr}	Thermal conductivity of frozen permafrost
k_i	Conductivity of ith bed
k_{ic}	Thermal conductivity of ice
k_l	Permeability from log
k_{ma}	Thermal conductivity of matrix
$k_{r,i}$	Real and imaginary parts of propagation factor
k_{th}	Thermal conductivity of thawed permafrost
L	Length
L	Detector spacing
L	Characteristic length
$\mathcal{L}$	Inductance
L_a	Tube length
L_d	Migration length
L_k	Total response of kth log in a combination of logs
ℓ_s	Slowing-down length
ℓ	Distance along hole
ℓ	Distance between two receivers
ℓ	Length
ℓ_m	Dipmeter search length
ℓ_{ab}	Depth shift between dipmeter pad responses
$\Delta\ell$	Distance between two measurements
$\Delta\ell$	Arc length in borehole
M	Variable in M-N plot
M	Volumetric rate of fluid injection
M	Number of γ's per cubic centimeter per second emitted by volume source
M	Mass
M	Magnetic polarization
M_i	Induced magnetic polarization
M_m	Minimum resolvable mass
M_n	Remanent magnetic polarization
m	Index
m	Mass
m	Exponent $(1/t)$; cementation exponent
m	Number of fractures
m	Mean number of disintegrations in time t
m_i	Magnitude of a single error in a directional survey
N	Variable in M-N plot
N	Noise intensity
N	Number of particles
N	Number
N	Neutrons per cubic centimeter
N_i	Number of nuclides of ith species
N_i	Number of turns in a coil
N_{i0}	Initial number of nuclides of ith species
N_0	Initial number of nuclides
n	Index
n	Saturation exponent
n	Number of atoms per cubic centimeter
n^*	Waxman-Smit saturation exponent
n_i	Ions (or atoms) of ith species per cubic centimeter
$[O]_{ma}$	Oxygen concentration of matrix
$[O]_o$	Oxygen concentration of oil
P	Pressure
P	Polarization charge
P	γ-Ray spectrum vector of an element
P_0	Probability of no disintegrations during dead time
P_c	Capillary pressure
P_e	Average photoelectric cross section per electron
P_f	Pore pressure
P^j	Descriptor spectrum of a given element
P_i^j	Counts in ith channel of spectrum j
P_x	Probability of x disintegrations per unit time
$P(E)$	Probability that a scattered neutron will have energy E

ΔP	Pressure difference
ΔP	Effective pressure
ΔP	Pressure drop
Q	Volumetric flow rate
Q	Geothermal heat flux
Q	Source strength in neutrons per second
Q	Quality factor (inverse of attenuation)
Q	Koenigsberger ratio
$Q_{p,s}$	Quality factor for compressional and shear waves
Q_v	Valence-concentration product
$\dot{q}$	Rate of heat removal
q	Electric charge
q	Slowing-down density
R	Resistivity
$\mathcal{R}$	Resistance
R	Radius of nucleus
R	Acoustic reflection coefficient
R_{21}	Electrical reflection coefficient
R_a	Apparent resistivity
R_m	Resistivity of mud
R_{mc}	Resistivity of mud cake
R_{mf}	Resistivity of mud filtrate
R_o	Resistivity of formation saturated with water
$\mathcal{R}_o$	Initial resistance of thermistor or resistance thermometer
$R_{r,i}$	Real and imaginary parts of resistivity
R_s	Resistivity of adjacent beds
R_t	True formation resistivity
$R_{v,h}$	Vertical and horizontal resistivity
R_w	Resistivity of water
R_{wa}	Apparent water resistivity
R_{we}	Equivalent water resistivity
R_{xo}	Resistivity of flushed zone
$R_{\gamma,\varphi,\theta}$	Dipmeter matrices
r	Radius
r	Correlation coefficient
r	Gravimeter dimension
r_{AM}	Distance between electrodes
r_{AN}	Distance between electrodes
r_{AO}	Distance between electrodes
r_{BM}	Distance between electrodes
r_{BN}	Distance between electrodes
r_e	Electrode radius
r_h	Mean hydraulic radius
$r_{m,n}$	Distances to electrodes
r_0	Casing radius
Δr_e	Accumulated directional survey error
S	Source term
S	Rate of energy generation
S	Fraction of pore space filled with liquid
S	Semblance
$\dot{S}$	Number of γ's produced per unit volume per unit time
S'_c	Fraction of fictional slurry porosity filled with clay
S_h	Saturation of pore space with hydrocarbon
S'_h	Fraction of fictional slurry porosity filled with hydrocarbon
S_{or}	Residual oil saturation
S_w	Fraction of pore space saturated with water
S'_w	Fraction of fictional slurry porosity filled with water
$S_{w1,w2}$	Each of two different saturations
S_{wi}	Irreducible water saturation
S_{xo}	Saturation of flushed zone
S/H	Ratio of low-energy to high-energy gammas
SP	Reading of SP, spontaneous potential or self-potential
SP_{sh}	SP reading opposite shale bed
SP_{ss}	SP reading opposite clean sand
SSP	Static SP reading
s	Distance (dummy variable of integration)
s	Spring length
T	Total vertical thickness
T	Temperature
T'	Second temperature
T	Bed thickness
T	Period of wave
T	Acoustic transmission coefficient
T_1	Observed NMR decay time for saturated sample

T_2	NMR relaxation time
T_a	Temperature of fluid in annulus
T_b	NMR decay time in fluid
T_f	Formation temperature
T_i	Injection temperature
T_m	Mud temperature
T_m	Thermal front temperature
T_0	Surface temperature
T_0	Initial temperature of thermistor or resistance thermometer
T_p	Temperature of fluid in pipe
T_s	NMR time constant for grain-fluid interface
T_s	Torque caused by gravimeter spring
T_w	Torque caused by gravimeter weight
t	Bed thickness
t	Time
t_l	Inverse phase velocity
$t_{+,-}$	Transport number for cations and anions
$t_{21,12}$	Transmitter-receiver transit time
t_d	Diffusion time
t_{form}	Neutron decay time in formation
t_{hole}	Neutron decay time in hole
t_i	Transport number
t_N	Neutron decay time or lifetime
t_0	Period of temperature variation
Δt	Transit time (inverse of velocity)
Δt_{cl}	Velocity log response in clay
Δt_f	Fluid transit time
$\Delta t_{m1,m2}$	Transit time in two matrices
Δt_{ma}	Matrix transit time
Δt_{shd}	Transit time in dispersed shale
Δt_{shl}	Transit time in laminar or structural shale
$(\Delta t_{ma})_a$	Apparent matrix transit time (MID plot)
$\Delta t_{p,s}$	*P* and *S* wave arrival times
$\delta\Delta t$	Uncertainty in measured transit time
$\delta\Delta t_f$	Uncertainty in fluid transit time
$\delta\Delta t_{m1,m2}$	Uncertainty in transit time in two matrices
$\delta\Delta t_{ma}$	Uncertainty in matrix transit time
U	Internal energy
U	Gravitational potential
U	Group velocity
U	Magnetic energy loss per cycle
U	Effective photoelectric cross section per unit volume
$(U_{ma})_a$	Apparent effective matrix photoelectric cross section per unit volume
u	Displacement in *x* direction
δu	Change in displacement in *x* direction
Δu	Error vector in directional survey
V	Volume
V	Electrical potential difference (voltage)
V	Variance of entropy *H*
V_b	Bulk volume
V_{cl}	Clay volume
V_f	Fluid volume
V_h	Volume of fluid other than water in pore space
V_i	Volume occupied by *i*th component
V_i	Dipmeter pattern vector
V_l	Liquid potential
V_m	Membrane potential
V_{ma}	Matrix volume
V_s	Volume of fluid near grain surface
V_w	Volume of water
V_0	Voltage amplitude
v	Flow velocity
v	Acoustic velocity (velocity of propagation of wave)
v_b	Acoustic velocity in bulk material
v_c	Plate velocity in casing
v_f	Acoustic velocity in fluid
v_i	Orthogonal unit vectors (directional survey)
v_l	Velocity of longitudinal wave
v_{ma}	Acoustic velocity in matrix
v_p	Velocity of compressional wave
$v_{pW,pD}$	Compressional wave velocity in saturated and dry material
v_r	Velocity of Rayleigh wave
v_s	Velocity of shear wave
$v_{sW,sD}$	Shear wave velocity in saturated and dry material

dW	Volume fraction of radioactive material
w	Width (horizontal extent) of bed
w	Displacement in y direction
w_i	Weighting factor
δw	Change in displacement in y direction
x	Cartesian coordinate
x	Number of disintegrations
x	Characteristic length
x_i	Sample of a waveform amplitude
x_i	Distance expressed in number of mean free paths
δx	Dimension of a volume element
δx_i	Uncertainty in ith quantity
Y	Average semblance
Y	Dimensionless grain size
y	Cartesian coordinate
δy	Dimension of volume element
Z	Atomic number
Z	Electric impedance
Z	Acoustic impedance
Z_c	Acoustic impedance of casing
Z_{eff}	Effective or averaged atomic number
Z_f	Acoustic impedance of solid
Z_i	Atomic number of ith species
Z_i	Terrain compartment height
Z_s	Acoustic impedance of solid (cement or formation)
z	Vertical Cartesian coordinate; depth
z_a	Actual depth
z_e	Equivalent depth
δz	Dimension of volume element
Δz	Difference between gravity station depths
α	Pore aspect ratio
α	Angular displacement of gravimeter beam
α	Angle
α	Azimuth angle
α	Attenuation coefficient
α	Dissociation factor
α	Direction cosine
β	γ-Ray absorption factor
β	Gravimeter spring angle
β	Angle
β	Direction cosine
Γ	Vertical air gradient (Free air gradient)
Γ	Number of γ-rays piercing surface of a sphere of unit area per unit time
γ	Temperature coefficient of thermal conductivity
γ	γ-Ray log response
γ	Dipmeter pad bearing angle; pad rotation
γ	Direction cosine
Δ	Diffusion effect on neutron decay time
Δ	Dilatation
δ	Skin depth
δ	Logarithmic acoustic amplitude decrement
δ	Kroneker delta function
ε	Dielectric permittivity
ε	Strain
ε_{ii}	Normal strain in arbitrary direction
ε_{ij}	Arbitrary shear strain
ε_{ma}	Permittivity of matrix
$\varepsilon_{r,i}$	Real and imaginary parts of permittivity
ε_w	Permittivity of water
ε_{xx}	Normal strain in x direction
ε_{xy}	Shear strain in xy plane ($= \varepsilon_{yx}$)
ε_{xz}	Shear strain in xz plane ($= \varepsilon_{zx}$)
ε_{yy}	Normal strain in y direction
ε_{yz}	Shear strain in yz plane ($= \varepsilon_{zy}$)
ε_{zz}	Normal strain in z direction
ζ	Displacement in z direction
ζ	Exponent
$\delta\zeta$	Change in displacement in z direction

η	Viscosity
η	Ellipse parameter
Θ	Induction number
θ	Azimuth angle
θ	Scattering angle
θ	Rotation
θ_1	Angle of incidence
θ_2	Angle of reflection
θ_3	Angle of refraction
$\theta_{a,b}$	Azimuth measured at each of two depths
θ_i	Rotation about arbitrary axis
θ_p	Angle normal to strike of bed
$\theta_{x,y,z}$	Rotation about x, y, z axes
κ	Thermal diffusivity
κ	Relative dielectric permittivity (dielectric constant)
$\kappa_{r,i}$	Real, imaginary parts of relative permittivity
κ_w	Relative permittivity of water
Λ	Grain size parameter
Λ_g	Ratio of pore surface area to grain volume
Λ_v	Ratio of pore surface area to rock volume
Λ_ϕ	Ratio of pore surface area to pore volume
λ	Radioactive decay constant
λ	Coefficient of anisotropy
λ	Mean free path
λ	Lamé constant
λ	Wavelength
λ_a	Absorption (or capture) mean free path
λ_i	Decay constant of ith nuclide
λ_s	Scattering mean free path
$\Delta\lambda$	Horizontal separation between two direction survey measurements
μ	Viscosity
μ	Linear attenuation coefficient
μ	Matrix attenuation coefficient
μ	Magnetic permeability
μ	Shear modulus
μ'	Second attenuation coefficient
μ_0	Magnetic permeability of free space
μ_C	Linear Compton attenuation coefficient
μ_E	Linear photoelectric attenuation coefficient
μ_P	Linear pair creation attenuation coefficient
μ_f	Fluid attenuation coefficient
μ_i	Mobility of ith species
ν	Poisson's ratio
ν	Ionic mobility
$\nu_{+,-}$	Mobility of cations, anions
ξ	Average logarithmic neutron energy decrement
ξ	Logarithmic acoustic energy decrement
ξ	Ellipse parameter
ρ	Density
ρ_l	Apparent density, long-spaced detector
ρ_{Dh}	Density log response to hydrocarbon
ρ_{ash}	Density log response to ash
ρ_b	Bulk density
ρ_c	Density contrast (between formation and surroundings)
ρ_{ca}	Apparent density contrast
ρ_{cl}	Density log response to clay
ρ_{coal}	Density log response to coal
ρ_{ct}	True density contrast
ρ_e	Electron density
ρ_f	Density of fluid
$\rho_{f1,f2}$	Density of two fluids
ρ_g	Density calculated from borehole gravity data
$\rho_{\log}$	Density from γ-γ density log
$\rho_{m1,m2}$	Density of two matrices
ρ_{ma}	Density of matrix
$(\rho_{ma})_a$	Apparent matrix density (for MID plot)
ρ_o	Observed density
ρ_s	Apparent density, short-spaced detector

ρ_{sur}	Density of surrounding formations
ρ_{sv}	Density of scattering volume
ρ_t	True density
ρ_w	Density of water
$\delta\rho$	Smallest measurable density contrast
$\delta\rho_b$	Uncertainty in bulk density
$\delta\rho_f$	Uncertainty in fluid density
$\delta\rho_{ma}$	Uncertainty in matrix density
Σ	Macroscopic cross section
$\Sigma_{1,2}$	Macroscopic cross section measured on each of two runs
Σ_a	Macroscopic cross section for absorption (or capture)
Σ_h	Macroscopic cross section of hydrocarbon
Σ_{ma}	Macroscopic cross section of matrix
Σ_{meas}	Measured value of macroscopic cross section
Σ_o	Macroscopic cross section in oil-saturated zone
Σ_s	Macroscopic cross section for scattering
Σ_t	Total macroscopic thermal neutron cross section
Σ_{true}	True value of macroscopic cross section
Σ_w	Macroscopic cross section of water
$\Sigma_{w1,w2}$	Macroscopic cross section of water for each of two different solutes
σ	Stress
σ	Mass per unit length
σ	Microscopic cross section
σ_a	Absorption (capture) cross section
σ_{el}	Cross section for elastic scattering
σ_{ii}	Normal stress in arbitrary direction
σ_{ij}	Shear stress in arbitrary plane
σ_{in}	Cross section for inelastic scattering
σ_{ne}	Cross section for nonelastic scattering
σ_s	Scattering cross section
σ_{xx}	Normal stress in x direction
σ_{xy}	Shear stress in xy plane ($= \sigma_{yx}$)
σ_{xz}	Shear stress in xz plane ($= \sigma_{zx}$)
σ_{yy}	Normal stress in y direction
σ_{yz}	Shear stress in yz plane ($= \sigma_{zy}$)
σ_z	Vertical component of stress
σ_{zz}	Normal stress in z direction
τ	Tortuosity
τ	Time constant
τ	Dead time
τ	Time delay (or lag) in cross-correlation
$\tau_{1/2}$	Half-life
Υ	Distance in dipmeter calculation
Φ	Neutron flux
Φ_{ab}	Cross-correlation product
Φ_{ri}	Flux from particle source at radius r_i
ϕ	Porosity
ϕ_2	Secondary porosity
ϕ_D	Porosity determined from density log
ϕ_{Dh}	Response of density log porosity to hydrocarbon
ϕ_f	"Free water" porosity
ϕ_m	Maximum porosity for which data are available (log-inject-log)
ϕ_{max}	Upper limit on porosity
ϕ_{min}	Lower limit on porosity
ϕ_N	Porosity measured by neutron log
ϕ_{Ncl}	Porosity that would be measured by neutron log in pure clay
ϕ_{Nf}	Porosity that would be measured by neutron log in fluid
ϕ_{Nh}	Porosity that would be measured by neutron log in hydrocarbon
ϕ_{N1}	Neutron log porosity in matrix 1
ϕ_{N2}	Neutron log porosity in matrix 2
ϕ_{Nma}	Porosity that would be measured by neutron log in pure matrix
ϕ_{Nsh}	Porosity that would be measured by neutron log in shale

ϕ_{Nw}	Porosity that would be measured by neutron log in water
ϕ_S	Porosity measured by sonic log
ϕ_{sd}	Sandstone porosity
ϕ_t	Total porosity (dual water model)
ϕ_z	Sum of true and fictional porosity (clay slurry model)
$\delta\phi$	Uncertainty in porosity
$\delta\phi_{Dsh}$	Uncertainty in porosity measured by density log caused by shale
$\delta\phi_{Dh}$	Apparent change in density porosity caused by hydrocarbon
$\delta\phi_{shd}$	Apparent change in porosity caused by dispersed shale
$\delta\phi_h$	Apparent change in porosity caused by hydrocarbon
$\delta\phi_{shl}$	Apparent change in porosity caused by laminar shale
$\delta\phi_{sh}$	Porosity uncertainty (or apparent change) caused by shale
$\Delta\phi_{Dsh}$	Apparent change in density porosity caused by shale
$\Delta\phi_{shl}$	Apparent sonic porosity caused by laminar shale
φ	Angle
φ	Angle of inclination of wellbore
φ	Sonde tilt
φ	Phase angle; phase shift
φ_p	Formation dip angle
χ	Magnetic susceptibility
Ψ	Electrical potential
Ψ	Arbitrary potential
ψ	Flux
ψ	Angle between tangents of angles measured at two depths (dogleg)
ψ	Solution to wave equation
ψ	Angle subtended by coils
ψ	Angle between formation plane and wellbore
ψ_m	Dipmeter search angle
ψ_0	Amplitude of solution to wave equation
ω	Angular frequency

Index

Page numbers in *italic* indicate tables and illustrations.

ABOUT THE AUTHORS

Joseph R Hearst, Ph.D., is Associate Program Leader for Advanced Development of the Nuclear Test Containment Program at Lawrence Livermore National Laboratory, with which he has been affiliated since 1960. As a member of the Plowshare Program at the Laboratory, he helped develop a number of new well-logging methods. In addition, he is a vice president of the Minerals and Geotechnical Logging Society. Dr. Hearst holds a B.A. degree in physics from Reed College, a B.S. in business and engineering administration from M.I.T., an M.A. in physics from Boston University, and a Ph.D. in physics (nuclear) from Northwestern University.

Philip H. Nelson, Ph.D., is a senior petrophysicist with the Sohio Petroleum Company, where he applies well-logging data to the development of producing oil fields. He was formerly with Kennecott Copper Company, where he helped initiate the application of logging methods for mineral exploration and development, and with the Lawrence Berkeley Laboratory, where he worked on the application of logging to problems of radioactive waste disposal. Dr. Nelson received his B.S. and Ph.D. degrees from M.I.T.